Frequency Synthesizer Design Handbook

For a complete listing of the *Artech House Microwave Library*,
turn to the back of this book

Frequency Synthesizer Design Handbook

James A. Crawford

Artech House
Boston • London

Library of Congress Cataloging-in-Publication Data
Crawford, James A.
Frequency synthesizer design handbook/James A. Crawford.
Includes bibliographical references and index.
ISBN 0-89006-440-7
1. Frequency synthesizers. I. Title
TK7872.F73C73   1994   94-7655
621.3815'486–dc20   CIP

A catalogue record for this book is available from the British Library

© 1994 ARTECH HOUSE, INC.
685 Canton Street
Norwood, MA 02062

All rights reserved. Printed and bound in the United States of America. No part of this book may be reproduced or utilized in any form or by any means, electronic or mechanical, including photocopying, recording, or by any information storage and retrieval system, without permission in writing from the publisher.

International Standard Book Number: 0-89006-440-7
Library of Congress Catalog Card Number: 94-7655

10 9 8 7 6 5 4 3 2

To Linda,
  Brian,
  James,
  Daniel, and
  Kristen

To God be all the glory
Hebrews 2:1–3

Contents

Preface		xiii
Chapter 1	Introduction to Computer-Aided Frequency Synthesizer Design	1
	References	2
Chapter 2	Building Blocks for Frequency Synthesis Using Phase-Locked Loops	3
2.1	The Design Equation	3
	2.1.1 Mutually Contradictory Requirements	4
	2.1.2 The Fuzzy Line That Separates RF/Analog and Digital Techniques	5
	2.1.3 Prudent System Design	6
2.2	Basic Componentry in Phase-Locked Loops	8
	2.2.1 Phase Detectors	8
	2.2.2 Loop Filters	23
	2.2.3 Voltage-Controlled Oscillators	27
	2.2.4 Feedback Dividers	27
2.3	Frequency-Locked Loops	31
2.4	Signal Contamination in Phase-Locked Loops	31
2.5	To Ground or Not to Ground	33
	References	39
	Selected Bibliography	40
Chapter 3	Phase Noise and Its Impact on System Performance	41
3.1	General Noise Theory	44
	3.1.1 Phase Noise First Principles	45
	3.1.2 Power Spectral Density Concept	46
3.2	Power Spectral Density Calculations in Discrete Sampled Systems	48

		3.2.1	Relationships Between Continuous-Time and Discrete-Time Fourier Transforms	48
		3.2.2	Power Spectral Density for Discrete Sampled Systems	50
		3.2.3	Example Computation With a Discrete System	51
		3.2.4	Phase Noise First Principles Continued	58
	3.3	Defining Standardized Phase Noise Quantities		60
	3.4	Creating Accurate Phase Noise PSDs in the Laboratory		63
	3.5	Creating Arbitrary Phase Noise Spectra in a Digital Signal Processing Environment		69
	3.6	Phase Noise in Devices		71
		3.6.1	Digital Frequency Dividers	71
		3.6.2	Phase Detector Phase Noise Performance	81
		3.6.3	Low-Noise Electronic Design	84
		3.6.4	Noise Sources in Lead-Lag Loop Filters	87
		3.6.5	Noise in Components	89
		3.6.6	Low-Noise Oscillator Design and Characterization	91
	3.7	Assessing System Performance in the Presence of Frequency Synthesizer Phase Noise		114
		3.7.1	Phase Noise Effects With Coherent Demodulation of Binary FSK	117
		3.7.2	Phase Noise Effects With Coherent Demodulation of BPSK	120
		3.7.3	Phase Noise Effects With Coherent Demodulation of QPSK	120
		3.7.4	Phase Noise Effects With Coherent Demodulation of 16-QAM	120
		3.7.5	M-ary Phase-Shift-Keyed Modulation (MPSK)	127
		3.7.6	Local Oscillator Phase Noise Impairment With Fast Changing Phase	135
		3.7.7	Phase Noise Impairments for More Advanced Waveforms	135
	3.8	Worst Case Bounding of Synthesizer Phase Noise Requirements		137
		References		140
		Selected Bibliography		145
Chapter 4	Phase-Locked Loop Analysis for Continuous Linear Systems			147
	4.1	Phase-Locked Loop Basics		148
	4.2	Pseudo-Continuous Phase Detector Models		152
		4.2.1	Tri-State (Output) Phase-Frequency Detector	153
		4.2.2	Simplified Phase-Frequency (Tri-Logic State) Detector	158
		4.2.3	Zero-Order Sample-and-Hold Phase Detector	163

	4.2.4	Inefficient Sample-and-Hold Phase Detector	165
	4.2.5	Digital Feedback Dividers	168
4.3		Modeling Time Delays in Continuous Systems	171
4.4		Frequency-Domain Analysis	174
4.5		Transient Response Evaluation for Continuous Systems	182
4.6		Summary	188

Appendix 4A Loop Bandwidth Considerations Using the
Phase-Frequency Detector 189
References 200
Selected Bibliography 201

Chapter 5 Frequency Synthesis Using Sampled-Data Control Systems 203
    5.1 Sampled-Data Control System Basics 203
        5.1.1 Frequency-Domain Representation for Sampled Signals 205
        5.1.2 $G_{OL}^*(s)$ and Its Relationship to z Transforms 207
    5.2 Frequency-Domain Analysis 208
    5.3 Stability 209
    5.4 Time-Domain Analysis 211
    5.5 Closed-Form Sampled-Data System Results 214
        5.5.1 Case 1: Ideal Type 1 214
        5.5.2 Case 2: Type 1 With Inefficient Phase Detector 216
        5.5.3 Case 3: Type 1 With Internal Delay 220
        5.5.4 Case 4: Ideal Type 2 225
        5.5.5 Case 5: Type 2 With Inefficient Phase Detector 228
        5.5.6 Case 6: Type 2 With Internal Delay 231
    5.6 Phase Noise 237
        5.6.1 Reference Phase Noise 238
        5.6.2 VCO-Related Phase Noise 242
    References 245
    Selected Bibliography 245

Chapter 6 Fast-Switching Frequency Synthesizer Design Considerations 247
    6.1 Type 1 Case Study 248
        6.1.1 Internal Design Details 250
        6.1.2 Post-Tuning Drift (Nemesis of Type 1 Designs) 264
    6.2 Initial Phase Error Impact on Switching Speed 264
    6.3 Sample-and-Hold Phase Detector Noise 270
    6.4 Type 2 Phase-Locked Loop With Simplified
Phase-Frequency Detector 273
        6.4.1 Phase-Frequency Detector Refinements 280
    6.5 Post-Tuning Drift With Type 2 Systems 282
    6.6 Choosing Between Sample-and-Hold and
Phase-Frequency Detectors 284

		References	285
Chapter 7		Hybrid Phase-Locked Loops	287
	7.1	Hybrid Architectures	287
	7.2	Dual-Loop Frequency Synthesis	289
		7.2.1 Frequency Channel Spacing	289
		7.2.2 Phase Noise Performance	289
		7.2.3 Offset Mixer Considerations	289
		7.2.4 Signal Isolation	291
		7.2.5 Gain Compensation Strategies	292
		7.2.6 Proper Sideband Selection	302
		7.2.7 Achievable Switching Speed	304
	7.3	Direct Digital Synthesizer Fundamentals	308
		7.3.1 General Concepts	309
		7.3.2 Representation of the Ideal DDS Output Spectrum	311
		7.3.3 Phase Truncation Related Spurious Effects	313
		7.3.4 Lookup Table Finite Word Length Effects on Spectral Purity	323
		7.3.5 Techniques for Mapping θ to $\sin(\theta)$	328
		7.3.6 Digital-to-Analog Converter Imperfections	335
		7.3.7 Spurious Suppression in Direct Digital Synthesizers Using Numerical Techniques	335
	7.4	Hybrid Phase-Locked Loop/DDS Frequency Synthesizers	343
	7.5	Pros and Cons of Direct Digital Frequency Synthesis	346
	7.6	Advanced Fractional-N Concepts	346
		7.6.1 Fractional-N Technique Based on Noise-Shaping Methods	347
		References	351
		Selected Bibliography	352
Chapter 8		MACSET: A Computer Program for the Design and Analysis of Phase-Locked Loop Frequency Synthesizers	353
	8.1	Time-Domain Modeling	354
	8.2	Numerical Integration Formulas	355
	8.3	Numerical Integration Algorithm Stability	365
	8.4	Discrete Circuit Models	369
	8.5	Circuit Macros	373
	8.6	Frequency-Domain Analysis	377
	8.7	Phase Noise Computations	382
	8.8	Summary	384
		References	384
		Selected Bibliography	385
Chapter 9		Fractional-N Frequency Synthesis	387
	9.1	Some Pros and Cons Behind Fractional-N Synthesis	393

9.2	Fractional-N Synthesis: General Concepts		395
9.3	Fractional-N Systems Aspects		401
	9.3.1	Spectral Performance: Discrete Spurious	401
	9.3.2	Loop Parameter Limitations	406
	9.3.3	An Alternative for Dealing With Truncation-Related Discrete Spurs	407
	9.3.4	System Noise Floor Degradations Arising in Fractional-N Synthesis	410
	9.3.5	API Gating Period Inaccuracies	410
	9.3.6	Phase Detector Ramp Nonlinearities	414
	References		419
About the Author			421
Index			423

Preface

The motivation for this text evolved primarily from my prior employment at Hughes Aircraft Company, Ground Systems Group, Fullerton, California, beginning in the late 1970s. During this time, it was my privilege to work with a number of excellent engineers and scientists. I want to specifically acknowledge Mr. Gary D. Frey, Mr. Jack Chakmanian, and Dr. Knut Konglebeck. From my later tenure at the Linkabit Corporation, San Diego, California, I want to acknowledge Mr. Fritz Weinert as well. In this company, the term "RF black magic" was largely put to rest, with rigorous analytical design methods and techniques favored instead. This text has been written with this same bent toward sound engineering design practice rather than ad hoc cut and try methods.

A number of excellent texts have been available for some time addressing the frequency synthesis area, but most if not all of these books are primarily circuit oriented. This text is intended to address the synthesis subject from a systems perspective instead. Special attention has also been given to the area of sampled-data control systems as they apply to modern phase-locked loop design and as well as other areas where clear design material was felt to be unavailable.

Finally, I want to give a special acknowledgment and word of thanks to Dr. William F. Egan who reviewed major portions of this text over the three years spent in this endeavor.

Chapter 1
Introduction to Computer-Aided Frequency Synthesizer Design

The phase-lock concept was originally described in a published work by de Bellescize in 1932 [1] but did not fall into widespread use until the era of television where it was used to synchronize horizontal and vertical video scans. One of the earliest patents showing the use of a phase-locked loop with a feedback divider for frequency multiplication appeared in 1970 [2]. The phase-locked loop concept is now used almost universally in many products ranging from citizens band radio to deep-space coherent receivers.

The material presented in this book largely grew out of the author's involvement with a high-performance frequency synthesizer, which is the subject of Chapter 6. Many of the concepts, particularly those related to sampled control systems, led to a number of valuable findings, which are discussed in Chapters 4 and 5. In Chapter 4, the transition between continuous and sampled control system perspectives is dealt with in some detail leading to Chapter 5, which enlarges the sampled control system concepts further.

Many of the underlying design guidelines pertaining to frequency synthesis for wireless communications are rooted in system-level performance requirements, particularly phase noise and discrete spurious requirements. These concepts are developed in detail in Chapter 3 with additional discussion purposely provided to allow individuals to simulate these system imperfections with a computer. Phase noise performance of key synthesis elements is also discussed at some length in Chapter 3.

Chapter 7 addresses a number of the most recent synthesis techniques including direct digital synthesizers (DDSs) and hybrid phase-locked loops. Fractional-N synthesis techniques based on sigma-delta modulation concepts are also discussed there.

Chapter 8 addresses the issue of phase-locked loop computer simulation, drawing in part on the results found in Chapters 4 and 5. Numerical techniques that are suitable for this application are developed from first principles.

Finally, Chapter 9 takes an in-depth look at traditional fractional-N frequency synthesis. Several analyses are considered, underscoring the need on the designer's part to be intimately aware of design details at the component level as well as at the circuit design level.

The material presented in this text should be viewed as a supplement to, rather than as a replacement for, a number of excellent texts that are currently available on this subject. Prior knowledge of basic phase-locked loop principles is generally assumed throughout the text. Stochastic process theory is employed sparingly in Chapters 3, 5, and 9. Fairly heavy use of z transforms is made in Chapter 5. Appropriate outside references for this material are provided in each respective chapter.

REFERENCES

[1] de Bellescize, H., "La Reception Synchrone," *Onde Electr.*, Vol. 11, June 1932, pp. 230–240.
[2] Sepe, R.B., R.I. Johnston, "Frequency Multiplier and Frequency Waveform Generator," U.S. Patent No. 3,551,826, December 29, 1970.

Chapter 2
Building Blocks for Frequency Synthesis Using Phase-Locked Loops

The explosion in commercial and consumer level communication products in the late 1980s and early 1990s dramatically revolutionized the availability of high-quality, low-cost componentry for RF system design. The frequency synthesis area has been one of the primary areas where this market pressure has resulted in unprecedented levels of high integration. The majority of consumer level frequency synthesis is currently being done with frequency synthesizers that are based on the single phase-locked loop (PLL) architecture. Designs that would have been multiloop in nature in years past are quickly giving way to hybrid loops, which often utilize a mixture of phase-locked loop technology along with direct digital synthesis. Even more bold progress is being made using, for example, delta-sigma modulator concepts with fractional-N synthesis to realize single-loop architectures that can deliver almost arbitrarily small channel spacings while maintaining excellent switching speed and spurious performance.

It seems that phase-locked loops are everywhere, from frequency synthesizer applications in radios to clock deskewing applications within high-speed digital processors. Dedicated hardware frequency multipliers followed by bulky filters are a thing of the past. The integration levels and cost achievable today in the phase-locked loop area make these techniques the solution of choice. As such, the contents of this chapter (and most of this text) focus primarily on the phase-locked loop for frequency synthesis.

2.1 THE DESIGN EQUATION

In about 1983, this author was asked to formulate a high-level "design equation" for general frequency synthesizer design that would factor in all of the normal performance measures plus size, complexity, risk, schedule length, and cost. The task was fairly meaningless because nearly every integral component, from feedback divider to voltage-controlled oscillator (VCO), had to be custom designed except in fairly benign situations where a few phase-locked loop components were available. The answer usually came back as "near infinite dollars, time, and risk" at least in the case of high-performance, military-type synthesizers. Technologies such as cellular telephone have changed this situation forever.

2.1.1 Mutually Contradictive Requirements

When designing frequency synthesis elements for standardized market areas such as analog cellular and digital cellular telephone, performance requirements have generally been arrived at that have taken into account the RF technology needed to perform the task. With the size of these markets being so large, most component vendors are ahead of the original equipment manufacturers (OEMs) in identifying clever design solutions or they are working directly with the OEMs. Because cost is such an overwhelming factor in consumer electronics, most hardware solutions for a given product are quite similar, having been optimized down to the minimally acceptable solution. Most of the system requirements, therefore, reflect prior consideration of the mutually contradictory requirements encountered in phase-locked frequency synthesizer design.

If starting from scratch, the normal trade-offs that must be made in synthesizer design include (1) phase noise performance, (2) switching speed, (3) discrete spurious performance, (4) frequency and tuning bandwidth, (5) size, (6) power consumption, and (7) cost. The last three items are by far the overwhelming factors in modern consumer electronics design.

Frequency Switching Speed

The precise definition of switching speed for a given system must be based on the modulation waveform employed (in the case of communications) and the system performance degradation incurred. These are system-level issues and are addressed in part in Chapter 3. For a quadrature phase-shift-keyed (QPSK) system, which uses phase shifts of 90-deg multiples to convey information, the settling time is normally defined as the time required for the loop to settle to within 0.1 radian of its steady-state final phase. In an analog FM system, such as an analog cellular telephone, switching speed might be defined as the time required for the loop to reach an output frequency within 100 Hz of the steady-state value. In receiving systems that estimate the incoming signal's carrier phase or frequency immediately after a frequency hop, system impairment will obviously result if the signal's parameters (i.e., phase and frequency) have not settled out because the presence of these transients will degrade the receiving system's estimates.

Increased switching speed always implies a larger phase-locked loop bandwidth, which normally results in higher discrete spurious levels. For larger loop bandwidths, frequency reference noise and phase detector noise sources lead to wider output phase noise pedestals about the carrier signal. Control loop stability margins normally degrade with increasing loop bandwidth as well, leading to increased "peaking" of noise sources present in the VCO. These issues are dealt with in greater detail in Chapters 3, 4, and 5.

A reasonably good rule of thumb for switching speed while maintaining low discrete spurs is simply 50 reference cycles. If a phase-locked loop is using a reference frequency of 30 kHz, achieving phase-lock will normally require a minimum of roughly 1.7 ms. If a switching speed requirement is at or below this guideline, more detailed assessment is mandated. This estimate is based in part on the fact that the switching transient error normally decays proportionally with respect to $\exp(-\zeta\omega_n t)$.

As discussed in Chapters 5 and 6, switching speeds on the order of a few reference periods are theoretically achievable, but not without extreme attention to detail. Normally, custom components will have to be designed if this performance level is needed, thereby driving cost issues dramatically.

Much of this text addresses phase-locked loop frequency synthesis in the context of a sampled control system. As developed in Chapter 5 at length, use of closed-loop bandwidths ≥ 0.1 F_{ref} must be done with care because sampling effects can significantly affect both system phase noise and transient performance.

A trade-off study between switching speed, loop filter attenuation of the first frequency reference sideband spurs, and system stability is given in Chapter 3, which should provide some design assistance. The final feasibility answers cannot normally be obtained without knowing a minimum of information, specifically: (1) phase detector leakage or glitch energy at the reference frequency when in steady state, (2) the phase detector and loop electronics anticipated noise floor, and (3) VCO tuning sensitivity and phase noise performance. Characterization of the phase detector area is normally the most difficult because few vendors provide adequately detailed information.

2.1.2 The Fuzzy Line That Separates RF/Analog and Digital Techniques

High-speed digital VLSI has dramatically changed the frequency synthesis landscape by making what were heretofore impossible concepts physically realizable. The highly integrated complementary metal-oxide semiconductor (CMOS) PLL devices that now operate at input frequencies as high as 2.5 GHz clearly fall into this category. Digital very large scale integration (VLSI) technology, however, has done more than simply brought faster PLL devices to the marketplace.

Perhaps the most dramatic impact of high-speed digital VLSI on frequency synthesis has been in the architecture area where truly new synthesis concepts have now become fairly commonplace. One of these new architectural concepts is the direct digital frequency synthesizer (DDS), which also relies on high-speed digital-to-analog converters for its implementation. DDS concepts are examined in considerable detail in Chapter 7. Since the VLSI technologies are capable of supporting high-frequency clock rates in CMOS (e.g., 100 MHz or more), intricate divider schemes and most recently delta-sigma modulation methods are being used for frequency synthesis. In communication receivers, where perhaps continuous frequency tuning is needed, much if not all of the extremely fine frequency resolution is realized either by using a DDS or by sampling the incoming signal (after amplification and translation to a reasonably low frequency) and eliminating the additional phase rotation mathematically in a digital signal processing (DSP) device. Digital VLSI techniques are significantly affecting the evolution of synthesis componentry from the highly integrated PLL devices to the highly accurate digitally controlled temperature compensated crystal oscillator (DCXO) references that are needed in every cellular phone. The impact of digital VLSI is affecting far more than basic synthesis componentry however. Its presence is dictating entire system architectures even in traditionally analog/RF areas.

2.1.3 Prudent System Design

Most frequency synthesizers that are designed are intended to be used as an integral element of a larger function or system. The prudent system design mitigates technical risk and development time by allocating performance requirements across the system in such a way as to make use of existing technologies wherever possible. An obviously flawed example would be the "perfect" receiver, which is perfect only if a 0.01% bandpass filter at 1.342 GHz can be built that has a 2-dB insertion loss and 80-dB stopbands. A good system design must evaluate the trade-offs between concepts as well as available componentry and, therefore, must cross all lines between RF/analog, signal processing, software, and digital hardware disciplines.

A little closer to home, the conceptual design of a frequency synthesizer is best thought of in terms of budgets, specifically, budgets that address the following areas:

Frequency resolution

Tuning bandwidth

Phase noise

Discrete spurious levels

Frequency switching speed

---

*Power supply contamination

*Interface related noise and isolation

Costs of material

Manufacturing costs

The items above the rule have always received substantial design scrutiny for inherent reasons. Items below the line have traditionally received less than their adequate share of attention as a result of schedule pressures, lack of information, or ignorance. With the dramatic level of product integration common now, these issues can no longer be ignored. Since the first five of these items are addressed elsewhere in this text, only the items marked with an asterisk are discussed here. The remaining two areas, although extremely important, are outside the intended scope of this text.

Power Supply Contamination Budgets

Power supply contamination in frequency synthesizers should be handled with a budget-like approach equivalent to that used to assess phase noise performance when multiple sources are being combined. Budgeting requires first a knowledge of the power supply contaminant levels that can be expected at the frequency synthesizer interface, plus knowledge of the permissible contaminant levels at important points within the frequency synthesizer. The latter information can be computed based on the output spectral purity requirements for the synthesizer. Given these two endpoints, it is possible to design the correct amount of power supply grooming within the synthesizer.

Ad hoc consideration of LC power supply filtering should be avoided. Dc current levels through inductors, particularly ferrite types, should be carefully considered to avoid any magnetic saturation effects that effectively render the inductor useless for filtering. Selection of LC values should also be done such that high-Q resonances are avoided. Otherwise, incoming power supply noise can actually be increased (on a voltage basis), thereby degrading rather than improving performance [1].

Special consideration must also be given to systems that employ switching regulators or digital systems that operate on a synchronous clock. Both of these situations result in strong spectral lines at discrete frequencies, which must be dealt with on an individual basis. Systems that use any amount of Schottky or FAST devices are nothing short of notorious for creating severe spectral components well into the gigahertz region.

The power source provided to the frequency synthesizer should ideally be separate from that used for digital circuitry elsewhere in the system but this is not always possible. The synthesizer's power input, however, should at a minimum be wired separately from other system elements in order to minimize common mode impedances between the synthesizer and other dynamically changing loads.

More detailed information about the effects of noise contamination on synthesizer performance can be found in [2–4] as well as later in this chapter.

Interface Related Noise and Contamination

Every interface signal to a frequency synthesizer should be considered to be a potential source of external contamination. Even simply lock-detect outputs can provide surprisingly effective conduits for external contamination largely because these signals easily find their way directly to the sensitive PLL chip area. All interface signals should be terminated as near to the frequency synthesizer perimeter as possible to prevent such outside contamination from leaching into the synthesizer interior.

2.2 BASIC COMPONENTRY IN PHASE-LOCKED LOOPS

The phase-locked loop can be reduced to a fairly small set of basic building blocks. The most common breakout of these building blocks is (1) the phase detector, (2) the loop filter, (3) the VCO, and (4) the feedback divider. This perspective is shown later in Chapter 4, Figure 4.2, where $K_d f(\ )$ represents the phase detector function and $H(s)$ represents the loop filter.

More complex frequency synthesis architectures that utilize multiple phase-locked loops can be considered of course, but for the purpose of this chapter, our focus will remain on essentially the single phase-locked loop case. Some discussion of multi-loop architectures is given in Chapter 3. More expansive discussions of this subject can be found in [2–5].

The basic building blocks we have identified will consume a substantial portion of the rest of this chapter. Since the primary focus of this text is to develop mathematical models for analyzing phase-locked loops rather than dwelling on detailed circuit design, outside references will be cited wherever possible when addressing these more detailed design areas.

2.2.1 Phase Detectors

The phase detector is often considered to be the heart of a phase-locked loop in that it is the origin of the control system dynamics, which supervise the desired phase-lock condition. When phase-locked loop techniques first became widely used, many patents were issued for different phase detector varieties (e.g., [6]). Additional concepts continue to evolve as device technologies allow new approaches to take realistic physical form [7]. The primary phase detector types encountered in practice include the following:

- Mixer (fundamental multiplicative types)
- Harmonic sampling
- Sample-and-hold
- Phase-frequency detectors

Given the importance of the phase detector topic to phase-locked loop synthesizers, each of these types is discussed further later in this chapter.

Other phase detector types are encountered in the context of phase-locked loops used for very weak signal detection and demodulation purposes (e.g., Costas loops [8]). These applications require a substantially different perspective for analysis and design, often requiring a deep understanding of renewal processes, Fokker-Planck techniques, etc. The interested reader should consult [9–12].

Mixer Phase Detectors

In early phase-locked loops, double-balanced and single-balanced mixers were often used as the phase detector element. Mixers are seeing increasingly limited service in more recent times because of the higher levels of integration achievable with monolithic digital phase detectors. An equal if not stronger reason for this trend is that the digital phase detector types that employ memory display an almost unlimited frequency pull-in range compared to the mixer-based detectors. *Pull-in range* refers to the initial frequency error over which a phase-locked loop is guaranteed to eventually establish phase-lock. Those unfamiliar with this concept may consult [3–5,13].

Mixers are still used as phase detectors in extremely high frequency applications or on occasions where the ultimate in phase noise floor performance is required. As mentioned in Chapter 3, the Schottky diode-based double-balanced mixer phase detector is still the top contender for achieving the ultimate in system sensitivity.

The mixer does have some additional advantages over digital approaches including its relative simplicity. Aside from the sum-frequency term, which results in the phase detector process, the output is also free of additional spurious outputs.

A simplistic application of a mixer used as a phase detector is shown in Figure 2.1. Signals must be in phase quadrature in order to realize the needed odd symmetry

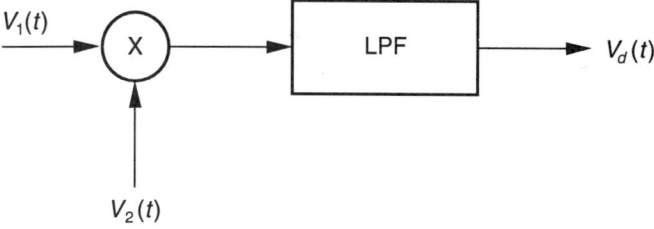

Figure 2.1 Mixer-based phase detector with output lowpass filter.

transfer function. Since the mixer has no signal memory mechanism, its pull-in range performance is very limited. For the situation shown, assume that

$$v_1(t) = V_A \sin[\theta_1(t)]$$

$$v_2(t) = V_B \cos[\theta_2(t)]$$

where (2.1)

$$\theta_1(t) = \omega_1 t + \phi_1$$

$$\theta_2(t) = \omega_2 t + \phi_2$$

in which case with application of the output low-pass filter, the output voltage is given by

$$\begin{aligned} V_d(t) &= \frac{V_A V_B}{2} \sin[(\omega_1 - \omega_2)t + (\phi_1 - \phi_2)] \\ &= \frac{V_A V_B}{2} \sin[\theta_e(t)] \\ &\approx \frac{V_A V_B}{2} \theta_e(t) \end{aligned} \qquad (2.2)$$

which becomes essentially linear for small phase error arguments. Since the phase detector gain is a function of V_A and V_B, measures must be taken to ensure that these levels are fairly constant.

The output voltage from a mixer-based phase detector is normally fairly small. Additional steps must be taken to deal with the mixer's output dc offset and its impedance termination sensitivity. The dc offset is directly related to the achievable pull-in range in the context of the so-called Richman voltage as discussed in [13]. The termination sensitivity issues generally require that isolation amplifiers be placed in both the RF and local oscillator (LO) paths to the mixer along with proper IF port termination at high frequencies.

The mixer phase detector characteristic can be taken to be linear only over a narrow range of phase error values θ_e. The percentage of deviation for the output voltage V_d from linear will be less than d percent provided that the phase error range is restricted to within [14]

$$|\theta_e| \leq \phi_L$$

where (2.3)

$$\phi_L = \sqrt{\frac{6d}{100}}$$

where the angular unit is radians. The dc offset at the mixer output may be estimated from the LO level expressed in dBm and the L to I port isolation expressed in decibels as [14]

$$V_{dc} = \frac{2\sqrt{50}}{\pi} \exp\left[\frac{LO - IS - 30}{20} \log_e(10)\right] \quad (2.4)$$

A thorough discussion of mixers used for low phase noise measurement can be found in [15]. To achieve the very best performance possible, the following guidelines should be observed:

1. Terminate the mixer output in 50 Ω at the sum product frequency.
2. Unload the mixer output at low frequencies in order to maximize its sensitivity.
3. Make incident powers at the L and R ports equal.
4. The careful choice of an additional capacitive reactance at the mixer output for low frequencies can further improve sensitivity.

A mixer-based phase detector that adheres to these guidelines is shown in Figure 2.2 [16]. A plot of the achievable phase detector gain versus input power levels is shown in Figure 2.3.

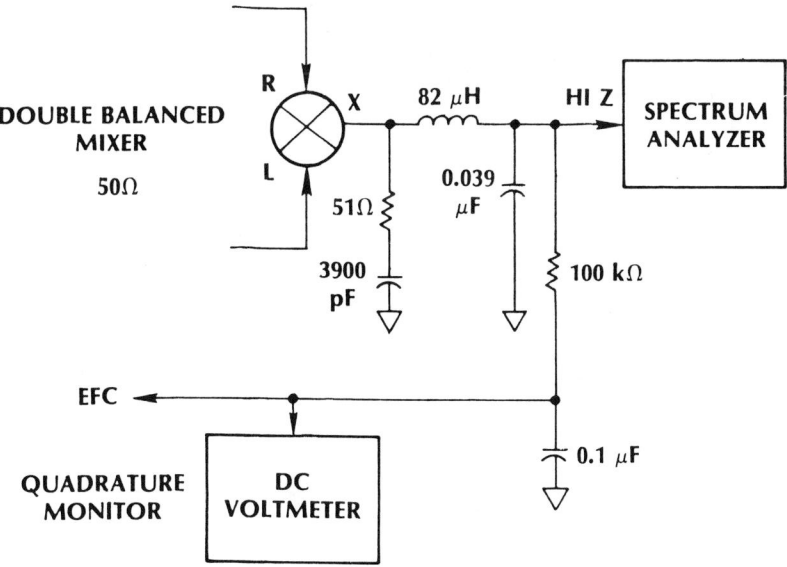

Figure 2.2 Recommended double-balanced mixer termination for phase detector operation. (©1978 IEEE; reprinted with permission). Source: [16], Figure 6.

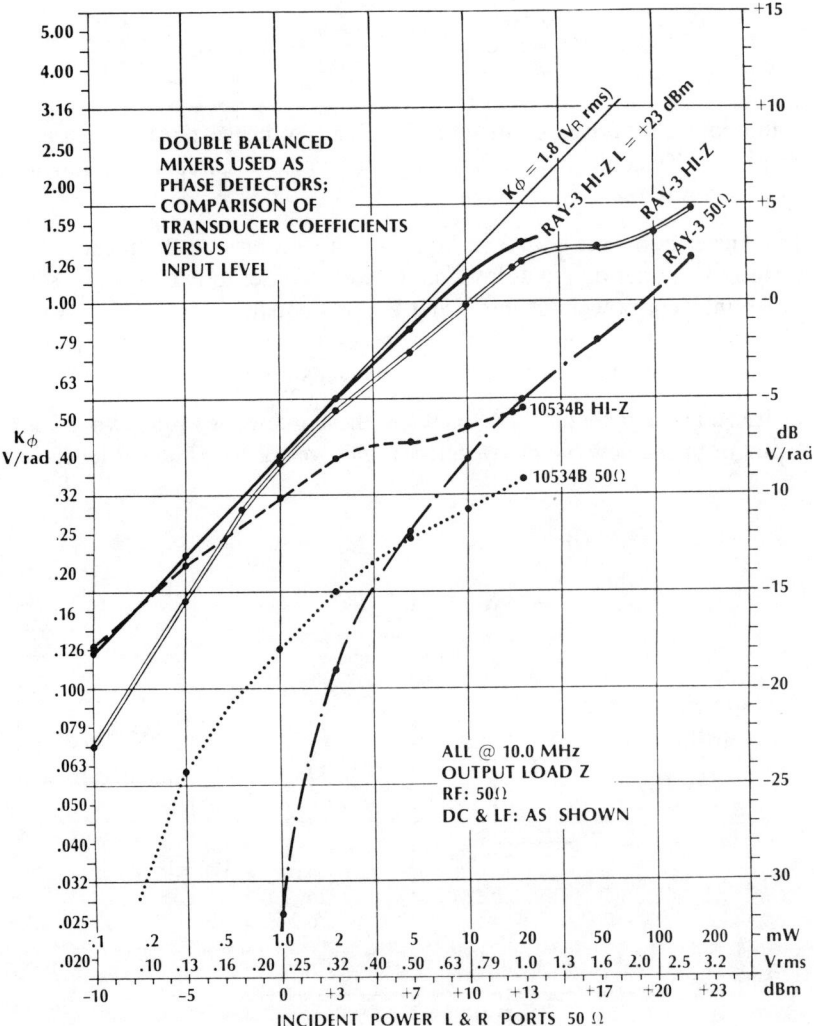

Figure 2.3 Comparison of various mixer sensitivities as a function of input signal level. (©1978 IEEE; reprinted with permission). Source: [16], Figure 7.

Finally, helpful information concerning isolation amplifiers and detailed low phase noise measurement can be found in [17].

Harmonic Sampling Phase Detectors

Harmonic sampling phase detectors take a number of different forms but all rely on essentially the same sampling concepts, which are developed at length in Chapter 5 where a band-limited signal is effectively aliased to baseband [18–20].

A first-order examination of the harmonic sampling process can be done based on Figure 2.4 where a simple gating circuit is assumed to be gated on for a time duration T. The sinusoid being sampled is assumed to have a phase argument of ϕ at the origin as shown. The Laplace transform for the gated portion of the one cycle is given by

$$F(s) = \frac{1 - e^{jN\omega_s T} e^{-sT}}{s - jN\omega_s} \tag{2.5}$$

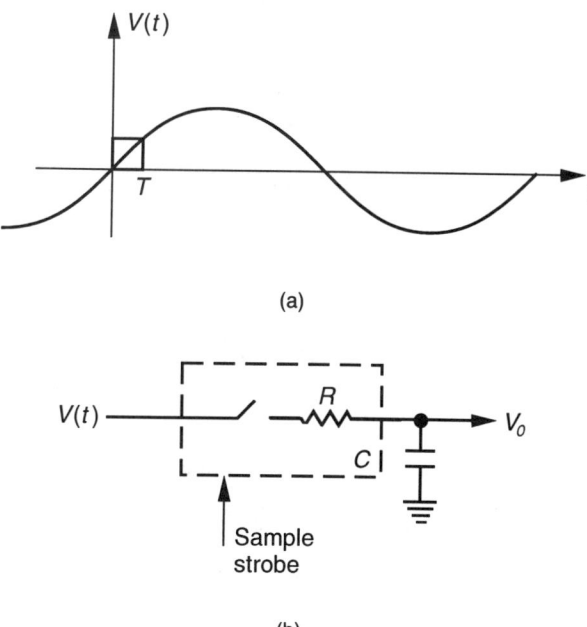

Figure 2.4 (a) Sampling gate interval for sampling a sinusoidal input signal and (b) harmonic sampler model.

where the sampled sinusoid has been assumed to have a frequency N times that of the radian sampling frequency ω_s and the input sine wave being sampled is represented by the imaginary part of the complex exponential. During the gating period, the output is passed through the RC low-pass filter as shown. This RC filter is implicit in the sampling process and represents the finite on-resistance and output capacitance of the sampling gate. The time response at the RC filter output to the single gated sinusoid sample is given by

$$V_0(t) = \left(\frac{1 - e^{jN\omega_s t} e^{-jN\omega_s t}}{1 + jN\omega_s RC} \right) e^{jN\omega_s t} \tag{2.6}$$

$$- \left(\frac{1 - e^{jN\omega_s t} e^{t/RC}}{\frac{1}{RC} + jN\omega_s} \right) \frac{e^{-t/RC}}{RC}$$

which simplifies nicely to

$$\frac{e^{jN\omega_s T} - e^{-T/RC}}{1 + jN\omega_s RC} e^{j\phi} \tag{2.7}$$

where the angle ϕ represents an arbitrary signal phase with respect to the origin in Figure 2.4. Taking only the imaginary portion corresponding to the real sin( ) input to the sampler, the output of the sampler at time $t = T$ given one sample is

$$V_o = V_{in} \sqrt{\frac{1 - 2\cos(N\omega_s T) e^{-T/RC} + e^{-2(T/RC)}}{1 + (N\omega_s RC)^2}} \sin(\theta) \tag{2.8}$$

where

$$\theta = \phi - \tan^{-1}(N\omega_s RC) + \tan^{-1}\left[\frac{\sin(N\omega_s T)}{\cos(N\omega_s T) - e^{-T/RC}}\right] \tag{2.9}$$

This value is meaningful only after the first sample has been taken. Subsequent output values are the superposition of this sample with an RC decay imposed and subsequent samples. After a fairly small number of samples, the precise number being dictated by the ratio T/RC, the output dc value will stabilize to a value given as

$$V_o = V_{in} \sqrt{\frac{1 - 2\cos(N\omega_s T) e^{-T/RC} + e^{-2(T/RC)}}{1 + (N\omega_s RC)^2}} \frac{\sin(\psi)}{1 - e^{-T/RC}} \tag{2.10}$$

where once again

$$\psi = \phi - \tan^{-1}(N\omega_s RC) + \tan^{-1}\left[\frac{\sin(N\omega_s T)}{\cos(N\omega_s T) - e^{-T/RC}}\right] \quad (2.11)$$

Inefficiency in the sampling operation, which is discussed further in Chapter 5 in the case of the sample-and-hold phase detector, introduces both a phase and amplitude error in the sampler output. As discussed further in Chapter 5, normally the additional phase introduced by inefficient sampling is the issue of most concern pertaining to phase-locked loop design.

A recent hardware product that employs harmonic sampling is discussed in [21].

Zero-Order Sample-and-Hold Phase Detector

The sample-and-hold phase detector normally consists of a linear sawtooth ramp voltage generator followed by a high-speed sample-and-hold, both of which are controlled by the reference and feedback divider inputs. A detailed discussion of this phase detector variety is given in Chapter 6.

This phase detector type is often considered to be more difficult to implement than the other detector types because it consists of considerably more analog circuitry. Unless a monolithic device is used, as shown in Chapter 6, a discrete implementation can require a substantial amount of circuitry.

Aside from complexity, the chief drawback of the sample-and-hold phase detector is that it lacks any frequency discrimination capability. Therefore, if the initial phase-locked loop frequency error is well outside the closed-loop bandwidth, acquisition will be extremely slow or simply may not occur. It is even possible for the loop to achieve lock on subharmonics of the reference frequency. In situations where this can arise, either a pretuning of the VCO or additional frequency steering must be provided.

Several significant advantages, however, make this phase detector quite attractive. First of all, a properly designed detector can exhibit extremely low reference spur sidebands. As discussed in Chapter 6, this is accomplished by designing a "doubly balanced" sampling bridge in which (1) sampling pulses are symmetrically proportioned about the nominal output voltage and (2) linear ramp leakage across the sampling gate is balanced out. Second, this phase detector type is capable of theoretically delivering complete phase-lock in as little as one reference period! This very interesting result is developed at length in Chapter 5.

The sample-and-hold phase detector can also display inefficient sampling similar to that discussed for the harmonic sampling phase detector. This phenomenon is analyzed in Chapter 5. Noise floor issues as they apply to the linear ramp generator are crucial for good phase noise performance and this issue is examined in some detail in Chapter 9.

The high-speed performance possible with the sample-and-hold phase detector makes the inclusion of sampling effects mandatory in any accurate analysis. Erroneous results for transient response, phase noise performance, and even stability can easily occur if the sampling nature of the underlying control system is not addressed with phase-locked loops that use bandwidths on the order of $0.1\ F_{\text{ref}}$ and higher. In Chapter 5, the departures from the continuous system analyses that occur with sampled control systems are developed in detail using the z transform and other techniques.

In the context of strictly continuous system analysis, the transfer function for the zero-order sample-and-hold phase detector in terms of Laplace transforms is given by

$$H(s) = \frac{K_d}{T} \frac{1 - e^{-sT}}{s} \tag{2.12}$$

where T is the length of the reference period in seconds and K_d is the phase detector gain in volts per radian. For Fourier frequencies that are small compared to T^{-1}, this may be approximated by

$$\begin{aligned} H(\omega) &= K_d\, e^{-j\omega T/2}\, \frac{\sin(\omega T/2)}{\omega T/2} \\ &\approx K_d\, e^{-j\omega T/2} \end{aligned} \tag{2.13}$$

where the $\sin(x)/x$ argument has been taken to be approximately unity. The appearance of the exponential delay term has led to considerable confusion in recent years. It is a direct result of the sampling process rather than any "transport delay" within the control system. This perspective is examined in considerable detail in Chapter 4 and from a different viewpoint in [22].

Before leaving this topic, it is instructive to evaluate the sample-and-hold transfer function represented by (2.12) for several measures. These quantities will be referred to later in our loop filter discussions that follow. The quantities of interest are

Impulse response peak value	K_d/T
Impulse response average value over T	K_d/T
Magnitude frequency response	$K_d\lvert\,\text{sinc}(\omega T/2)\rvert$
dc gain	K_d
3-dB bandwidth	$\omega T/2 \approx 1.39$
Apparent group delay	$T/2$

Phase-Frequency Detectors

In the general sense, phase-frequency detectors come in many varieties. All phase-frequency detectors exhibit an ability (1) to operate in a frequency discriminator mode for large initial frequency errors and (2) to perform as a coherent phase detector once the system transient response is within the pull-in range of the phase-locked loop. The distinction between these two operational modes during any given transient response is the point at which phase detector cycle slips cease to occur. All such detectors require that some form of memory be present within the detector.

In an analog frequency discriminator, the needed memory element takes the form of the finite group delay through a frequency selective element. Frequency discrimination is possible by noting that different amounts of phase shift occur due to this specific group delay as a function of input frequency. Phase-frequency detectors use similar concepts to achieve frequency discrimination but the memory element is almost always a digital element such as a flip-flop.

The phase-frequency detector was popularized by the onset of devices from Motorola and RCA in the early 1970s. The RCA CD4046A used four internal flip-flops and a state table of 12 states in order to perform the complete phase detector function [3]. The Motorola MC4344 was quite similar and also made use of 12 states. Diagrams of the CD4046 and MC12040 phase-frequency detectors are shown in Figures 2.5 and 2.6, respectively.

Many phase-frequency detectors suffer from a phenomenon known as the *zero phase error dead zone* in which the phase detector gain is very erratic. This phenomenon was first recognized by Egan [4]. A typical plot of phase detector gain versus time delay between feedback divider and reference divider inputs is shown in Figure 2.7. Large phase detector gain differences obviously lead to dramatically different phase-locked loop performance than anticipated including instability. Additional discussion of the dead-zone phenomenon may be found in [23–25].

A number of factors can contribute to the dead-zone problem but the most significant factor is architectural rather than semiconductor in nature, particularly at the phase detector output(s). Early phase-frequency detectors directed pulse outputs to one of two outputs depending on whether the reference edge or the feedback divider edge was observed first. If the input edges occurred very close together in time, occasionally the wrong decision would be made, which could cause the device to momentarily reenter its frequency discriminator mode. Other limitations due simply to finite rise and fall times associated with very narrow pulses also led to phase detector gains that were much smaller than normal.

To avert this dead-zone problem, early designers normally introduced an offset error at the loop filter input purposely to cause the phase detector to operate away from its otherwise zero phase error condition (in the case of type 2 and higher loops). This method, although effective, was also undesirable because it led to higher reference spurs at the synthesizer output.

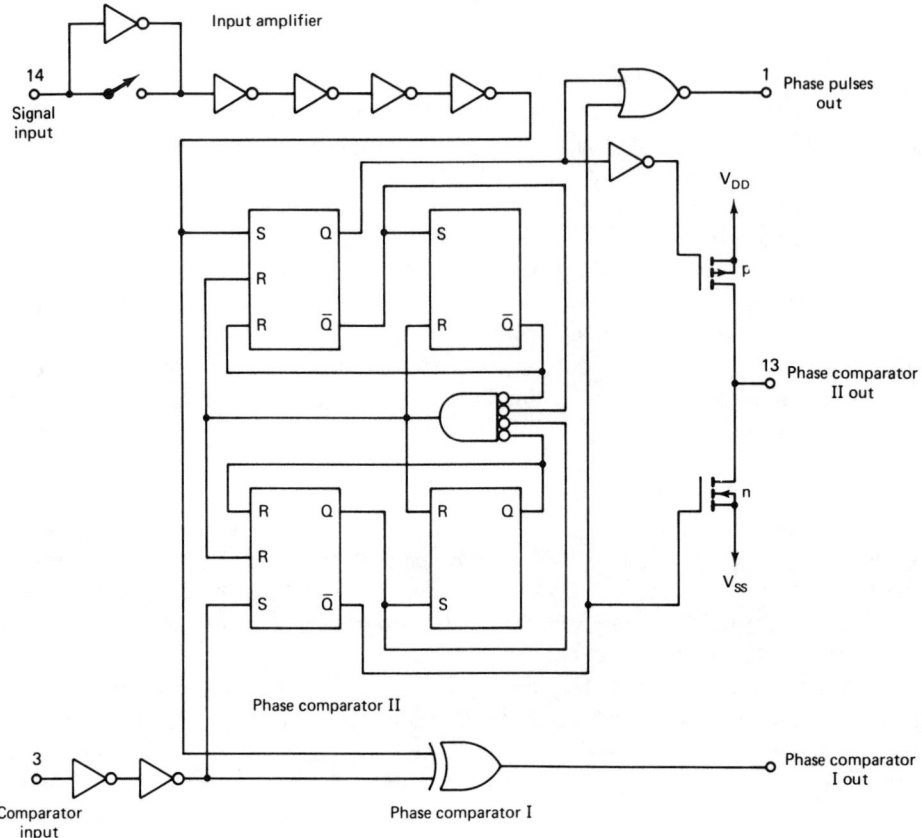

Figure 2.5 CD4046 phase/frequency comparator. Source: [3], Figure 4–74, p. 224. (Reprinted with permission.)

Other designers have resorted to other methods to alleviate the dead-zone problem. One effective approach has been to marry a traditional digital frequency discriminator with a separate analog or digital phase detector that displays no dead zone but lacks frequency discriminator capability. Devices from Plessey have incorporated a sample-and-hold type phase detector to avoid the dead-zone problem whereas a recent device from Analog Devices [26,27] (AD9901) makes use of an EXCLUSIVE-OR type phase detector for the same purpose. In principle, both should be effective in eliminating the dead-zone problem.

By far the simplest approach to avoid the dead-zone problem is to use the three-state phase-frequency detector, which was first identified by Sharpe [28]. Rather than attempt to make a digital determination about which divider input occurred first, this

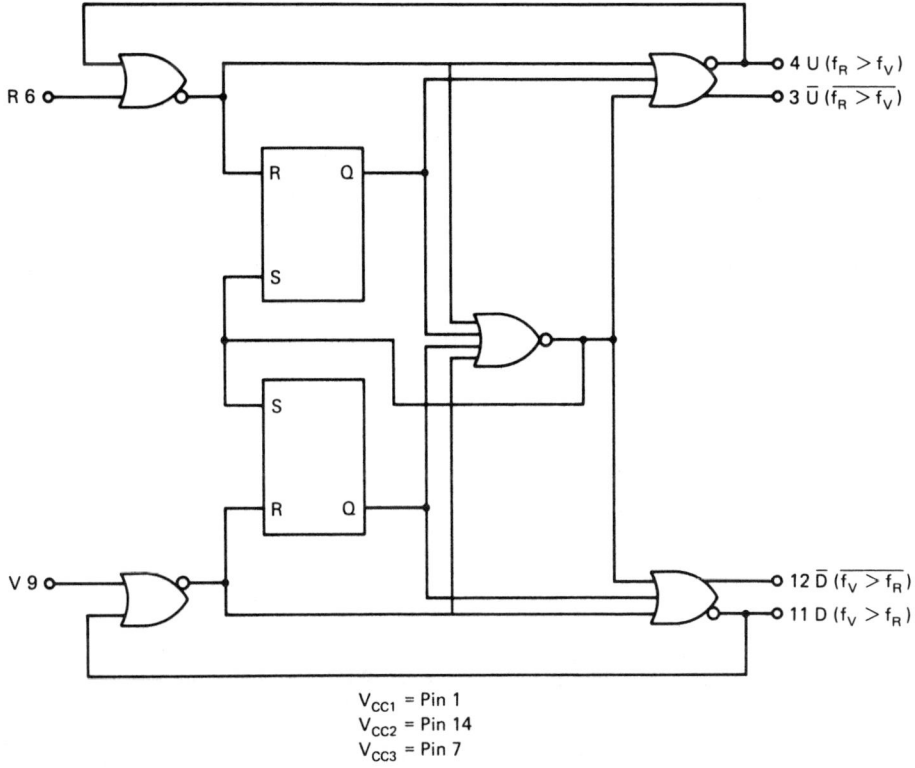

Figure 2.6 Block diagram of Motorola MC12040 phase/frequency comparator. (Source: [3], Figure 4–75, p. 225. Reprinted with permission.)

phase detector deals with both divider outputs in a parallel manner where the analog voltage outputs are combined in the differential analog electronics that follow. Since each flip-flop in the detector exhibits a finite output pulse width, finite rise and fall time issues are essentially eliminated [29]. This phase detector type is shown in Figure 2.8 and is referred to as the "simplified phase-frequency detector" throughout Chapter 4.

The presence of finite pulse widths at the phase detector outputs can create potentially large reference sidebands on the synthesizer output unless the gains for each phase detector arm are well balanced. The phase-locked loop architecture under consideration is shown in Chapter 7, Figure 7.15. The RC sections represented by R_1, C_1 and R_3, C_2 form what will be referred to as the *postdetection filtering*. Since the phase detector outputs are pulse-like in nature, it is mandatory that these low-pass filters be in place in order (1) to reduce the voltage slew rate as seen at the

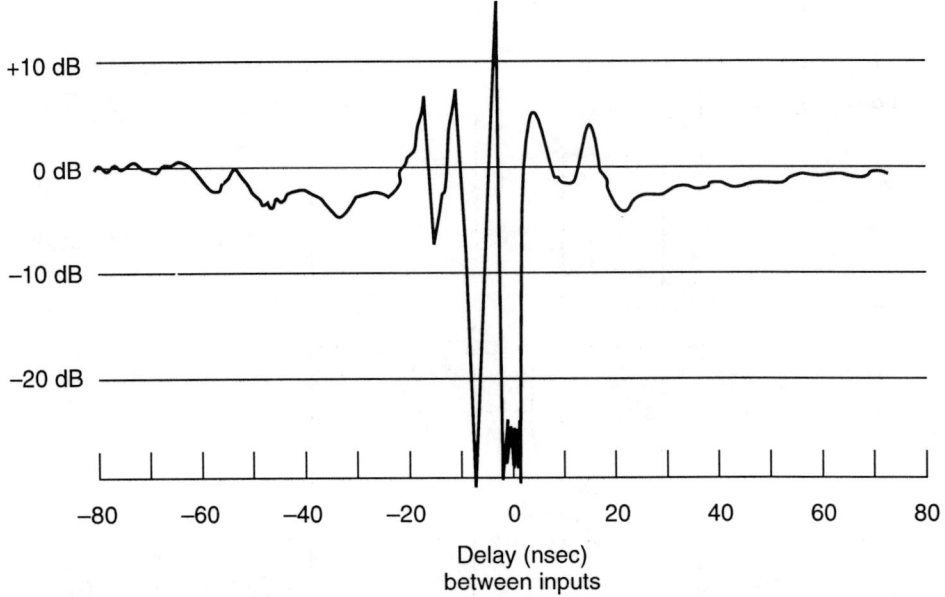

Figure 2.7 Measured gain (K_ϕ) for a phase/frequency detector near crossover. Source: [4], Figure 5.23, p. 119. Reprinted with permission. (Copyright © 1981 John Wiley & Sons, Inc. Reprinted by permission [4].)

lead-lag filter inputs (i.e., op-amp inputs) and (2) to extract only the baseband fundamental spectral information rather than pass along higher frequency alias spectra. A fairly large number of factors can contribute to the issue of gain balance between the two phase detector arms. If the input to the lower arm of the postdetection filter in Figure 7.15 (e.g., input to resistor R_3) is given by $V_2(s)$ and for the time being C_4 and C_3 are taken to be short circuits, the op-amp output is given by

$$V_o(s) = V_2(s) \frac{R_5}{R_1 + R_2} \frac{1}{1 + sC_1 \frac{R_1 R_2}{R_1 + R_2}} \gamma(s) \qquad (2.14)$$

where $\gamma(s)$ represents the balance error factor, which is given by

$$\gamma(s) = \frac{R_5}{R_6} \frac{R_1 + R_2 + R_6}{R_3 + R_4 + R_5} \left[\frac{1 + sC_1 \dfrac{R_1(R_2 + R_6)}{R_1 + R_2 + R_6}}{1 + sC_2 \dfrac{R_3(R_4 + R_5)}{R_3 + R_4 + R_5}} \right] \qquad (2.15)$$

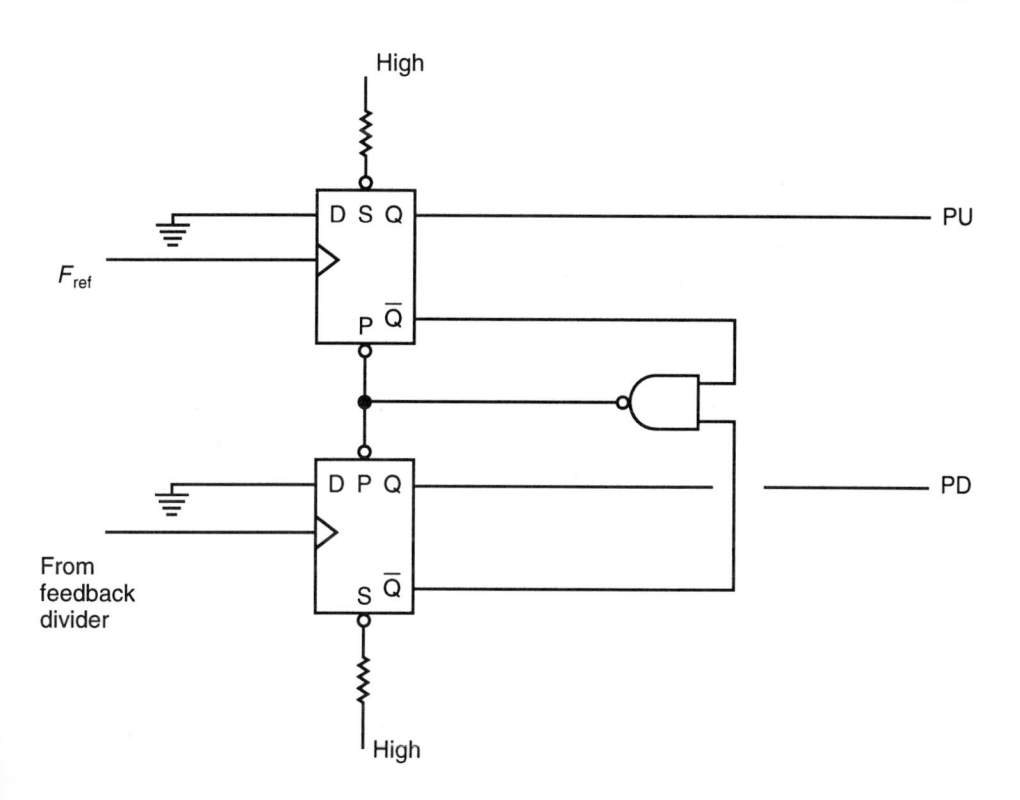

Figure 2.8 Three-state phase detector. (Reprinted with permission from *EDN Magazine*, Sept. 20, 1976. © Cahners Publishing Company 1994; a division of Reed Publishing USA [28].)

The tolerances of many components effect the degree to which $\gamma(s)$ is made identical to unity as desired. Imbalances in the postdetection filtering can also lead to unexpected results in the loop transient performance as is discussed shortly.

Selection of the RC time constant for the postdetection filter has rarely been addressed other than in the context of system phase margin. Normally, the corner frequency of this filtering has been placed at five times the closed-loop bandwidth or larger. Since the sample-and-hold phase detector is the optimum choice for high-speed frequency switching, a better design methodology (when appropriate) is to select the postdetection filter time constant such that the filter approximates the behavior of a zero-order sample-and-hold. Contrasting the impulse response of the postdetection filter to that of the sample-and-hold, for a simple RC low-pass filter with $\tau = RC$,

Impulse response peak value	K_d/τ
Impulse response average value over T	$K_d[1 - \exp(-T/\tau)]/T$
Magnitude frequency response	$K_d[1 + (\omega\tau)^2]^{-0.5}$

dc gain	K_d
3-dB bandwidth	$\omega\tau \approx 1.0$
Apparent group delay	$\approx \tau$

Recognizing that the loop transfer function phase is the most critical item to be matched with the sample-and-hold, choosing $\tau = T/2$ is then recommended. In narrowband phase-locked loops where speed is much less an issue than is reference spur reduction, a larger postdetection filter time constant should of course be selected.

Additional modifications to this simplified (three-state) phase detector were also suggested in [28] for enhancing switching speed for large initial frequency errors. This phase detector is named the quad-D phase-frequency detector owing to its use of 4D flip-flops. It is shown in Figure 2.9. When operating in frequency discriminator mode, the voltage output of this phase detector as compared to the simplified three-state phase frequency detector is as shown in Figure 2.10. The quad-D detector output increases to its full output value for frequencies that are in error by a factor of 2.

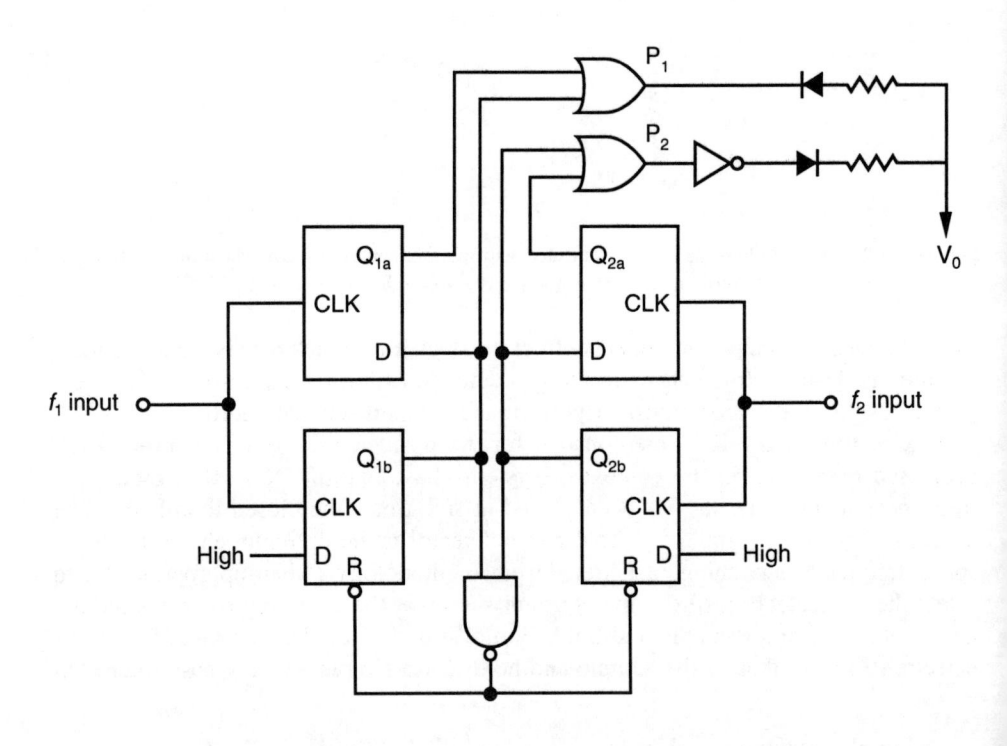

Figure 2.9 Example of a quad-D phase/frequency detector. (Source: [3], Figure 4–73, p. 223. Reprinted with permission.)

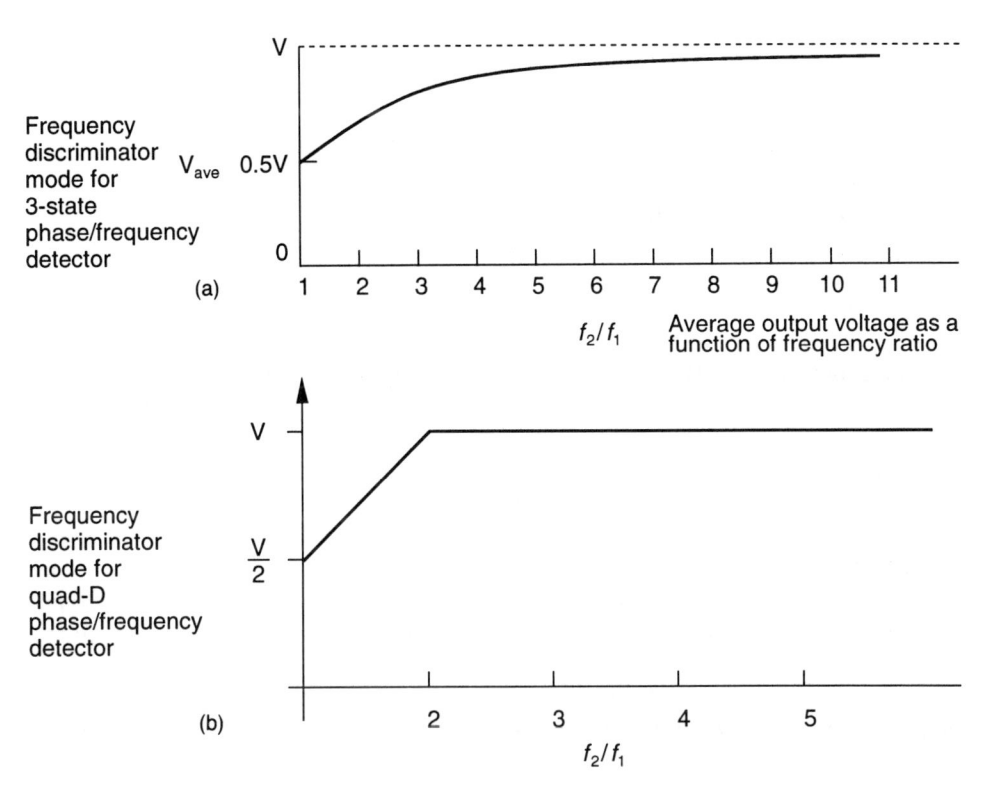

Figure 2.10 Average phase detector output voltage in frequency discriminator mode: (a) Three-state phase detector. Source: [3], Figure 4–72, page 222. Reprinted with permission. (b) Quad-D phase/frequency detector. (Reprinted with permission from *EDN Magazine*, Sept. 20, 1976. © Cahners Publishing Company 1994; a division of Reed Publishing USA [28].)

Phase-frequency detectors can in principle output a voltage or a current that is proportional to the instantaneous phase error in the system. Although the voltage variety has been of primary focus in this chapter, the current output variety, which are often called *charge pumps*, are actually just as common if not more so. The interested reader is referred to [30].

More exhaustive discussions of phase detector types may be found in [3] and [5].

2.2.2 Loop Filters

The complexity of the loop filter area can range from a complete absence in the case of an ideal type 1 phase-locked loop to the systematically gain-compensated lead-lag filter shown in Chapter 7, Figure 7.11, combined with an elliptic filter for improved spurious performance as shown in Figure 7.15. Several configurations are discussed in Chapters 4 and 5 with several others presented in [3]. The author's approach in this area has been to rely on back-of-the-envelope analysis based on concepts presented in this text combined with detailed computer analysis as discussed in part in Chapter 8.

The loop filter subject is discussed in a number of excellent texts [2–5,12] as well as others so the basic concepts are not reiterated here. Rather, several concepts that have not appeared in the general literature to date are presented instead.

The first of these concepts is to pattern use of the postdetection filters that are used with most phase/frequency detectors after the (optimal) sample-and-hold. This was discussed earlier.

A second useful concept is the use of linear programming techniques to synthesize a loop filter structure that exhibits a constant phase margin over a wide range of closed-loop bandwidths. The general theory behind this approach is discussed in detail in Chapter 7.

Another useful concept that at first may appear unadvised is the use of an additional pole-zero pair immediately before the VCO to reduce the effective tuning sensitivity of the VCO. Such a network is shown in Figure 2.11. The pole and zero frequencies are purposely chosen to lie well within the closed-loop bandwidth and therefore do not affect the loop stability. In the network shown, the VCO tuning sensitivity is then reduced by a factor of $R_2/(R_1 + R_2)$. This approach is particularly helpful in situations where a very narrow loop bandwidth is needed, thereby leading to extreme component values in the lead-lag filter. The chief drawback of this technique is that the loop's transient response (for large frequency steps) is dictated by

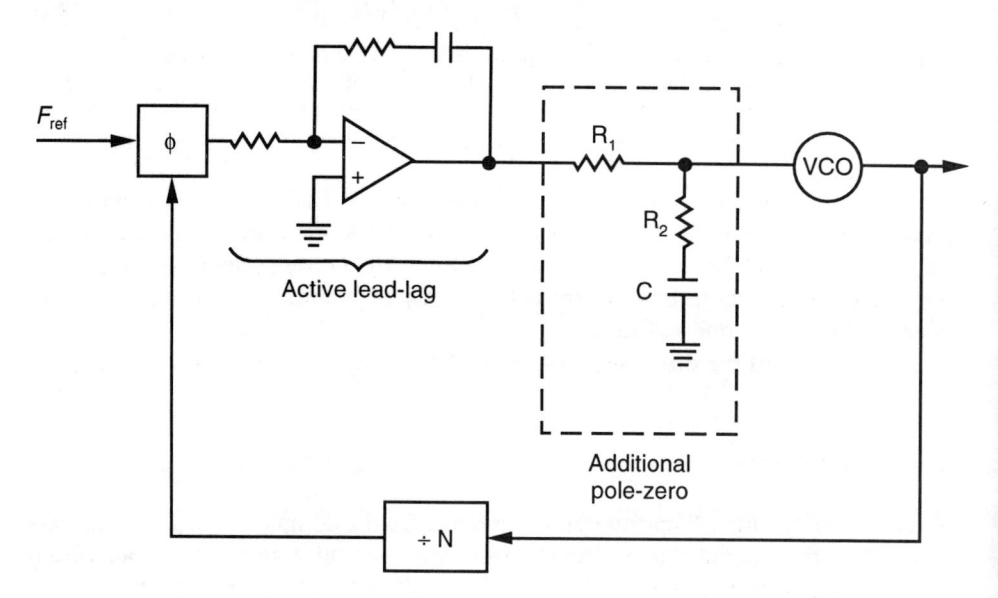

Figure 2.11 An additional pole-zero pair preceding the VCO can be used to reduce the effective VCO tuning sensitivity.

the low-frequency pole and zero introduced by this network rather than the closed-loop bandwidth. This approach can also be used to decrease the source resistance level as seen by the VCO tuning port, thereby reducing resistor Johnson noise, which would otherwise modulate the VCO.

In phase-locked loop situations where an acquisition aid is required because the phase detector does not have a frequency discrimination capability, an ingenious method that has gone largely unnoticed is the positive feedback method reported in [13]. In this approach, a positive feedback network causes the active lead-lag loop filter to oscillate at some low frequency when the loop is unlocked. Once the oscillation brings the VCO within the pull-in range of the control loop, the much greater negative feedback present within the control loop when approaching phase lock overcomes the locally applied positive feedback, thereby quenching the oscillation and achieving lock. In the case of the normal active lead-lag loop filter, such a filter is as shown in Figure 2.12. The oscillation frequency must be chosen such that the maximum VCO frequency slew rate does not result in large cycle slips once the VCO is within the pull-in range of the control loop. Given a linear VCO slew rate of $\Delta\omega$ rad/s/s and a loop natural frequency of ω_n rad/s, the phase error observed at the phase detector is given by

$$\varphi = \frac{\Delta\omega}{\omega_n^2 N} \qquad (2.16)$$

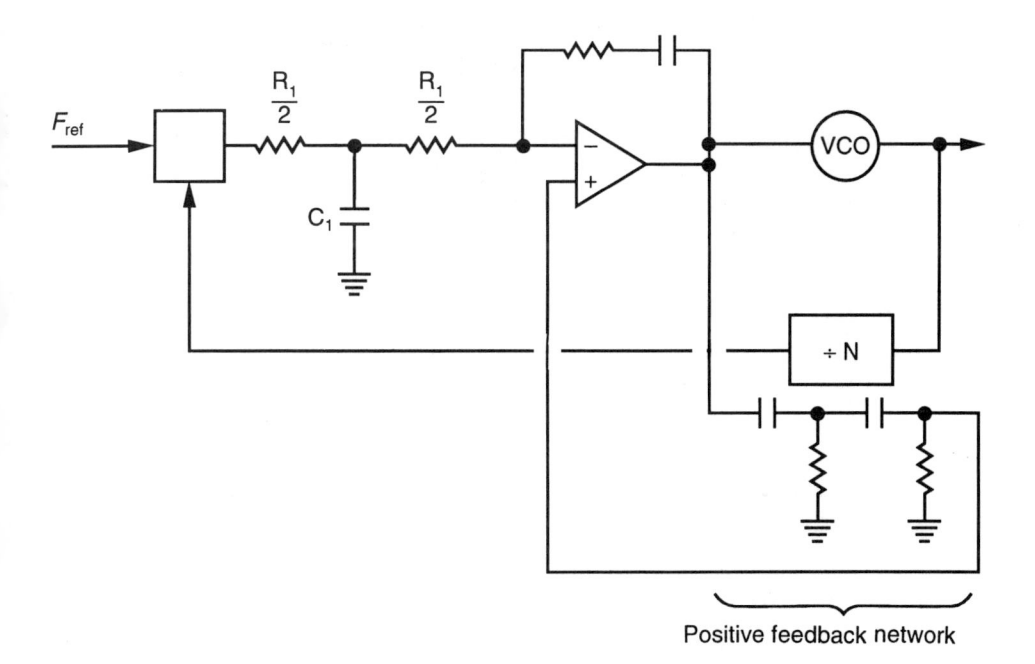

Figure 2.12 A typical type 2 phase-locked loop augmented with automatic acquisition circuitry.

in radians where N is the feedback divider ratio being used. The VCO slew rate used should not cause this phase error to exceed the phase detector's dynamic range.

One of the more interesting phenomenon reported lately arises in the simplified phase-frequency detector due to gain imbalances as described earlier with the $\gamma(s)$ factor of (2.15). As reported in [31], gain imbalances between the two postdetection filter arms can lead to unexpectedly long phase-locking times. A different explanation than that given in [31] seems appropriate. Ideally, the lead-lag loop filter maintains a single state variable in the form of the voltage residing on a single integration capacitor. In the differential filter form, two capacitors are present with each experiencing slightly different time constants owing to these imbalancing effects. As the active outputs from the differential phase detector alternate, the system time constants are in effect modulated in a nonlinear manner. Closed-form analysis of this situation would of course be quite difficult if not impractical.

The MACSET computer program described in Chapter 8 was used to examine the gain imbalance question further using the simple phase-locked loop shown in Figure 2.13.

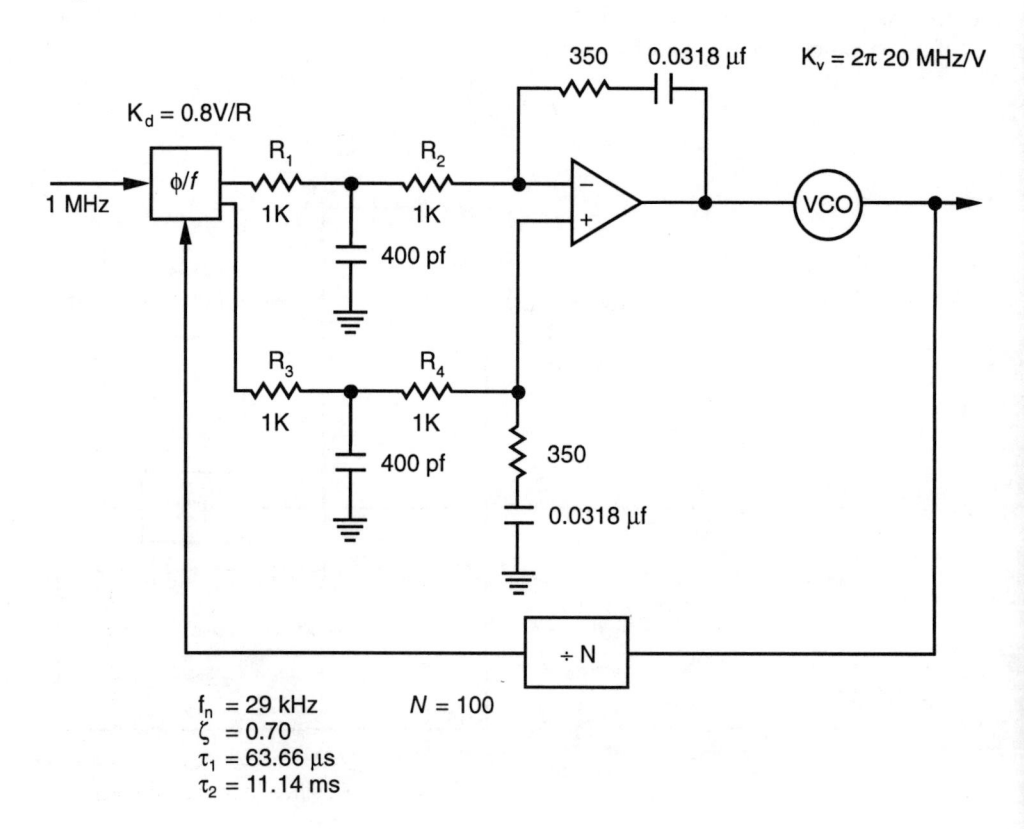

Figure 2.13 Phase-locked loop structure used to investigate differential phase detector arm imbalance.

The nominal loop parameters selected were as follows:

$\omega_n = 2\pi\ 20$ kHz
$\zeta = 0.70$
$N = 150$

Each transient response considered was based on changing the feedback divider from 150 to 151 in a synchronous manner. Three cases were then considered:

Case 1: Frequency hop $N = 150$ to 151 with perfectly matched arms.
Case 2: Frequency hop $N = 150$ to 151 with R_1 and R_2 in error by -10%.
Case 3: Frequency hop $N = 151$ to 150 with R_1 and R_2 in error by -10%.

Case 2 as compared to case 1 is shown in Figure 2.14 and shows little if any impact due to the imbalance. Case 3 as compared to Case 1 is shown in Figure 2.15 and a definite phase tail is clearly visible. Although it was possible to reduce the phase tail somewhat by using a higher damping factor for the loop, the tail presence could not be completely eliminated. Further investigation of this phenomenon is certainly warranted.

2.2.3 Voltage Controlled Oscillators

The principle purpose for employing phase-locked loop technology is to improve the frequency and phase stability of an otherwise free-running oscillator. The control that is exerted on the oscillator is in the form of a control voltage that normally controls the oscillator's output frequency using a varactor. Throughout most of this text, the VCO is assumed to be an integrator of phase (Chapter 4). Detailed design issues pertaining to VCOs are considered in Chapter 3.

As high-frequency monolithic phase-locked loops become more and more common place, new techniques for dealing with the difficulties imposed by work in silicon will become more widely known. Special design considerations are normally necessary for dealing with issues such as ground-bounce in mixed-mode devices and high $1/f$ noise in GaAs devices. A number of representative examples can be found in [32] where monolithic phase-locked loops operating as high as 28 GHz are considered.

2.2.4 Feedback Dividers

The feedback divider plays an integral role in modern phase-locked frequency synthesis in that aside from the VCO itself, only a portion of the feedback divider op-

(a)

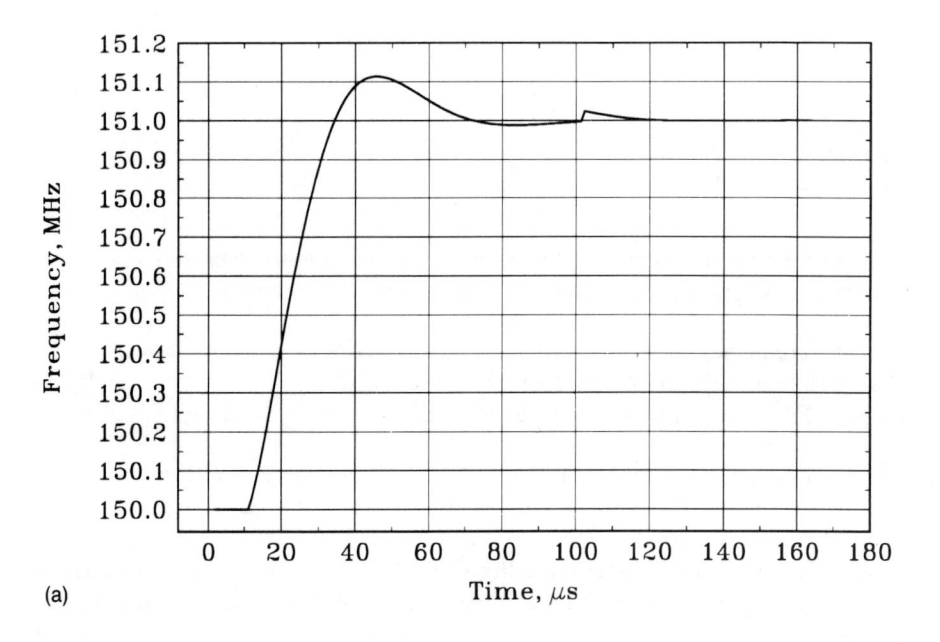

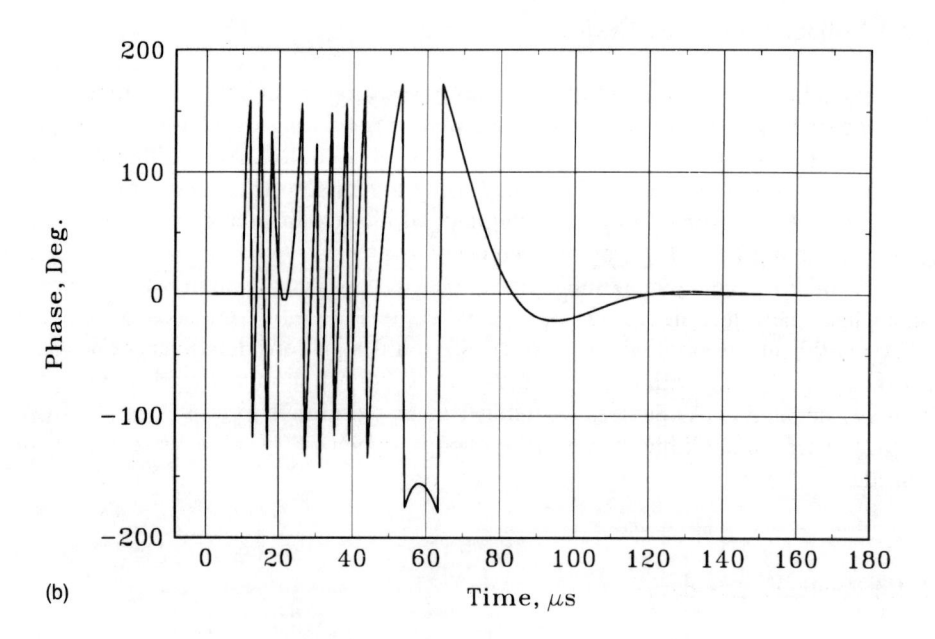

(b)

Figure 2.14 (a) Frequency and (b) phase settling for case 1, which has no gain imbalance.

(c)

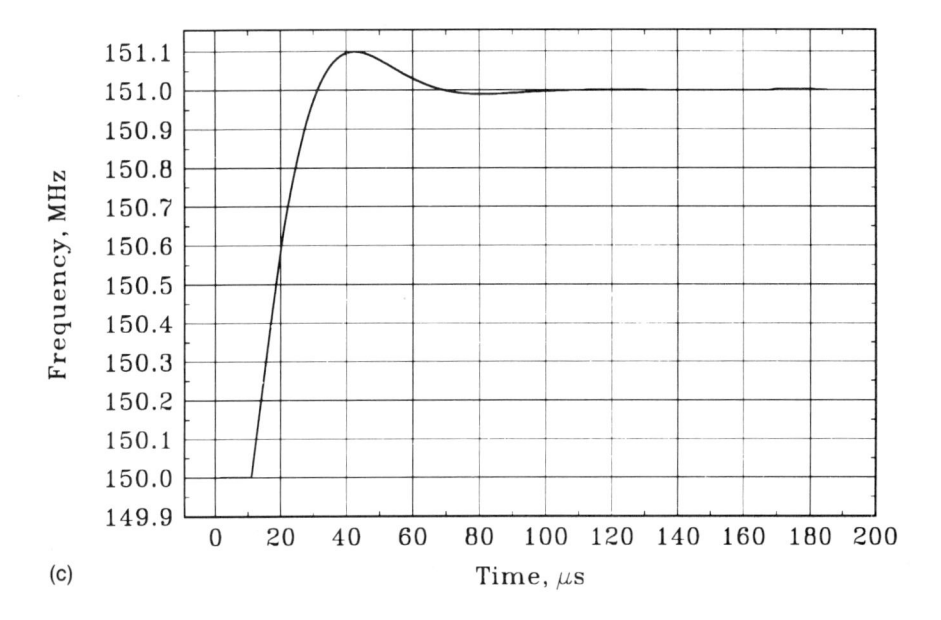

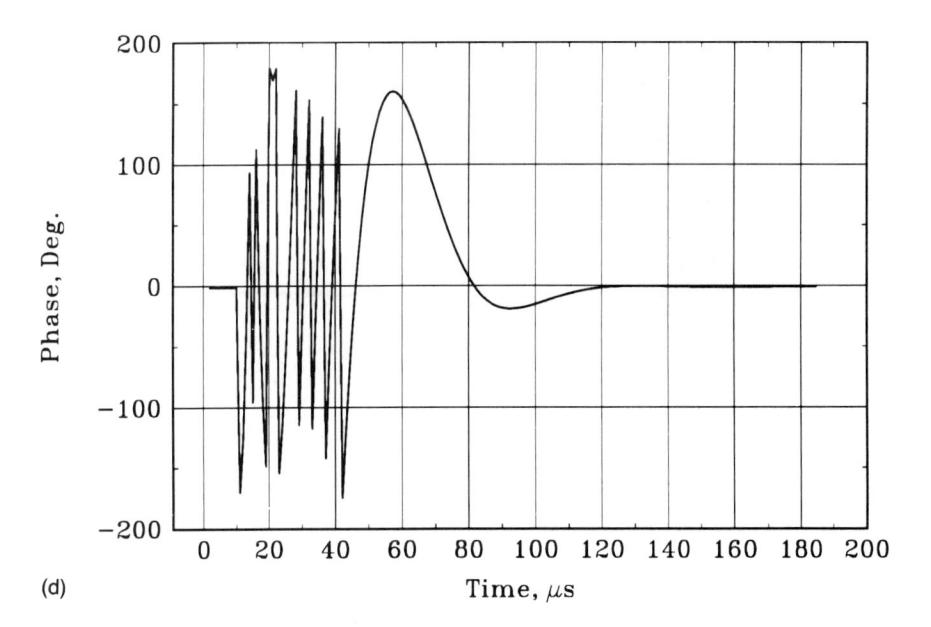

(d)

Figure 2.14 (Continued). (c) Frequency and (d) phase settling for case 2.

(a)

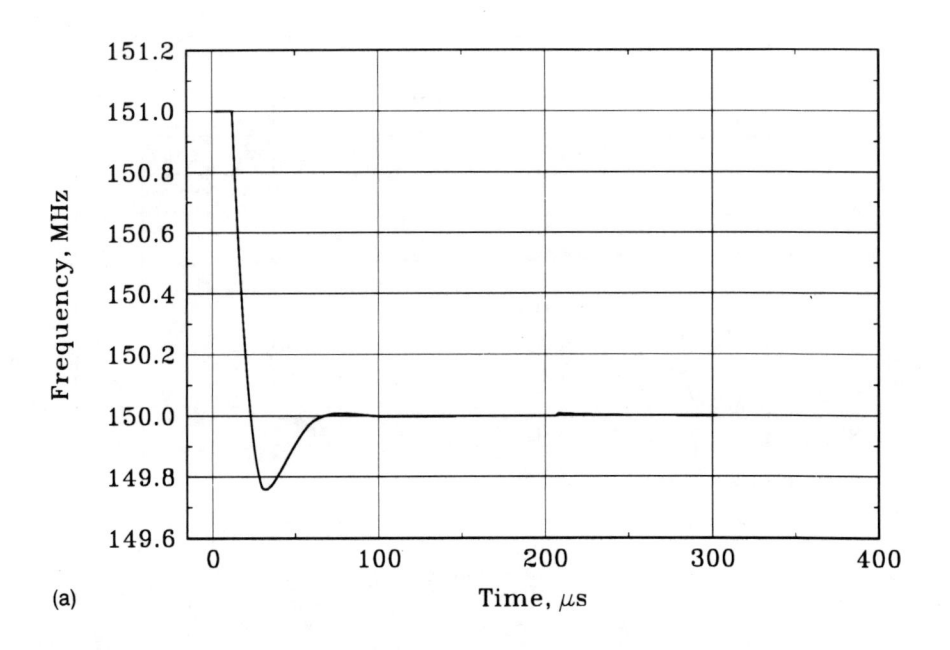

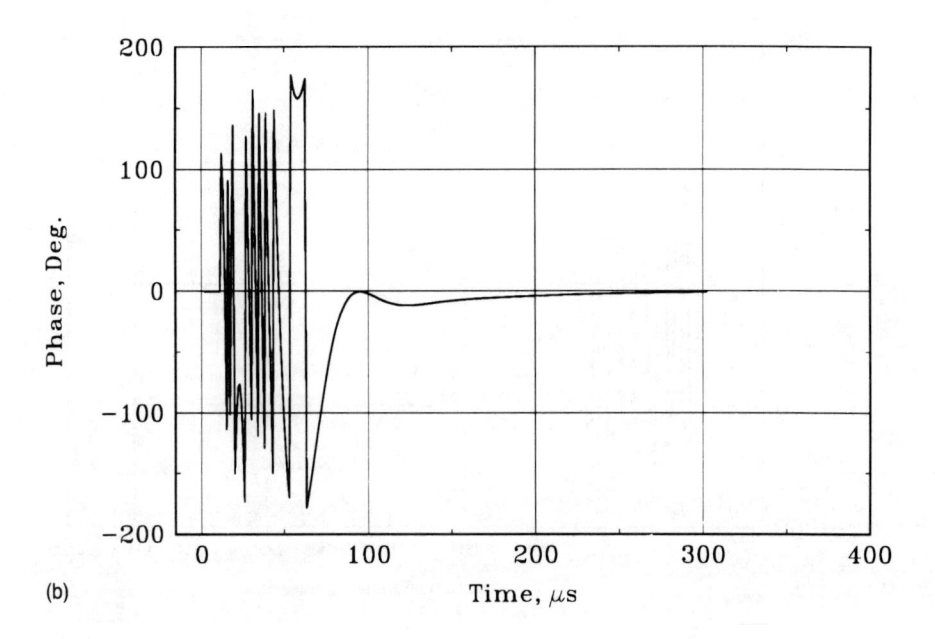

(b)

Figure 2.15 (a) Frequency and (b) phase settling for case 3.

erates at the full frequency rate of the VCO output. All other control functions within the loop are essentially handled at much lower baseband frequencies.

The concepts of the feedback divider as a "zero-crossing selector gate" and as a zero-crossing "sampler" are introduced in Chapter 4. Phase noise performance of digital divider elements is discussed in Chapters 3 and 5. Divider signal contamination issues are briefly examined in Sec. 2.4.

2.3 FREQUENCY-LOCKED LOOPS

Aside from this brief section, there will be no further attention given to frequency-locked loops within this text. Since most modern systems are striving to obtain the last ounce of performance (which normally implies coherent processing), frequency-locked loops see much less demand than their phase-locked counterparts. Three references [33–35] for these loops are cited in the bibliography.

2.4 SIGNAL CONTAMINATION IN PHASE-LOCKED LOOPS

Signal contamination in phase-locked loops is perhaps the greatest design fear encountered in synthesizer design aside from simply having the design "work." Only a modest amount of material addressing circuit detail issues is presented in this text but that in no way should detract from the importance of this subject in actual design practice.

Aside from normal design practices, which should be fairly commonplace, one additional perspective added by this text is the necessity to consider sampling effects during the design process. As discussed at length in Chapters 4 and 5, wherever there is a digital element within a phase-locked loop, the system is in effect a sampled control system. Therefore, any (and all) contaminant signals that may be present will be aliased to baseband by the sampling action.

Take for instance a situation where the entire phase-locked loop is operating at frequencies that are strictly (coherent) multiples of 3 MHz except for the reference frequency in the system which is at 10 MHz. Any 10 MHz present near the dividers will automatically be aliased to baseband and appear as potential 1-MHz spurs (10 MHz − 3 · 3 MHz). Because frequency dividers are inherently wideband, a contaminant at 303.0123 Mhz could in principle be aliased to baseband and create spurious elements at 12.3 kHz. This perspective clearly widens one's definition of "acceptable contaminant signals."

The same considerations, of course, are applicable to power supply-related contamination [36]. As stated earlier in this chapter, every signal entering or leaving a frequency synthesizer is a potential source of signal contamination. Power supply-related spurious contaminants for digital devices are directly related to (1) the state transition slew rate of the digital device and (2) the amplitude of the power supply induced contaminant as shown in Figure 2.16. In this figure, voltage supply contaminations were introduced from frequencies ranging from 10 to 200 kHz and the dBc spurious sideband levels measured. In both circuit cases, the measurements were essentially independent of the contamination frequency. This statement is true for contaminant frequencies well into the megahertz region.

Serious sideband contamination can also result in RF amplifiers that are run into output signal compression. Since VCOs must inherently contain some form of self-limiting, power supply contamination can easily result in undesired discrete sideband spurs there also. A wealth of knowledge concerning the effects of small contaminating signals on nonlinear elements used in frequency synthesis and conversion can be found in [37].

(a) Test circuit considered

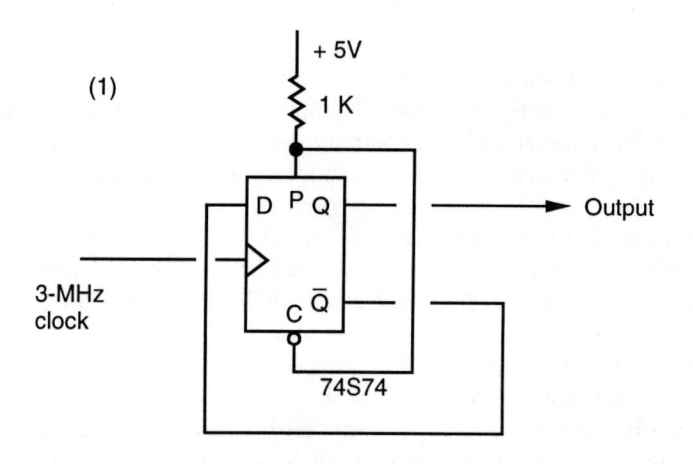

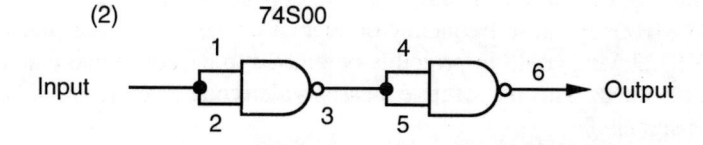

Figure 2.16 Spurious sideband levels (dBc) versus power supply contamination level (mV pk-pk): (a) Divide-by-2 flip-flop circuit (74574) and (b) cascaded NAND (74500) gate circuit.

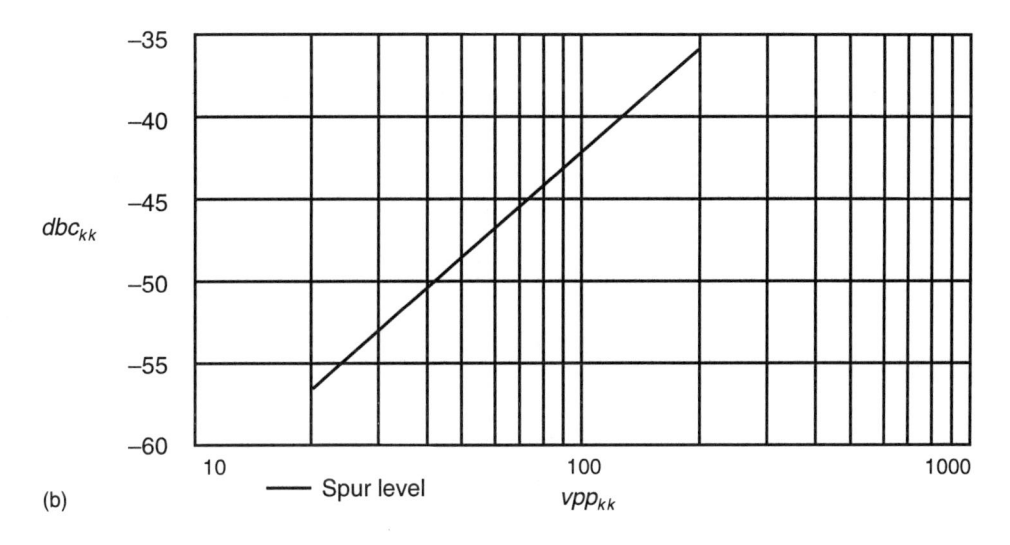

Figure 2.16 (Continued).

2.5 TO GROUND OR NOT TO GROUND

The most important signal in RF design in general is ground. Without proper signal grounding, the best laid plans of man run aground! Without proper grounding and shielding, no amount of ingenuity can rescue the hardware design. All too frequently, however, consistent grounding guidelines are left as an afterthought rather than being identified at the beginning of the overall design process. Several first principles for proper RF circuit grounding are presented within this section for the simple reason that grounding is a system issue.

Ohm's Law

Shielding and grounding issues are often lumped into the category of RF "black magic" where discussions of skin depth and circuit radiation effects abound. Although these issues are certainly at play, at its most basic level, grounding techniques must derive from Ohm's law.

Consider for example the simple low-pass filter shown in Figure 2.17. Depending on the grounding techniques employed, due to the finite ground plane distributed resistance and inductance, the circuit that is actually realized is as shown in Figure 2.18.

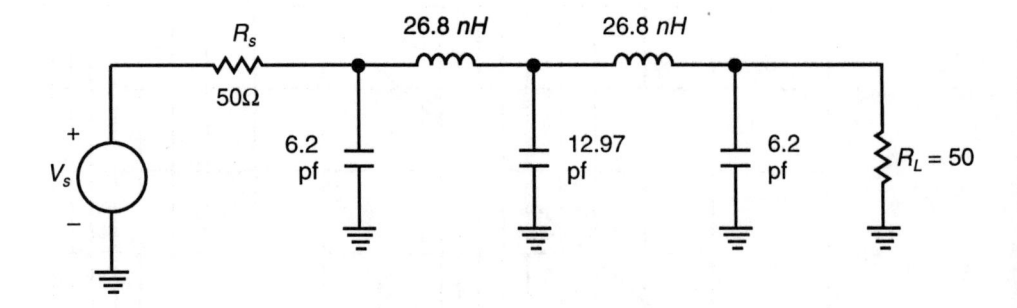

Figure 2.17 Idealized $N = 5$ Chebyshev lowpass filter ($f_c = 500$ MHz, 0.1-dB ripple).

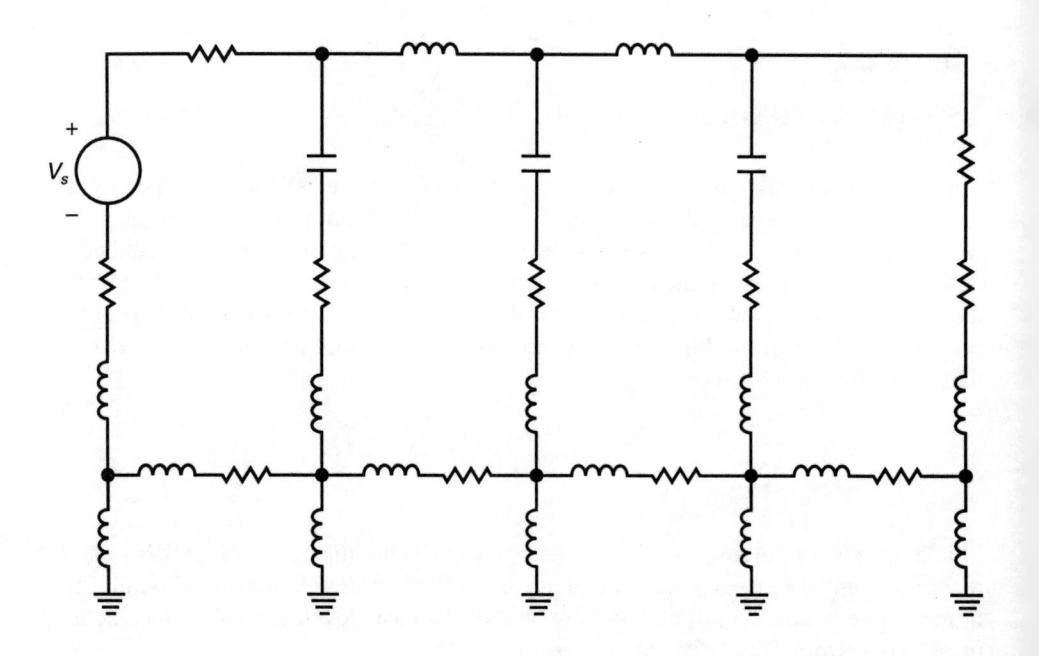

Figure 2.18 Actual Chebyshev filter circuit with primary circuit parasitics shown.

To amplify this perspective further, consider the stray inductance effects of the circuit board vias on the performance of this same filter as shown in Figure 2.19. Anticipating this problem, further assume that the designer has laid a microstrip length of copper along the top side of the circuit board, which connects the grounded side of the shunt components as shown.

The attenuation performance of this filter was analyzed using Touchstone[1] for copper strip widths of 20, 100, 250, and 500 mils. The results are shown in Figure 2.20. Note how the stopband attenuation performance is actually degraded by the added microstrip line for some width values. In effect, it is as if the introduction of this additional grounding provides an additional path to the filter output, thereby worsening the overall performance. This is precisely the point that must be made; ill-conceived grounding measures can impair rather than improve performance.

Grounding in the context of Ohm's law is as simple as recognizing that just as much care must be taken with the current (signal) flow out of every electrical components as with current flow into the component. The microstrip example just considered is not unlike attempting to carry two grounding paths for the circuit and therein lies the fundamental problem. There should be no confusion as to how a signal current is returned to its source ground.

Grounding Guideline No. 1: In RF circuit design, use one ground plane. There should be no confusion as to how each signal current is returned (through ground) to its source.

---

[1]Touchstone is a product from Eesof, Inc. Westlake, CA.

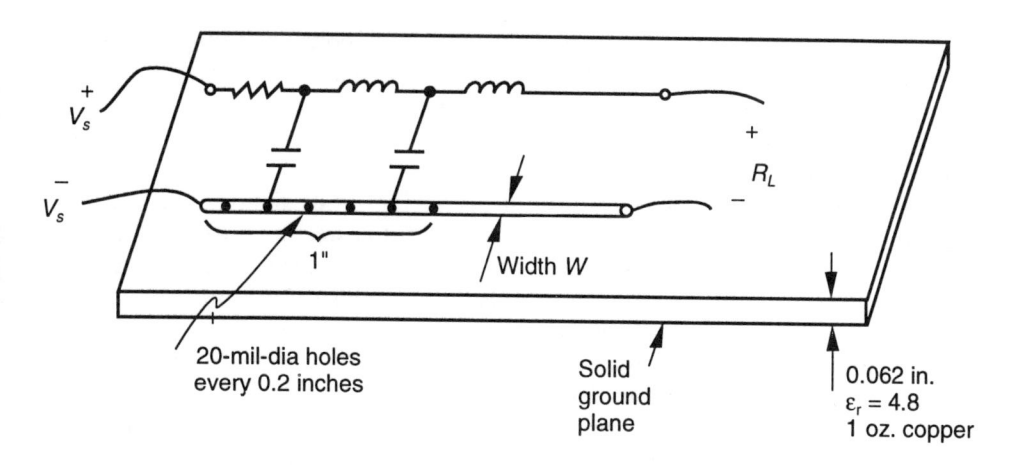

Figure 2.19 Grounding effectiveness when carrying an additional top-layer microstrip ground length.

- 20-mil-dia holes every 0.2 inches
- Solid ground plane
- 0.062 in.
- $\varepsilon_r = 4.8$
- 1 oz. copper
- Width W
- 1"

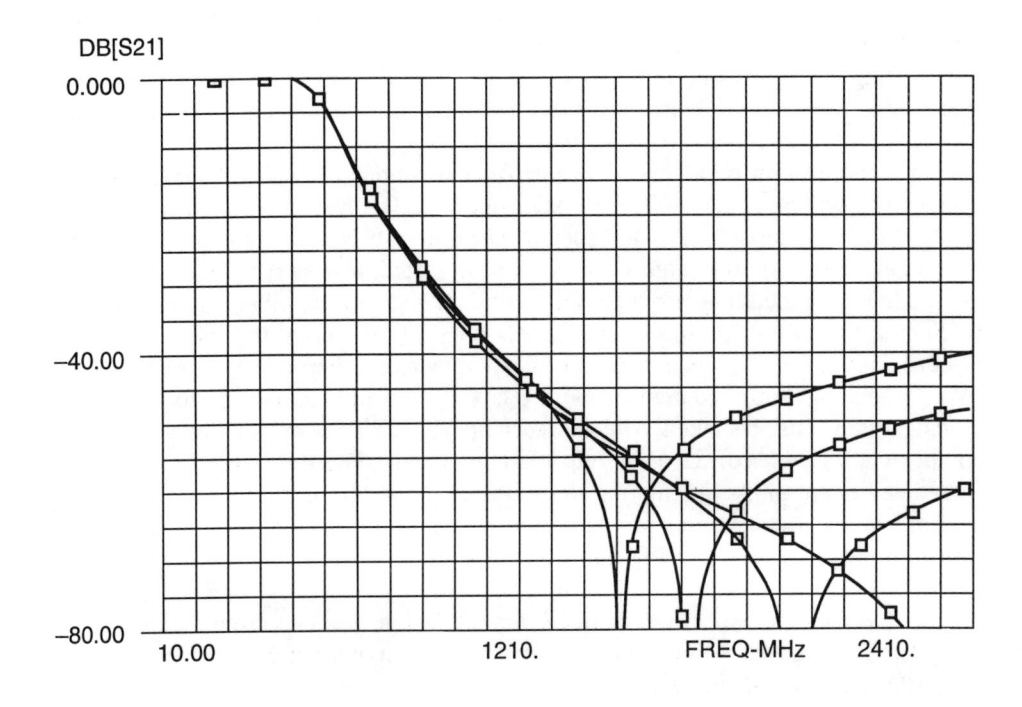

Figure 2.20 Filter stopband degradation as a function of increasing microstrip width (20, 100, 250, and 500 mils).

The ideal grounding and shielding situation is as shown in Figure 2.21 where each electrical components is shielded and contained within its own cavity. Clearly, the flow of every signal and its return current are easily identifiable as being constrained by the cavity walls. Such an approach is obviously inappropriate for small, low-cost products such as cellular phones.

The important concept to grasp from Figure 2.21 is its similarity with a classical transmission line. Current flow within an ideal transmission line is, of course, well defined and no signal leakage occurs. Each cavity shown in Figure 2.21 can be envisioned as a bubble along a transmission line. Expanding further on this thought, an amplifier-filter circuit could appropriately be built as shown in Figure 2.22. This form of circuit shielding and grounding is more reasonable and also retains excellent integrity. Although it is still unsuitable for many cases based on its size and cost, this form of construction has been used for decades within the electronics industry with excellent results. Here again, signal and return currents are easily identifiable.

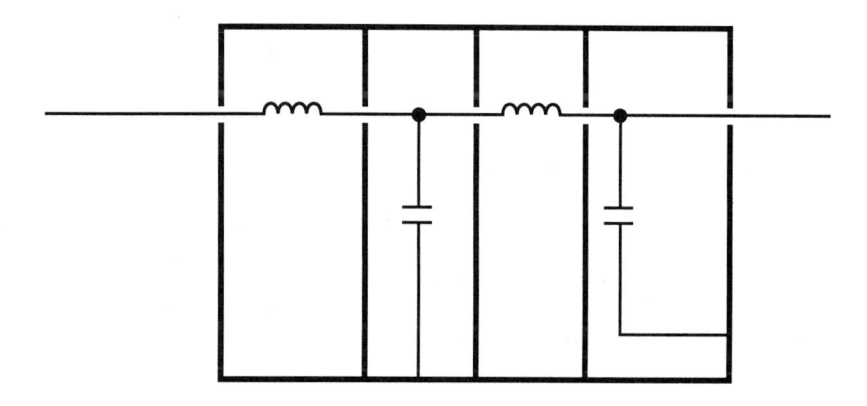

Figure 2.21 Idealized cavity-based electronic shielding.

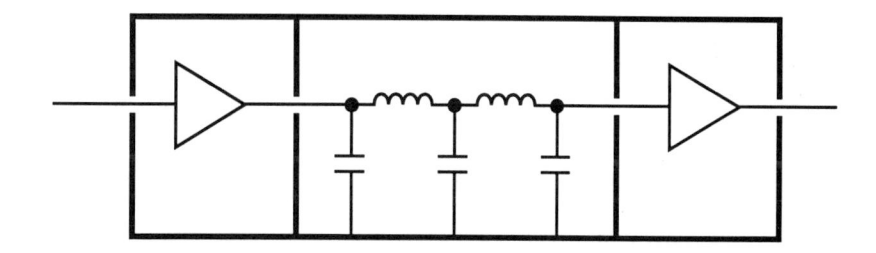

Figure 2.22 More reasonable cavity shielding based on proper functional partitioning.

The performance attainable in Figure 2.22 can be approached with less cost by using the grounding approach shown in Figure 2.23. The cavity construction shown in Figure 2.22 is approximated here by using separate coplanar ground plane areas for each circuit area that is to be self-contained. Power is provided to the first circuit area, again using transmission line reasoning with the local grounding brought to a single point. This approach ideally eliminates signal current flow in the power distribution circuitry between modules yet maintains a coaxial type connection between modules by way of the microstrip structures used to connect functional areas. If more serious shielding is required for a single circuit area such as a filter, an appropriate shield can be installed for that one area alone.

Grounding Guideline No. 2: Apply the transmission line concept to signal and power distribution thereby eliminating signal flow except where it is intended.

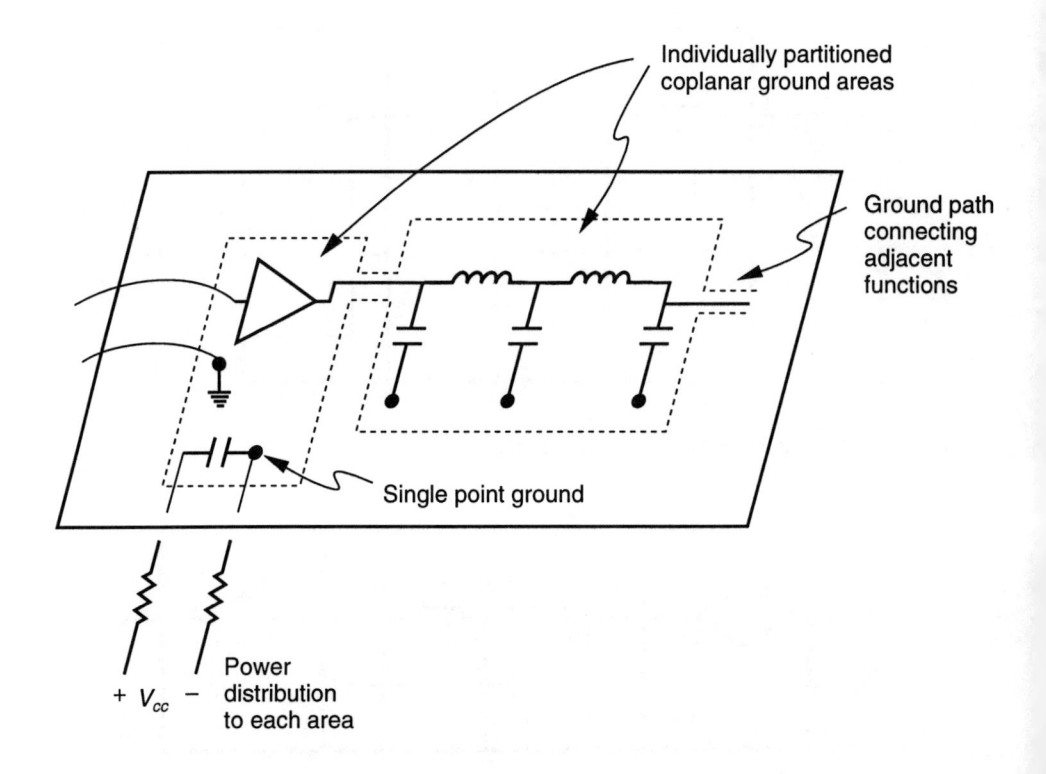

Figure 2.23 Cost-effective grounding methodology.

A closely related guideline pertains to situations in which a signal (digital, RF, or otherwise) must be routed some distance away. Since the characteristic impedance of any transmission line increases with the distance from its center conductor to its shield, which increases signal radiation, conductor-shield spacing should be kept to a minimum (or equivalent to 50 Ω between functional areas).

Grounding Guideline No. 3: When transporting signals between functional circuit areas, the accompanying signal return path should be kept physically close in order to keep the line's characteristic impedance at a proper level.

In mixed-signal applications, separate ground areas such as those shown in Figure 2.23 are generally mandatory. Digital signals with their fast edges quickly contaminate most RF circuits. In these situations, differential signals are highly recommended for signaling between the different circuit areas, particularly if physical separations are large.

In keeping with the focus of this text, explicit grounding details such as shielding effectiveness, skin depth, etc., have not been addressed in detail. A more detailed introduction into these matters can be found in [2]. As a closing note, with almost 20 years in this industry, the author has observed only a handful of contamination issues that were caused by signal radiation problems; the remainder have been conducted signal problems that were explainable by simply employing Ohm's law.

REFERENCES

[1] Middlebrook, R.D., "Design Techniques for Preventing Input-Filter Oscillations in Switched-Mode Regulators," *Fifth National Solid-State Power Conversion Conference,* San Francisco, CA, May 4–6, 1978.
[2] Manassewitsch, V., *Frequency Synthesizers Theory and Design,* 2nd ed., John Wiley & Sons, New York, 1980.
[3] Rohde, U.L, *Digital PLL Frequency Synthesizers Theory and Design,* Prentice-Hall, Englewood Cliffs, NJ, 1983.
[4] Egan, W.F., *Frequency Synthesis by Phase Lock,* John Wiley & Sons, New York, 1981.
[5] Wolaver, D.H., *Phase-Locked Loop Circuit Design,* Prentice-Hall, Englewood Cliffs, NJ, 1991.
[6] Grossimon, H.P., et al., "Phase Comparator Circuit," U.S. Patent No. 3,551,808, December 29, 1970.
[7] Wulich, D. "Fast Frequency Synthesis by PLL Using a Continuous Phase Divider," *Proc. IEEE,* Vol. 76, No. 1, January 1988, pp. 85–86.
[8] Halgren, R.C., et al., "Improved Acquisition in Phase-Locked Loops with Sawtooth Phase Detectors," *IEEE* COM-30, No. 10, Oct. 1982, pp. 2364–2375.
[9] Lindsey, W.C., C.M. Chie, *Phase-Locked Loops,* IEEE Press, New York, 1986.
[10] Lindsey, W.C., M.K. Simon, *Telecommunication Systems Engineering,* Prentice-Hall, Englewood Cliffs, NJ, 1973.
[11] Meyr, H., G. Ascheid, *Synchronization in Digital Communications, Volume 1,* John Wiley and Sons, New York, 1990.
[12] Blanchard, A., *Phase-Locked Loops Applications to Coherent Receiver Design,* John Wiley & Sons, New York, 1976.
[13] Gardner, F.M., *Phaselock Techniques,* 2nd ed., John Wiley & Sons, New York, 1979.
[14] Kurtz, S.R., "Specifying Mixers as Phase Detectors," *Microwaves,* January 1978.
[15] Walls, F.L., S.R. Stein, "Accurate Measurements of Spectral Density in Devices," *31st Annual Frequency Control Symposium,* 1977.
[16] Fischer, M.C., "Frequency Domain Measurement Systems," *10th Annual Precise Time and Time Interval Applications and Planning Meeting,* December 1978.
[17] Walls, F.L., S.R. Stein, J.E. Gray, "Design Considerations in State-of-the-Art Signal Processing and Phase Noise Measurement Systems," *30th Annual Frequency Control Symposium,* 1976.
[18] Gilchrist, B., et al., "Sampling Hikes Performance of Frequency Synthesizers," *Microwaves & RF,* January 1984.
[19] Gilchrist, B., et al., "The Use of Sampling Techniques for Miniaturized Microwave Synthesis Applications," *IEEE MTT-S Digest,* 1982, pp. 431–433.
[20] Weisskopf, P.A., "Subharmonic Sampling of Microwave Signal Processing Requirements," *Microwave J.,* May 1992.
[21] Vella, P., et al., "Novel Synthesizer Cuts Size, Weight, and Noise Levels," *Microwaves & RF,* May 1991.
[22] Egan, W.F., "Sampling Delay—Is It Real?" *RF Design,* February 1991.

[23] *Electronics*, May 5, 1983, p. 137.
[24] Osafune, K., et al., "High-Speed and Low-Power GaAs Phase Frequency Comparator," *IEEE Trans.* Vol. MTT-34, January 1986, pp. 142–146.
[25] Gavin, D.T., R.M. Hickling, "A PLL Synthesizer Utilizing a new GaAs Phase Frequency Detector," INS EEA 89–000943.
[26] Hill, A., J. Surber, "The PLL Dead Zone and How to Avoid It," *RF Design*, March 1992.
[27] Analog Devices, "Ultrahigh Speed Phase/Frequency Discriminator," Data Sheet AD9901.
[28] Sharpe, C.A., "A 3-State Phase Detector Can Improve Your Next PLL Design," *EDN*, September 20, 1976, pp. 55–59.
[29] O'Leary, M., "Practical Approach Augurs PLL Noise in RF Synthesizers," *Microwaves & RF*, September 1987.
[30] Gardner, F.M., "Charge-Pump Phase-Lock Loops," *IEEE Trans.* Vol. COM-28, November 1980, pp. 1849–1858.
[31] Endres, T.J., J.B. Kirkpatrick, "Sensitivity of Fast Settling PLLs to Differential Loop Filter Component Variations," *IEEE 47th Annual Symposium on Frequency Control*, Hershey, PA, May 27–29, 1992.
[32] Digest of Technical Papers, IEEE International Solid-State Circuits Conference, Session 5, 1992.
[33] Walls, F.L., S.R. Stein, "A Frequency-Lock System for Improved Quartz Crystal Oscillator Performance," *IEEE Trans. Instrum. Measurement*, September 1978, pp. 249–252.
[34] Puglia, K.V., "Frequency Locking a Microwave Source," *Appl. Microwave*, November/December 1989.
[35] Rao, J.B.L., W.M. Waters, "Voltage-Controlled Oscillator with Time-Delay Feedback," Naval Research Laboratory, Washington D.C., date unknown.
[36] Martin, M., "RF Sidebands Caused by DC Power Line Fluctuations," *Microwave J.*, September 1991.
[37] Ward, C.J., "Delay Reduction Techniques in Phase-Locked Loop Amplifiers," *Electron. Lett.*, April 2, 1981, pp. 253–255.

SELECTED BIBLIOGRAPHY

Egan, W.F., "The Effects of Small Contaminating Signals in Nonlinear Elements Used in Frequency Synthesis and Conversion," *Proc. IEEE*, July 1981, pp. 797–811.

McNab, K., "Measure Open-Loop Gain Using Floating Voltage (Open-loop Gain Measurements, Part 2)," *Microwaves & RF*, April 1987.

"Method of Detecting Frequency Offset," Section 3.5.2, *Bell Syst. Tech. J.*, December 1983, pp. 3386–3389.

Shen, E.K., et al., "Monolithic GaAs Variable Modulus Divider for Frequency Synthesis Applications," *Microwave J.*, August 1985.

Chapter 3
Phase Noise and Its Impact on System Performance

All practical frequency sources inherently have undesirable phase (or frequency) modulation imposed on the otherwise ideal carrier due to internal noise within the source [1]. Thermal noise is also the ultimate factor limiting the performance of highly sensitive systems. The free electrons that are in random motion within a conductor due to thermal agitation create an open-circuit noise voltage at the ends of the conductor as shown in Figure 3.1. Since a physical conductor has losses with an associated resistance R, which is positive, the conductor may be modeled as a noise generator and ideal noiseless resistance as shown in Figure 3.1. The broadband noise spectrum represented by e_o in Figure 3.1 is known as Johnson noise [2]. The magnitude of the mean-squared thermal noise voltage in a conductor having resistance R is given by [3,4]

$$e_o^2 = 4kTR \int_{f_1}^{f_2} \frac{hf}{kT(e^{hf/kT} - 1)} df \qquad (3.1)$$

where

k = Boltzmann's constant 1.38×10^{-23} J/K
T = resistor temperature Kelvin
R = resistance ohms
h = Plank's constant 6.62×10^{-34} J/sec

and f_1 and f_2 represent the frequency limits of interest in hertz. This relationship is intimately involved with an area of physics known as blackbody radiation [2].

Phase noise is particularly important in receiving and transmitting systems [5]. Undesirable phase noise levels can lead to (1) increased bit error rates (BERs) in digital communication systems, (2) contamination of adjacent frequency channels,

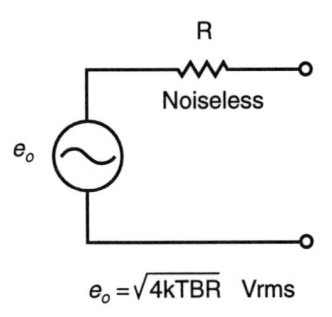

Figure 3.1 Noise source Thevenin equivalent circuit.

$$e_o = \sqrt{4kTBR} \text{ Vrms}$$

and (3) receiver desensitization in the presence of other strong signals due to reciprocal mixing. These different performance degradations are depicted in Figure 3.2.

In Figure 3.2(a), phase noise present in the transmitter and receiver local oscillators cause the otherwise stationary signal constellation to rotate in a random fashion, thereby degrading BER performance. In Figure 3.2(b) the spectral sidelobes of the transmitted waveform are shown falling into adjacent frequency channels, thereby increasing the noise level on these channels for any receivers located nearby. Finally in Figure 3.2(c), the local oscillator phase noise sidebands in the receiver are shown directly transferred to the signals at the IF port due to the mixing process. The presence of strong adjacent signals mixing with the local oscillator phase noise sidebands results in a corresponding reduction in the receiver's sensitivity.

The importance for a sound understanding of phase noise basics in frequency synthesizer design cannot be understated. All too frequently, system designers are not aware of the cost versus performance trade-offs involved in synthesizer design, whereas RF hardware designers are often not conversant with system performance measures. In the end, this information gap must be bridged in order to realize a cost-effective, high-quality final product.

Phase noise requirements vary dramatically depending on the modulation waveform being used (e.g., FSK, MPSK) and the system interoperability concerns that apply. A number of different modulation types and the impact of phase noise on system performance are considered toward the end of this chapter.

In high-speed data communications, phase noise creates clock jitter, which affects the clock recovery process in bit synchronizers. The same type of noise phenomena causes degraded BER performance for these systems.

The content of this chapter spans a large range of topics from basic phase noise concepts through device noise contributions, to the actual impact on communication system performance. The chapter begins with an introduction to basic phase noise concepts beginning with a deterministic [sinusoidal phase modulation (PM)] example. Then power spectral density concepts are developed along with a detailed dis-

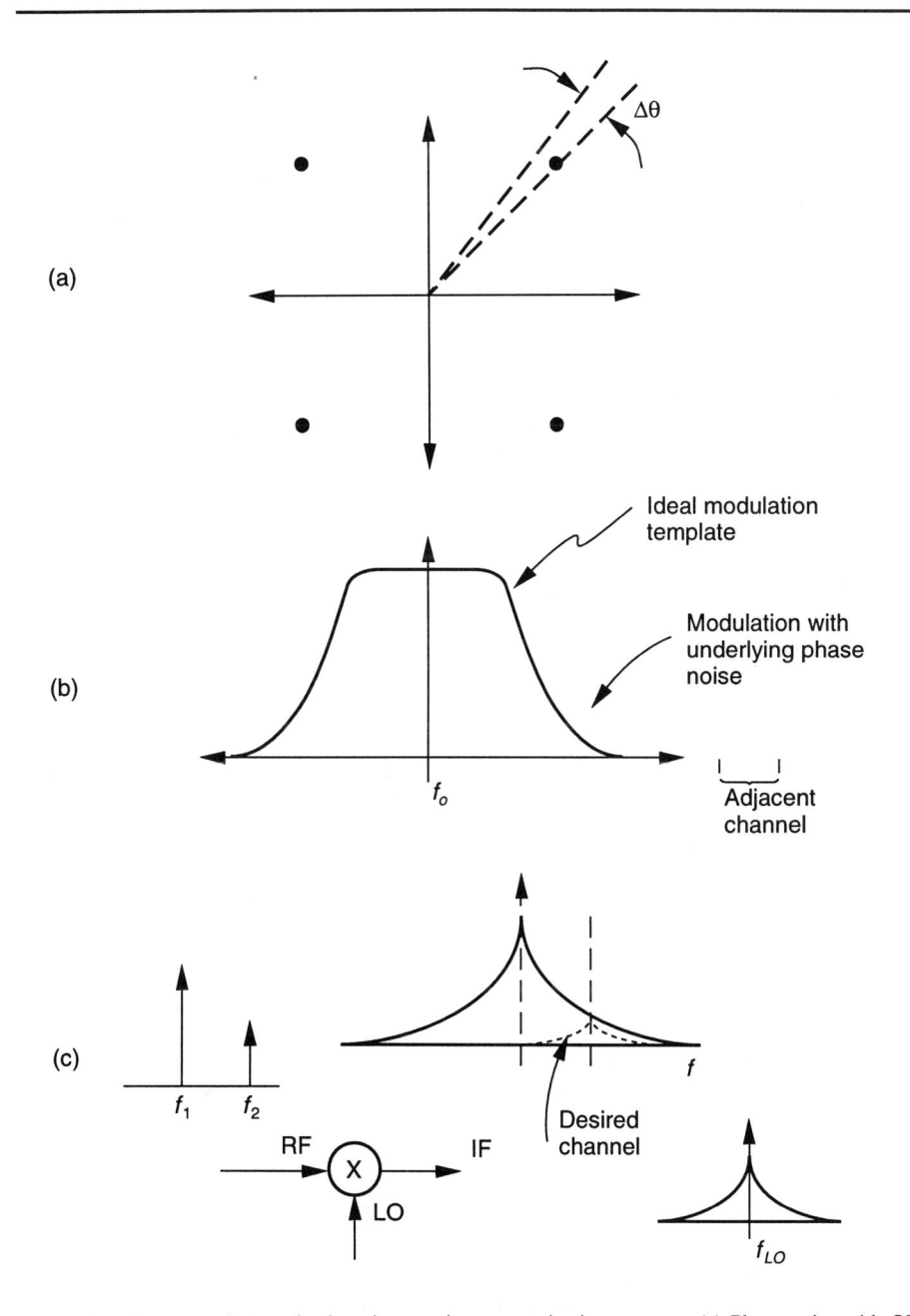

Figure 3.2 Common phase noise impairments in communication systems: (a) Phase noise with QPSK causes the constellation points to be randomly rotated by $\Delta\theta$, thereby increasing bit errors. (b) Transmitted sidelobe energy can fall into adjacent channels. (c) Reciprocal mixing due to high phase noise sidebands can lead to out-of-band signals being translated into the IF.

cussion about simulating random signal quantities in the time domain. Phase noise contributions that arise in different phase-locked loop (PLL) components are then discussed. Based on this material, attention is finally given to quantifying the impact of phase noise on the BER performance of a number of digital communication systems. This latter material is intended as only an introduction to this very broad area of communication system engineering.

3.1 GENERAL NOISE THEORY

The theory relating to phase noise has received substantial attention in recent years, with much of the work addressing the development of precise atomic clocks. Noise processes are normally categorized with respect to their power spectral density (PSD) behavior, but with the appearance of fractals [6] and chaos—to name only two active research areas—other characterization techniques can also be encountered. Primary attention in this text is given to the PSD characterization.

Thermal noise or Johnson noise is associated with the thermodynamic properties of a simple resistance as described earlier. The statistics of the thermal noise voltages are Gaussian [7]. The PSD of the noise is white in that all frequencies display the same spectral density of kT W/Hz.

Shot noise results from the discrete nature of electronic current and also displays a white PSD. Shot noise is generally described by the Poisson distribution.

Power law noise sources are sources whose PSD varies as f^α over a wide range of Fourier frequencies f. Obviously, some bounds on f must apply near 0 and ∞ depending on the sign of α. The most common power law noise sources encountered in communications work are given in Table 3.1.

Table 3.1
Power Law Noise Sources

Name	PSD Shape
White PM	Flat
Flicker PM	$\propto f^{-1}$
White FM	$\propto f^{-2}$
Flicker FM	$\propto f^{-3}$
Random Walk FM	$\propto f^{-4}$

Development of the PSD concept is presented later in this chapter.

Flicker noise is a particularly interesting phenomena within nature. It can be found almost everywhere, from the daily height of the Nile river to the music of Bach and the Beatles [7]. In fact, the $1/f$ noise process provides a remarkably good starting point for stochastic music composition [8,9].

3.1.1 Phase Noise First Principles

An intuitively attractive means for representing the output of an oscillator is given by

$$V(t) = [V_o + \epsilon(t)] \sin[2\pi \nu_o t + \varphi(t)] \tag{3.2}$$

where the amplitude variation is represented by $\epsilon(t)$, the nominal output frequency is given by ν_o, and $\phi(t)$ is a random process representing phase noise. Throughout all of this material, $\epsilon(t)$ is assumed to be identically equal to zero and only the phase noise contribution is considered.

In initiating this discussion about phase noise, consider first the deterministic signal (sinusoidal phase modulation) case where

$$\varphi(t) = \Delta\theta \sin(\omega_m t) \tag{3.3}$$

Substituting (3.3) into (3.2) and expanding,

$$V(t) = V_o \sin(2\pi\nu_o t) + \frac{V_o \Delta\theta}{2} \{\sin[(2\pi\nu_o + \omega_m)t] - \sin[(2\pi\nu_o - \omega_m)t]\} \tag{3.4}$$

The resulting spectrum is shown in Figure 3.3. In arriving at this result, the small angle assumptions applicable when $|\Delta\theta| \ll \pi/2$ have been used. Otherwise, the more rigorous derivation given in Chapter 2 of [10] must be followed.

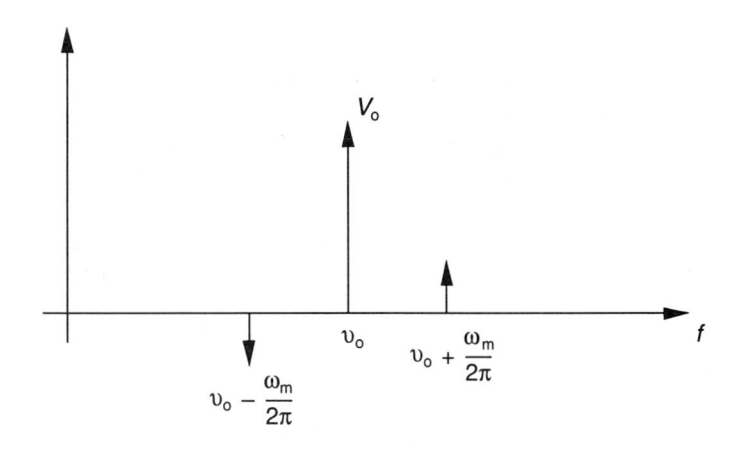

Figure 3.3 First-order spectrum for sinusoidal PM.

The key point to be made in this example is that the level of each sideband tone with respect to the carrier is given by

$$\mathscr{L}(f_m) = 20 \log_{10}\left(\frac{\Delta\theta}{2}\right) \qquad (3.5)$$

where once again, $|\Delta\theta| \ll \pi/2$ has been assumed.

In taking the next logical step where the phase perturbations represented by $\phi(t)$ are assumed to arise from random noise processes, we must employ several basic concepts from stochastic processes, specifically the concept of PSD. Since this is a very important subject central to the remainder of this chapter, some time is devoted to developing this important topic.

3.1.2 Power Spectral Density Concept

The PSD concept is central to the study of phase noise. The traditional definition for the PSD $P_x(f)$ of a random signal $x(t)$ is given by

$$P_x(f) = \lim_{T \to \infty} \left\{ \frac{1}{T} \mathbf{E}[|X_T(f)|^2] \right\} \qquad (3.6)$$

where $X_T(f)$ is the finite time Fourier transform of $x(t)$ defined as

$$X_T(f) = \int_{-T/2}^{T/2} x(\tau) e^{-j2\pi f \tau} d\tau \qquad (3.7)$$

and $\mathbf{E}$ denotes statistical (ensemble) expectation [11]. In more tangible terms, if x represents the random source phase perturbations as given in (3.2), then (3.6) closely represents the phase noise spectrum that would be seen on a spectrum analyzer. In this form, $P_x(f)$ has units of watts per hertz. This specific viewpoint is developed shortly.

Although the definition of PSD (3.6) involves statistical expectation, the basic operation of a classical spectrum analyzer provides a clear indication that the PSD can be obtained (for most signals encountered in everyday practice) by using time averages exclusively rather than ensemble averages. This perspective has been revived in several popular recent texts [12–14]. This viewpoint is advantageous since the theory of random processes based on time averages is considerably more developed than is the theory based on statistical ensemble averages. The time averages more closely represent the physical world for most communication phase noise problems as well. The time-averaged-based concepts have historically been known as generalized harmonic analysis and were first explored by Wiener.

In developing the definition of PSD based on time-averaged quantities, first consider the (ergodic) random process $x(t)$ and its finite time segment defined by

$$x_T(t) = \begin{cases} x(t) & |t| \leq T/2 \\ 0 & |t| > T/2 \end{cases} \tag{3.8}$$

and its Fourier transform representation

$$X_T(f) = \int_{-\infty}^{\infty} x_T(t) \, e^{j2\pi ft} \, dt \tag{3.9}$$

which is the same starting point used in (3.6). As developed in [12,13], the PSD based on time averages is given by

$$P_x(f) = \lim_{T \to \infty} \lim_{U \to \infty} \left[\int_{-U/2}^{U/2} \frac{1}{T} |X_T(v,f)|^2 \, dv \right] \tag{3.10}$$

where

$$X_T(u,f) = \int_{u-T/2}^{u+T/2} x(t) \, e^{-j2\pi ft} \, dt \tag{3.11}$$

The order of limit taking in (3.10) cannot in general be interchanged. Furthermore, it can be shown that the PSD $P_x(f)$ generally does not converge but is instead very erratic if only the limit over T is used in (3.10) [12–14].

In the case of widesense stationary random processes (of which ergodic processes are a subset), the autocorrelation function and the PSD are Fourier transform pairs as given by

$$R_x(\tau) = \int_{-\infty}^{\infty} P_x(f) \, e^{j2\pi f\tau} \, df \tag{3.12}$$

This relationship is often employed to calculate the PSD density indirectly from the autocorrelation function and is known as the Wiener-Khinchine relation.

In the case of a uniformly sampled random process ($T_s = 1$), which gives rise to a random sample sequence $x(nT_s)$, the PSD of the sample sequence is given by [15]

$$P_x(f) = \sum_{n=-\infty}^{\infty} R_x(n) \, e^{-j2\pi fn} \quad \text{for } |f| < \frac{1}{2} \tag{3.13}$$

where it can be shown that

$$R_x(n) = \int_{-1/2}^{1/2} P_x(f)\, e^{j2\pi fn}\, df \qquad (3.14)$$

With this albeit brief discussion of PSDs, we can proceed with the development of the more computationally oriented aspects of this important topic.

3.2 POWER SPECTRAL DENSITY CALCULATIONS IN DISCRETE SAMPLED SYSTEMS

Although the preceding section primarily discussed PSDs as applied to continuous systems, with the rapid evolution of digital signal processing in modern systems, discrete sampled systems bear increasing importance in actual hardware implementations compared to traditional analog techniques. This is particularly true where discrete computer models are employed to simulate continuous-time systems.

3.2.1 Relationships between Continuous-Time and Discrete-Time Fourier Transforms

Consider a deterministic signal $x(t)$ whose Fourier transform is given by

$$X(f) = \int_{-\infty}^{\infty} x(t)\, e^{-j2\pi ft}\, dt \qquad (3.15)$$

If the continuous signal is sampled every T_s seconds, the sample value at time $t = nT_s$ is given by

$$x(nT_s) = \int_{-\infty}^{\infty} X(f)\, e^{j2\pi fnT_s}\, df \qquad (3.16)$$

This may be expanded as

$$\begin{aligned} x(nT_s) &= \sum_{r=-\infty}^{\infty} \int_{(2r-1)/2T_s}^{(2r+1)/2T_s} X(f)\, e^{j2\pi fnT_s}\, df \\ &= \int_{-1/2T_s}^{1/2T_s} \left[\sum_{r=-\infty}^{\infty} X\!\left(f + \frac{r}{T_s}\right) \right] e^{j2\pi fnT_s}\, df \end{aligned} \qquad (3.17)$$

A similar equation may be developed for the discrete Fourier transform (DFT) of the sampled sequence $y_n = x(nT_s)$. This may be done by recognizing that the DFT is equivalent to the z transform of the sample sequence evaluated on the unit circle in the complex z plane. The z transform of y_n is given by

$$Y(z) = \sum_{n=-\infty}^{\infty} y_n z^{-n} \qquad (3.18)$$

and its inverse is given by

$$y_n = \frac{1}{j2\pi} \oint Y(z) z^{n-1} \, dz \qquad (3.19)$$

The z transform $Y(z)$ exists provided that the sample sequence y_n is absolutely summable. Additionally, if the sequence is taken as causal, all of the z-transform poles must lie within the unit circle and we may choose the unit circle as the contour of integration leading to

$$y_n = \frac{1}{j2\pi} \int_{-1/2T_s}^{1/2T_s} Y(e^{j2\pi fT_s}) \, e^{j2\pi fnT_s} \, df \qquad (3.20)$$

Assuming now that y_n and $x(nT_s)$ are identical sample sequences, it must be true that

$$Y(e^{j2\pi fT_s}) = \frac{1}{T_s} \sum_{r=-\infty}^{\infty} X\left(f + \frac{r}{T_s}\right) \qquad (3.21)$$

Assuming further that $x(t)$ is sufficiently band-limited with respect to the sampling frequency (i.e., Nyquist criterion is satisfied), the DFT can then be approximated by

$$Y(e^{j2\pi fT_s}) \approx \frac{1}{T_s} X(f) \quad \text{for } |T_s| \leq \frac{1}{2} \qquad (3.22)$$

This convention then requires a slight modification to the normal discrete-time Fourier transform pair as

$$X_k = T_s \sum_{n=0}^{N-1} x_n \, e^{-j2\pi(kn/N)}$$

$$x_n = \frac{1}{NT_s} \sum_{k=0}^{N-1} X_k \, e^{j2\pi(kn/N)} \qquad (3.23)$$

This definition for the DFT will be used throughout the discussions of PSD in this chapter. We now need to extend these relationships from deterministic signals to random signals.

3.2.2 Power Spectral Density for Discrete Sampled Systems

Although the pertinent results for sampled systems have already been stated in (3.13) and (3.14), the importance of these results warrants a closer look at the derivation details. As introduced at the beginning of this chapter, time averages for statistical quantities are much more tangible and accessible than ensemble probability treatments. It is desirable to estimate all pertinent statistical properties from a single sample waveform $x(t)$ by substituting time averages for ensemble averages. The property required to accomplish this is *ergodicity*:

> A random process is said to be ergodic if all of its statistics can be predicted from a single waveform of the process ensemble via time averaging [14]. Ergodicity, which requires subject sample sequences to be stationary up to fourth-order moments will be assumed throughout the discussions that follow.
>
> The equivalent time average definition for the PSD of a random process $x(t)$

is given by

$$P_x(f) = \lim_{M \to \infty} \mathbf{E}\left[\frac{1}{(2M+1)T_s} \left| T_s \sum_{n=-M}^{M} x_n \, e^{-j2\pi f n T_s} \right|^2 \right] \quad (3.24)$$

To show that this is equivalent to the Wiener-Khinchine relation, the squared magnitude in (3.24) can be expanded in terms of the embedded DFT, and interchanging the order of expectation (averaging) and summation results in

$$P_x(f) = \lim_{M \to \infty} \frac{T_s}{2M+1} \sum_{m=-M}^{M} \sum_{n=-M}^{M} R_x(m-n) \, e^{-j2\pi f(m-n)T_s} \quad (3.25)$$

Recognizing that since x_n is widesense stationary (ergodic),

$$\sum_{m=-M}^{M} \sum_{n=-M}^{M} R(m-n) = \sum_{m=-2M}^{2M} M(2M+1-|m|) R_x(m) \quad (3.26)$$

we finally obtain

$$P_x(f) = T_s \sum_{m=-\infty}^{\infty} R_x(m)\, e^{-j2\pi f m T_s} \qquad (3.27)$$

In this context, $R_x(m)$ is the autocorrelation function of the sample sequence as given earlier. Therefore, the PSD is simply the scaled DFT of the discrete autocorrelation function $R_x(m)$.

3.2.3 Example Computation With a Discrete System

Consider the low-pass filtered noise process shown in Figure 3.4 where $n(t)$ is an ideal Gaussian noise source and $H(f)$ is an $N = 2$ Butterworth low-pass filter having a 3-dB corner frequency of 1 Hz. Although it is quite simple to calculate the resulting output PSD in the frequency domain, it is more instructive to perform the calculations in the time domain instead by making use of the PSD concepts just developed.

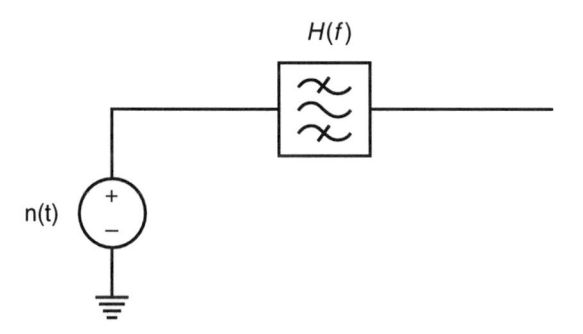

Figure 3.4 Low-pass filtering of a continuous Gaussian noise source.

The transfer function for the Butterworth low-pass filter is given by

$$H(s) = \frac{\omega_c^2}{s^2 + \sqrt{2}\omega_c s + \omega_c^2} \qquad (3.28)$$

where in this case, $\omega_c = 2\pi\, 1$ Hz will be assumed. If the analysis were to be done in the frequency domain, we would make use of the power transfer function of the low-pass filter, which in this case simplifies nicely to

$$|H(f)|^2 = \frac{1}{1 + (f/f_c)^4} \qquad (3.29)$$

Assuming that the Gaussian noise source has a flat (white) noise spectrum with a density of $N_o/2$ W/Hz bandwidth for frequencies from $-\infty$ to ∞ Hz, the PSD at the output of the filter is simply given by

$$P_o(f) = \frac{N_o}{2} \frac{1}{1 + (f/f_c)^4} \tag{3.30}$$

This result will be used to verify the result we obtain by computing the PSD in the time domain.

Continuing now with the time-domain analysis, the impulse response of the low-pass filter is given by

$$h(t) = \frac{\omega_c}{\sqrt{1-\zeta^2}} e^{-\zeta \omega_c t} \sin(\omega_c \sqrt{1-\zeta^2} \, t) \tag{3.31}$$

where $\zeta = 0.5\sqrt{2}$. The z transform for this time-domain-sampled filter impulse response is given by

$$H(z) = K \frac{z \, e^{-aT_s} \sin(\omega_c T_s)}{z^2 - 2ze^{-aT_s} \cos(\omega_c T_s) + e^{-2aT_s}} \tag{3.32}$$

where

$$K = \zeta \omega_c = 2\pi \frac{\sqrt{2}}{2}$$

$$\omega_o = \omega_c \sqrt{1-\zeta^2} = 2\pi \frac{\sqrt{2}}{2} \tag{3.33}$$

$$K = \frac{\omega_c}{\sqrt{1-\zeta^2}} = 2\pi \frac{2}{\sqrt{2}}$$

Assuming that the sampling rate is 64 Hz in the discrete representation for the continuous system, the z transform for the filter impulse response is given by

$$H(z) = \frac{0.008992 \, z}{z^2 - 1.861375 \, z + 0.870367} \tag{3.34}$$

The DFT representation for $H(s)$ can be found by evaluating (3.34) for $\omega_k = 2\pi k/(T_s N)$ where N is the number of points used in the DFT.

To proceed, the random Gaussian noise source must be modeled in the time domain. This may be accomplished by using the concept shown in Figure 3.5. Rather than attempt to model the infinitely wide bandwidth noise source $n(t)$, it suffices to band-limit the wideband noise to a bandwidth much larger than the system bandwidth of interest. Since the Butterworth filtering in this example has a 3-dB bandwidth of only 1 Hz, band-limiting the input noise to 4 or 5 Hz would probably be adequate. A closer examination of Figure 3.5 will provide the needed insight.

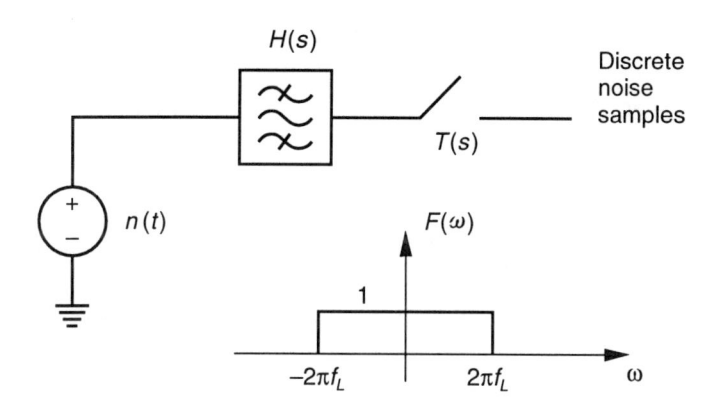

Figure 3.5 Methodology for creation of discrete random noise samples for simulation purposes.

Assume that $F(s)$ in Figure 3.5 is an ideal low-pass filter having a bandwidth of f_L Hz and that the filter output is sampled once every T_s seconds. Using the Wiener-Khinchine relation, the autocorrelation function at the filter output is given by

$$R_n(\tau) = \int_{-f_L}^{f_L} \frac{N_o}{2} e^{-j2\pi f\tau} \, df$$

$$= N_o f_L \frac{\sin(2\pi f_L \tau)}{2\pi f_L \tau}$$

(3.35)

From this calculation, it is clear that the variance of each individual filtered noise sample is $\sigma^2 = R_n(0) = N_o f_L$. Furthermore, samples that are separated by an integral multiple of $(2f_L)^{-1}$ sec are uncorrelated [i.e., $R_n(\tau) = 0$] making them statistically independent in the case of Gaussian noise. Therefore, if $2f_L$ is taken to be equal to the sampling rate used in our discrete-time simulation, this approach can be used to model the white Gaussian noise source very accurately. The variance of the discrete Gaussian random number generator then required is given simply as $\sigma_g^2 = N_o f_s/2$.

The complete time-domain simulation can now be assembled for this simple case as shown in Figure 3.6. The Gaussian random number generator just described

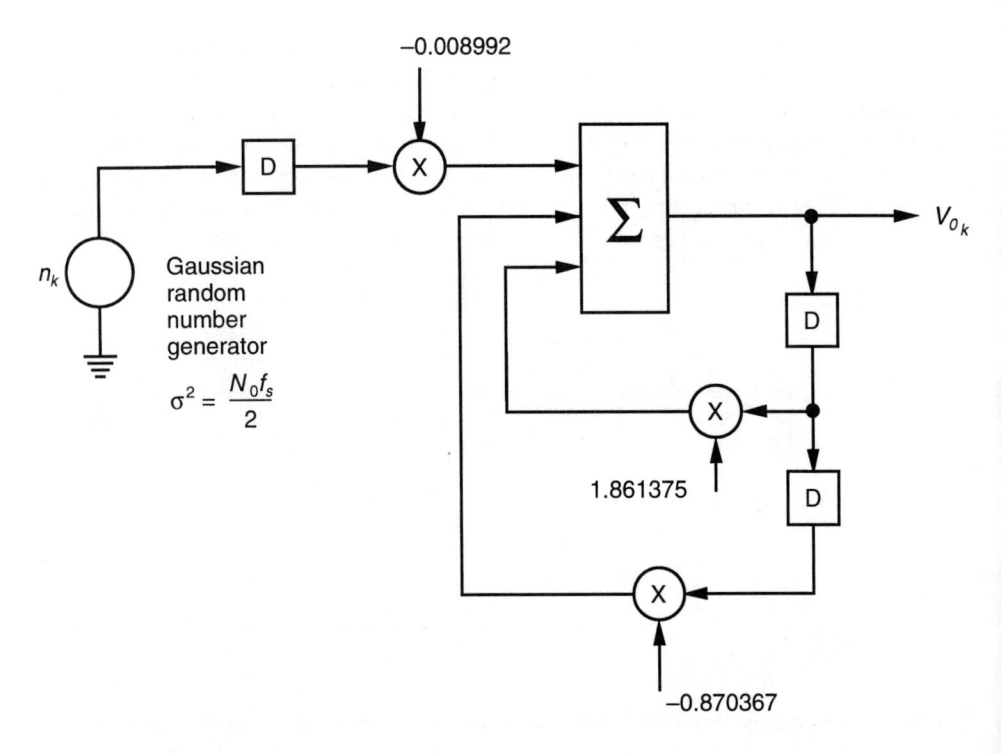

Figure 3.6 Discrete model for creating low-pass filtered white Gaussian noise.

represents the noise source in the system, and the z transform of the filter impulse response is expanded into the finite difference equation represented here graphically. Once a sufficiently long sequence of output samples V_{o_k} have been computed numerically, the previous numerical techniques may be used to calculate the resulting PSD at the filter output.

Before continuing, some additional material is needed for creating the Gaussian random number generator. Although most computer languages provide functions for calculating uniformly distributed random variables based on linear congruential techniques [16], generally, the programmer must create his or her own Gaussian random number generator. The most common approach to do this is to make use of a transformation on two random variables, which allows us to compute the Gaussian random sample as

$$n_k = \sqrt{-2\sigma_g^2 \log_e(u_k)} \cos(2\pi v_k) \qquad (3.36)$$

where u_k and v_k are statistically independent uniformly distributed random variables spanning the range (0,1]. Gaussian random variables can also be accurately produced by properly summing M independent uniformly distributed random variables [17].

In the spirit of (3.24), three types of numerical techniques have historically been used to compute the PSD using DFTs [14]. In the Daniell method, the smoothing represented by the expectation operator in (3.24) is done by averaging the finite DFT calculation over adjacent frequency bins [18]. The Bartlett method averages multiple DFTs computed from segments of the original sample sequence in order to affect the needed expection [19]. The Welch method extends the Bartlett approach by overlapping the sample segments and applying data windowing functions in order to improve the spectral leakage, which is normally observed between adjacent frequency bins [20]. A Fortran computer program is provided in [14] that uses this latter approach for PSD calculation.

A modified PSD calculation technique is used here for illustrative purposes. Assume that for each group of N output samples, the DFT is computed using a triangular-shaped window function. Let the kth DFT calculated in this manner be denoted by

$$\hat{P}_k^N \qquad (3.37)$$

The desired spectral smoothing can then be accomplished by recursively filtering the individual DFTs so calculated as described by

$$P_k = P_{k-1}\frac{k-1}{k} + \hat{P}_k^N \frac{1}{k} \qquad (3.38)$$

where P_0 is the starting value for the smoothed spectrum computation, which is taken to be zero (for all N vector values).

Carrying out these computations, the PSD for $k = 2$ is shown in Figure 3.7. The erratic behavior of the PSD is due to the poor conditioning of individual DFTs mentioned earlier. The estimated PSD using $k = 128$ is shown in Figure 3.8 where it can be shown that the variance of the PSD is reduced by a factor of 64 compared to that shown in Figure 3.7. Finally, a nearly perfect PSD is shown in Figure 3.9 for this example using $k = 2048$.

In these computational examples, a value of 0.02448 W/Hz was arbitrarily assumed for $N_o/2$ corresponding to $\sigma_g^2 = 1.567$ given the sampling rate of 64 Hz. On a per hertz basis, the spectral density for this example at dc should then be 10 $\log_{10}(N_o/2) = -16.11$ dBW/Hz. The discrepancy in the figures for the spectral density at dc arises from the use of the triangular window function, which reduces the measurement noise bandwidth of the system (simulation). If no data windowing had been used, the equivalent noise bandwidth of each DFT frequency bin would have been given by

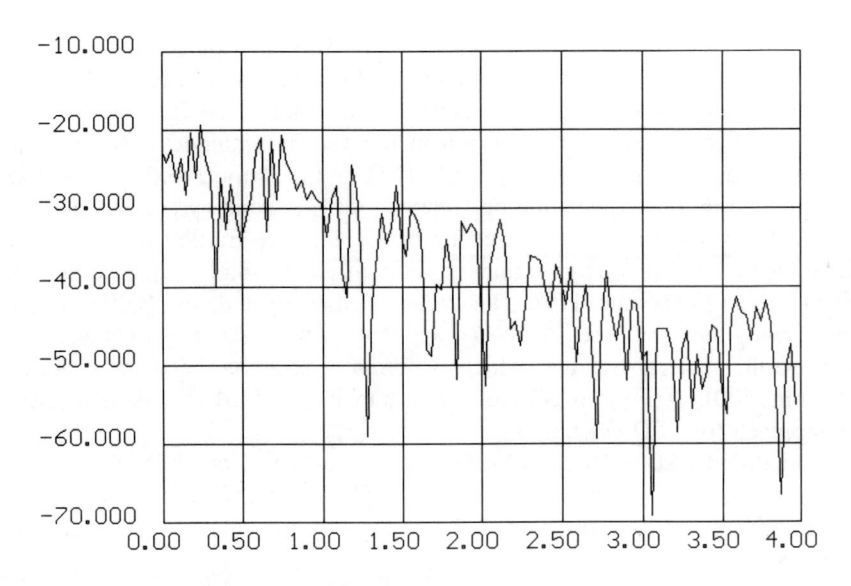

Figure 3.7 Power spectral density for smoothing case $k = 2$.

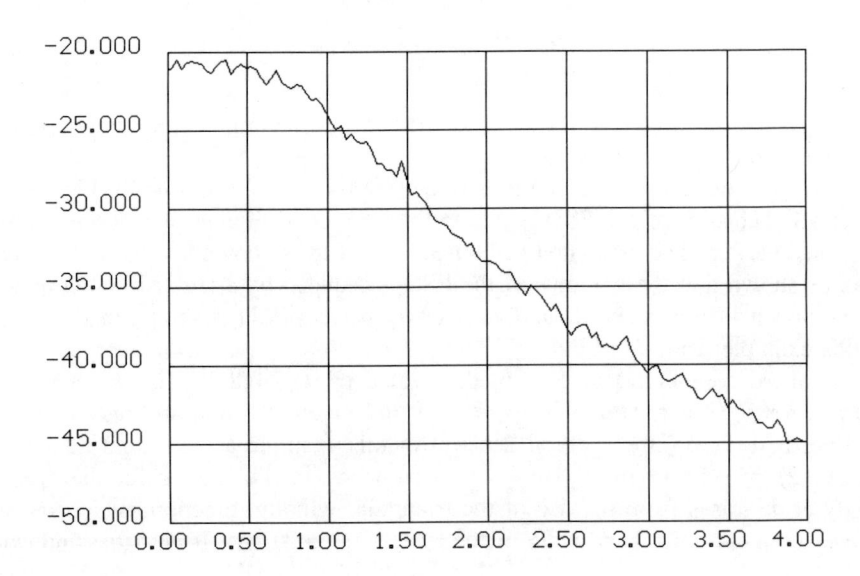

Figure 3.8 Power spectral density with reduced variance for case $k = 128$.

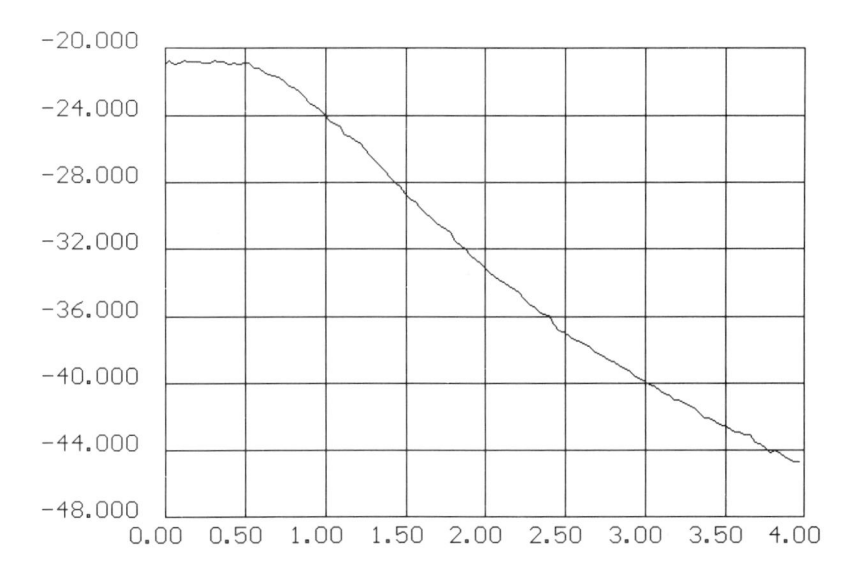

Figure 3.9 Nearly ideal power spectral density for case $k = 2048$.

$$B_m = \int_{-\infty}^{\infty} |W(f)|^2 \, df \tag{3.39}$$

$$W(f) = \frac{e^{j\pi fT_s}}{M} \frac{\sin(2\pi fMT_s)}{\sin(\pi fT_s)}$$

Evaluating (3.39) in the nonwindowed case (i.e., weighting function identical to 1), $B_m = 0.125$ Hz with $M = 1024$ and $T_s^{-1} = 64$ Hz. For the triangular-shaped window function,

$$W(f) = \left| \frac{\sin(2\pi fMT_s)}{M \sin(\pi fT_s)} \right|^4 \tag{3.40}$$

in which case, $B_m = 0.041666$ Hz. Since the equivalent noise bandwidth for the windowed case is 3.0 times smaller than for the nonwindowed case, the plotted PSD in Figures 3.7, 3.8, and 3.9 is no longer per Hertz bandwidth but is rather per 0.33-Hz bandwidth in which case the difference of $10 \log_{10}(3) = 4.77$ dB is seen between the theoretical and plotted results. Restoring the plotted values to a per hertz basis, the spectral density computed at dc is $-21 + 4.77 = -16.23$ dBW/Hz, which is within 0.12 dB of theory. The small remaining discrepancy can be easily attributed to the remaining variance in the PSD calculations, finite frequency warping in the discrete representation for the continuous low-pass filter, and potential departures from ideal in the Gaussian numerical noise source used in the simulation. Additional discussion of this analysis method and the variance of the PSD thus estimated are addressed in greater detail in [21].

The concept of data windowing is fundamental to digital signal processing. In the context used here, windowing is used to allow multiple data segments to be combined without incurring appreciable spectral splattering, which would otherwise result from the signal discontinuities at the sample sequence boundaries. Thorough discussions of windowing fundamentals can be found in [22,23].

3.2.4 Phase Noise First Principles Continued

Having developed the PSD concepts in some detail, this ground work permits a reexamination of the phase noise situation in which $\varphi(t)$ in (3.3) is now a stochastic random process. Ignoring the AM component in (3.3) and using (3.7) and (3.8),

$$P_v(f) = \lim_{T \to \infty} \left\{ \frac{1}{T} \mathbf{E}[|V_T(f)|^2] \right\} \tag{3.41}$$

where

$$V_T(f) = \int_{-T/2}^{T/2} \left\{ \frac{e^{j[2\pi v_o t + \varphi(t)]} - e^{-j[2\pi v_o t + \varphi(t)]}}{j2} \right\} e^{-j2\pi f t} \, dt \tag{3.42}$$

and $P_v(f)$ again has units of watts per hertz. Performing a transformation of variables and taking the expectation and limit over T,

$$P_v(f) = \frac{1}{4}[Z(f - v_o) + Z(-f - v_o)] \tag{3.43}$$

where

$$z(\tau) = \mathbf{E}[e^{j\Delta\varphi(\tau)}] \tag{3.44}$$

$$Z(f) = \int_{-\infty}^{\infty} z(\tau) \, e^{-j2\pi f \tau} \, d\tau$$

which is simply another restatement of the Wiener-Khinchine relation where $z(\tau)$ is the correlation function of the random phase noise process expressed in complex exponential form. The quantity $P_v(f)$ represents the PSD seen on a spectrum analyzer exactly.

If the small angle assumption for $|\Delta\varphi(\tau)|$ can again be employed, it is possible to relate $P_v(f)$ back directly to the PSD of the phase noise process $\varphi(t)$ itself. Considering only the positive frequency term in (3.43),

$$P_v^p(f) \approx \int_{-\infty}^{\infty} \frac{1}{4} \mathbf{E}\left\{1 + j\Delta\varphi(\tau) - \frac{[\Delta\varphi(\tau)]^2}{2}\right\} e^{j2\pi(f-\nu_o)\tau} d\tau$$

$$\approx \int_{-\infty}^{\infty} \frac{1}{4} [1 - \sigma_\varphi^2 + R_\varphi(\tau)] e^{-j2\pi(f-\nu_o)\tau} d\tau$$

$$\approx \frac{1}{4}[1 - R_\varphi(0)] \delta(f - \nu_o) + \frac{1}{4} P_\varphi(f - \nu_o) \qquad (3.45)$$

The first term in (3.45) represents a delta function in the frequency domain with a slightly reduced amplitude due to the random phase noise present, whereas the second term represents the continuous phase noise spectrum centered at ν_o. Therefore, the PSD of $v(t)$ can be made directly equivalent to the PSD of the underlying phase noise process $\varphi(t)$ provided that the small-angle approximation for $\Delta\varphi(\tau)$ can be made.

At this point, it is important to highlight an important distinction between the PSD given by (3.45) and the sideband phase noise level at a specific carrier offset which is normally expressed in dBc/Hz (sideband phase noise level in 1-Hz bandwidth with respect to the carrier expressed in dB). From (3.2), it should be clear that the idealized carrier power in this representation is $1/4$ at $f = \nu_o$ and $1/4$ at $f = -\nu_o$. Hence, the continuous phase noise sideband level expressed as dBc/Hz is given from (3.45) as

$$\mathcal{L}(f) = P_\varphi(f) \qquad (3.46)$$

where f is the frequency offset from the carrier frequency ν_o. Both $\mathcal{L}(f)$ and $P_\varphi(f)$ are two-sided spectra, being defined for both positive and negative frequencies. Therefore, the phase noise dBc/Hz value observed on the spectrum analyzer is numerically equivalent to $10 \log_{10}[P_\varphi(f)]$ where $P_\varphi(f)$ has units of rad^2/Hz.

Two-sided spectra have been used throughout the discussions thus far in this chapter because they are a natural representation for use with the Weiner-Khinchine relation. The whole subject of one-sided (defined for only positive frequencies) versus two-sided PSDs is a troublesome issue, which has caused factors of 2 to be lost or added erroneously in many systems analyses. The definitions used throughout this chapter (which have also been adopted in most of the technical literature) for the primary phase noise quantities of importance are therefore addressed in the next section.

3.3 DEFINING STANDARDIZED PHASE NOISE QUANTITIES

Thus far in this chapter, a minimal amount of standardized terminology has been used concerning phase noise in order that the focus could be maintained on the underlying fundamental principles. It is now appropriate, however, to introduce some of the phase noise quantities that have been standardized in the field.

A number of excellent papers have been published in recent years that discuss phase noise characterization fundamentals [24–30]. The updated recommendations of the IEEE are provided in [31] and those of the CCIR in [32].

In the discussion that follows, the nominal carrier frequency is denoted by v_o in hertz and the frequency offset from the carrier by f, also in hertz. Some authors term the quantity f the *Fourier frequency* [33].

One of the most prevalent measures of phase noise used by manufacturers and users is $\mathcal{L}(f)$, which we have previously encountered. This important quantity is defined as follows from [33]:

> $\mathcal{L}(f)$: The normalized frequency domain representation of phase fluctuations. It is the ratio of the PSD in one phase modulation sideband, referred to the carrier frequency on a spectral density basis, to the total signal power, at a frequency offset f. The units for this quantity are Hz^{-1}. The frequency range for f ranges from $-v_o$ to ∞. Therefore, $\mathcal{L}(f)$ is a two-sided spectral density. It is also called single-sideband phase noise [5].

Another quantity of equal importance is the one-sided PSD of the phase fluctuations $S_\varphi(f)$. The two-sided equivalent of $S_\varphi(f)$ was represented by $P_\varphi(f)$ in (3.46). The working definition for $S_\varphi(f)$ is given as follows [33]:

> $S_\varphi(f)$: The one-sided spectral density of phase fluctuations. The range of frequencies f span from 0 to ∞, and the dimensions are rad^2/Hz. The value of $S_\varphi(f)$ is measured by passing the signal through a phase detector and measuring the PSD at the detector output. Normally, the approximation

$$\mathcal{L}(f) \approx \frac{1}{2} S_\varphi(|f|) \qquad (3.47)$$

is made, but this is only valid as long as

$$\int_{f_1}^{\infty} S_\varphi(f)\, df \ll 1 \text{ rad}^2 \qquad (3.48)$$

It is not uncommon to see $\mathcal{L}(f)$ redefined as exactly equal to $S_\varphi(f)/2$ in order to avoid erroneous use of $\mathcal{L}(f)$ in situations where the small-angle approximation is not valid [30].

One of the more troublesome issues in spectral characterizations is the use of one-sided versus two-sided measures. Two-sided spectral densities are defined such that the frequency range of integration is from $-\infty$ to ∞. Normally, the one-sided PSDs are taken to be simply twice as large as the corresponding two-sided spectral density. Two-sided spectral densities are primarily useful in pure mathematical analyses involving Fourier transforms. We note that the terminology for single-sideband signals versus double-sideband signals is totally distinct from the one-sided and two-sided spectral density terminology. They are totally different concepts [27].

The simple phase noise representation given by (3.2) and the ensuing analysis assumes *a priori* the stationarity (at least to second-order) of $\varphi(t)$ and to fourth order for time-averaged-based analyses [26]. In the rigorous sense, the theoretical analysis of internal white noise in oscillators leads to a phase diffusion process analogous to Brownian motion, which is not stationary. In this case, it is impossible to use the autocorrelation/PSD theory developed in this chapter. As a rule, however, stationary random processes can be assumed for the work undertaken in this chapter, and further concern with noncompliant cases will not be addressed further at this time. The reader is referred to [26] for an excellent review of other important characterization quantities including finite data length effects, use of the two-sample Allan variance, Hadamard variance concepts, structure functions, and more.

Other characterization quantities are often computed in order to more clearly represent the impact of phase noise on different communication systems. In phase-modulated systems, the impact of phase noise is often quantified in terms of the total rms phase jitter of a given source. The rms phase jitter is the square root of the variance of the phase noise processes and is given by

$$\sigma_\varphi = \sqrt{\int_{f_1}^{f_2} S_\varphi(f)\, df} \tag{3.49}$$

where f_1 is normally taken to be from $0.001R$ to $0.05R$ and f_2 is normally taken to be equal to R where R is the symbol rate in hertz for digital communication systems work. The precise values for f_1 and f_2 really depend on other elements of the system design, specifically, the carrier recovery loop dynamics and the receiver's matched filtering. In lieu of this additional information, σ_φ^2 can be used to characterize the spectral quality of the source, but it is not suitable for estimating the achievable BER performance of the system.

In analog FM systems, the dominant characterization quantity for phase noise performance is called residual FM. It may be computed as

$$\sigma_f = \sqrt{\int_{f_1}^{f_2} f^2 S_\varphi(f)\, df} \tag{3.50}$$

where f_1 and f_2 represent the range of offset frequency for which the quantity applies.

Both the phase and frequency variance quantities can be easily calculated assuming that the phase noise quantity $S_\varphi(f)$ can be represented as a straight-line template as shown in Figure 3.10.

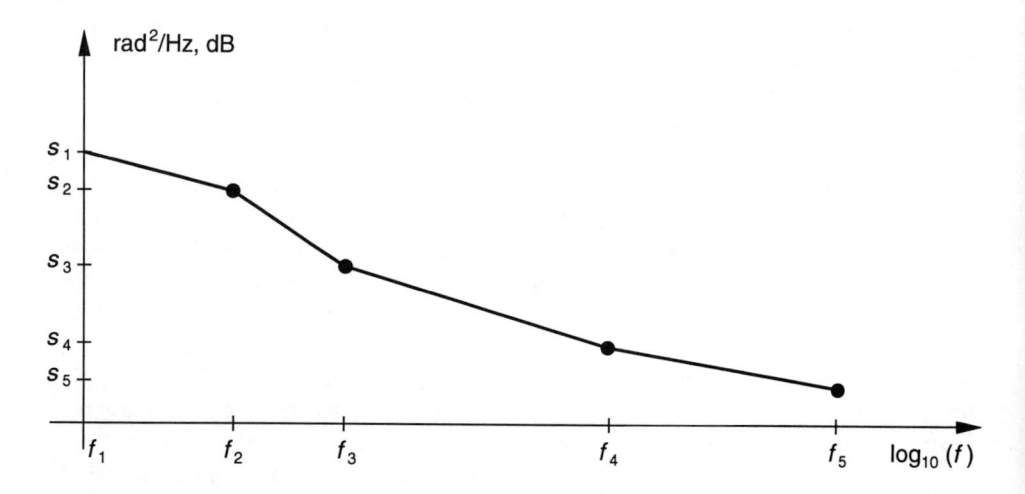

Figure 3.10 Phase noise template description.

In this form,

$$S_\varphi(f) = \sum_{m=1}^{M} K_m f^{\alpha_m} W_m(f) \qquad (3.51)$$

where the $M + 1$ line segment data points are represented by (f_k, S_k) and the template points (f_k, S_k) have units of (Hz, rad^2/Hz in dB). In this form,

$$W_m(f) = \begin{cases} 1 & f_m \leq f \leq f_{m+1} \\ 0 & \text{otherwise} \end{cases}$$

$$\alpha_m = \frac{S_{m+1} - S_m}{10 \log_{10}[(f_{m+1})/f_m]} \qquad (3.52)$$

$$K_m = \frac{10^{0.1 S_m}}{(f_m)^{\alpha_m}}$$

$$\sigma_\varphi^2 = \int_{f_1}^{f_{M+1}} S_\varphi(f)\, df$$

$$= \sum_{m=1}^{M} \begin{cases} \alpha_m = -1 & 10^{0.1 S_m} f_m \log_{10}\left(\dfrac{f_{m+1}}{f_m}\right) \\[1em] \alpha_m \neq -1 & \dfrac{f_m}{\alpha_m + 1}\left[\left(\dfrac{f_{m+1}}{f_m}\right)^{\alpha_m+1} - 1\right] 10^{0.1 S_m} \end{cases} \quad (3.53)$$

The residual FM may be calculated by using (3.52) with the α_m replaced with $\alpha_m' = \alpha_m + 2$. Detailed calculation examples for both of these quantities may be found in [10] and [34,35].

The performance impact of phase noise in differential phase-shift-keyed (DPSK) systems that is most useful is the variance of the excess phase accumulated over one data symbol time interval. This variance is easily calculated as

$$\sigma_{\Delta\varphi}^2 = \mathbf{E}\{[\varphi(t + T) - \varphi(t)]^2\}$$

$$= 2[R_\varphi(0) - R_\varphi(T)] \quad (3.54)$$

$$= 4\int_0^\infty S_\varphi(f) \sin^2(\pi f T)\, df$$

From the integrand in (3.54), it is clear that close-in phase noise components near the carrier whose offset is much less than the symbol rate T^{-1} are suppressed by the $\sin^2(\ )$ quantity significantly. This fact alone provides significant insight into what spectral components are most damaging to DPSK systems.

3.4 CREATING ACCURATE PHASE NOISE PSDS IN THE LABORATORY

Although the measurement of phase noise quantities has received much attention in the literature, creation of accurate phase noise PSDs in the laboratory for equipment and system evaluation has generally not been addressed. One means by which this gap may be filled is the NLO-100 test equipment family, which is briefly discussed in this section [36]. The NLO-100 utilizes a mixture of digital and RF techniques to accurately produce arbitrary phase noise PSDs using the simple phase noise model given by (3.2).

To accurately utilize the suggested model (3.2), it is necessary to understand both the statistics and spectral content of the random quantity $\varphi(t)$. We will assume throughout this section that the source being modeled is a simple phase-locked loop frequency synthesizer.

The signal-to-noise ratio (SNR) within a frequency synthesizer is normally tremendous. In this respect, the more complicated phase error diffusion processes that take place under very low SNR conditions are not at issue here. As discussed earlier, the mechanism limiting the SNR within the PLL control loop electronics is thermal noise that is Gaussian distributed. If we think of this Gaussian noise as interfering in an additive manner with the finite edge-rate digital signal shown in Figure 3.11 and assume a constant transition slew rate, there is little question but that the phase error of the actual zero crossing will also be Gaussian distributed. More elegant arguments are certainly possible, but they lead to this same general result [10,37–39]. Well outside the phase-locked loop closed-loop bandwidth, the phase noise is dominated by the self-noise of the locked oscillator. It has long been assumed that the oscillator self-noise displays Gaussian statistics as well [40,41]. Hence, the Gaussian assumption for the statistics of $\varphi(t)$ is used here.

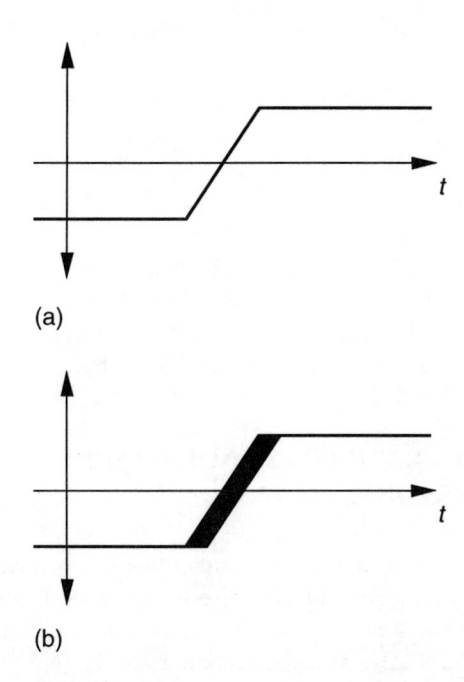

Figure 3.11 Impact of additive noise on digital signal zero-crossings: (a) Noiseless zero crossing and (b) jittered zero crossings caused by additive noise.

In general then, the desired NLO-100 output is a constant envelope signal having a nominal frequency of ν_o, Gaussian phase noise statistics, and a phase noise spectrum that can be described as shown in Figure 3.10. The example calculation based on Figure 3.4, which has already been presented, contains all of the essential ingredients required to create a user-defined output phase noise spectrum. Rather than address the general case here, a classical type 2 phase-locked loop is assumed for this discussion.

As developed in Chapter 4, the transfer function from the reference phase noise disturbance to the output phase of the loop is given by

$$H_1(s) = \frac{N\omega_n^2\left(1 + s\frac{2\zeta}{\omega_n}\right)}{s^2 + 2\zeta\omega_n s + \omega_n^2} \tag{3.55}$$

and the transfer function from the voltage-controlled oscillator (VCO) self-noise to the output (under closed-loop conditions) is given by

$$H_2(s) = \frac{s^2}{s^2 + 2\zeta\omega_n s + \omega_n^2} \tag{3.56}$$

where ω_n is the loop natural frequency in radians per second, ζ is the damping factor, and N is the feedback divider ratio as shown in Figure 3.12.

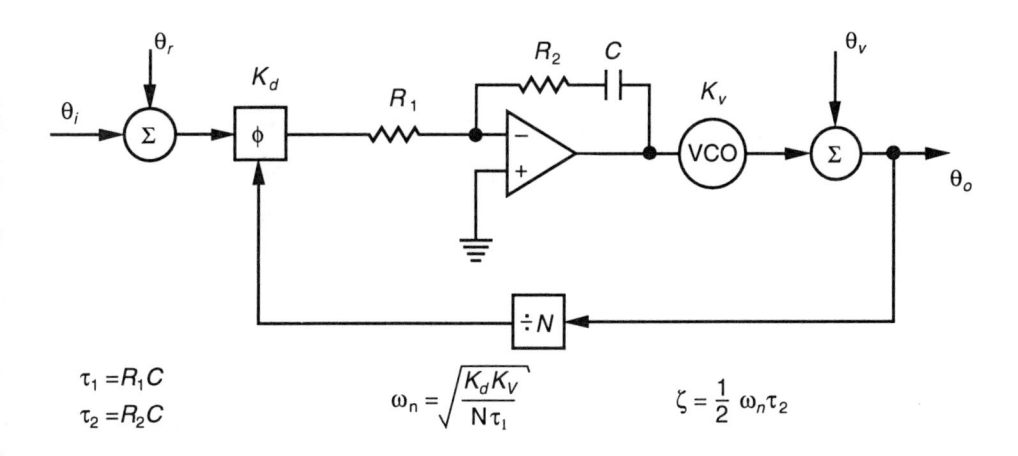

$\tau_1 = R_1 C$
$\tau_2 = R_2 C$
$\omega_n = \sqrt{\dfrac{K_d K_v}{N\tau_1}}$
$\zeta = \dfrac{1}{2}\omega_n \tau_2$

Figure 3.12 Classical type 2 phase-locked loop including modeling for dominant noise sources.

The reference and VCO self-noise are normally independent random processes. Although both noise sources can of course be modeled, for simplicity only the reference noise source is addressed here.

Bilinear transformation techniques may be used to convert the system frequency domain impulse response (3.55) into an equivalent discrete time model as

$$\frac{\theta_o(z)}{\theta_v(z)} = \frac{(\omega_n T/2)^2 (a_1 z^2 + a_2 z + a_3)}{b_1 z^2 + b_2 z + b_3} \tag{3.57}$$

where

$$a_1 = 1 + \frac{4\zeta}{\omega_n T}$$

$$a_2 = 2$$

$$a_3 = 1 - \frac{4\zeta}{\omega_n T}$$

$$b_1 = 1 + \zeta\omega_n T + \left(\omega_n \frac{T}{2}\right)^2 \tag{3.58}$$

$$b_2 = -2 + 2\left(\omega_n \frac{T}{2}\right)^2$$

$$b_3 = 1 - \zeta\omega_n T + \left(\omega_n \frac{T}{2}\right)^2$$

Recognizing z^{-1} as the simple unit delay operator (see Chapters 4 and 5), (3.57) may be rewritten in the time domain as the recursion

$$\theta_o[n+1] = \theta_v[n+1]\frac{a_1}{b_1} + \theta_v[n]\frac{a_2}{b_1} + \theta_v[n-1]\frac{a_3}{b_1} - \frac{b_2}{b_1}\theta_o[n] - \frac{b_3}{b_1}\theta_o[n-1] \tag{3.59}$$

This simple discretized model for creating a noisy phase-locked loop output is shown graphically in Figure 3.13.

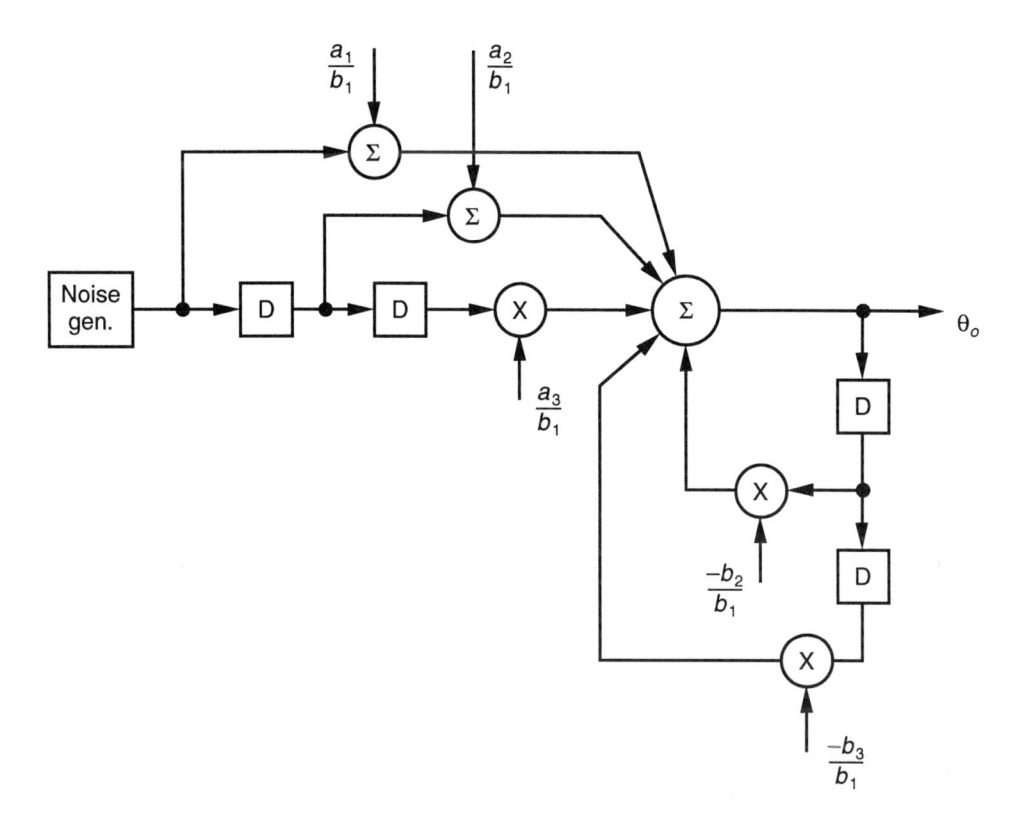

Figure 3.13 Discrete model for a noise corrupted phase-locked loop.

Using this type of formulation, the NLO-100 outputs a noisy constant amplitude sine wave per the model given by (3.2) with repeatable digital precision. Phase noise as well as discrete spurs may both be synthesized in this manner. Two sample spectrum analyzer measurements of the NLO-100 output for different loop parameters are shown in Figures 3.14 and 3.15. In general, the NLO-100 can create any constant-envelope signal as long as (1) the maximum information bandwidth is limited to roughly 100 kHz, and (2) the output waveform is periodic with a period not exceeding the buffer storage capacity of the instrument (corresponds to approximately 0.3 sec). RF center frequencies ranging from approximately 800 to 1600 MHz can be accommodated in the basic unit with additional coverage available in other models. The NLO-100 utilizes many other similar but generally more sophisticated techniques to synthesize arbitrary output waveforms under high-level user control. A top-level block diagram of the NLO-100 is shown in Figure 3.16.

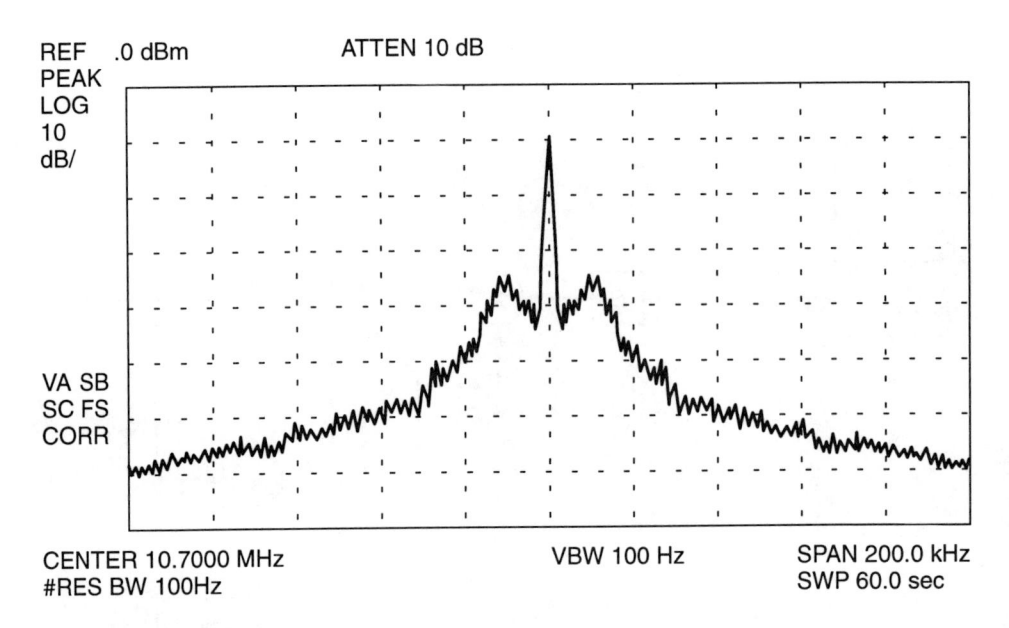

Figure 3.14 NLO-100 output noise spectrum with clearly apparent noise peaking.

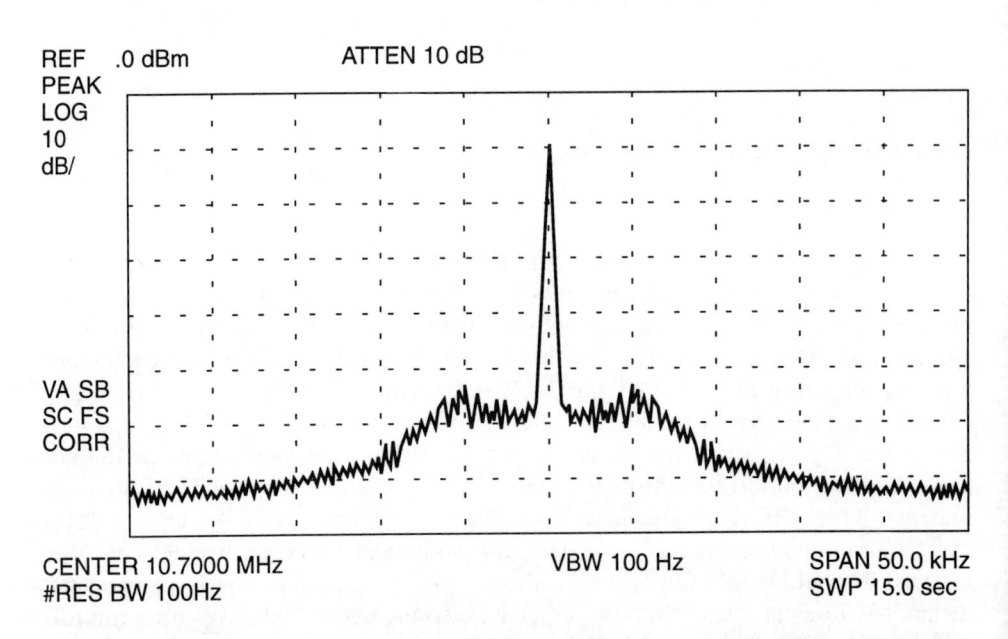

Figure 3.15 NLO-100 output noise spectrum representing the phase noise spectrum of a well-designed system.

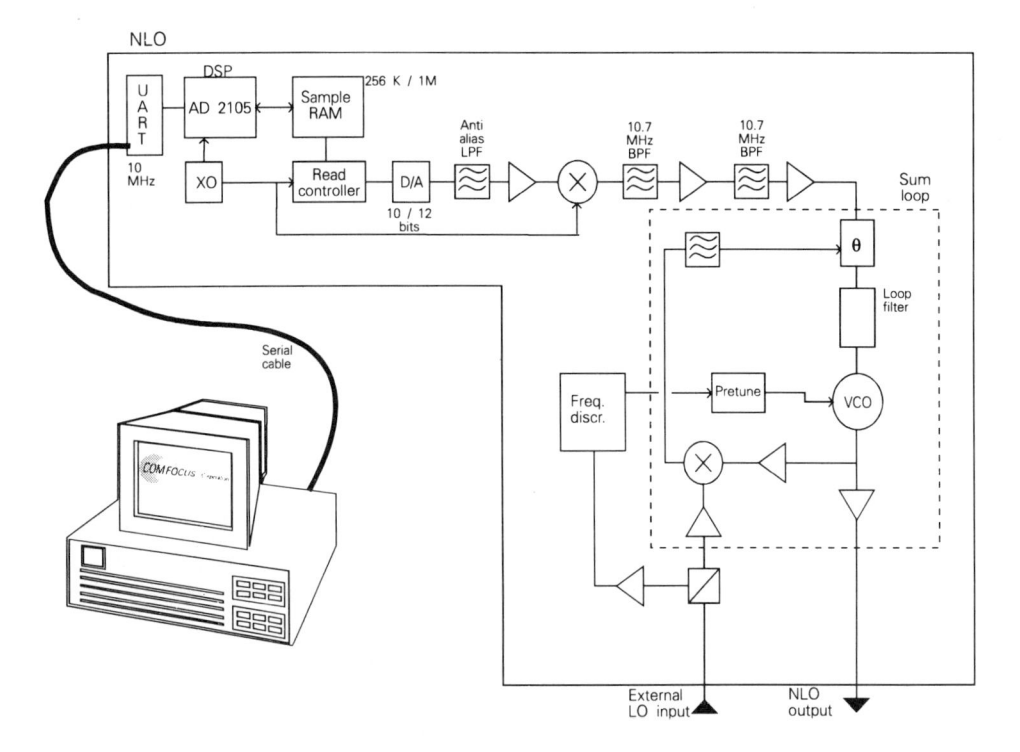

Figure 3.16 NLO-100 top-level block diagram.

3.5 CREATING ARBITRARY PHASE NOISE SPECTRA IN A DIGITAL SIGNAL PROCESSING ENVIRONMENT

Creation of phase noise spectra using parametric methods like that just described for the classical type 2 phase-locked loop is fairly straightforward. In contrast, a suitable approach for an arbitrary spectrum template is much less intuitive. Fortunately, it is possible to borrow from modern signal processing techniques in order to handle this case in an expedient manner.

Much of modern digital signal processing concerns itself with the estimation of signal spectra. In the present case, the desired signal spectrum is known exactly (user defined) and we desire to create a (digital) sample sequence that faithfully represents the specified spectrum. Assuming that only widesense stationary random processes are to be used, a solution to this problem is to make use of the maximum entropy (all-pole) method from power spectrum estimation theory, which is also known as *autoregressive modeling* [42].

In this approach, we assume that the user-specified PSD can be accurately represented by the all-pole model

$$P(f) = \frac{a_0}{\left|1 + \sum_{k=1}^{M} a_k z^k\right|^2} \quad (3.60)$$

for some adequately large modeling order M. Since the Wiener-Khinchine theorem states that the Fourier transform of the autocorrelation function is equal to the PSD, and the Fourier transform of $P(f)$ is just the Laurent series expansion of (3.60) in z, it is possible to write

$$\frac{a_0}{\left|1 + \sum_{k=1}^{M} a_k z^k\right|^2} \approx \sum_{i=-M}^{M} \phi_i z^i \quad (3.61)$$

where the $\approx$ is meant to imply that the series expansion of the left-hand side of the equation agrees term by term with the right-hand side for terms z^{-M} to z^M [42]. It can be shown that the coefficients in (3.61) satisfy the matrix equation

$$\begin{bmatrix} \phi_0 & \phi_1 & \phi_2 & \cdots & \phi_M \\ \phi_1 & \phi_2 & \phi_3 & \cdots & \phi_{M-1} \\ \cdots & \cdots & \cdots & \cdots & \cdots \\ \phi_M & \phi_{M-1} & \phi_{M-2} & \cdots & \phi_0 \end{bmatrix} \begin{bmatrix} 1 \\ a_1 \\ a_2 \\ \cdots \\ a_M \end{bmatrix} = \begin{bmatrix} a_0 \\ 0 \\ 0 \\ \cdots \\ 0 \end{bmatrix} \quad (3.62)$$

where the matrix, because of its special symmetry, is known as a Toeplitz matrix. This fact may be exploited to solve (3.62) for the a_i coefficients in a much more efficient manner than classical Gaussian elimination or LU decomposition techniques. The well-known LPC-10 speech compression techniques are based on this mathematical solution as it turns out.

Once the a_i are known in (3.60), it should be clear that the desired PSD can be obtained by recursively filtering a unit-variance Gaussian noise sample stream with the IIR filter represented by

$$H(f) = \frac{\sqrt{a_0}}{1 + \sum_{k=1}^{M} a_k z^k} \quad (3.63)$$

Although finite precision, modeling order, and other issues remain open to further discussions, this technique provides a reliable means for accurately creating a user-specified PSD. Source code for a class of C++ routines that perform these mathematical techniques is provided on disk in the companion software program.

Additional helpful references for sampled data systems are [43–46].

3.6 PHASE NOISE IN DEVICES

At the beginning of this chapter, thermal noise was mentioned as the root cause of phase noise in communication systems. Although this perspective is helpful, accurate modeling and prediction of phase noise and its impairment on system performance are an absolute necessity. An introduction into this important design area is provided in this section.

Assuming that our discussions are limited to linear time-invariant systems, calculation of phase noise quantities is primarily driven by ascertainment of the PSDs of the underlying noise sources themselves. Given this information, the calculation of the noise spectrum at the output of any linear network is straightforward.

In this section, we address the noise contributions that arise in (1) digital frequency dividers, (2) phase detectors, (3) loop filter components, and (4) oscillators. A thorough look at the phase noise performance of direct digital synthesizers is developed in Chapter 7.

3.6.1 Digital Frequency Dividers

Digital devices, specifically digital dividers, have played an important role in phase-locked loop frequency synthesis and frequency synthesis in general for some time now. Even so, manufacturers rarely if ever provide adequate phase noise performance information that would allow *a priori* prediction of phase noise performance. The most notable exceptions to this statement at the time of this writing are the phase-locked loop chip families from Qualcomm and National Semiconductor. In sharp contrast to most of the industry, important phase noise quantities are explicitly provided in most of their device literature. This historical trend is gradually changing fortunately, and independent researchers are also contributing to the device characterization database for existing devices.

As developed at some length in other references [10,38], phase noise levels at the input to an ideal digital divider of ratio N are reduced by $20 \log_{10}(N)$ at the divider output. In this respect, the phase noise floor requirements for divider elements near the divider output can be more critical than at the divider input, which is directly opposite to the more familiar concept of a receiver cascaded noise figure. If, for instance, several digital divider stages are cascaded as shown in Figure 3.17,

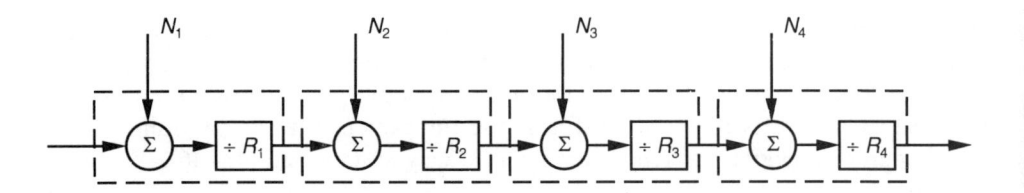

Figure 3.17 Model for assessing cascaded divider phase noise performance.

each having a divide ratio of R_k and input-referenced noise floor of N_k rad^2/Hz, the overall input-referred noise floor for the cascade is given roughly by

$$N_T = N_1 + \sum_{m=2}^{M} N_m \left(\prod_{n=1}^{m-1} R_n \right) \qquad (3.64)$$

This reasoning should make it clear why it is possible to use high-frequency prescaler devices, which may have worse phase noise performance and yet still be limited by the performance of the later divider elements. To make accurate phase noise performance comparisons between different dividers, device noise characterizations should be normalized to the same input frequency.

It is also important to make the distinction between synchronous counters and ripple counters in discussing phase noise performance [47]. Additional phase noise degradation beyond the 20 $\log_{10}(N)$ rule can occur in ripple counters because noise sources in each stage independently contribute noise components.

Some researchers report their measurement results at the divider outputs. Given this orientation, the output phase noise spectrum from a divide-by-N digital divider is normally expressed as [38]

$$S_{\varphi,D,N}(f) = \frac{S_{\varphi,\text{in}}(f)}{N^2} + S_{\varphi,\text{add}}(f) \qquad (3.65)$$

Normalizing this result with respect to output frequency, and expressing $S_{\varphi,\text{in}}(f)$ as a power series [e.g., (3.51)],

$$\frac{S_{\varphi,D,N}(f)}{f_{\text{out}}^2} = \sum_{k=-2}^{2} h_k f^{k-2} + \frac{h_{1,\text{add}}}{f} + h_{2,\text{add}} \qquad (3.66)$$

For frequencies of less than approximately 1 MHz, the best reported noise performance is given by [38]

$$S_{\varphi,\text{add}}(f) \approx \frac{10^{-14.7}}{f} + 10^{-16.5} \qquad (3.67)$$

For an output frequency greater than 1 MHz, input noise dominated the results reported in [38].

It is important to remember that in a phase-locked loop, the feedback divider action causes noise components that are far removed in frequency to be aliased to within F_{ref} Hz of dc where F_{ref} is the reference frequency used by the phase detector. Therefore, care must be taken to ensure that the feedback divider input is free of unwanted spurious signals.

Accurate measurement of divider phase noise requires substantial attention to design details. The same similar precautions must of course be followed in using the dividers in frequency synthesizers if the same performance levels are to be achieved. Recent measurements with several families of ECL devices have resulted in reported measures that are 6 to 20 dB better than earlier published results [48]. The phase noise results referred to here apply to several divide-by-20 devices that were evaluated, the results for which are shown in Figure 3.18.

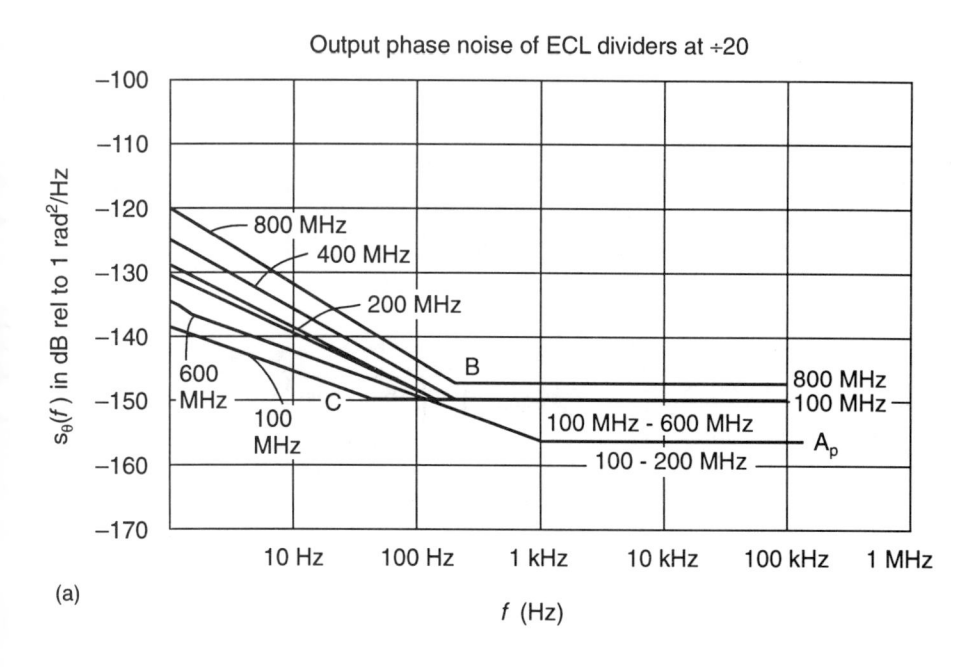

Figure 3.18 Phase noise performance of different divide-by-20 components. (©1987 IEEE; reprinted with permission). Source: [48].

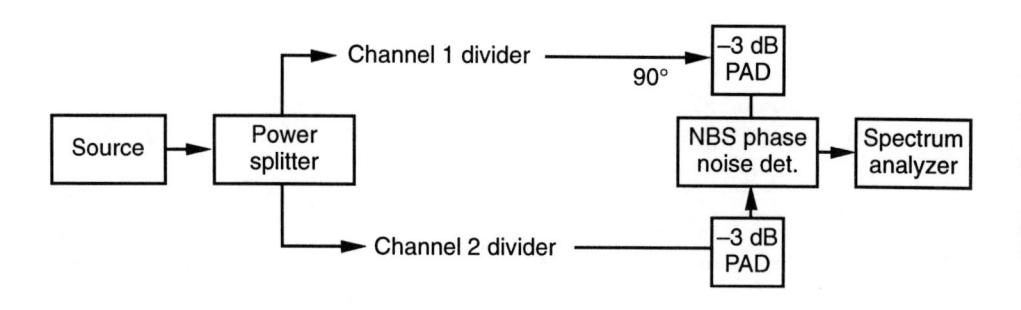

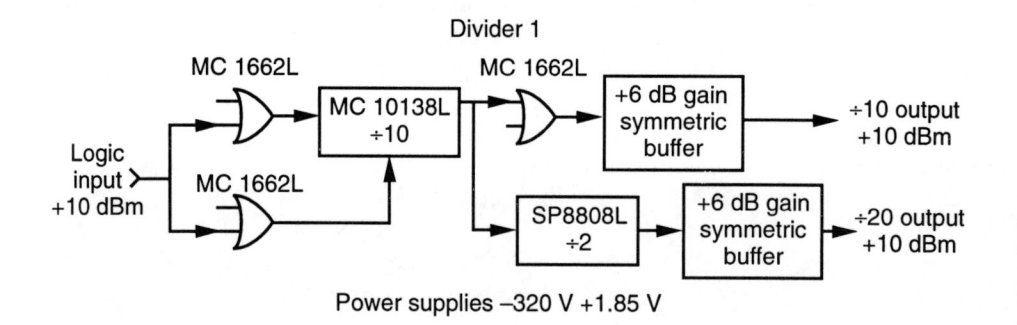

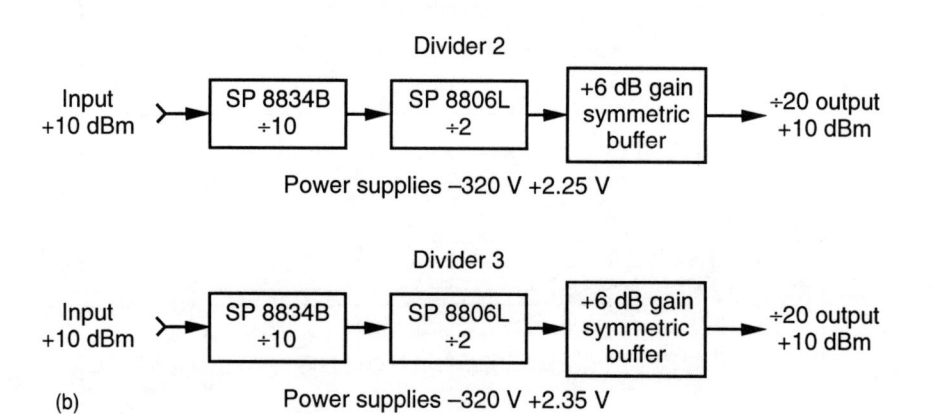

Figure 3.18 (Continued).

Special attention was given when making the measurements to eliminating the phase noise of the driving source, and in special cases where the divider output was not symmetrical, additional divide-by-2 blocks were added. Quite surprisingly, this step improved the measured phase noise performance by 10 to 30 dB rather than the theoretically anticipated 6 dB. Phase noise was measured using the three different configurations shown in the lower portion of Figure 3.18. Curves A, B, and C were obtained using the divider configurations 1, 2, and 3, respectively, for both channels input to the NBS phase noise test set. Additional details are given in [48].

Care must of course be taken to provide adequately clean power sources for the devices. In the case of Schottky TTL for instance, a voltage supply ripple of 20-mV peak-to-peak is sufficient to create spurious sidebands of -57 dBc at the divider output. Extrapolating this result to a target divider noise floor of -165 dBc/Hz would then require the voltage supply noise to be less than 28 nV/$\sqrt{\text{Hz}}$. In contrast, output noise levels for many three-terminal regulators are in the tens of $\mu V/\sqrt{\text{Hz}}$.

One of the earliest reports given on phase noise performance of digital dividers was provided in [49] and reprinted in [50]. The measured results are shown here in Figure 3.19. Curve A in Figure 3.19 has motivated many a synthesizer designer to attempt to achieve a noise floor on the order of -170 dBc/Hz, but such performance levels require extreme attention to detail and are in practice rarely obtained.

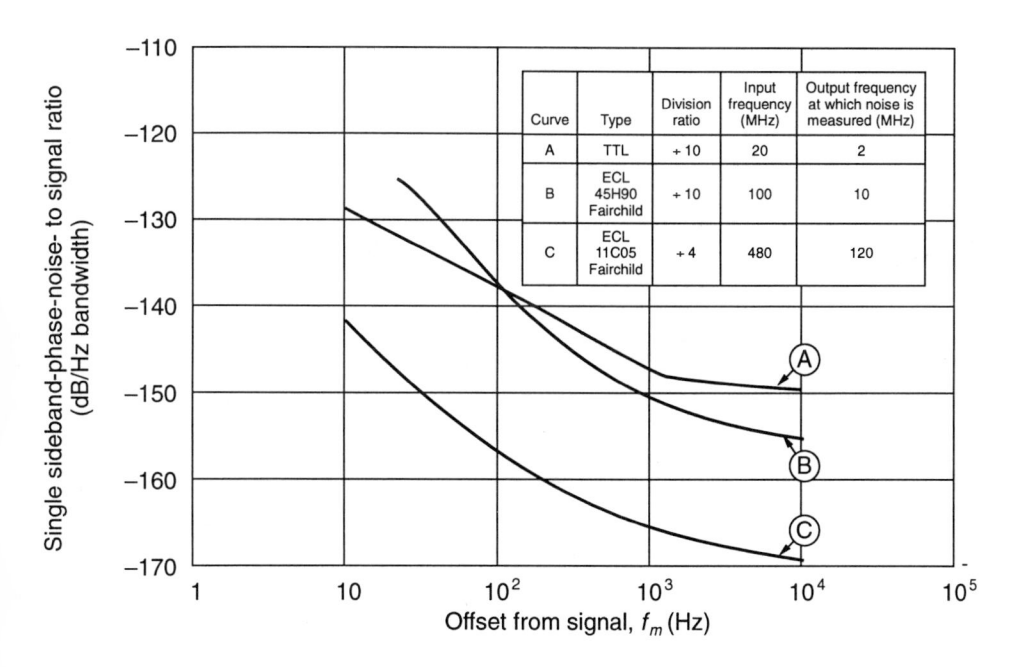

Curve	Type	Division ratio	Input frequency (MHz)	Output frequency at which noise is measured (MHz)
A	TTL	÷10	20	2
B	ECL 45H90 Fairchild	÷10	100	10
C	ECL 11C05 Fairchild	÷4	480	120

Figure 3.19 Phase noise of digital frequency dividers. Source: [50], Appendix B, page 574. Reprinted with permission.

Divider phase noise performance is strongly dependent on device technology, particularly when GaAs devices are also considered. Generally, the phase noise levels of GaAs devices are substantially higher than for silicon based devices. This is clearly the case presented in Figure 3.20.

Additional results for several different digital technologies are shown in Figures 3.21, 3.22, and 3.23 [52]. In general, the best overall system phase noise performance can only be obtained when competing divider technologies are first evaluated and selected based on a normalization of measurement data to the same input (or output) frequency, from which detailed design efforts evolve.

To date, some of the latest divider phase noise results are reported in [47]. The TTL and ECL divider test sets that were used are shown in Figure 3.24 and the noise measurement setups are shown in Figure 3.25.

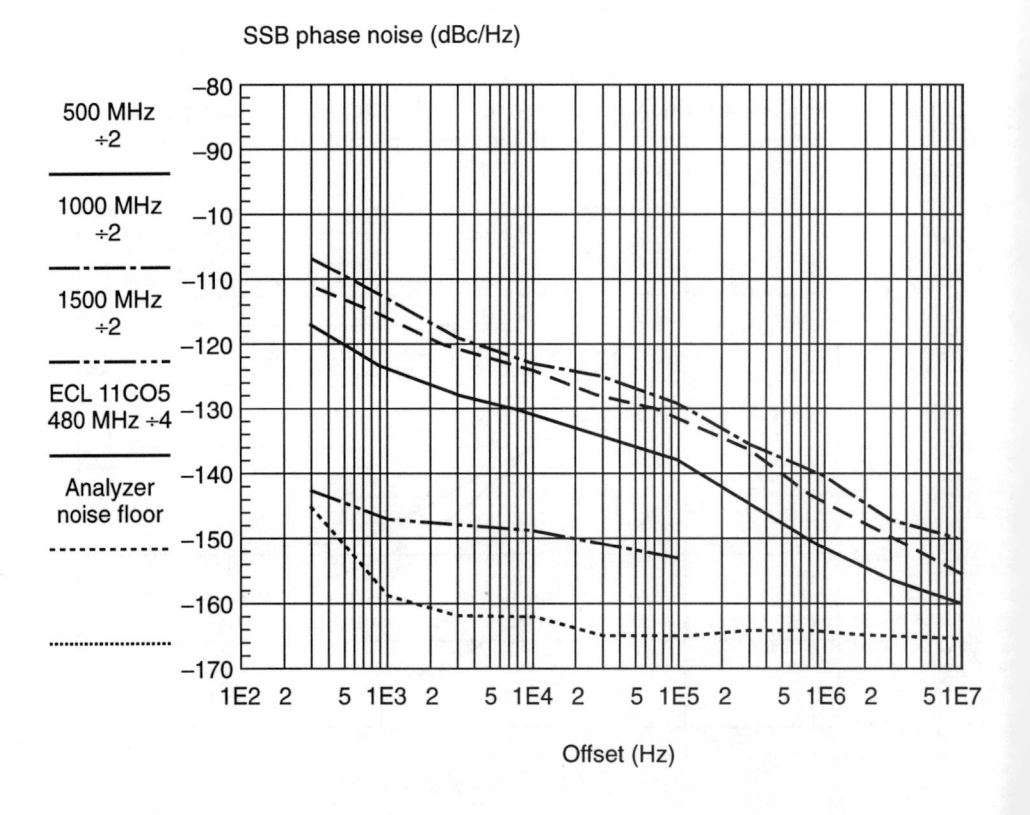

Figure 3.20 Phase noise performance comparison of frequency dividers. Source: [51]. Reprinted with permission.

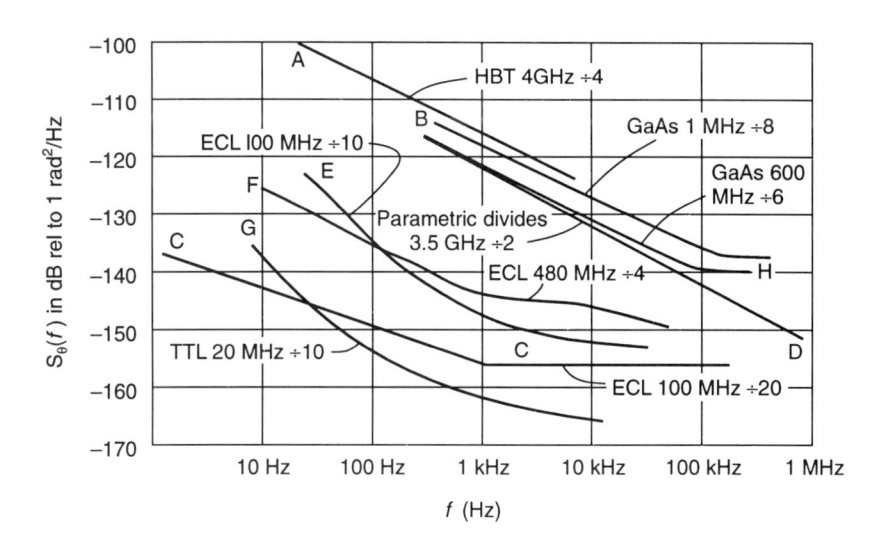

Figure 3.21 Phase noise performance of various frequency dividers. (©1987 IEEE; reprinted with permission). Source: [48].

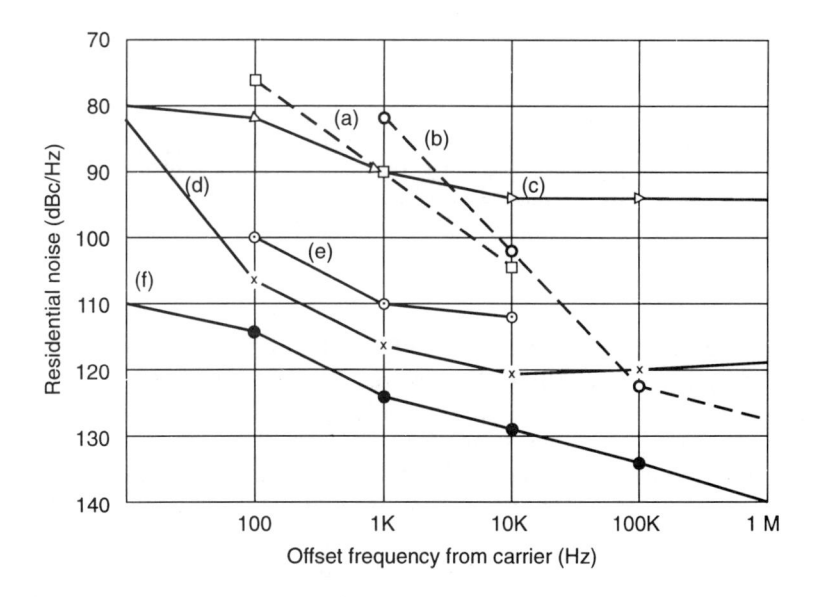

Figure 3.22 Residual noise of various frequency divider types measured at a specific frequency for each type: (a) Digital GaAs ÷4 at 300 MHz, (b) analog parametric divider ÷4 at 3 GHz, (c) digital ECL ÷4 at 125 MHz, (d) digital ECL ÷4 at 120 MHz, (e) digital ECL ÷10 at 10 MHz, and (f) digital TTL ÷2 at 2 MHz. Source: [52]. Reprinted with permission.

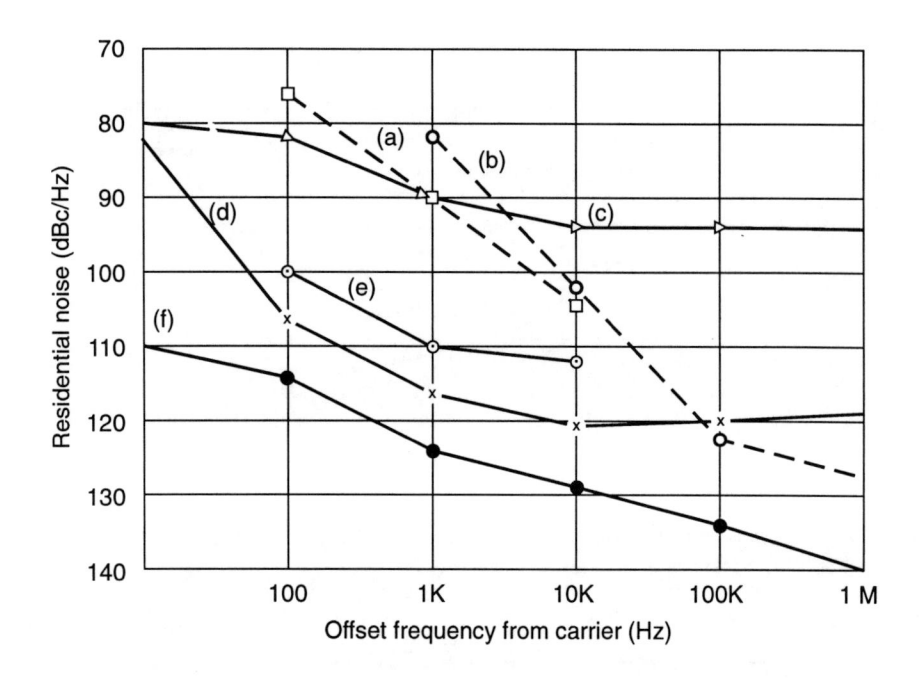

Figure 3.23 Frequency divider residual noise normalized to 10 GHz: (a) Digital ECL ÷10 at 10 MHz, (b) digital GaAs ÷4 at 300 MHz, (c) digital TTL ÷10 at 2 MHz, (d) digital ECL ÷4 at 125 MHz, (e) digital ECL ÷4 at 120 MHz, and (f) analog parametric divider ÷2 at 3 GHz. Source: [52]. Reprinted with permission.

In each test case, the divider under test performed a divide-by-4 function using an input signal frequency of 24 MHz. The test results are shown here in Figure 3.26. Under these conditions, the ACT family clearly demonstrated the best phase noise performance of the logic families evaluated.

With all of the data presented on divider phase noise performance, the obvious question remaining is what design guidelines should be used for phase-locked loop design? If the divider is to be used stand-alone, we recommend that the results presented in Figure 3.26 be adopted. On the other hand, if the divider(s) are used within a phase-locked loop, many other factors contribute to the overall phase noise performance in many cases, masking the lower noise floor performance of the divider element(s).

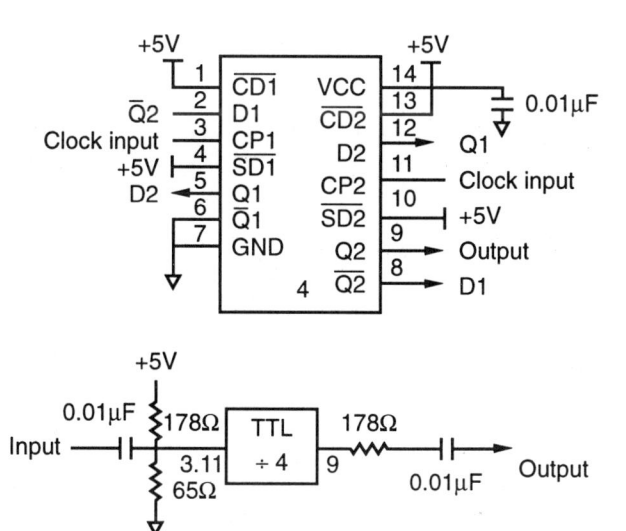

The TTL frequency divider test fixture

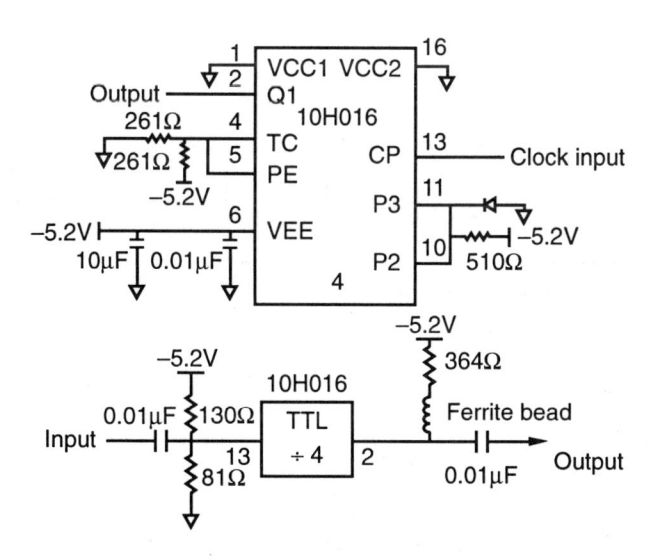

The ECL frequency divider test fixture

Figure 3.24 TTL and ECL frequency divider test fixtures used in characterization efforts. Source: [52]. Reprinted with permission.

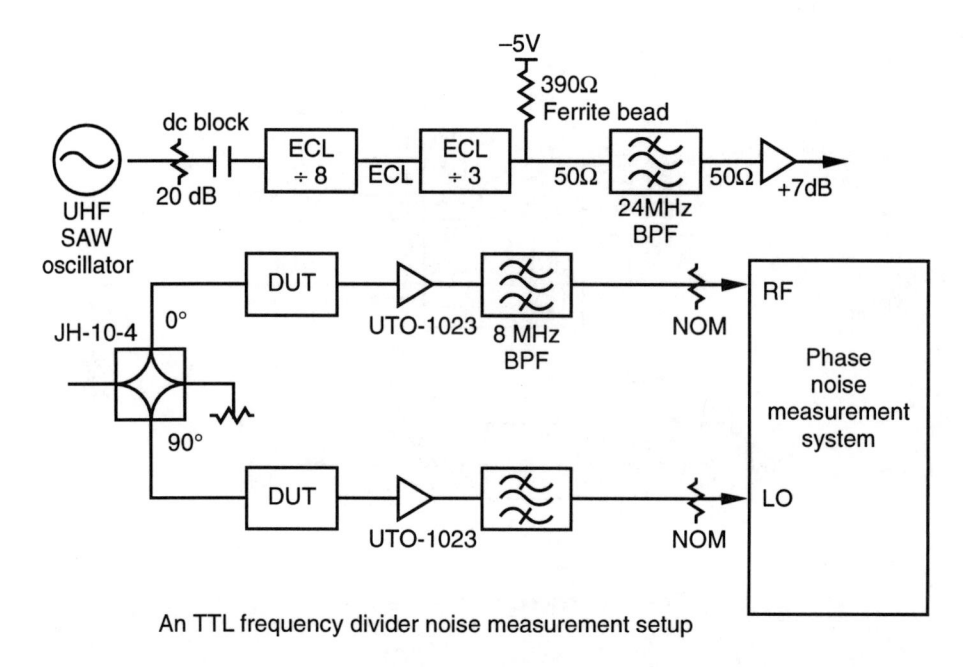

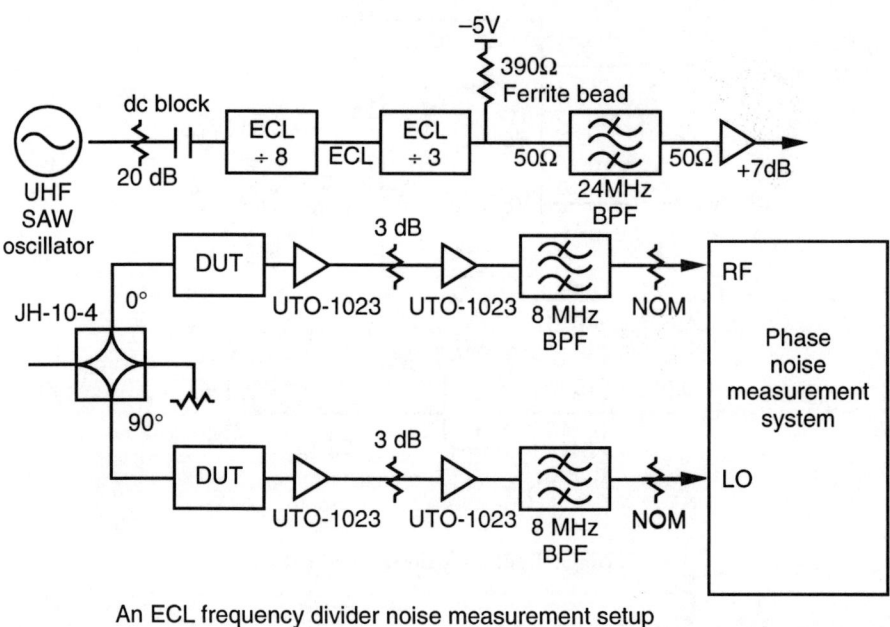

Figure 3.25 TTL and ECL frequency divider noise measurement setups. Source: [52]. Reprinted with permission.

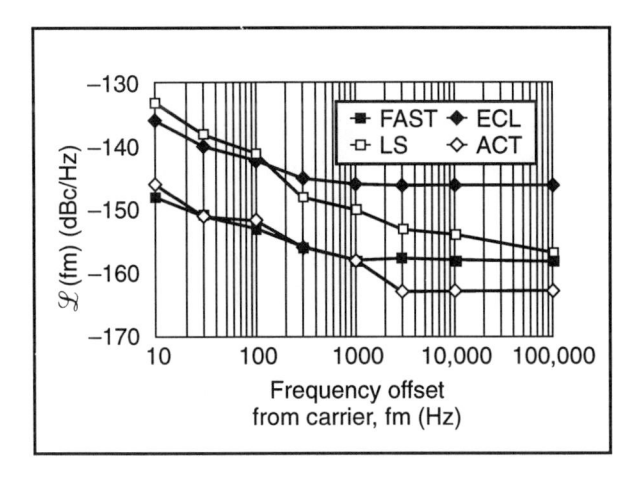

Figure 3.26 Digital frequency divider residual phase noise. Source: [47]. Reprinted with permission.

3.6.2 Phase Detector Phase Noise Performance

The phase noise performance of the phase detector element used in phase-locked loops is at present one of the more unquantified noise sources that must be considered in low-noise frequency synthesizer design. The phase noise performance of digital dividers has received considerably more attention.

Although phase detectors are discussed in a number of texts [10,50,53–56], phase noise performance is only briefly addressed from a theoretical perspective in [10]. More extensive research into this area seems appropriate at this time.

In proceeding, we want to differentiate between phase detector noise phenomenon arising from normal linear operation and noise resulting from the so-called "dead-zone" behavior exhibited by some digital phase detectors. The dead-zone issue was addressed briefly in Chapter 2. Since this phenomenon is well understood and easily avoided, this potential noise source is not germane to our present discussions.

Double-Balanced Mixer Phase Detectors

Although rarely used in modern synthesizer design, the double-balanced mixer (DBM) still represents the best phase detector noise performance achievable [38,57–59]. The best phase detectors are double-balanced mixers that utilize Schottky barrier diodes in the ring configuration [38]. Measurements performed by different researchers put the achievable noise floor attainable at

$$S_\varphi(f) \approx \frac{10^{-14\pm 1}}{f} + 10^{-17} \qquad (3.68)$$

Substantial care must be paid in using DBM phase detectors as evidenced in discussions presented in [57–59].

Although not limited solely to DBM-type phase detectors, it is interesting to note that correlation techniques may be used to improve phase noise measurement limits by as much as 20 dB or more as shown in Figure 3.27. The correlation-based

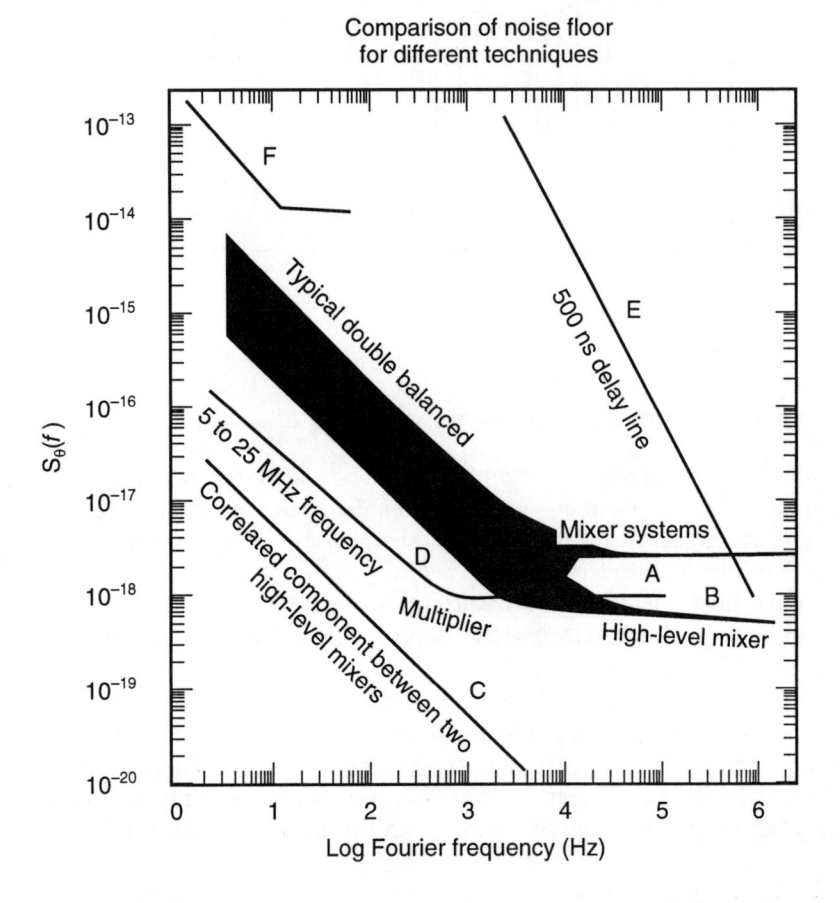

Figure 3.27 Comparison of noise floor for different phase noise measurement techniques. Curve A: The noise floor $S_\phi(f)$ (resolution) of typical DBM systems at carrier frequencies from 0.1 MHz to 26 GHz; similar performance possible to 100 GHz. Curve B: The noise floor $S_\phi(f)$ for a high-level mixer. Curve C: The correlated component of $S_\phi(f)$ between two channels using high-level mixers. Curve D: The equivalent noise floor $S_\phi(f)$ of a 5- to 25-MHz frequency multiplier. Curve E: Approximate phase noise floor using a discriminator with 500-ns delay line. Curve F: Approximate phase noise floor using a specialized optical fiber technique to realize a delay line of 1 ms. (©1989 IEEE; reprinted with permission). Source: [59], Figure 4.

measurement system used is shown in Figure 3.28. Unfortunately, there does not appear to be a simple means to extend this performance level to the actual phase detector component as used within a phase-locked loop.

Phase noise performance of digital phase detectors has gone largely unpublished in recent years even though the subject noise contributions can be substantial. Digital phase detectors that were evaluated in the frequency range of 0.1 to 1.0 MHz were found to exhibit a phase noise floor of [38]

$$S_\varphi(f) = \frac{10^{-10.6 \pm 0.3}}{f} \qquad (3.69)$$

ECL and complementary metal-oxide semiconductor (CMOS) logic families were found to exhibit better phase noise performance by up to 22 dB in the flicker noise region [38]. More recently, the phase noise floor performance of the Motorola 12040 ECL phase detector has been reported at -144 dBc/Hz at 10 MHz while that of an

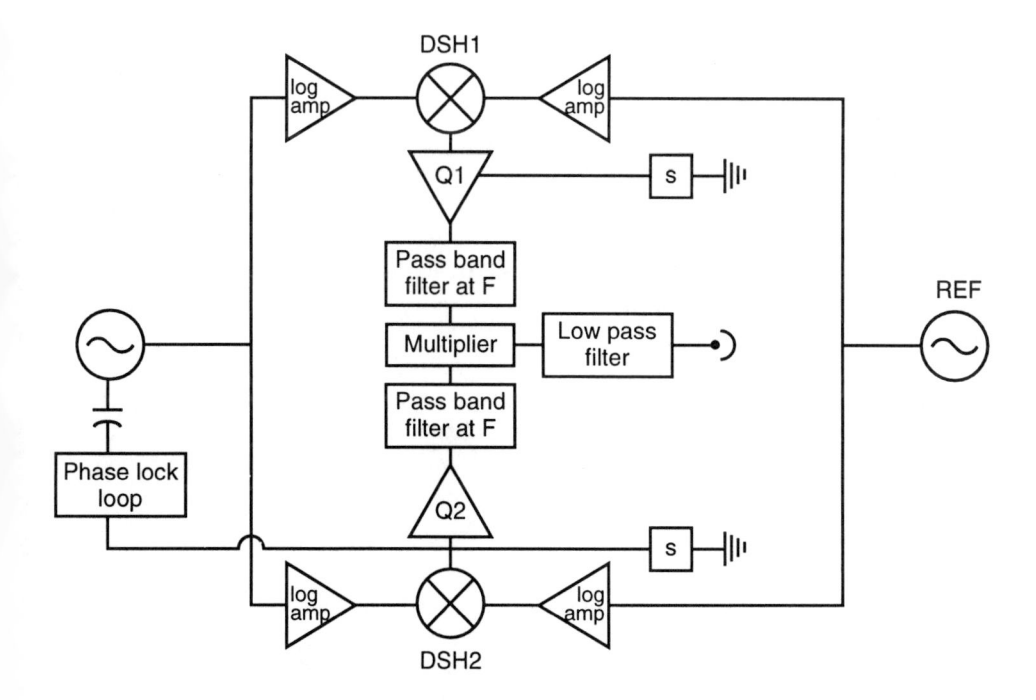

Figure 3.28 Correlation phase noise measurement system. (©1989 IEEE; reprinted with permission). Source: [59], Figure 5.

HCMOS-type phase-frequency detector operating at 1 MHz was reported at roughly −160 dBc/Hz [60].

The best phase noise floor performance normally observed with highly integrated phase-locked loop chips is approximately −150 dBc/Hz. Some of these devices may be as much as 15 to 20 dB poorer however. Therefore, before committing a component choice to a design, it is important to determine the anticipated phase noise of the subject device in advance.

3.6.3 Low-Noise Electronic Design

In the context of individual phase-locked loop design, aside from the digital elements and the VCO, the remaining circuitry is comprised of baseband analog circuitry. To realize the best possible phase noise performance achievable, low-noise analog design guidelines must be observed.

Active as well as passive circuit components used in the loop filter, VCO, and accompanying RF/analog circuitry can and will contribute noise to the overall output phase noise spectrum. In a well-designed phase-locked loop, the overall output phase noise spectrum should only be limited by the inherent phase detector/feedback divider phase noise floor well within the closed loop bandwidth, and by the VCO self-noise well outside this bandwidth. This statement assumes of course an ideal frequency reference and somewhat arbitrary frequency step size.

Noise and its impact on systems has received much attention during the past decades, and no attempt will be made to condense all of this material. Beyond the brief material presented here, the interested reader is referred to a number of excellent references on the subject [61–77], particularly [65] for a broad discussion of noise phenomena and [63] for an extensive discussion on low-noise electronic circuit design practices.

The material that follows was originally used as the basis for an extensive system-level noise characterization computer program. Modern spreadsheet software can be used to streamline the computational elements if desired. The use of spreadsheets for system-level noise analysis is now a widely accepted design practice [78].

Filters and Noise

Noisy linear two-port models are commonly used to describe electrical networks over a wide range of complexity levels ranging from simple resistor models (e.g., Figure 3.1) to complete amplifiers, oscillators, and subsystems. Normally, individual noise sources are statistically (independent) uncorrelated, which, when true, greatly simplifies the ensuing mathematical analysis.

We will begin by considering the noise performance of an ideal lossless LC filter as shown in Figure 3.29. The methodology employed is applicable to generalized noise analysis.

```
                    Source                              Load

                   ┌─────┐  ┌──────────┐  ┌─────┐
              R_S ─┤     ├──┤ Lossless ├──┤     ├─ R_L
                   │     │  │ network  │  │     │
                   └─────┘  └──────────┘  └─────┘
```

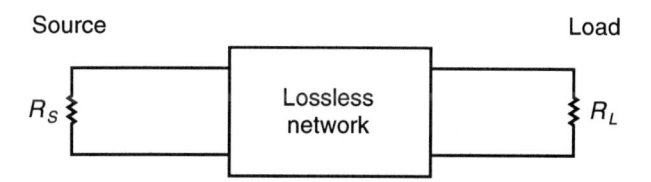

Figure 3.29 Lossless two-port network model for filter noise considerations.

The noise power that is *available* from each resistor is kT W/Hz bandwidth. Scattering parameter analysis, which is commonly used in RF and microwave design, defines the transducer gain for the lossless filter as

$$G_T = \frac{\text{Power delivered to load}}{\text{Power available source}} \tag{3.70}$$

which is mathematically given as

$$G_T = \frac{|S_{21}|^2 (1 - \Gamma_S^2)(1 - \Gamma_L^2)}{|(1 - S_{11}\Gamma_S)(1 - S_{22}\Gamma_L) - S_{12}S_{21}\Gamma_S\Gamma_L|^2} \tag{3.71}$$

where the S_{ij} represent the scattering parameters for the lossless network and Γ_S and Γ_L represent the source and load reflection coefficients, respectively. This relationship can be used to calculate the amount of input noise power delivered to the load and vice versa.

Assuming now that the source and load are perfectly matched, Γ_S and Γ_L are both zero and $G_T = |S_{21}|^2$. In terms of standard parameter notation, the incident, transmitted, and reflected waves are as shown in Figure 3.30.

Summing the total power incident at the load,

$$kT|S_{21}|^2 + kT|S_{22}|^2 = kT \tag{3.72}$$

and at the source

$$kT|S_{12}|^2 + kT|S_{11}|^2 = kT \tag{3.73}$$

For a passive lossless network, it is also true that

$$[S^T]^* \, S = I \tag{3.74}$$

where the $*$ denotes complex conjugate, T denotes matrix transposition, and I is the identity matrix. From this relationship,

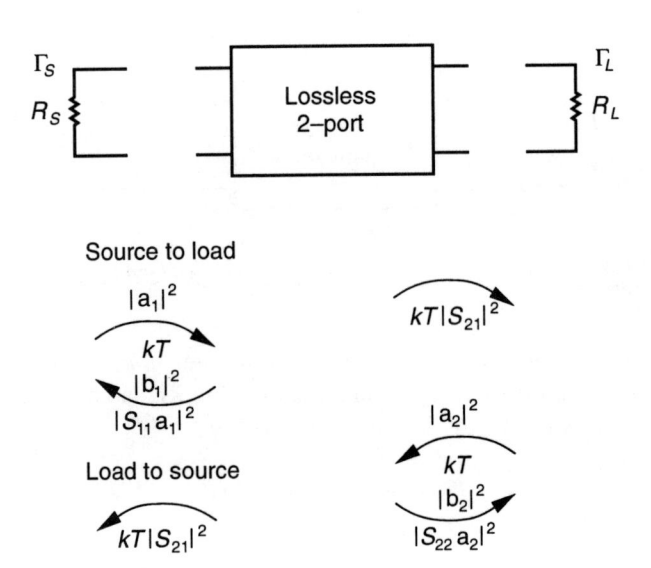

Figure 3.30 Incident and reflected noise power terms for the lossless linear two-port model.

$$[S^T]* S = \begin{bmatrix} |S_{11}|^2 + |S_{21}|^2 & S_{11}^*S_{21} + S_{21}^*S_{22} \\ S_{21}^*S_{11} + S_{22}^*S_{21} & |S_{21}|^2 + |S_{22}|^2 \end{bmatrix} \quad (3.75)$$

which leads to the identities

$$|S_{11}|^2 + |S_{21}|^2 = 1$$
$$|S_{22}|^2 + |S_{12}|^2 = 1 \quad (3.76)$$
$$|S_{22}|^2 + |S_{21}|^2 = 1$$

We note that these equalities are simply restatements of the famous Feldtkeller energy equation for reactance two ports that arises in modern filter design based on insertion loss techniques [79]. Using the last equality in (3.76)

$$kT|S_{21}|^2 + kT[1 - |S_{21}|^2] = kT \quad (3.77)$$

This is a very interesting result in that it states that even if the power available from the source is attenuated by the lossless filter, the same noise power is delivered to the load regardless.

Using this result and the definition of $\mathscr{L}(f)$ given earlier in this chapter, the $\mathscr{L}$ transfer function for the lossless filter can be written as

$$\mathcal{L}_o(f) = \frac{P(f)}{P_{in}}|S_{21}|^2 + \frac{kT}{P_{in}}[1-|S_{21}|^2]$$
$$= \mathcal{L}_i(f)|S_{21}|^2 + \frac{kT}{P_{in}}[1-|S_{21}|^2] \qquad (3.78)$$

where $P(\omega)$ is the input power spectral density in watts per hertz (AM and PM components) and P_{in} is the carrier power present at the filter input.

In the case of an attenuation pad with numerical power loss $L \geq 1$, the output power spectral density is given by

$$\mathcal{L}_o(f) = \mathcal{L}_i(f) + \frac{kT}{P_{in}}[L-1] \qquad (3.79)$$

This result can be used to include the effects of filter insertion loss.

Results for other synthesizer elements such as mixers and amplifiers can be derived by means of similar arguments. In the case of a frequency mixer, it is insightful to consider the local oscillator (LO) port further. In normal operation, the LO port is driven into saturation, which effectively removes any AM noise components. The discussion thus far does not deal with this circumstance properly because linear networks were assumed throughout. This deficiency can be corrected by carrying through separate $\mathcal{L}(f)$ terms for AM and PM noise components and replacing kT in each of the earlier equations with $kT/2$ since the additive noise components result in equal measures of AM and PM noise in the system. A recent discussion about cascaded noisy two-port models that may be helpful is given in [80].

3.6.4 Noise Sources in Lead-Lag Loop Filters

As developed in Chapters 4 and 5, it is standard practice to reflect all noise sources in a phase-locked loop back to one of two points, either as an equivalent phase noise source at the reference input, or as an additive phase noise contribution to the VCO output. The impact of the circuit noise on the phase-locked loop output noise spectrum can then be readily ascertained.

Operational amplifier noise modeling is covered in detail in [62–64] and [69,75]. In this discussion, all noise sources are assumed to be statistically independent and we want to reflect the noise of all of the individual sources back in series with V_s as shown in Figure 3.31.

In the case of e_n, the output noise observed is given by

$$V_o = e_n \frac{1 + s(\tau_1 + \tau_2)}{s\tau_1} \qquad (3.80)$$

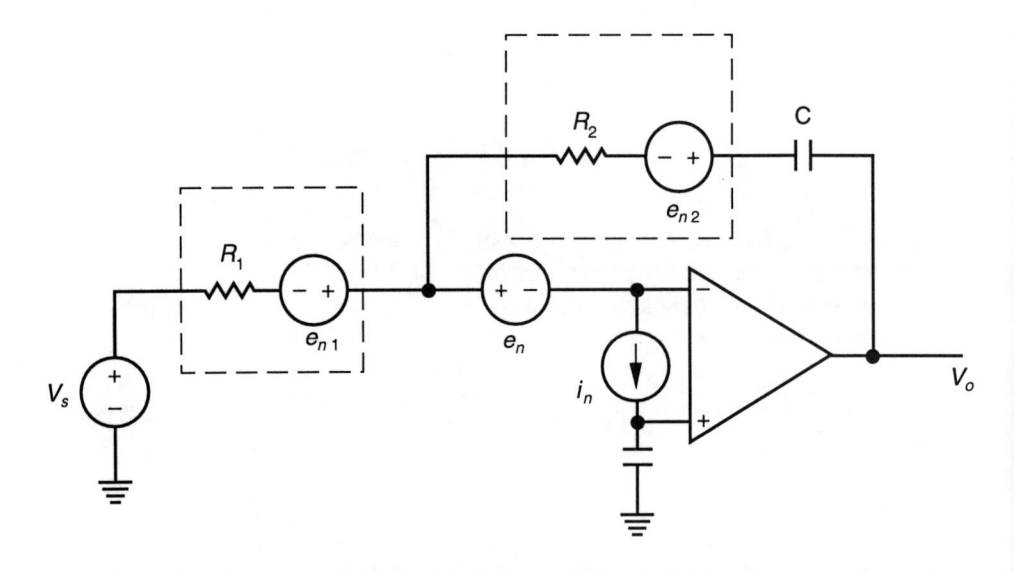

Figure 3.31 Active lead-lag filter with principle noise sources.

Since the voltage gain from V_s to the output is given by

$$-\frac{1 + s\tau_1}{s\tau_1} \tag{3.81}$$

the equivalent input noise represented by e_n reflected back to the input is given by

$$e_{s1} = e_n \frac{1 + s(\tau_1 + \tau_2)}{1 + s\tau_2} \tag{3.82}$$

Similar expressions may be computed for the i_n and e_{n2} noise sources, respectively, as

$$e_{s2} = i_n R_1 \tag{3.83}$$

$$e_{s3} = e_{n2} \frac{s\tau_1}{1 + s\tau_2}$$

Since all of the equivalent sources are assumed to be statistically independent, the resulting rms noise voltage reflected to the input as a function of frequency f is given by

$$V_T^2(f) = e_{n1}^2 + e_{s1}^2 + e_{s2}^2 + e_{s3}^2$$

$$= 4kTR_1 + e_n^2(f) \frac{1 + (2\pi f)^2(\tau_1 + \tau_2)^2}{1 + (2\pi f \tau_2)^2}$$

$$+ [i_n(f)R_1]^2 + 4kTR_2 \frac{(2\pi f \tau_1)^2}{1 + (2\pi f \tau_2)^2} \quad (3.84)$$

Normally, the $V_s(t)$ source is used to represent the phase detector output. Therefore, assuming that the phase detector gain is given by K_d volts/rad, the phase noise floor due to these noise sources is then given by $V_T(f)/K_d$. Thorough discussion of these techniques for phase-locked loops may be found in [81–83].

3.6.5 Noise in Components

The noise contributions from operational amplifiers, transistors, and field-effect transistors (FETs) are fairly obvious areas for focused attention, but passive components can also exhibit unfavorable noise behavior. Most zener diodes are particularly bad for creation of excess electrical noise. As shown in Figure 3.32, the noise density of the 1N941 diode family from Motorola is particularly poor in that high noise levels exist at low audio frequencies, which are difficult to remove with filtering. In contrast, there are notable exceptions such as the 1N935 and 1N821 zener families from Motorola, which are shown in Figures 3.33 and 3.34.

For these diodes, the low-frequency noise density is very low and only becomes appreciable at higher frequencies where it is more easily removed with filtering. More recent data book publications do not always provide this information.

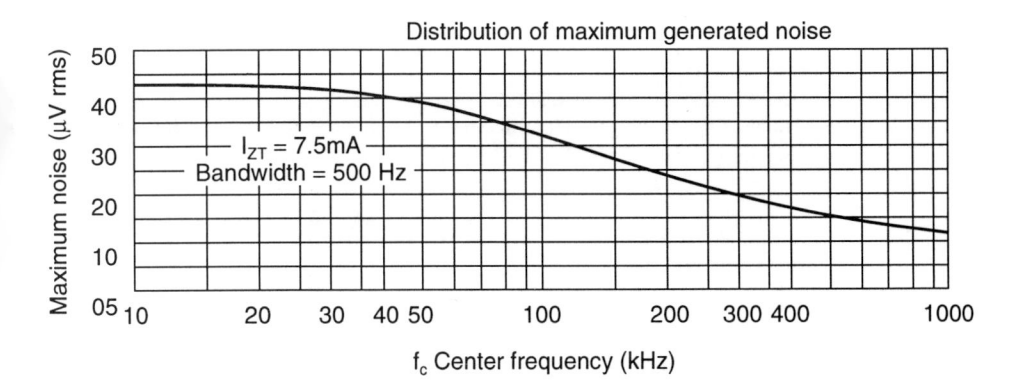

Figure 3.32 Distribution of maximum generated noise for the Motorola 1N941. (Copyright of Motorola, Inc. Used by permission). Source: [84].

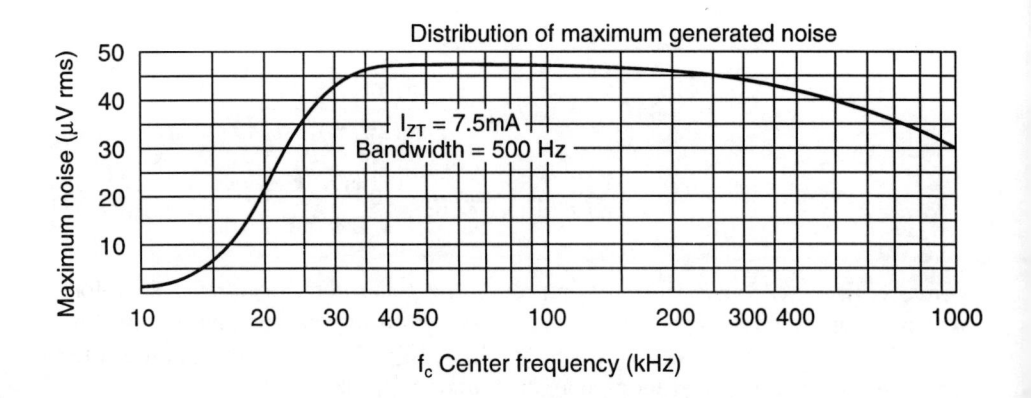

Figure 3.33 Distribution of maximum generated noise for the Motorola 1N935. (Copyright of Motorola, Inc. Used by permission). Source: [84].

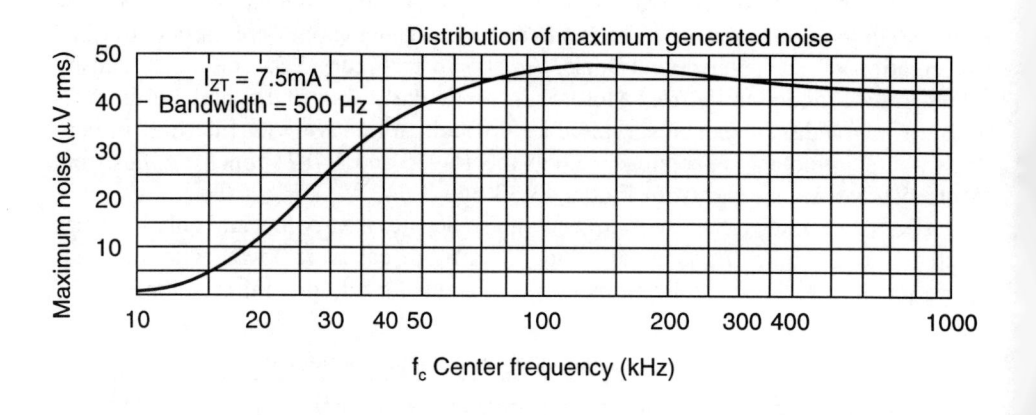

Figure 3.34 Distribution of maximum generated noise for the Motorola 1N821. (Copyright of Motorola, Inc. Used by permission). Source: [84].

Somewhat more subtle, carbon composition resistors are particularly bad for low noise design and should almost never be used. It has also been found that using transistor amplifier stages without emitter degeneration can result is $1/f$ noise performance that is as much as 40 dB inferior if local dc feedback is used. Electrolytic, ceramic, and silver mica capacitors can also be sources of excessive flicker noise and should only be used in noncritical circuit locations [57].

3.6.6 Low-Noise Oscillator Design and Characterization

If it were possible to construct the "perfect" oscillator, frequency synthesizer design would be much less difficult indeed. In many respects, frequency synthesizer design must begin with the oscillator.

In this section, we want to develop oscillator concepts in a systematic way in order to eventually arrive at a PSD description for the oscillator output. This characterization is one of the most important quantities needed for predicting overall synthesizer phase noise performance. We will primarily concern ourselves with lumped LC oscillators in this chapter, avoiding more specialized oscillator design areas such as YIG-based [85] and SAW-based [86] technologies.

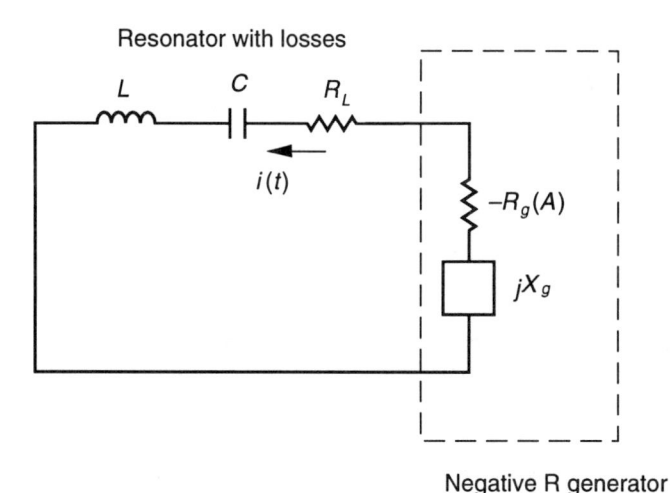

A Simple Negative Resistance Oscillator Model

Most microwave oscillators are designed using negative resistance concepts. This is particularly advantageous when computer-aided design software for linear network analysis is available. The existence of a negative resistance in an RLC network implies instability and hence oscillation. The negative resistance oscillator model we will consider is shown in Figure 3.35.

Figure 3.35 Simple negative resistance oscillator model.

A popular negative resistance generator is the Colpitts topology shown in Figure 3.36, which exhibits an input impedance of

$$Z_{in} = r_1 \frac{\left(\frac{1}{R_e} + \frac{1}{r_1} + g_m\right) + j\omega(C_\alpha + C_\beta)}{\left(\frac{1}{R_e} - \omega^2 C_\alpha C_\beta r_1\right) + j\omega\left(C_\beta + \frac{C_\alpha r_1}{R_e}\right)} \quad (3.85)$$

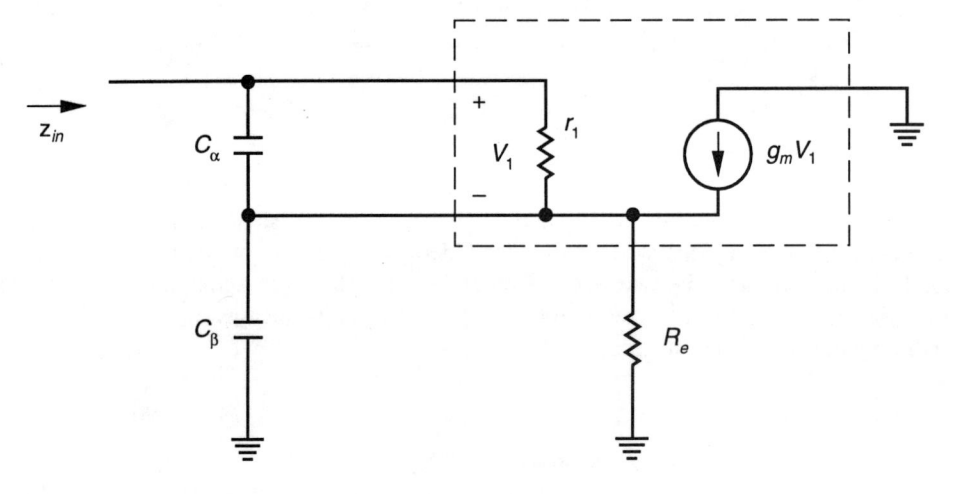

Figure 3.36 Common-collector Colpitts negative resistance generator.

The Z_{in} term displays a negative real part provided that

$$\omega \geq \frac{1}{\sqrt{C_\alpha C_\beta r_1 R_e}} \quad (3.86)$$

In Figure 3.35, the negative resistance generator is shown as a function of the oscillator signal amplitude $A(t)$. Once steady-state operation has been achieved, oscillator losses are precisely compensated for by oscillator gains implying that $R_L = R(A_\infty)$.

Assuming that the resonator Q in Figure 3.35 is reasonably high, the current waveform will be essentially sinusoidal and it is possible to represent the resonator current $i(t)$ as

$$i(t) = A(t) \cos[\omega_o t + \varphi(t)] \quad (3.87)$$

where both the phase and amplitude functions are slowly varying with respect to time compared to $\cos(\omega_o t)$ (i.e., definition of high-Q). The sum of the voltage drops across the RLC elements is given by

$$i(t)R_L + L\frac{di(t)}{dt} + \frac{1}{C}\int i(t)\, dt \qquad (3.88)$$

Substituting (3.87) for $i(t)$, employing integration by parts, and dropping higher order terms, (3.88) may be simplified to

$$\frac{A}{\omega_o}\left(1 - \frac{\varphi'}{\omega_o}\right)\sin(\omega_o t + \varphi) + \frac{A'}{\omega_o^2}\cos(\omega_o t + \varphi) \qquad (3.89)$$

where the primes represent differentiation with respect to time. Similarly,

$$\frac{d}{dt}[A\cos(\omega_o t + \varphi)] \approx A'\cos(\omega_o t + \varphi)$$
$$- A\omega_o \sin(\omega_o t + \varphi) \qquad (3.90)$$
$$- A\varphi' \sin(\omega_o t + \varphi)$$

Using these results in (3.88) and equating the voltage drop across the RLC elements to the negative of the voltage drop across the negative resistance generator, upon equating like sin( ) and cos( ) terms, we obtain

$$\frac{d\varphi}{dt} \approx -\frac{1}{2L}X_g$$
$$\frac{dA}{dt} \approx \frac{A}{2L}[R_g(A) - R_L] \qquad (3.91)$$

As shown by these results, the oscillator build-up to steady state is governed by these two equations, which, under the present assumptions, are uncoupled. This same observation was made earlier by Edson in the context of oscillator amplitude perturbations arising from noise [87]. This result allows us to consider oscillator phase perturbations independently. More extensive discussions concerning oscillator build-up as well as steady-state operation in this context can be found in [88,89]. A detailed look at using shunt and series feedback with active devices in order to create negative resistance generators is provided in [90].

Leeson's Phase Noise Model

The oscillator phase noise model, which has historically been used to describe oscillator phase noise, was heuristically deduced by Leeson in [91]. This model is widely used throughout industry. Based on this model, the oscillator single-sideband phase noise spectrum is given by

$$\mathscr{L}(f) = \frac{FkT}{2P_o}\left[1 + \left(\frac{f_o}{2Q_L f}\right)^2\right] \tag{3.92}$$

where

F = active device noise factor
k = Boltzmann constant (J/K)
T = temperature (K)
P_o = output power (W)
f_o = oscillator center frequency (Hz)
Q_L = loaded resonator quality factor, Q
f = frequency offset from carrier (Hz).

Leeson's model was later substantiated more rigorously in [92]. As later pointed out in [93], this model does have some deficiencies in its present form. A nearly equivalent statement of Leeson's model can be derived by considering the arguments presented next, which are based on the oscillator model shown in Figure 3.37. The analysis approach used here bears strong resemblance to that employed in [10,94].

Referring to Figure 3.37, the transfer function for the RLC resonator is given by

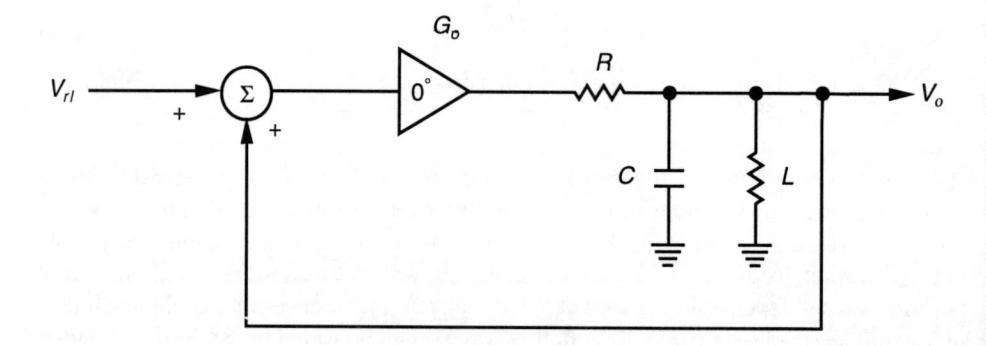

Figure 3.37 Linear oscillator analysis model.

$$F(\omega) = \frac{\omega RC}{j(\omega^2 LC - 1) + \omega RC} \qquad (3.93)$$

$$\approx \frac{1}{1 + j\dfrac{2Q(\omega - \omega_o)}{\omega_o}}$$

where $Q = (\omega_o RC)^{-1}$. The bandwidth of the resonator can be defined as $B_N = f_o/Q$. Hypothesizing linear steady-state operation of the network in Figure 3.37, the steady-state gain G_o exactly compensates for the losses represented by resistor R. The closed-loop voltage transfer function for the oscillator is given by

$$H(\delta) \approx \frac{G_o}{1 - G_o} \frac{1}{1 + j\dfrac{2Q\delta}{f_o(1 - G_o)}} \qquad (3.94)$$

where δ represents the frequency offset from the center frequency $f - f_o$. Given $H(\delta)$, the effective noise bandwidth of the oscillator model is

$$BW = \int_{-\infty}^{\infty} \left|\frac{H(\delta)}{H(0)}\right|^2 d\delta$$

$$= \int_{-\infty}^{\infty} \left|\frac{1}{1 + j\dfrac{2Q\delta}{f_o(1 - G_o)}}\right|^2 d\delta$$

$$= (1 - G_o)\frac{\pi}{2} B_N \qquad (3.95)$$

Oscillator feedback causes the normal noise bandwidth of the RLC network ($\pi B_N/2$ Hz) to be reduced substantially. It is therefore reasonable to define the closed-loop bandwidth of the oscillator as

$$B_c = (1 - G_o)B_N \qquad (3.96)$$

Assume now that the PSD represented by V_n in Figure 3.37 is that of white broadband noise and is therefore given by kT W/Hz. The PSD at the oscillator output is then

$$P(\delta) = kT\left(\frac{G_o}{1 - G_o}\right)^2 \frac{1}{1 + \left(\dfrac{2\delta}{B_c}\right)^2} \text{ W/Hz} \qquad (3.97)$$

The total signal power is simply

$$P_o = \int_{-\infty}^{\infty} P(\delta) \, d\delta \approx \frac{kT}{(1 - G_o)^2} \frac{\pi}{2} B_c \qquad (3.98)$$

Therefore, on inclusion of an additional factor of 0.5 to reflect that only phase noise is of concern (AM noise assumed to be removed), the normalized single-sideband noise spectrum is finally

$$\mathcal{L}(\delta) = \frac{P(\delta)}{2P_o}$$

$$= \frac{1}{\pi B_c} \frac{1}{1 + \left(\dfrac{2\delta}{B_c}\right)^2} \text{ Hz}^{-1} \qquad (3.99)$$

where, upon combining (3.96) and (3.98) and including a nonunity noise factor F for the active device,

$$B_c = \frac{\pi}{2} \frac{FkT}{P_o} \left(\frac{f_o}{Q}\right)^2 \qquad (3.100)$$

In comparing this model with Leeson's, it is interesting to observe that this model remains defined for $\delta = 0$ but does not properly predict the ultimate limit for $\mathcal{L}(\infty)$, which is given by $FkT/2P_o$. On the other hand, Leeson's model is not defined for $\delta = 0$ but does properly predict the ultimate limit for $\mathcal{L}(\infty)$.

A computational comparison between the linear model and Leeson's model is shown in Figure 3.38 where the following oscillator parameters have been assumed:

$F = 3$
$k = 1.38 \times 10^{-23}$ J/K
$T = 290$ K
$P_o = 5$ mW
$Q = 10$
$f_o = 1$ GHz

For all practical purposes, the two oscillator phase noise models are identical in the frequency range of normal interest. Continued use of the more familiar model of Leeson is therefore advocated.

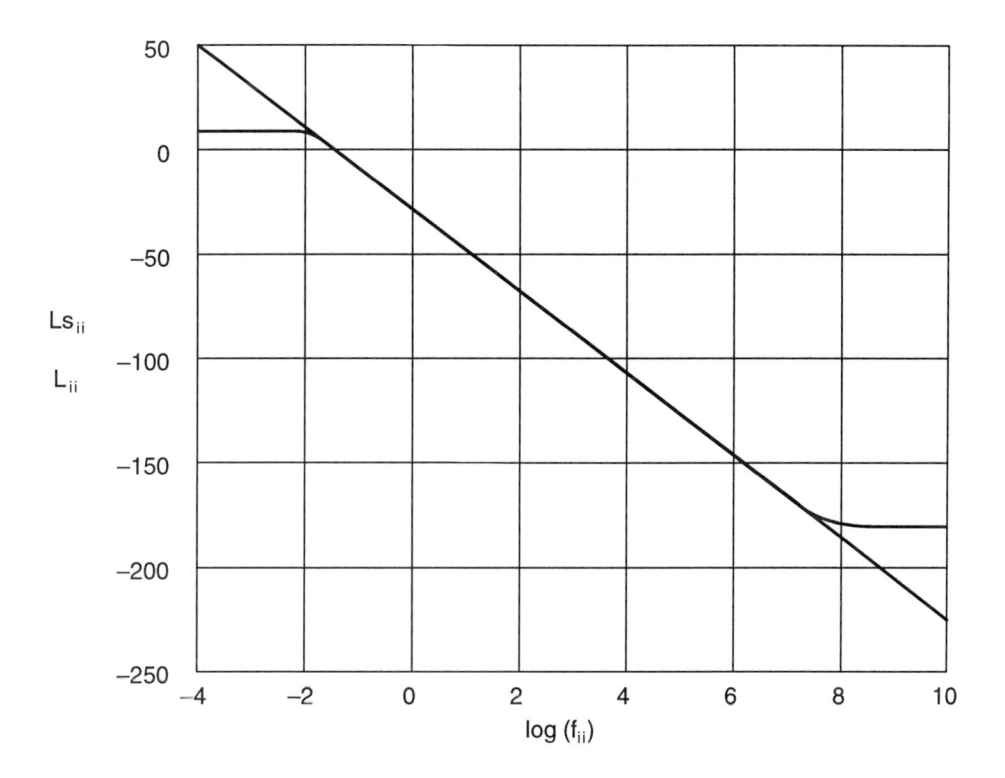

Figure 3.38 Comparison of Leeson's phase noise relationship with derivation in the text (3.99).

Extensions and Implications of Leeson's Model

As alluded to earlier in this chapter, a number of different noise processes can be observed with oscillator PSDs. Scherer provides an extension to Leeson's model, which includes most of the causes of phase noise in oscillators [49] which is given by

$$\mathscr{L}(f_m) = \frac{FkT}{2P_o}\left[1 + \left(\frac{f_o}{2Q_L f_m}\right)^2\right]\left(1 + \frac{f_c}{f_m}\right) \qquad (3.101)$$

The factor involving f_c represents flicker noise in the oscillator. If the $1/f$ phase noise modulation is in the resonator or active device, then an increase in the resonator Q will improve the oscillator phase noise. On the other hand, if the $1/f$ noise mechanism results in frequency modulation of the resonator center frequency, no improvement in resonator Q will lower the observed phase noise in the $1/f$ region.

Minimization of oscillator phase noise can be achieved by observing the following design guidelines [49]:

- Maximize the unloaded resonator Q.
- Maximize the resonator reactive energy within the constraints of device breakdown voltages. Forward bias of the varactor due to high RF voltage swing should be avoided.
- Active devices with low noise figures should be used.
- Active devices that display low $1/f$ noise should be used.
- Local dc feedback should be used with the active device to minimize flicker noise.
- Built-in resonator losses and loss represented by extracted power from the resonator (oscillator) should be made roughly equal as shown later.
- Signal power should be coupled directly out of the resonator portion of the oscillator.

In many VCOs, the spectral purity is dominated by AM to FM conversion mechanisms near the carrier frequency. One method to predict the AM-to-FM conversion effect in a varactor-tuned VCO is based on a simple observation of the VCO output frequency oscillator signal amplitude. As discussed later, a change in the RF voltage amplitude across the tuning varactor normally affects the observed tuning capacitance value in the resonator, thereby providing one substantial AM-to-FM conversion mechanism in the oscillator. Since the collector current in a bipolar transistor displays full shot current, the minimum percent of AM that will be observed is given by

$$\frac{\sqrt{2}\sqrt{2qI_c}}{I_c} 100\% \tag{3.102}$$

Extensive discussions of low-noise oscillator design, which develops these points, can be found in [95–98].

Although previous theories reveal the dependence of $\mathcal{L}(f)$ on P_o and Q, it is not clear what dependence between unloaded resonator $Q(Q_u)$ and loaded resonator $Q(Q)$ should be maintained for best phase noise performance. From (3.92), the phase noise for offset frequencies that appear in the -20 dB/decade regime is closely approximated by

$$\mathcal{L}(f) \approx \frac{FkT}{2P_o}\left(\frac{f_o}{2Qf}\right)^2 \tag{3.103}$$

Assume further that the loaded resonator Q is given by

$$\frac{1}{Q} = \frac{1}{Q_u} + \frac{1}{Q_L} \tag{3.104}$$

where Q_L represents the decrease in resonator Q, which occurs when power is removed from the resonator as the output signal. Minimizing $\mathcal{L}(f)$ with respect to Q_L, the optimal choice for Q_L can be shown to be $Q_L = Q_u$. Therefore, the power actually extracted from the resonator should be equivalent to the other circuit losses present in the resonator, which is synonymous with impedance matching for maximum power transfer.

In a similar derivation based on slightly different assumptions, it can be shown that the optimal ratio of loaded to unloaded $Q(Q/Q_u)$ should be 0.67 rather than 0.5 [94]. As shown in Figure 3.39, a range of Q/Q_u ratios can be accommodated with minimum performance impact, thereby making the actual ratio a fairly noncritical design consideration.

In the context of negative resistance oscillators from the Kurokawa theory perspective, several other insightful observations are possible. Let $Z(\omega)$ represent the input impedance seen looking into the passive network portion of the oscillator model

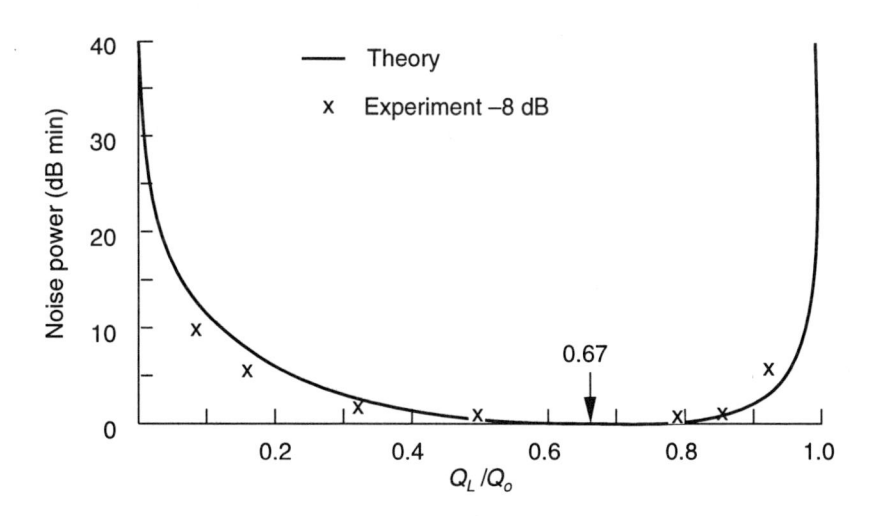

Figure 3.39 Oscillator sideband noise versus Q_L/Q_o. (©1986 IEEE; reprinted with permission). Source: [94].

shown in Figure 3.35 including jX_g leaving $-R(A)$ as the frequency-independent negative resistance generator as shown. In steady-state operation, it must be true that

$$R(A) = \text{Re}[Z(\omega_o)] \tag{3.105}$$
$$\text{Im}[Z(\omega_o)] = 0$$

This situation is shown graphically in Figure 3.40 [99]. If a small noise $e(t) = a\, e^{j(\omega_o t + \phi)}$ appears in series with the negative resistance element having frequency ω_o and phase ϕ with respect to the noise-free signal current, assuming that the noise-free current $i(t)$ in Figure 3.35 is represented by $A_o e^{j\omega t}$,

$$Z(\omega) = R(A) + \frac{a\, e^{-j\varphi}}{A_o} \tag{3.106}$$

Figure 3.40 shows how the noise voltage determines the instantaneous amplitude and frequency of the oscillating current, which is represented by

$$i'(t) = [A_o + \Delta A(t)]\, e^{j[\omega_o + \Delta\omega(t)]t} \tag{3.107}$$

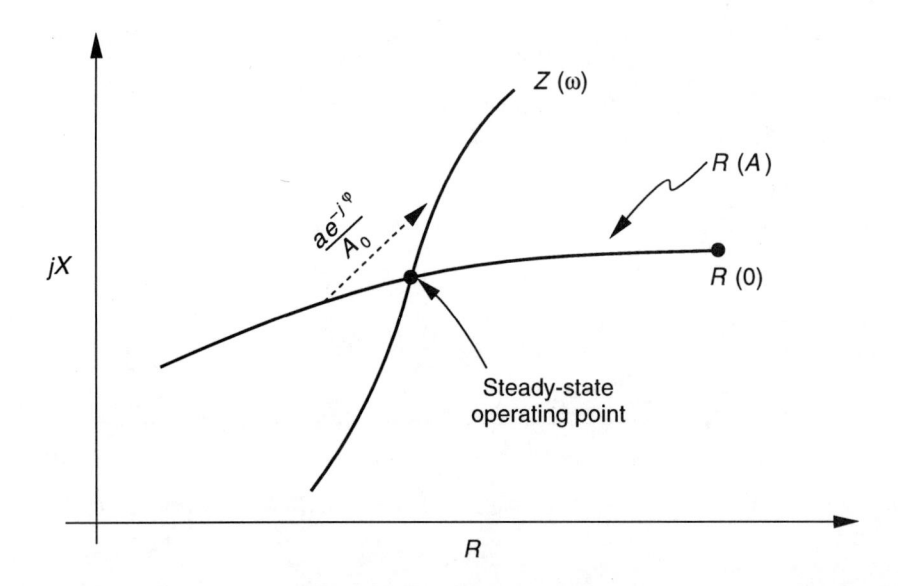

Figure 3.40 Negative resistance oscillator impedance conditions for steady-state operation. After: [99]. Reprinted with permission.

If the noise source $e(t)$ can be characterized in this manner, $R(A)$ and $Z(\omega)$ can be used to determine the AM and FM noise spectra of the oscillating current [99].

Equivalent reasoning in terms of reflection coefficients is given in [100]. If $Z(\omega)$ is represented by Γ_L and $-R(A)$ by Γ_T, steady-state conditions that are necessary for oscillation to occur are given by $\Gamma_L \Gamma_T = 1$. The loci of reflection coefficient trajectories is shown graphically in Figure 3.41. Esdale and Howes [101] show that the angle between $d\Gamma_{T^{-1}}/dA$ and $d\Gamma_L/d\omega$ must make an angle α of less than 180 deg. For optimum oscillator phase noise performance,

- $d\Gamma_{T^{-1}}/dA$ and $d\Gamma_L/d\omega$ should intersect with $\alpha = 90$ deg
- $d|\Gamma_{T^{-1}}|/dA$ should be minimized
- $d\Gamma_L/d\omega$ should be maximized.

Before leaving the topic of oscillator design based on linear methods, the extensive work done by Janning [102] is worthy of comment. A similar derivation of this work is more accessible as given in [10].

Design Guidelines From Nonlinear Oscillator Theory

Generalized nonlinear oscillator analysis leads immediately to sophisticated mathematical techniques including nonlinear differential equations and Volterra series techniques. Although we do not explore any of these methods here, several important

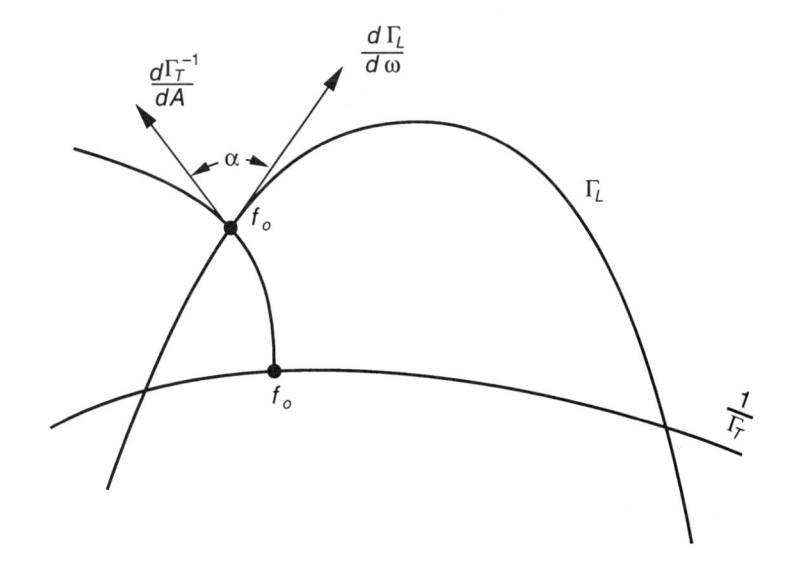

Figure 3.41 Plots of Γ_L and Γ_T^{-1} in the complex plane. After: [100]. Reprinted with permission.

results are presented. An overview of a number of nonlinear analysis techniques can be found in [103].

All oscillators require some form of limiting or signal level control once steady-state operation is achieved. While it is true that automatic gain control could be used to control the oscillator signal level, this is rarely done in RF and microwave oscillators because of the circuit complexity that would normally result. Most of these oscillators rely on nonlinear limiting mechanisms for determining the steady-state oscillator signal levels.

van der Pol Oscillator Model

The van der Pol oscillator model addresses the negative resistance oscillator model shown in Figure 3.35 where $X_g = 0$ and $-R_g(A) = -A^3/3$ [104]. The nonlinear differential equation that represents this model is

$$v'' + \frac{1}{C}\left(\frac{1}{R} - v^2\right)v' + \frac{1}{LC}v = 0 \qquad (3.108)$$

where v is the voltage across the negative resistance generator as a function of time, and the primes denote differentiation with respect to time. In the special case where $L = C^{-1} = \epsilon$ and $R = 1$, this reduces to

$$v'' + \epsilon(1 - v^2)v' + v = 0 \qquad (3.109)$$

which for small ϵ leads to a sinusoidal steady-state solution having frequency $\omega_o \approx 1$ and amplitude ≈ 2. Using Chua's second-order determining equation approach [104], a closer approximation for ω_o is given by

$$\omega_o \approx 1 - \frac{\epsilon^2}{16} \qquad (3.110)$$

The third-order determining equation approximation would have a correction term proportional to ϵ^4.

In this form, if $\omega_o \approx 1$ is assumed, the resonator Q is given by ϵ^{-1}. The steady-state sinusoidal frequency is then given by

$$\omega_o \approx 1 - \left(\frac{1}{4Q}\right)^2 \qquad (3.111)$$

It is simple then to conclude that $Q > 1.12$ and $Q > 2.5$ are adequate to reduce the shift in ω_o from unity to less than 5% and 1%, respectively. Therefore, unless the resonator Q is extremely poor, the observed resonance frequency is given very accurately by $\omega_o = (LC)^{-0.5}$.

Classical Methods of Nonlinear Oscillator Analysis

An extensive treatment of nonlinear oscillator analysis based on more traditional mathematical techniques can be found in [105]. One of the more insightful discussions presented there examines nonlinear oscillator limiting using phase plane techniques.

The phase plane approach is strictly limited to second-order dynamic systems, represented by Figure 3.37, where the gain block is replaced by a generalized nonlinearity in which case the output is given by

$$V_o = \sum_{i=1}^{M} g_i V_{in}^i = G(V_o) \qquad (3.112)$$

The integro-differential equation representing this system is given by

$$C\frac{dV_o}{dt} + \frac{V_o - G(V_o)}{R} + \frac{1}{L}\int V_o\, dt = 0 \qquad (3.113)$$

Letting

$$y = \frac{1}{L}\int V_o\, dt \qquad (3.114)$$

$$\frac{dV_o}{dt} = \frac{dV_o}{dt} y'$$

on substitution into (3.113), the state of the system is completely characterized by the differential equation

$$\frac{dy}{dV_o} = \frac{1}{L}\frac{V_o RC}{G(V_o) - V_o - yR} \qquad (3.115)$$

Once this differential equation has been solved, quantities such as steady-state signal amplitude and harmonic distortion can be computed as shown in [105].

Modern oscillator designers have a wide variety of computer-aided design tools at their disposal. The best techniques tend to utilize a mixture between small and large signal analysis tools [110]. A number of papers have appeared addressing large signal oscillator design [106–110] but phase noise performance is rarely addressed. Even though (3.92) clearly predicts that increasing P_o will improve $\mathcal{L}(f)$, clearly this cannot be done at the expense of resonator Q or other important design guidelines.

Varactor Diode Nonlinear Effects

One of the principle nonlinear oscillator elements, particularly in wideband VCOs, is the varactor diode. The potentially large voltage swing across the varactor(s) leads to departures in the frequency tuning curve from nominal and upconversion of low-frequency noise components that contribute to oscillator phase noise sidebands.

Nonlinear varactor diode effects are considered at length in [111] using Volterra series analysis techniques. It can be shown that minimal varactor distortion occurs when the oscillator tuning varactors are used in the back-to-back topology as shown in Figure 3.42. Second-order distortion is theoretically reduced to zero when matched abrupt junction varactor diodes are used in this configuration.

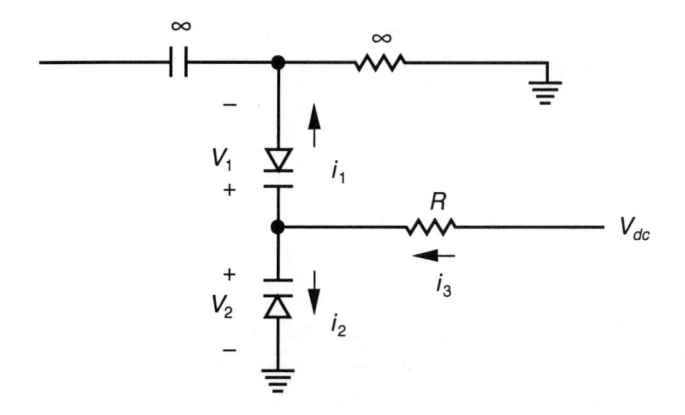

Figure 3.42 Back-to-back varactor diode configuration for improved balance.

An intuitive look at the varactor nonlinear effects in the back-to-back configuration can be had by assuming that the varactors are used in a fairly high-Q resonant circuit in which case the total voltage across the two varactors can be taken to be $A\cos(\omega_o t)$. In this case, the circuit shown in Figure 3.42 is completely described by

$$i_1 = \frac{dQ_1}{dV_1}\frac{dV_1}{dt} = \frac{C_o}{\sqrt{V_1 + \varphi}}\frac{dV_1}{dt}$$

$$i_2 = \frac{C_o}{\sqrt{V_2 + \varphi}}\frac{dV_2}{dt} \tag{3.116}$$

$$V_2 = V_{dc} - (i_1 + i_2)R$$

$$V_2 = A\cos(\omega_o t) + V_1$$

where the individual varactor diode C versus V curve has been assumed to be

$$C(v) = \frac{C_o}{\sqrt{V + \varphi}} \tag{3.117}$$

and ϕ is the junction contact potential, which is normally about 0.7V. Simple algebra can be used to write the solution for this example as the differential equation

$$\frac{dV_o}{dt} = -\frac{V_2 - V_{dc} + ARC_o \sin(\omega_o t) f(t)}{RC_o \left[\dfrac{1}{\sqrt{V_2 + \varphi}} + f(t) \right]} \tag{3.118}$$

where

$$f(t) = [V_2 - A\cos(\omega_o t) + \phi]^{-1/2} \tag{3.119}$$

This differential equation may be easily solved using numerical integration software packages. Once $V_2(t)$ has been found, the calculation of the input varactor current $-i_1(t)$ is straightforward. This computation has been carried out in Figure 3.43 for the case where $A = 0.5$, $V_{dc} = 4.0$, $\omega_o = 1.0$, $R = 4.0$, $C_o = 1.0$, and $\phi = 0.70$ where the current i_1 appears essentially sinusoidal. Increasing the input signal voltage peak value A to 10.0V, nonsinusoidal current flow is readily apparent as shown in Figure 3.44. The degree of harmonic distortion introduced by the varactor diodes can be readily ascertained using these results.

Computer-aided design and analysis can also be used to actually design the varactor diode doping profiles for linear frequency tuning even in the presence of large signals [112]. Any reduction in sensitivity of the tuning curve to signal amplitude A is desirable because this leads to a corresponding decrease in AM to FM conversion within the oscillator.

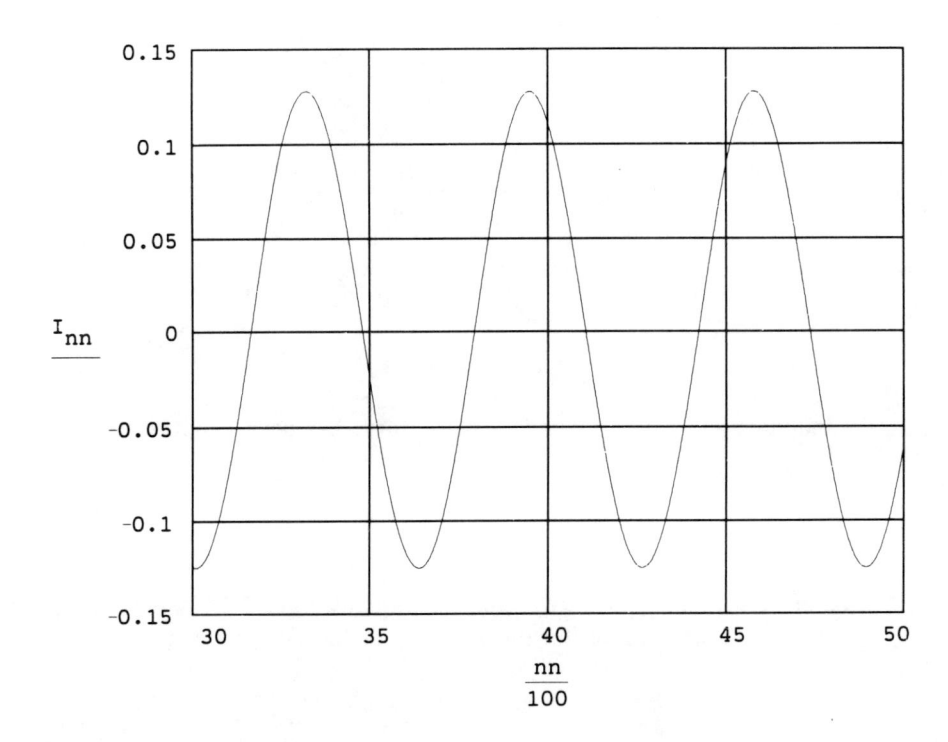

Figure 3.43 Varactor current i_1 for $A = 0.5$, $V_{dc} = 4.0$, $\omega_o = 1$, $R = 4.0$, and $\phi = 0.70$ versus time.

Additional Nonlinear Effects in Oscillators

Oscillator circuit nonlinearities cause low-frequency noise components to be upconverted and to appear as noise sidebands on the oscillator output. Although this statement is intuitively obvious, quantifying this mechanism is much more complex. Second-order nonlinear distortion determines the degree of $1/f$ noise contamination of the oscillator output for instance [100,113]. Therefore, second-order distortion in the oscillator should be minimized. The degree to which any oscillator accomplishes this goal can be judged based on the second harmonic output level of the oscillator. A good oscillator should exhibit second harmonic levels on the order of -40 dBc [100].

Another useful indicator of good oscillator design is the change in oscillation frequency versus dc bias reduction. A slow reduction of the supply voltage from nominal to the point at which oscillation just ceases should result in a very small frequency change (on the order of 20 kHz for a well-designed 2-GHz oscillator [100]). Solution of the van der Pol equation in [113] also shows that the oscillator excess gain (which is necessary for initial oscillator build-up) should be minimized in order

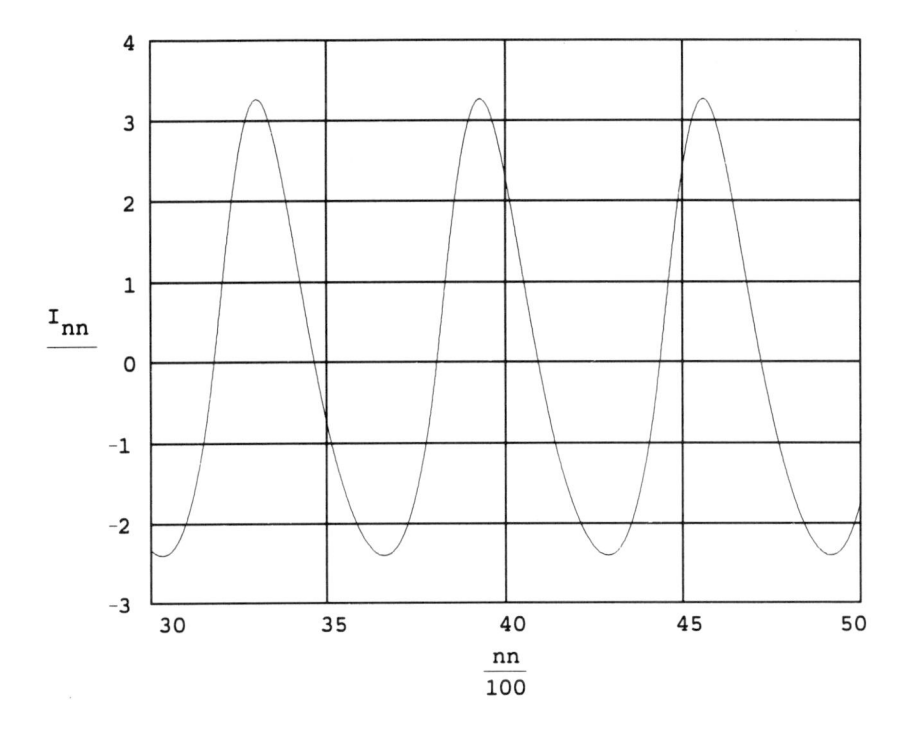

Figure 3.44 Varactor current i_1 for $A = 10.0$, $V_{dc} = 4.0$, $\omega_o = 1$, $R = 4.0$, and $\phi = 0.70$ versus time.

to prevent amplitude fluctuations from being converted into significant frequency fluctuations.

Healey [113] also reported that second harmonic currents in the oscillator sustaining stage can appear in phase quadrature with the fundamental current, thereby worsening the conversion of AM noise to PM noise. Ideally, the second and third harmonic frequencies are well below the beta cutoff frequency of the oscillator sustaining stage transistor thereby minimizing this effect.

Other Oscillator Impairments

Other impairments including injection locking, load pulling, and power supply pushing can cause serious oscillator performance degradation, particularly in phase-locked systems. If the induced impairments fall within the closed-loop bandwidth of the system, potentially chaotic spectral behavior can result. Design margins must be identified and held for each of these potential problem areas.

Injection Locking

Injection locking has been a recognized oscillator phenomenon for many years. It can be shown that when a signal of sufficient amplitude and sufficiently small frequency error is impressed on a free-running oscillator, over time, the free-running oscillator changes its frequency to that of the impressed signal with a corresponding change in its signal phase and amplitude [114,115]. Normally, injection locking is a very undesirable situation, but it has been used to advantage on occasion such as in narrowband bit synchronizers [116]. Adler [115] showed that the injection locking frequency range is given by $\pm \Delta \omega_m$ with

$$\Delta \omega_m = \frac{\omega_o}{Q} \sqrt{\frac{P_i}{P_o}} \qquad (3.120)$$

where

Q = loaded resonator Q
ω_o = free-running oscillator radian frequency
P_o = oscillator resonator power level
P_i = injected signal power of interfering signal.

Within the injection locking range, if a signal is applied at time $t = 0$, the phase transient of the oscillator, which occurs during the locking process based on Adler's concepts, is given by

$$\varphi = 2.0 \tan^{-1} \left[-\frac{\Delta \omega_m}{\Delta \omega} - \sqrt{1 - \left(\frac{\Delta \omega_m}{\Delta \omega}\right)^2} \tan\left\{ \frac{\Delta \omega t}{2} \sqrt{1 - \left(\frac{\Delta \omega_m}{\Delta \omega}\right)^2} \right\} \right] \qquad (3.121)$$

These results provide some guidelines for predicting potential injection locking problems.

This phenomenon can be particularly troublesome in type 2 control systems where multiple signals may be present that are all integrally related to a fixed reference frequency. In this case, interferers may be present that have the same frequency, but different phase, as the oscillator being locked. Since the loop electronics attempt to drive the steady-state phase error ideally to zero, it may be very difficult to achieve zero oscillator phase and frequency error simultaneously within the loop. In this situation, either the contaminating signal must be eliminated or provision must be made for allowing a static phase error within the phase-locked loop (e.g., utilize a leaky integrator).

A more comprehensive look at injection locked oscillators based on the inhomogeneous van der Pol equation for forced self-sustained oscillations is available in [117].

Load Pulling

Oscillator load pulling refers to the change in oscillator frequency that occurs when the oscillator load impedance is changed. If this impedance change is dynamic in nature, load pulling of the oscillator leads to direct frequency modulation of the oscillator. Obviously, if this oscillator is contained within a phase-locked loop and the frequency of modulation lies within the closed-loop bandwidth, strange interactions can result.

In this context, one of the most serious load pulling situations that can occur in practice arises in modulators where the modulation signal causes (low-frequency) baseband frequency modulation of the load as seen by the oscillator such as in the modulator situation shown in Figure 3.45. In this figure, Γ_m is the load reflection coefficient, which is assumed to be a function of the modulation signal $m(t)$.

In the special case where $m(t)$ is a sinusoid with frequency ω_m, assume that

$$\Gamma_m(t) = \Gamma_{mo} + \Delta\Gamma \sin(\omega_m t) \tag{3.122}$$

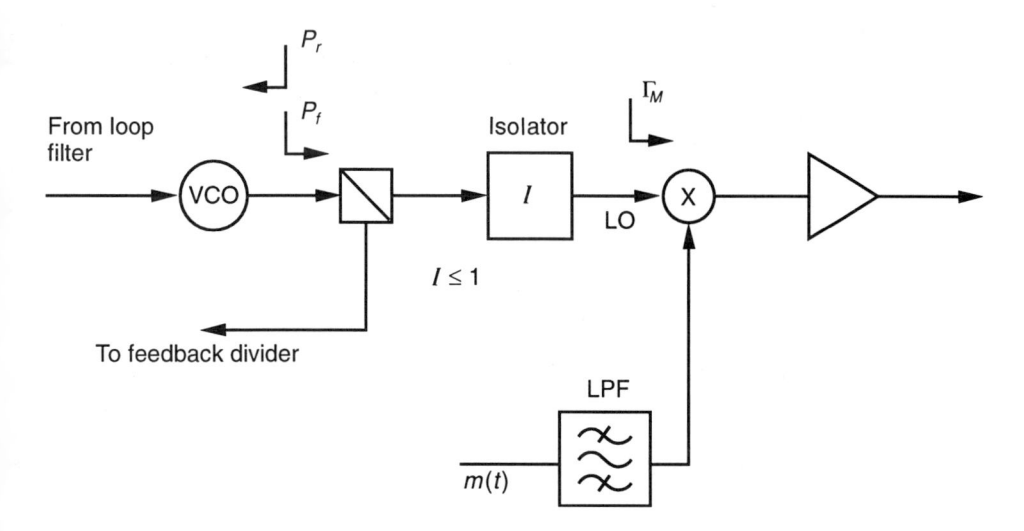

Figure 3.45 Baseband-modulation-induced load pulling.

If P_f is the forward power output from the oscillator, the reflected signal power is then given by

$$P_r = P_f I[\Gamma_{mo} + \Delta\Gamma \sin(\omega_m t)]^2 I \qquad (3.123)$$

In terms of voltage phasers at the oscillator output then, the reflected signal creates a peak phase shift of

$$\theta_p = \tan^{-1}(I\Delta\Gamma) \approx \Delta\Gamma I \qquad (3.124)$$

which translates into an approximate peak-to-peak change in oscillator frequency of

$$\Delta\omega_{pp} \approx \theta_p \frac{\omega_o}{2Q} 2 \qquad (3.125)$$

where Q is the loaded-Q of the oscillator and ω_o is the nominal oscillator radian frequency.

To defeat this damaging frequency modulation, in the case of baseband signals $m(t)$, the isolation in Figure 3.45 must be made quite high. It is not unusual in the case of DBMs for the required isolation to be equivalent to 70 to 90 dB.

An exact derivation of the (nonlinear) oscillator pulling figure is given in [118]. Although instructive, use of the results requires knowledge of two oscillator quantities that are rarely known in actual practice.

A measurement-based assessment of oscillator load pulling is described in [119]. This method is based on using a variable attenuator that is terminated in a sliding short. This permits both the magnitude and phase of the load reflection coefficient to be easily changed. The test configuration is shown in Figure 3.46

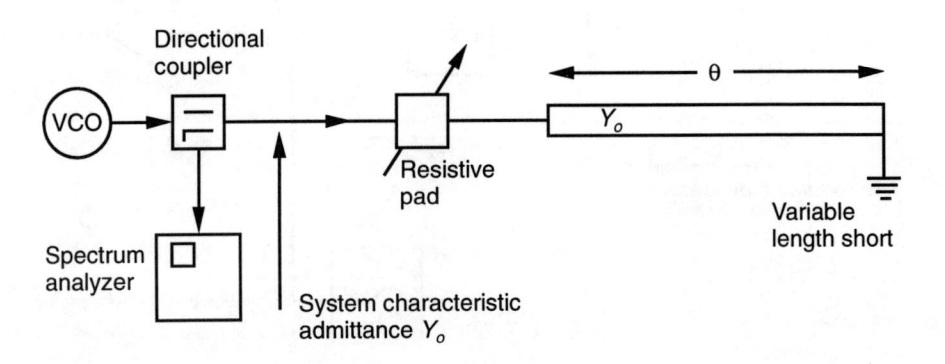

Figure 3.46 Oscillator load-pulling assessment. After: [119].

where Y_o represents the system characteristic admittance, the resistive pad represents the variable attenuator, and θ represents the electrical length of the variable short.

In the case of a variable attenuator with characteristic admittance Y_o,

$$R_1 = \frac{1-\alpha}{Y_o(1+\alpha)}$$

$$R_2 = \frac{2\alpha}{Y_o(1-\alpha^2)} \qquad (3.126)$$

$$\alpha = 10^{A_{dB}/20}$$

where A_{dB} is the attenuation in decibels. The peak-to-peak maximum variation of the oscillator frequency due to the mismatched load can then be shown to be

$$\Delta f = \frac{f_o}{Q_e}\frac{\gamma^2-1}{\gamma^2+1} \qquad (3.127)$$

where Q_e is the oscillator's external (loaded) Q and γ is the VSWR of the load being considered. If the external Q of the oscillator is not known *a priori*, it can be computed as

$$Q_e = \frac{f_o}{\Delta f}\frac{2\alpha^2}{1+\alpha^4} \qquad (3.128)$$

Oscillator Pushing

Oscillator pushing is the technical term applied to the oscillator frequency perturbations that result from small changes in the oscillator's supply voltage(s). These perturbations can result from a number of factors including (1) changes in device capacitance values caused by modified reverse biased junction capacitances, (2) changes in the oscillator self-limiting signal mechanism, and (3) changes in the sustaining stage gain. Oscillator pushing can lead to substantial phase noise degradation because any power supply noise directly can lead to frequency modulation of the oscillator.

Oscillator Design Example

One of the most impressive low-noise oscillator designs appearing in the recent literature is that reported in [120]. Typical design parameters for such an oscillator are given in Table 3.2.

Table 3.2
Typical VCO Design Parameters

Typical VCO Design Parameters	
Supply Voltage	12 V
Supply Current	20 mA
Tuning Range	1–1.4 GHz
Output Power	+5 dBm
Tuning Voltage Range	1–9 V
Tuning Port Time Constant	$\leq$50 ns
Phase Noise Performance	
10 kHz	$<$−110 dBc/Hz
100 kHz	$<$−130 dBc/Hz
1 MHz	$<$−150 dBc/Hz
Harmonic Content	
Second	$<$−35 dBc
Third	$<$−20 dBc
Load Pulling	
VSWR 1:1 to ∞:1	$<$100 kHz
Pushing	$<$50 kHz/V

A representative oscillator schematic for such an oscillator based largely on [120] is shown in Figure 3.47. Additional measures have also been included in this schematic to further improve performance beyond that reported in [127]. With reference to Figure 3.47, the key design methodologies are as outlined:

1. The push-pull design ideally eliminates even-mode distortion terms thereby reducing $1/f$ noise impact.
*2. Output power coupling is chosen to match the internal resonator losses with the power actually extracted from the oscillator.
*3. "Noiseless" biasing [63] is used in the transistor base circuits. Absence of any significant dc current in the base resistors minimizes shot current, which has a magnitude of $(2qI_{dc})^{1/2}$ A rms/$\sqrt{\text{Hz}}$
4. Local emitter (dc) feedback present to reduce $1/f$ noise.
*5. Oscillator self-limiting mechanism implemented such that the resonator Q is not degraded. Actual limiting occurs across the base-emitter junction rather than the base-collector junction. Aside from the feedback level, which is set by the collector-emitter capacitance dividers, the emitter biasing time constant R_1C_1 should be kept greater than $2.5/f_{ol}$ where f_{ol} is the lowest oscillator output frequency being tuned.
6. The amount of feedback to the emitter from the collector must not result in base-emitter junction voltage breakdown (critical point for most microwave transistors).
7. The collector supply currents are inserted with negligible resonator loading.

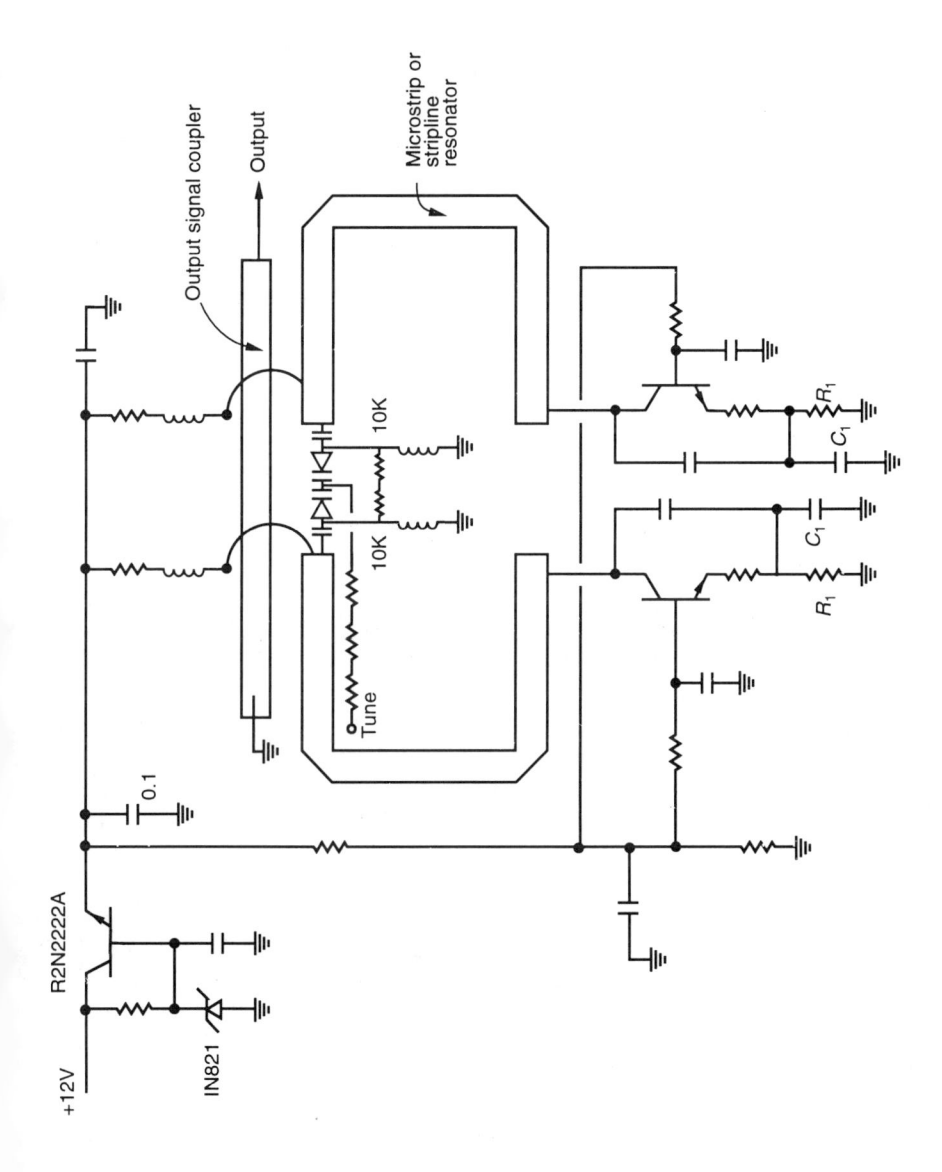

Figure 3.47 State-of-the-art balanced VCO concept. (Reprinted with permission. After: [120].)

8. Back-to-back varactors are used to minimize even-order varactor nonlinearities.
*9. Varactors are placed at minimum voltage swing in the resonator.
*10. Resistors placed about the varactors in order to maintain reverse bias have their noise voltage short-circuited through an inductor.
*11. Tuning line resistance is kept very low, thereby keeping Johnson noise effects minimal.
12. Power supply cleaning is provided to reduce pushing effects and noise contamination.

The asterisks in the preceding list denote significant design enhancements over [120].

3.7 ASSESSING SYSTEM PERFORMANCE IN THE PRESENCE OF FREQUENCY SYNTHESIZER PHASE NOISE

The discussion presented thus far in this chapter satisfies most system-level analysis needs for the stand-alone design of frequency synthesizers. In the case of communication systems work, however, it is important to understand how synthesizer phase noise contributions directly impact the end performance of the complete system. Usually, this performance impact is quantified in terms of BER performance degradation with respect to an ideal phase noise free system.

The advent of high-throughput digital communication techniques has accentuated the need for high-quality, low-cost signal sources. Although older binary and quadrature phase-shift-keyed (BPSK, QPSK) systems were certainly subject to phase noise considerations, the higher throughput waveforms such as 16-QAM[1] (QAM stands for quadrature amplitude modulation), 64-QAM, and M-PSK trellis-coded modulation are much more sensitive to phase noise related impairments. In contrast, the mobile communications industry often employs noncoherent waveforms such as DQPSK in order to combat the severe fading nature of the mobile channel, and these waveforms are inherently immune to close-in phase noise effects. Therefore, phase noise degradation in any given modern system cannot generally be accurately predicted without factoring in many other system attributes. Even the type of forward error correction (FEC) being used in the system and the data interleaving depth (if used) can significantly affect the system sensitivity to nonideal synthesizer phase noise behavior [121]. This perspective motivated the earlier discussion about simulating arbitrary phase noise spectra on the computer because such modeling can be appended to existing system simulation analysis tools if desired.

Recognizing the voluminous nature of complete communication systems performance characterization, we still want to develop a basic understanding of phase noise performance as it relates to system BER performance with simple communi-

---

[1]First presented to the author by Gary D. Frey.

cation systems. The balance of this chapter is committed to that endeavor. First, a number of basic digital communication waveforms are considered followed by a more general viewpoint where phase noise is equivalenced with a degradation in the received signal carrier-to-noise ratio (CNR). We will conclude with a more theoretical treatment of phase noise degradation as it applies to lost channel capacity.

Phase Noise Related Impairments With Traditional Digital Waveforms

The presence of local oscillator phase noise in coherent receiving systems results in a correlation-related processing loss as well as causing the recovered estimate of the carrier phase to be noisy [33]. In general, the criticality of synthesizer phase noise performance increases rapidly as the modulation waveform complexity is increased.

In lieu of exhaustive computer simulation to assess phase noise impact on a given system, simplifying assumptions can be made in order to allow analytical design guidelines to be established. One of two assumptions is normally made in this context concerning the behavior of the carrier phase error over a symbol period, namely, (1) the phase error is assumed to be constant over each symbol interval T or (2) the phase error is assumed to vary rapidly over each symbol period [122, Chapter 6]. The degree to which each assumption is appropriate depends on many system aspects as well as on the received signal's SNR. The most conservative design methodology is to levy phase noise requirements, which result in acceptable BER performance regardless of which assumption is made. Generally speaking, the most severe system impact normally arises under assumption (1) because the irreducible error problems associated with (1) do not normally appear with (2) [122].

To investigate these issues further, consider the QAM situation where the transmitted signal is given by

$$s(t) = \sqrt{2}\left[\sum_k a_k p(t - kT)\right] \cos(\omega_o t) \\ - \sqrt{2}\left[\sum_k b_k p(t - kT)\right] \sin(\omega_o t) \quad (3.129)$$

The a_k and b_k values represent the in-phase and quadrature channel (i.e., I and Q channel) information bits being sent, $p(t)$ is the symbol pulse shape used for both channels, and ω_o is the carrier radian frequency. This may be rewritten as

$$s(t) = \text{REAL PART}\left[\sqrt{2}\, e^{j\omega_o t} \sum_k (a_k + jb_k)\, p(t - kT)\right] \quad (3.130)$$

Assume now that the receive signal processing is as shown in Figure 3.48 where the receiver's estimates of the incoming signal center frequency and phase are given by ω and θ, and the filters $h(t)$ are perfectly matched to the transmitted pulse shape $p(t)$ with

$$\int_{-\infty}^{\infty} |H(f)|^2 \, df = 1 \tag{3.131}$$

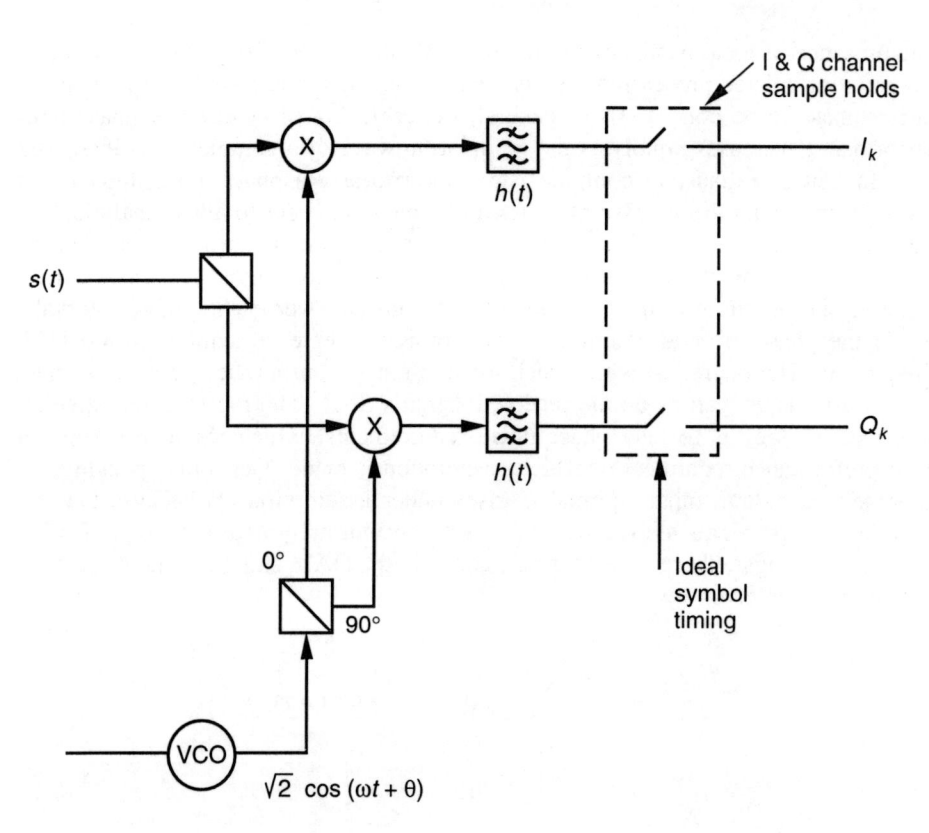

Figure 3.48 Two-channel coherent receiver.

If $p(t)$ has been appropriately chosen such that the matched filter outputs are free of intersymbol interference (e.g., Nyquist condition) and the data symbol timing is known exactly, the recovered information bits are given simply by

$$\begin{aligned} I_k &= a_k \cos(\Delta\omega t + \theta) + b_k \sin(\Delta\omega t + \theta) \\ Q_k &= b_k \cos(\Delta\omega t + \theta) - a_k \sin(\Delta\omega t + \theta) \end{aligned} \tag{3.132}$$

where $\Delta\omega$ represents the frequency error between the incoming signal and the receiver's frequency estimate. Assuming that this frequency error is made zero by the receiver, the recovered I and Q sample values are given by

$$I_k = a_k \cos(\theta) + b_k \sin(\theta)$$
$$Q_k = b_k \cos(\theta) - a_k \sin(\theta) \quad (3.133)$$

Any phase error θ in the receiver's local oscillator results in (1) a loss of correlation, which is represented by the cosine terms, and (2) an undesirable coupling between the I and Q channels, which is represented by the sine terms. This latter effect is particularly damaging to QAM systems. In the discussion that follows, the BER performance degradation that results when θ is constant over each symbol period T will be examined first, followed by considerations when θ varies rapidly over each symbol period.

3.7.1 Phase Noise Effects With Coherent Demodulation of Binary FSK

The BER for coherent demodulation of binary FSK is given by

$$P_{b\,\text{BFSK}} = Q\left(\sqrt{\frac{E_b}{N_o}}\right) \quad (3.134)$$

where

$$Q(x) = \int_x^\infty \frac{e^{-u^2/2}}{\sqrt{2\pi}}\,du \quad (3.135)$$

for the simple additive white Gaussian noise (AWGN) channel [123]. A convenient approximation for evaluating $Q(x)$ is [123]

$$Q(x) = \begin{bmatrix} \dfrac{e^{-x^2/2}}{\sqrt{2\pi}} \sum_{i=1}^{5} b_i t^i & \text{for } x \geq 0 \\ 1 - \dfrac{e^{-x^2/2}}{\sqrt{2\pi}} \sum_{i=1}^{5} b_i t^i & \text{for } x < 0 \end{bmatrix} \quad (3.136)$$

where

$$t = \frac{1}{1 + p|x|} \tag{3.137}$$

and

$p = 0.2316419$
$b_1 = 0.319381530$
$b_2 = -0.356563782$
$b_3 = 1.781477937$
$b_4 = -1.821255978$
$b_5 = 1.330274429.$

Nonideal frequency synthesizer phase noise performance is assumed to introduce a phase error θ, which is assumed to remain constant over individual symbol intervals. This error is assumed to have a Tikhonov probability distribution function [33,121,123], which is given by

$$p(\theta) = \frac{e^{\alpha \cos(\theta)}}{2\pi I_0(\alpha)} \tag{3.138}$$

where α is the SNR for the carrier recovery phase-locked loop and $I_0(\ )$ represents the zeroth-order modified Bessel function. For large α arguments, numerical evaluation of the Bessel function can become numerically ill conditioned and it is advisable to use the asymptotic approximation for the density function

$$P(\theta, \alpha) \approx \sqrt{\frac{\alpha}{2\pi}} \, e^{\alpha[\cos(\theta) - 1]} \tag{3.139}$$

In frequency synthesizer terminology where the total integrated phase noise is given by σ_θ^2, $\alpha = \sigma_\theta^{-2}$.

Employing Bayes rule, the BER for coherent BFSK with phase noise impairment constant over each symbol period is given by

$$\begin{aligned}
P_{b\,\text{BFSK}}\left(\frac{E_b}{N_o}, \alpha\right) &= \int_{-\pi}^{\pi} P_b\left(\frac{E_b}{N_o} \bigg| \theta\right) p(\theta) \, d\theta \\
&= \int_{-\pi}^{\pi} Q\left[\sqrt{\frac{E_b}{N_o}} \cos(\theta)\right] \frac{e^{\alpha \cos(\theta)}}{2\pi I_0(\alpha)} \, d\theta
\end{aligned} \tag{3.140}$$

which is most easily evaluated numerically. The results of this calculation for loop SNRs of 8, 10, 12, 15, and 20 dB are presented in Figure 3.49.

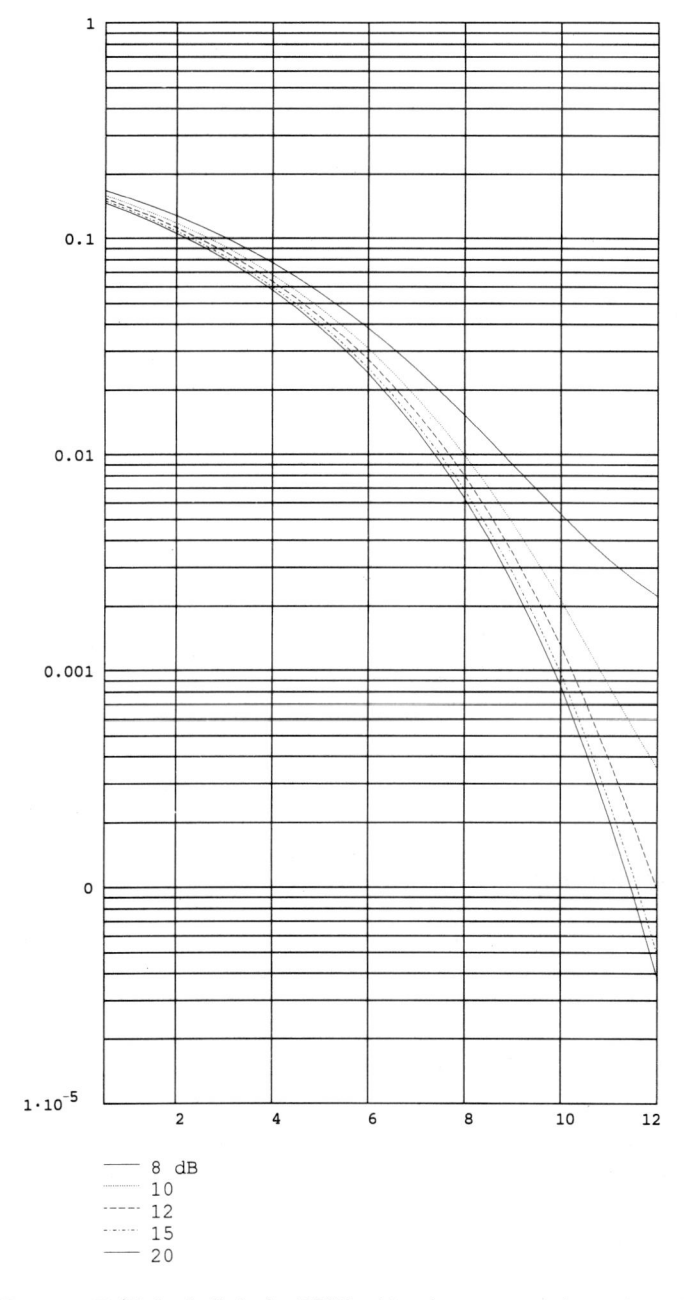

Figure 3.49 BER versus E_b/N_o in decibels for BFSK with coherent processing and specified carrier loop SNR.

3.7.2 Phase Noise Effects With Coherent Demodulation of BPSK

In the case of coherent BPSK demodulation with the AWGN channel and carrier phase constant over each symbol interval, the BER is given by

$$P_{b\,\text{BPSK}}\left(\frac{E_b}{N_o}, \alpha\right) = \int_{-\infty}^{\infty} Q\left[\sqrt{\frac{2E_b}{N_o}} \cos(\theta)\right] p(\theta)\, d\theta \tag{3.141}$$

The results for this case are shown in Figure 3.50 for loop SNRs of 8, 10, 12, 15, and 20 dB.

3.7.3 Phase Noise Effects With Coherent Demodulation of QPSK

The BER performance in the case of coherent demodulation of QPSK over the AWGN channel is given by

$$\begin{aligned} P_{b\,\text{QPSK}}\left(\frac{E_b}{N_o}, \alpha\right) = &\frac{1}{2}\int_{-\pi}^{\pi} Q\left\{\sqrt{\frac{2E_b}{N_o}}\left[\cos(\theta) + \sin(\theta)\right]\right\} p(\theta)\, d\theta \\ &+ \frac{1}{2}\int_{-\pi}^{\pi} Q\left\{\sqrt{\frac{2E_b}{N_o}}\left[\cos(\theta) - \sin(\theta)\right]\right\} p(\theta)\, d\theta \end{aligned} \tag{3.142}$$

where the I and Q channel cross-coupling adds additional complications. The BER curves for this waveform are shown in Figure 3.51 for loop SNRs of 8, 10, 12, 15, 20, and 25 dB. It is fairly clear that considerably better phase noise performance is needed for QPSK than for either of the previous waveforms in order to incur the same departure from ideal performance without phase noise.

3.7.4 Phase Noise Effects With Coherent Demodulation of 16-QAM

The next square signal constellation above QPSK is 16-QAM. In this waveform, the I and Q channel constellation points are allowed to each take on four different values $\pm a$ and $\pm 3a$. A representative ideal 16-QAM signal constellation with Gray coding is shown in Figure 3.52.

The Gray coding assigns the specified (I, Q) bit patterns to the constellation points in such a way that any nearest neighbor reception errors only result in a single bit error. This coding results in the system BER performance being very nearly equivalent to the more simply computed symbol error rate of the system.

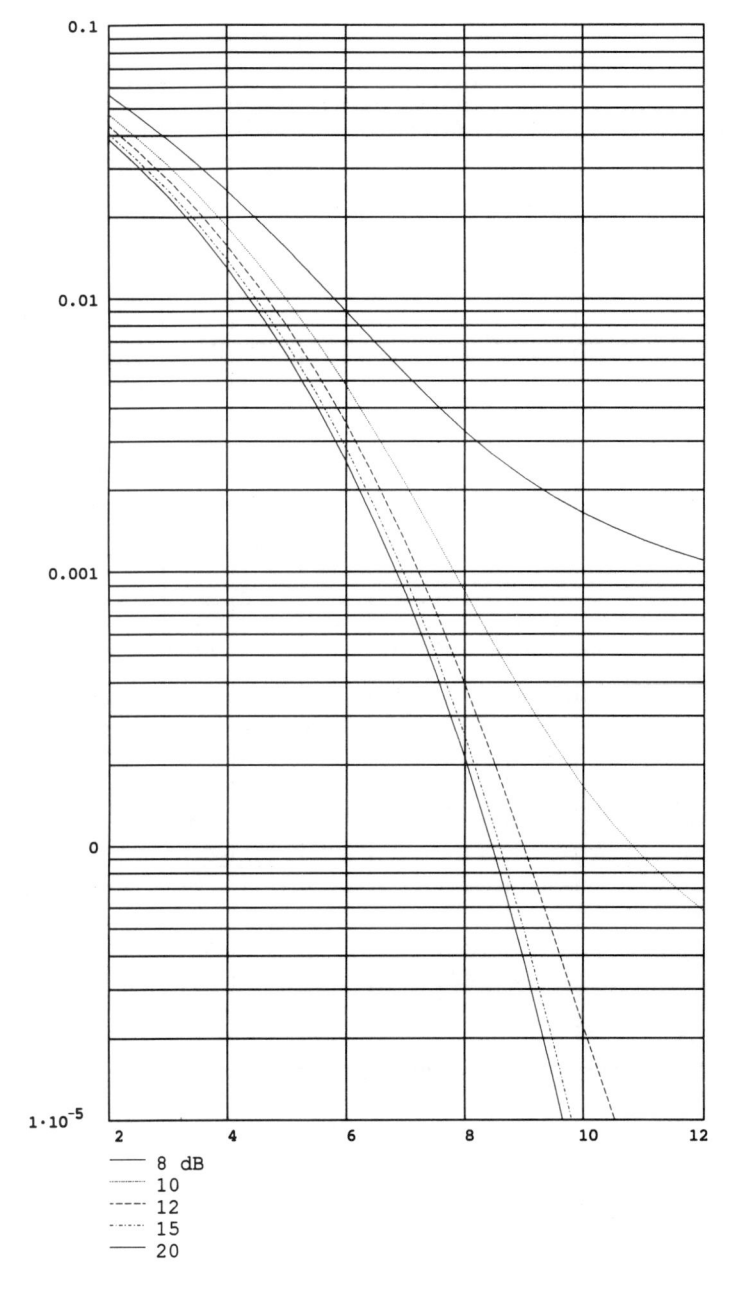

Figure 3.50 BER for BPSK versus E_b/N_o in decibels with coherent processing and specified carrier loop SNR.

122

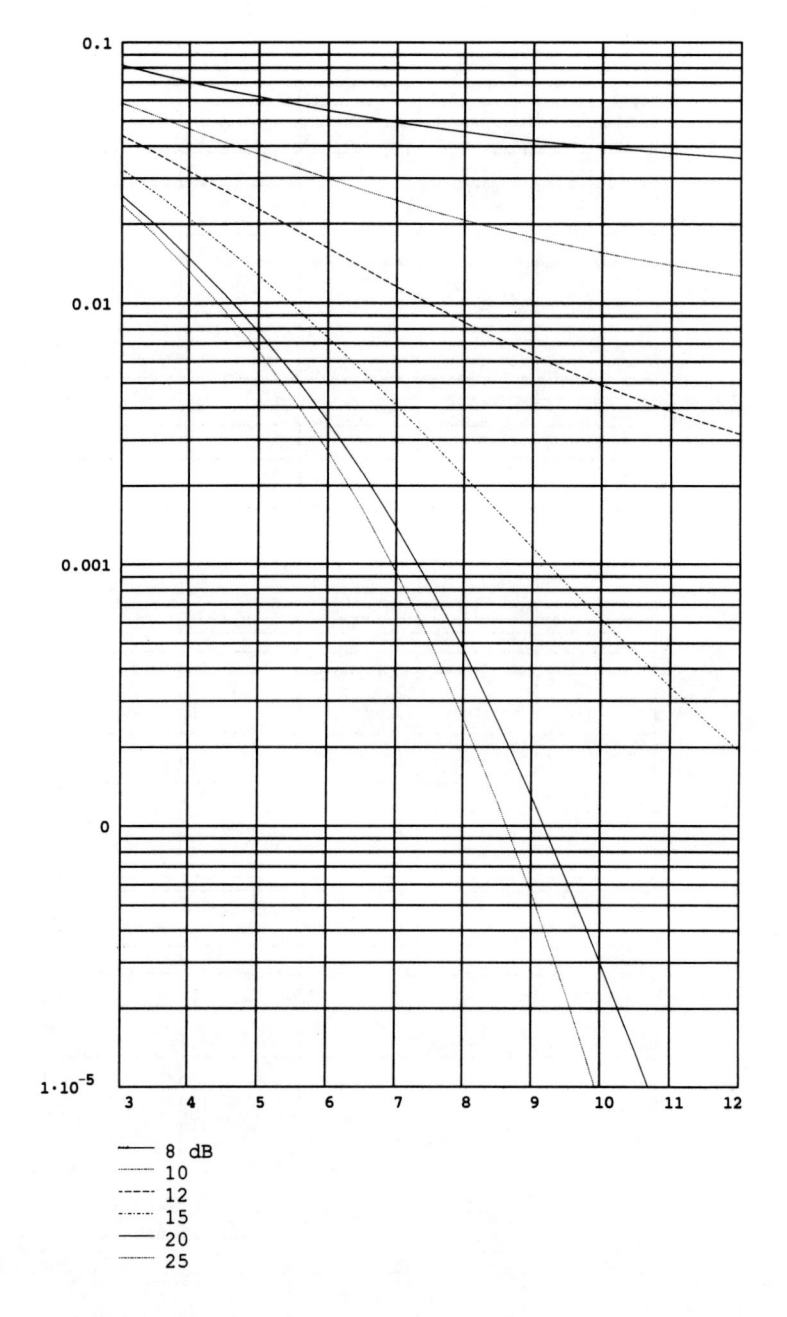

```
——— 8 dB
········· 10
----- 12
·-·-· 15
——— 20
········· 25
```

Figure 3.51 BER for QPSK versus E_b/N_o in decibels with coherent processing and specified carrier loop SNR.

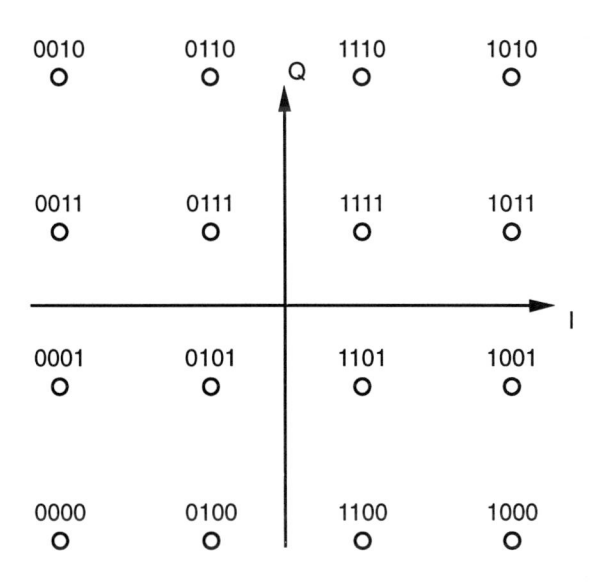

```
0010      0110        1110      1010
 O          O     Q    O          O
                  ↑
                  |
0011      0111   |  1111      1011
 O          O    |   O          O
                 |
─────────────────┼──────────────────▶ I
                 |
0001      0101   |  1101      1001
 O          O    |   O          O
                 |
                 |
0000      0100   |  1100      1000
 O          O    |   O          O
```

Figure 3.52 Gray-coded 16-QAM signal constellation.

A lucid derivation of the symbol error rate for Gray-coded 16-QAM signals over an AWGN channel is given in [123, Chapter 4]. In the absence of any local oscillator phase noise impairments, the symbol error rate is given by

$$Ps = 1 - \left[\frac{4}{16}P(C|I) + \frac{8}{16}P(C|II) + \frac{4}{16}P(C|III)\right] \quad (3.143)$$

where the $P(|)$ functions represent the probability of a symbol error for the regions of symmetry identified in Figure 3.53 by Roman numerals. These quantities are given by

$$P(C|I) = [1 - Q(\gamma_o)]^2$$

$$P(C|II) = [1 - 2Q(\gamma_o)][1 - Q(\gamma_o)] \quad (3.144)$$

$$P(C|III) = [1 - Q(\gamma_o)]^2$$

where the Q function was given in (3.136) and γ_o is related to the average SNR per bit as

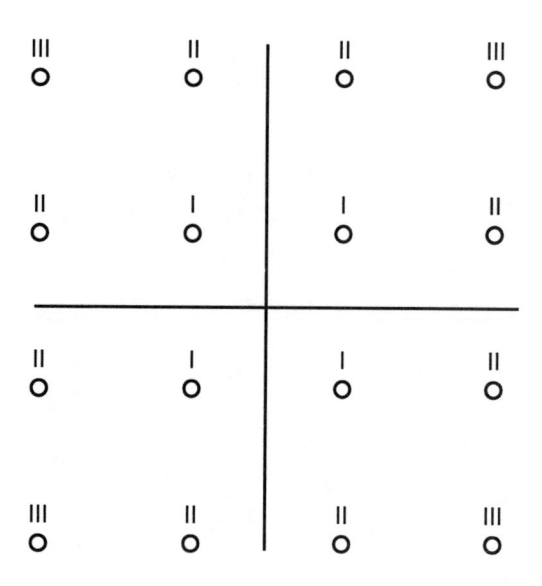

```
III    II     |  II    III
 o      o     |   o     o


 II     I     |   I     II
  o     o     |   o      o
              |
──────────────┼──────────────
 II     I     |   I     II
  o     o     |   o      o


III    II     |  II    III
 o      o     |   o      o
```

Figure 3.53 Constellation symmetries useful for BER assessment of 16-QAM.

$$\gamma_o = \sqrt{\frac{2a^2}{N_o}}$$

$$= \sqrt{\frac{4}{5}\frac{\bar{E}_b}{N_o}} \qquad (3.145)$$

With the introduction of LO phase noise impairment to the system, the symmetry conditions utilized in (3.143) must be modified leading to a predicted symbol error rate upper bound given by

$$P_s(\alpha, \gamma_o) \leq \sqrt{\frac{\alpha}{2\pi}} \int_{-\pi}^{\pi} \left[1 - \sum_{k=1}^{12} Qm_k(\gamma_o, \phi)\right] e^{\alpha[\cos(\phi)-1]}\, d\phi \qquad (3.146)$$

where

$$Qm_1(\gamma_o, \theta) = Q\{\gamma_o[\cos(\theta) - \sin(\theta)]\}$$

$$Qm_2(\gamma_o, \theta) = Q\{\gamma_o[2 - \cos(\theta) + \sin(\theta)]\}$$

$$Qm_3(\gamma_o, \theta) = Q\{\gamma_o[\sin(\theta) + \cos(\theta)]\}$$

$$Qm_4(\gamma_o, \theta) = Q\{\gamma_o[2 - \sin(\theta) - \cos(\theta)]\}$$

$$Qm_5(\gamma_o, \theta) = Q\{\gamma_o[3\cos(\theta) - \sin(\theta) - 2]\}$$

$$Qm_6(\gamma_o, \theta) = Q\{\gamma_o[\cos(\theta) + 3\sin(\theta)]\} \qquad (3.147)$$

$$Qm_7(\gamma_o, \theta) = Q\{\gamma_o[2 - \cos(\theta) - 3\sin(\theta)]\}$$

$$Qm_8(\gamma_o, \theta) = Q\{\gamma_o[-2 + 3\cos(\theta) + \sin(\theta)]\}$$

$$Qm_9(\gamma_o, \theta) = Q\{\gamma_o[2 - \cos(\theta) + 3\sin(\theta)]\}$$

$$Qm_{10}(\gamma_o, \theta) = Q\{\gamma_o[\cos(\theta) - 3\sin(\theta)]\}$$

$$Qm_{11}(\gamma_o, \theta) = Q\{\gamma_o[3\cos(\theta) - 3\sin(\theta) - 2]\}$$

$$Qm_{12}(\gamma_o, \theta) = Q\{\gamma_o[3\sin(\theta) + 3\cos(\theta) - 2]\}$$

The results of this analysis for phase-locked loop SNRs of 20, 25, 27, 30, and 35 dB are shown versus average E_b/N_o in Figure 3.54. The ideal case represented by (3.143) is also shown. At a 10^{-7} BER, performance varies dramatically with phase-locked loop SNRs between 25 and 30 dB as evidenced here. Clearly, operation at any SNR in this vicinity would result in highly varied results due to this sensitive dependency.

A partial summary of the detection performance loss for several digital modulation waveforms is shown in Figure 3.55. There is little question but that the minimum-shift-keying (MSK) waveform exhibits the most severe degradation with respect to recovered carrier phase of the waveforms considered.

The 16-QAM waveform would of course exhibit the greatest sensitivity of all the waveforms mentioned thus far, but from Figure 3.54, it is clear that discussions at an E_b/N_o of 7 dB are fairly meaningless since the BER in this SNR region is quite high. At a BER of 10^{-6} for 16-QAM, a loop SNR of 25 dB results in a loss with respect to ideal of nearly 6 dB, whereas a mere 2-dB additional loop SNR reduces the performance loss to approximately 2 dB. If additionally complex signal constellations are employed, correspondingly more severe carrier phase recovery requirements must be achieved. In the case of Gray-coded 256-QAM at a BER of 10^{-6} for instance, a static carrier reference phase error of only 0.5° will result in a performance loss of 0.6 dB [124].

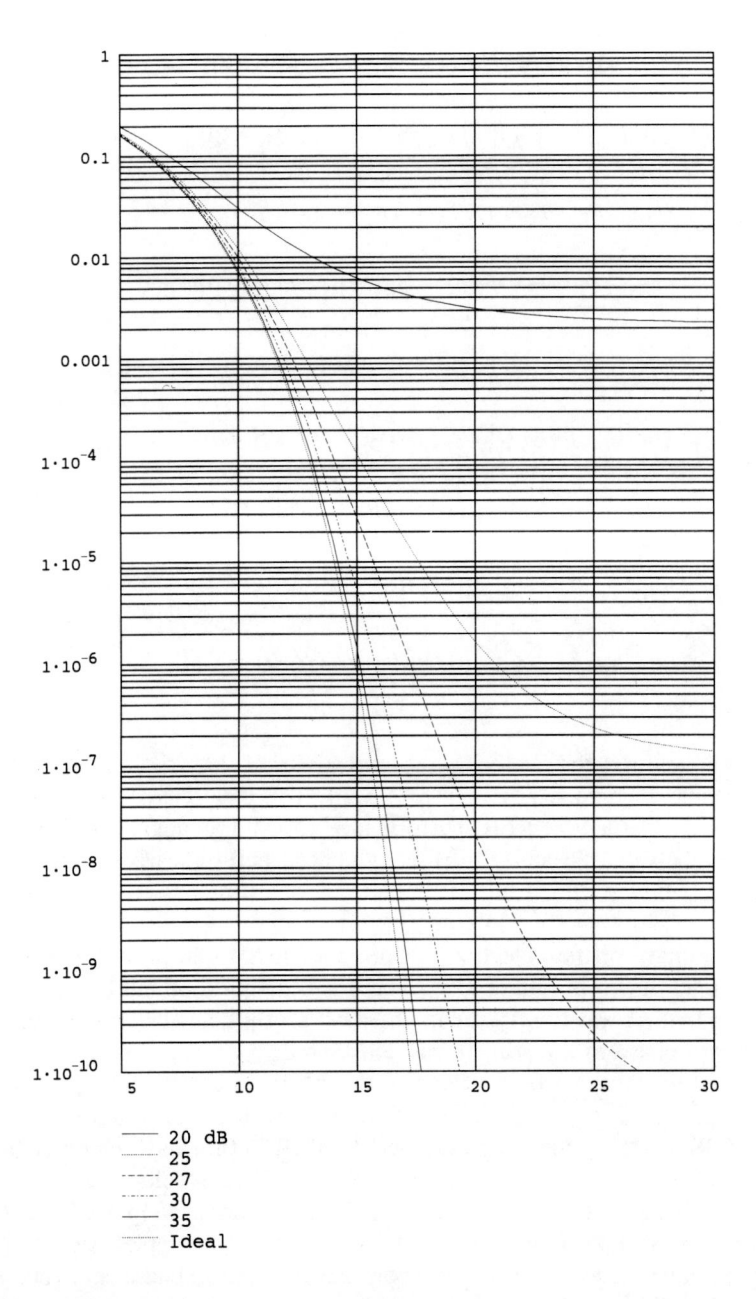

Figure 3.54 Symbol error rate for Gray-coded 16-QAM versus E_b/N_o in decibels with coherent processing and specified carrier loop SNR.

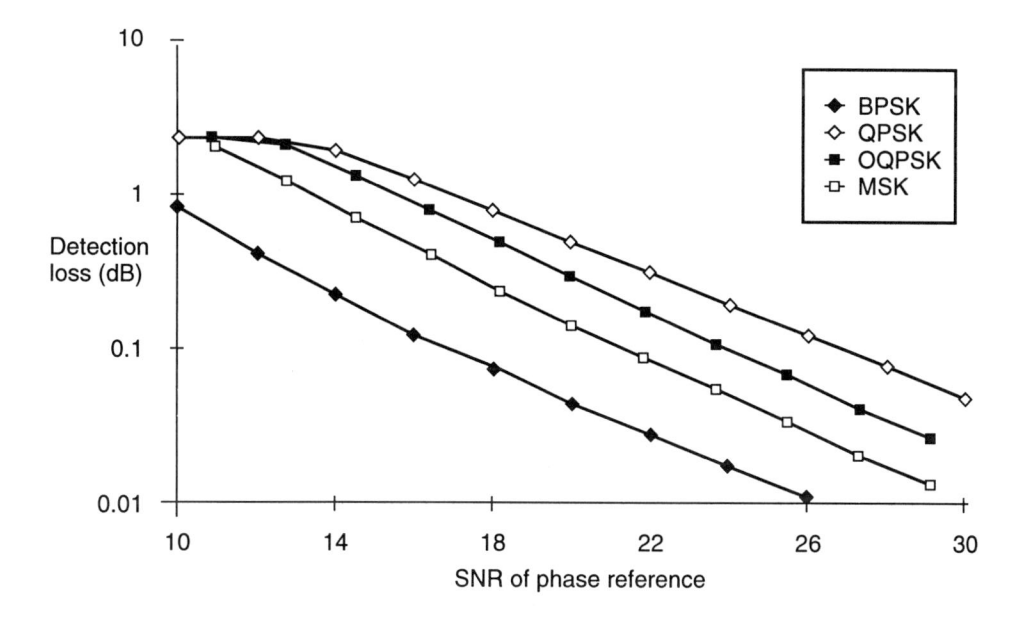

Figure 3.55 Detection loss in decibels versus SNR of phase reference at $E_b/N_o = 7$ dB. Source: [33]. Reprinted with permission.

3.7.5 *M*-ary Phase-Shift-Keyed Modulation (MPSK)

In the context of phase noise, it is only appropriate that some consideration of MPSK waveforms be given. MPSK waveforms utilize constant envelope signaling with constellation points that are normally equally spaced in angle. (Some consideration has also been given to unequally spaced constellations for mobile communications, e.g., [125].) The signal constellation for 8-PSK is shown in Figure 3.56 and the impact of local oscillator phase noise for this waveform is examined here.

The information detection process for MPSK is degraded by (1) the phase error introduced by the additive Gaussian noise of the communication channel and (2) phase perturbations introduced by the nonideal frequency synthesizer. Although the foregoing discussions have only addressed random noise contributions, discrete spurs on the synthesizer output that result from phase modulation are a separate important consideration in any demodulator and will not be studied further here.

A thorough treatment of MPSK BER performance in the ideal case (i.e., no local oscillator phase noise impairment) is given in [126]. The performance of the MPSK system may be assessed by analyzing the case of a sinusoid immersed in the AWGN channel. The coherent MPSK demodulator used to determine the phase of

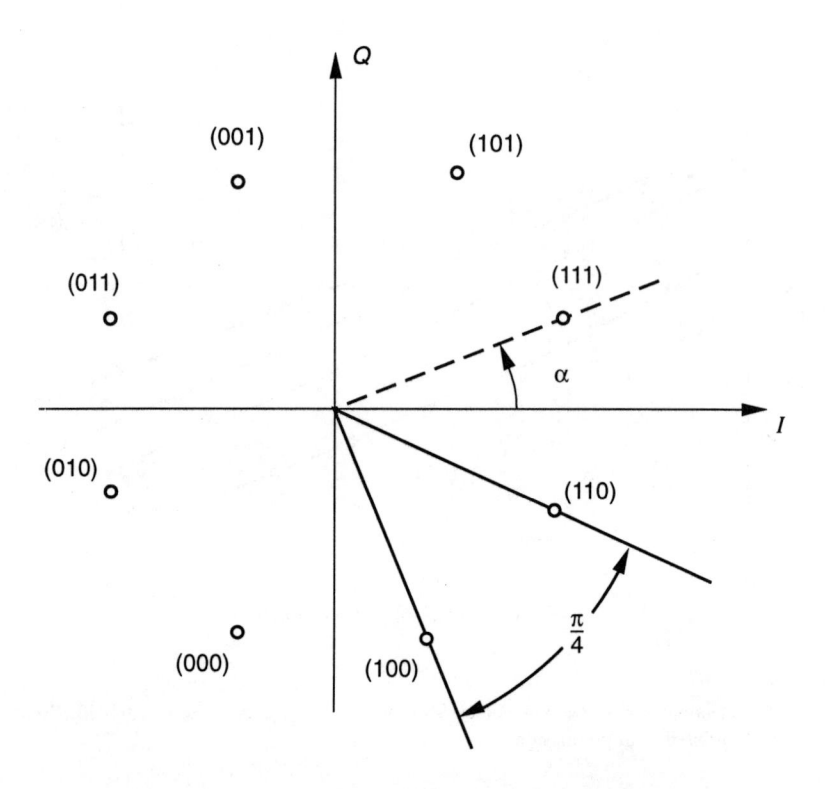

Figure 3.56 Representative Gray-coded 8-PSK signal constellation. After: [128].

the received signal is shown in Figure 3.57. In this configuration, the incoming signal $s(t)$ plus noise $n(t)$ is observed for T_s seconds, which corresponds to a single symbol time interval. It can be shown [127] that under these conditions, the PDF of the computed phase estimate θ is given by

$$p(\varphi|s_m) = \frac{1}{2\pi} e^{-E_s/N_o} \left(1 + \sqrt{\frac{4\pi E_s}{N_o}} \cos(\varphi)\, e^{(E_s/N_o)\cos^2(\varphi)} \left\{ 1 - Q\left[\sqrt{\frac{2E_s}{N_o}} \cos(\varphi)\right] \right\} \right)$$

(3.148)

where s_m represents the signal transmitted with phase θ_m, E_s represents the signal energy per symbol, and ϕ represents the estimation error, which is given by

$$\varphi = \hat{\theta} - \theta_m$$

(3.149)

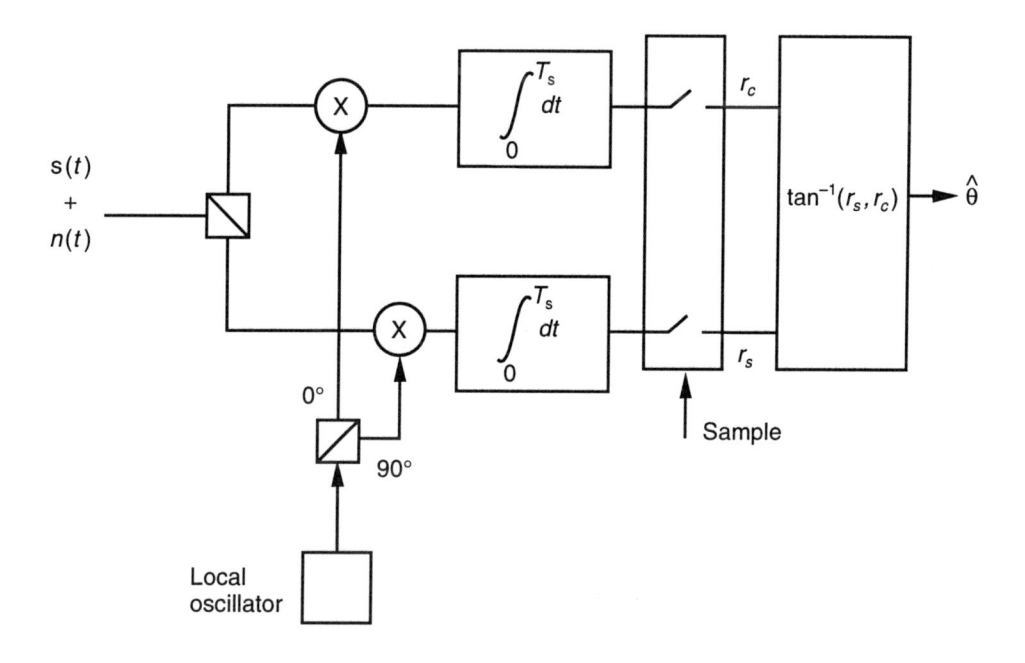

Figure 3.57 Coherent processor for calculation of received signal phase.

This probability density function is computed for E_s/N_o ratios starting from -3 dB and moving up in 2.5-dB steps in Figure 3.58.

In the absence of any local oscillator phase noise impairments, the probability of a correct symbol decision is equal to the probability that ϕ falls within the region given by $\pm \pi/M$ where M is the number of discrete signal phases used in the signal constellations. This probability may be computed as

$$P_c = \int_{-\pi/M}^{\pi/M} p(\varphi|s_m) \, d\varphi \qquad (3.150)$$

For large E_s/N_o ratios, the $Q(\ )$ function may be approximated as

$$Q\left[\sqrt{\frac{2E_s}{N_o}} \cos(\varphi)\right] \approx \frac{1}{\sqrt{4\pi \frac{E_s}{N_o} \cos(\varphi)}} e^{-(E_s/N_o)\cos^2(\varphi)} \qquad (3.151)$$

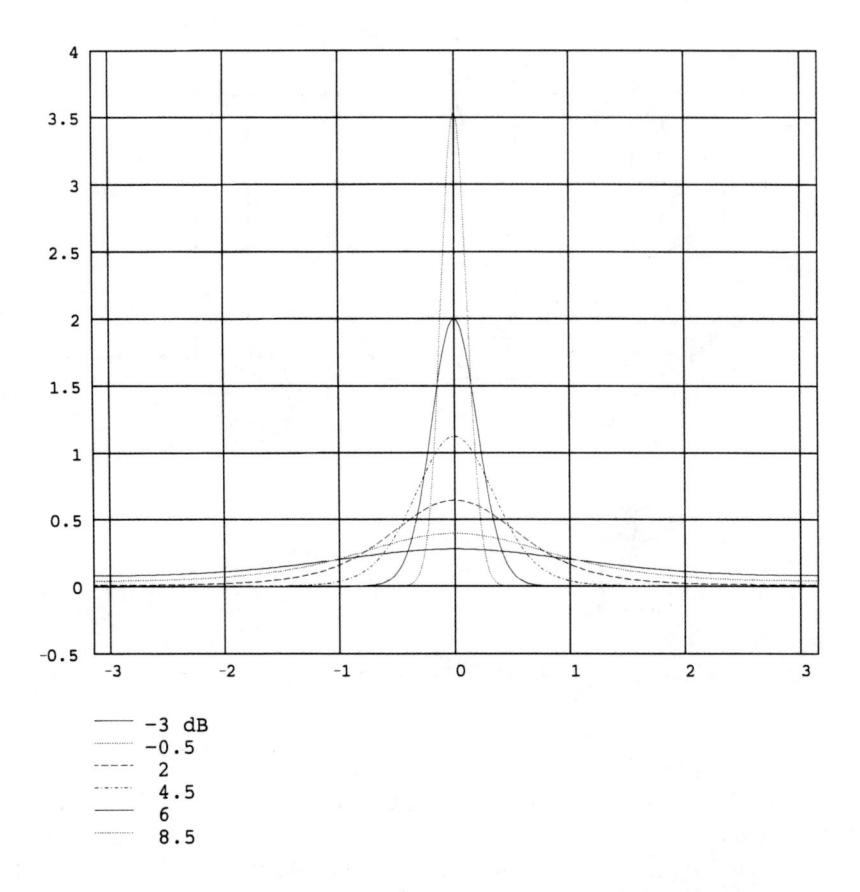

Figure 3.58 PDF for the computed phase estimate as a function of E_s/N_o in decibels.

Legend:
— −3 dB
⋯ −0.5
--- 2
-⋅- 4.5
— 6
⋯ 8.5

The probability of an 8-PSK symbol error is then given approximately by

$$P_{se} = 1 - \int_{-\pi/M}^{\pi/M} p(\varphi|s_m)\, d\varphi$$

$$\approx 1 - \int_{-\pi/M}^{\pi/M} \sqrt{\frac{E_s}{\pi N_o}} \cos(\varphi)\, e^{-(E_s/N_o)\sin^2(\varphi)}\, d\varphi$$

$$\approx 2Q\left[\sqrt{\frac{2E_s}{N_o}} \sin\left(\frac{\pi}{M}\right)\right] \quad \text{for } M \geq 4 \tag{3.152}$$

When Gray coding is used to assign actual information bits to the constellation points, any two symbols that correspond to adjacent constellation points are made to differ in only one bit value. Hence, the average BER for MPSK can be approximated by

$$P_b \approx \frac{P_{se}}{\log_2(M)} \qquad (3.153)$$

If additional phase noise is present in the receiving system due to frequency synthesizer phase noise and the phase error can be considered to be constant over each symbol interval, the situation is akin to the QAM discussions presented earlier. Once again, the Tikhonov density function can be used to describe the synthesizer phase noise in which case the PDF of the phase error ϕ must be modified as

$$q(\varphi|s_m) = \int_{-\pi}^{\pi} p(\varphi - \psi|s_m) p_{\text{PLL}}(\psi) \, d\psi \qquad (3.154)$$

where $p_{\text{PLL}}(\psi)$ represents the Tikhonov phase error PDF. Strictly speaking, the implied linear convolution must be modified since ϕ is limited to the interval $(-\pi, \pi)$. Given this form, the symbol error probability is then

$$P_{se} = 1 - \int_{-\pi/M}^{\pi/M} q(\varphi|s_m) \, d\varphi \qquad (3.155)$$

Since this result requires the evaluation of a double integral, numerical evaluation is exceedingly slow. Fortunately, an alternative approach can be used to compute the P_{se} result more quickly without sacrificing precision.

Returning to Figure 3.57, assume that the transmit signal phases are given by

$$\theta_i(t) = \begin{cases} \alpha + (i-1)\dfrac{\pi}{4} & \text{for } i = 1, 3, 5, 7 \\ i\dfrac{\pi}{4} - \alpha & \text{for } i = 2, 4, 6, 8 \end{cases} \qquad (3.156)$$

over the time interval $0 \leq t \leq T_s$ where normally $\alpha = \pi/8$. As shown in [128,129], it is possible to form signal constellation decision regions that do not make use of the traditional polar signal representations. It is particularly advantageous that the final result for 8-PSK can be expressed solely in terms of the Q function. A block diagram of this two-channel direct detection receiver is shown in Figure 3.59. Since it can be shown that this receiver is optimal and also equivalent to Figure 3.57 for

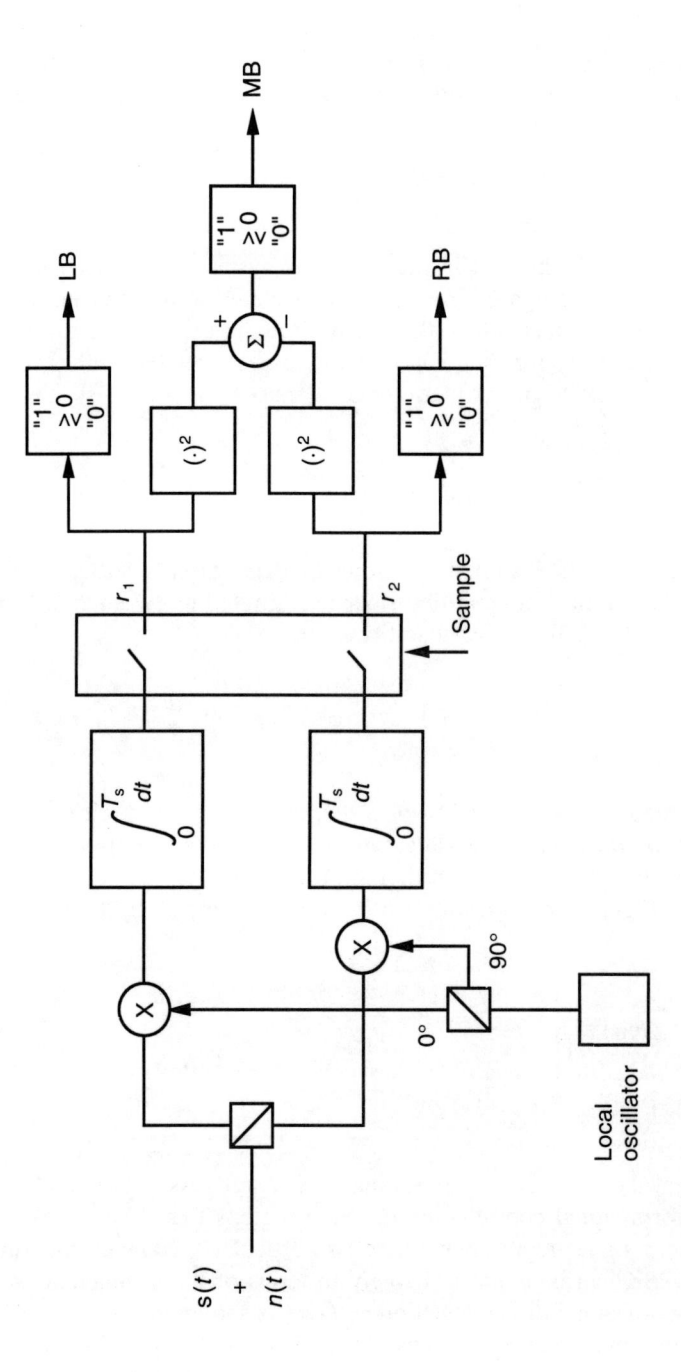

Figure 3.59 Two-channel direct detection receiver for 8-PSK assuming Gray-coded three-tuples (LB, MB, RB). After: [128].

Gray-coded constellations, calculated phase noise impairments on BER should be generally applicable.

It can be shown that the BER for this system when the synthesizer phase noise contribution is constant over each symbol interval and given by the Tikhonov PDF is

$$P_b(\gamma, \alpha) = \int_{-\pi}^{\pi} \frac{e^{\alpha \cos(\theta)}}{2\pi I_o(\alpha)} \frac{1}{3} \left[\sum_{k=1}^{5} Qnx_k(\theta, \gamma) \right] d\theta \qquad (3.157)$$

where

$$Qnx_1(\theta, \gamma) = Q\left[2\sqrt{\gamma} \cos\left(\theta + \frac{\pi}{8}\right) \right]$$

$$Qnx_2(\theta, \gamma) = Q\left[2\sqrt{\gamma} \sin\left(\theta + \frac{\pi}{8}\right) \right]$$

$$Qnx_3(\theta, \gamma) = Q\left\{ \sqrt{2\gamma} \left[\cos\left(\theta + \frac{\pi}{8}\right) + \sin\left(\theta + \frac{\pi}{8}\right) \right] \right\} \qquad (3.158)$$

$$Qnx_4(\theta, \gamma) = Q\left\{ \sqrt{2\gamma} \left[\cos\left(\theta + \frac{\pi}{8}\right) - \sin\left(\theta - \frac{\pi}{8}\right) \right] \right\}$$

$$Qnx_5(\theta, \gamma) = -2Qnx_3(\theta, \gamma)Qnx_4(\theta, \gamma)$$

Although this result is lengthy, it is exact and does not require a double integration given the Q function approximation provided by (3.136). The two key parameters in (3.158) are α and γ where α is the SNR of the synthesizer phase-locked loop defined earlier and γ is defined as

$$\gamma = \frac{3E_b}{2N_o} \qquad (3.159)$$

The computed BER for the 8-PSK waveform including synthesizer phase noise is shown in Figure 3.60 for phase-locked loop SNRs of 15, 20, 25, 27, and 30 dB, and for the nonimpaired case, which was computed using (3.152).

Synthesizer phase noise impairment results in a very similar irreducible error probability as observed with the other waveforms, and at a BER of 10^{-6}, a loop SNR of at least 27 dB is required in order to realize negligible performance impact.

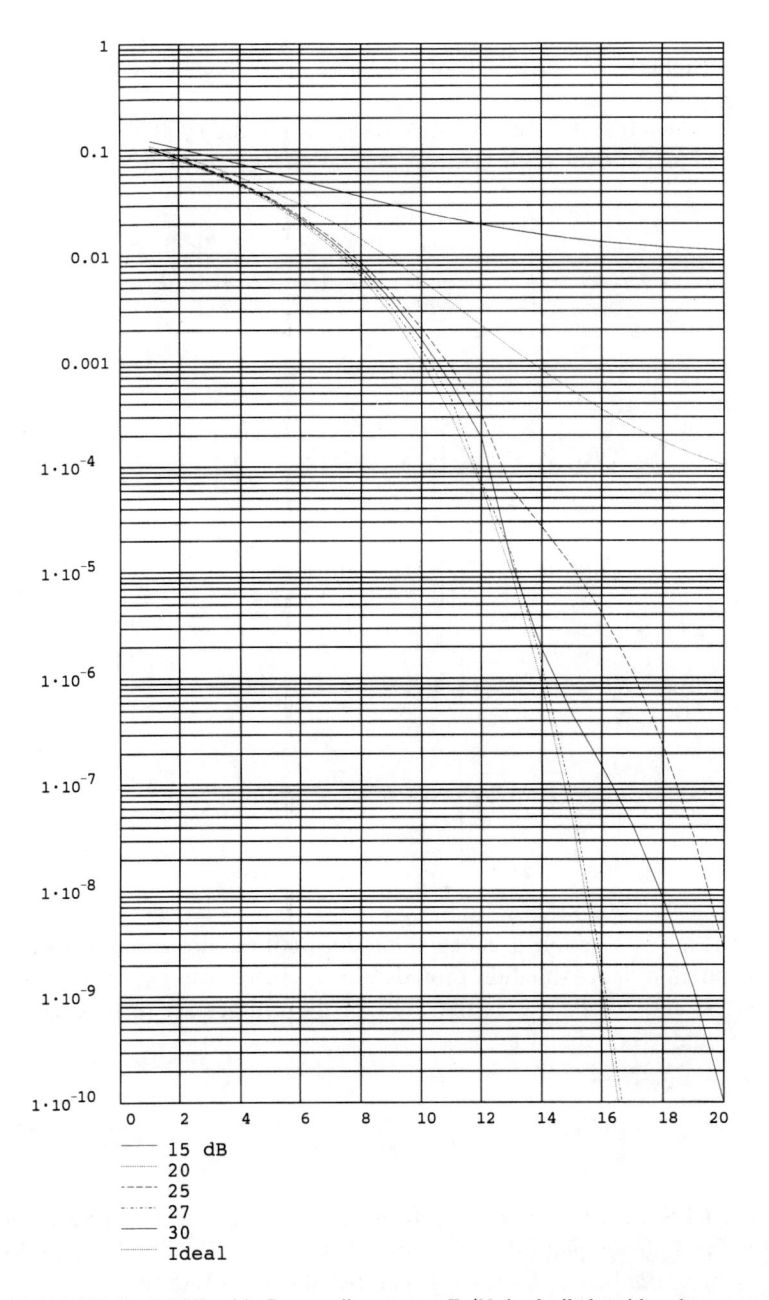

Figure 3.60 BER for 8-PSK with Gray coding versus E_b/N_o in decibels with coherent processing and specified carrier loop SNR.

3.7.6 Local Oscillator Phase Noise Impairment With Fast Changing Phase

As substantiated by (3.133), the presence of LO phase noise leads to (1) a correlation processing loss and (2) undesirable cross-coupling between the in-phase and quadrature-phase channels. When the phase noise process is slowly changing with respect to an individual symbol period, substantial irreducible BER degradation occurs as shown earlier. In contrast, the system impairment that occurs when the phase noise processes are rapidly changing with respect to a symbol interval are much less severe.

As shown in [122], the primary degradation that occurs under rapid phase change conditions is a loss of correlation. The rapid phase changes cause cross-coupled channel terms to essentially average to zero over each symbol period. The correlation loss in symbol E_s/N_o is closely approximated by

$$L = \mathbf{E}\left[\int_0^{T_s} \cos(\theta)\, d\theta\right] \qquad (3.160)$$

where $\mathbf{E}[\ ]$ denotes statistical expectation and losses cause L to be less than unity.

3.7.7 Phase Noise Impairments for More Advanced Waveforms

Fast-paced developments in mobile and satellite communications have resulted in an explosion of advanced signaling and processing techniques during the past decade. In this section, a small portion of this innovative work is touched on briefly in the context of phase noise related impairments.

In strictly coherent or noncoherent systems, the methods described earlier for assessing phase noise impairments are quite effective. However, difficulties presented by the mobile communications channels have caused considerably more sophisticated schemes to be developed where essentially noncoherent processing is extended over multiple received symbols [125,130–132]. This automatically causes more severe requirements to be imposed on the frequency synthesizer compared to the strictly symbol-by-symbol noncoherent case. In these types of cases, a more complete examination of phase noise related losses must be made.

Spread-spectrum communications lead to substantially relaxed synthesizer phase noise requirements, particularly if differential detection is used and proper forward error correction (FEC) techniques are employed. In general, synthesizer phase noise requirements in terms of BER performance are generally more severe for narrowband systems than for wideband systems. In narrowband systems, close-in phase noise performance is of particular concern compared to wideband systems where larger frequency offset noise contributions dominate.

Continuous phase modulation (CPM) is one class of modulation waveforms that has received considerable attention in recent years [133]. The fundamental assessment quantity in CPM is called the minimum (Euclidean) distance $d_{\min}$. Euclidean distance measures also play an important role in trellis-coded modulated systems. To define this quantity, assume that two constant envelope signals $s_i(t)$ and $s_k(t)$ differ over a period of N symbol periods. The distance between these two signals is given by

$$\int_0^{NT_s} [s_i(t) - s_k(t)]^2 \, dt \tag{3.161}$$

which can be recast in a normalized Euclidean distance measure $d(,)$ as

$$2E_b d^2[s_i(t), s_k(t)] \tag{3.162}$$

where

$$d^2[s_i(t), s_k(t)] = \log_2\left(\frac{M}{T_s}\right) \int_0^{NT_s} \{1 - \cos[\Delta\varphi(t)]\} \, dt \tag{3.163}$$

and M is the symbol alphabet size. The quantity $\Delta\phi(t)$ denotes the phase difference between the two signals as a function of time. Clearly, the presence of local oscillator phase noise directly contributes to $\Delta\phi(t)$ in a negative manner not unlike the correlation loss discussed earlier. In CPM, it is common practice to define $d_{\min}$ as

$$d_{\min} = \min_{\substack{i,k \\ i \neq k}} \{d^2[s_i(t), s_k(t)]\} \tag{3.164}$$

The significance of $d_{\min}$ is that the probability of a symbol error is on the order of

$$P_e \approx Q\left[\sqrt{d_{\min}^2 \frac{E_b}{N_o}}\right] \tag{3.165}$$

Distance measures like $d_{\min}$ are foundational to modern digital communication theory. Another important system-level concept is the channel cutoff rate parameter R_o channel capacity as defined by Shannon [121,123,125]. In the case of 16-QAM over an AWGN channel, the cutoff rate is given by [125]

$$R_o = -\log_2\left(\frac{1}{M^2} \sum_{i=1}^{\sqrt{M}} \sum_{j=1}^{\sqrt{M}}\right) e^{-|s_i - s_j|^2/4N_o} \tag{3.166}$$

where equally likely symbols have been assumed and $M = 16$. The significance of R_o in the communication system performance sense is that the probability of a symbol error with FEC applied can be upper bounded as

$$P_e \leq e^{-N(R_o-R)} \tag{3.167}$$

where the code rate is given by $R = k/N$, the length of each code word is N, and the number of information bits per code word is k [125]. Clearly, the same type of distance function appears in the R_o expression and the presence of local oscillator phase noise will reduce R_o.

The use of R_o is particularly attractive when overall system performance including the benefit of FEC techniques is sought [134,135]. With local oscillator phase noise present, in principle, the reduced R_o must be compensated for by an increase in E_b/N_o. A plot of R_o versus E_b/N_o for the 16-QAM case is shown in Figure 3.61.

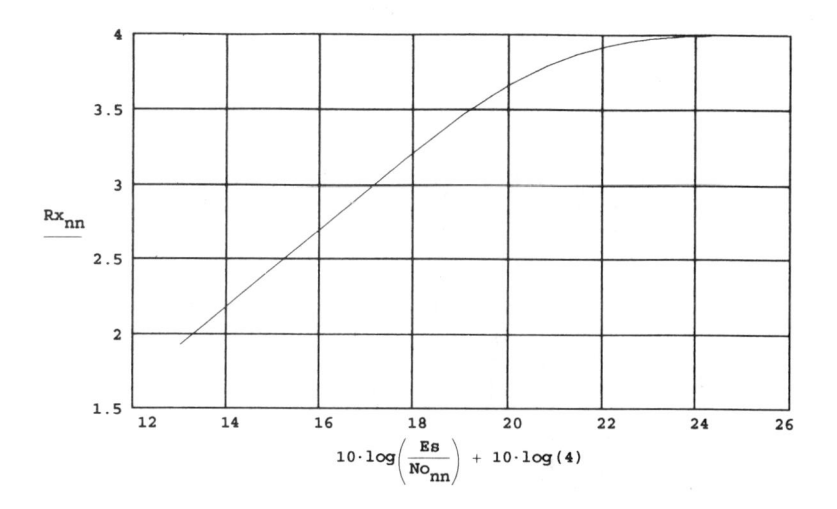

Figure 3.61 Channel cutoff rate versus E_b/N_o in decibels for 16-QAM.

3.8 WORST CASE BOUNDING OF SYNTHESIZER PHASE NOISE REQUIREMENTS

Where possible, it is highly desirable to uncouple the phase noise performance requirements of local oscillator sources from the overall system performance requirements while at the same time guaranteeing minimal system degradation from these sources. In communications work, normally the three degradation issues described in Figure 3.2 must be treated separately. Only BER performance degradation has been considered thus far.

In coherent communications, the best overall rule of thumb to employ is the total integrated phase noise concept, which was discussed in the context of Figure 3.10. The standard deviation of the integrated phase noise should be small compared to the minimal constellation angle as substantiated in (3.152) for an angular separation of $2\pi/M$. In the case of noncoherent communications, the standard deviation of the phase accumulated over a symbol interval (3.54) should be similarly small in order to achieve negligible performance impact.

In situations where the bounded phase noise quantities do not lead to adequate design margins, some assessment of the benefit resulting from suppressed carrier recovery (in coherent systems) should be quantified in order to avoid potentially overdesigning the frequency synthesizer. As a rule, proper application of FEC techniques and interleaving can dramatically reduce phase noise impairments to a negligible level and at reduced cost.

In MPSK and M-QAM systems, Feher has advocated a first-order approximation to local oscillator phase noise impairments which is very straight forward to

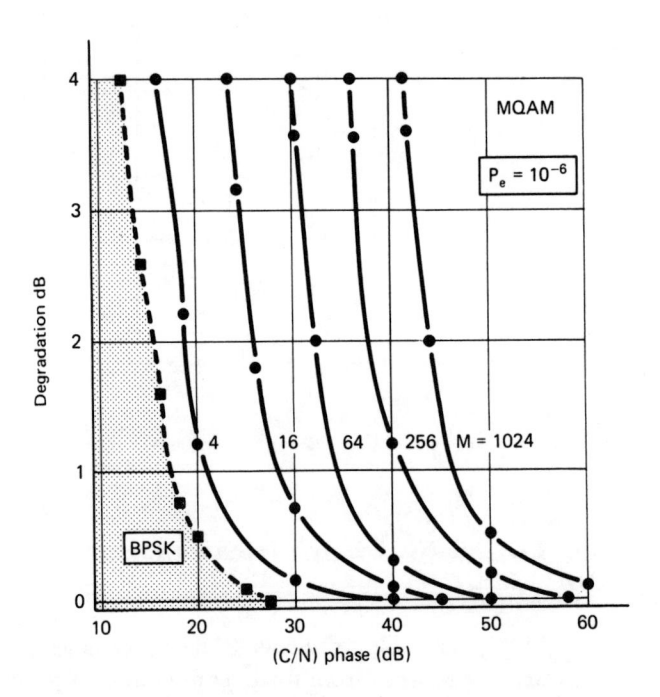

Figure 3.62 Degradation of M-QAM systems versus carrier-to-phase noise ratio for $P_e = 10^{-6}$. Double-sided Nyquist bandwidth. Source: [136], Figure 7.69. Reprinted with permission.

compute [136,137]. In this approximation, the channel noise and LO phase noise are both assumed to have Gaussian PDFs. The total CNR power for the system is calculated as

$$\frac{C}{N} = \frac{1}{\left(\frac{C}{N}\right)_{CH}^{-1} + \left(\frac{C}{N}\right)_{P}^{-1}} \quad (3.168)$$

where $(C/N)_{CH}$ is the C/N ratio of the channel caused by AWGN, and $(C/N)_P$ is the carrier-to-phase noise ratio represented by the LO. This latter quantity is calculated as the total integrated phase noise within the receive bandwidth. The computed degradations from theory for QAM and PSK systems using this approach are shown in Figures 3.62 and 3.63. These results represent easily accessible starting points for allocating performance measures for the frequency synthesizer.

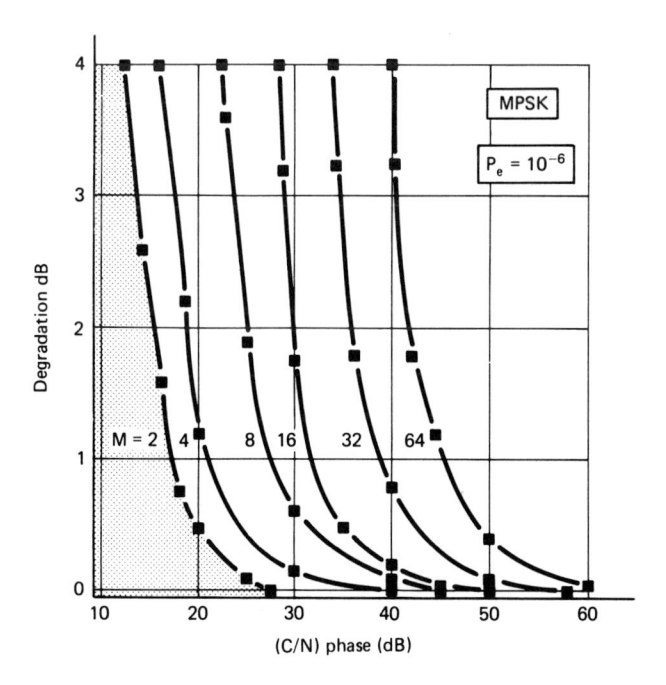

Figure 3.63 Degradation of M-PSK systems versus carrier to phase noise ratio for $P_e = 10^{-6}$. Double-sided Nyquist bandwidth. Source: [136], Figure 7.70. Reprinted with permission.

REFERENCES

[1] Paczkowski, H.C., J. Whelehan, "Understanding Noise: Part I," *IEEE MTTS Newsletter*, Winter 1988.
[2] Charschan, S.S., *Lasers in Industry*, New York, Van Nostrand Reinhold, 1972, Chap. 2.
[3] Johnson, J.B., "Thermal Agitation of Electricity in Conductors," *Phys. Rev.*, Vol. 32, July 1928, pp. 97–109.
[4] Nyquist, H., "Thermal Agitation of Electricity in Conductors," *Phys. Rev.*, Vol. 32, July 1928, pp. 110–113.
[5] "Local Oscillator Phase Noise and Its Effect on Receiver Performance," Technical notes, Watkins-Johnson Co., Vol. 8, No. 6, November/December 1981.
[6] Gagnepain, J., J. Groslambert, R. Brendel, "The Fractal Dimension of Phase and Frequency Noises: Another Approach to Oscillator Characterization," IEEE CH2186–0/85/0000–0113, 1985.
[7] Barnes, J.A., "Noise and Time and Frequency—A Potpourri," *42nd Annual Frequency Control Symposium*, 1988.
[8] Voss, R.F., "$1/f$ (Flicker) Noise: A Brief Review," IBM Thomas J. Watson Research Center, Yorktown Heights, NY, est. 1970.
[9] Dodge, C., C.R. Bahn, "Music Fractals," *Byte*, June 1986.
[10] Robins, W.P., *Phase Noise in Signal Sources*, Peter Peregrinus Ltd., London, 1982.
[11] Blachman, N.M., *Noise and Its Effect on Communication*, 2nd ed., Robert E. Krieger Publishing, Malabar, FL, 1982.
[12] Gardner, W.A., *Introduction to Random Processes with Applications to Signals and Systems*, 2nd ed., McGraw-Hill, New York, 1990.
[13] Gardner, W.A., *Statistical Spectral Analysis: A Nonprobabilistic Theory*, Prentice-Hall, Englewood Cliffs, NJ, 1988.
[14] Marple, S.L., *Digital Spectral Analysis with Applications*, Prentice-Hall, Englewood Cliffs, NJ, 1987.
[15] Shanmugan, K.S., A.M. Breipohl, *Random Signals, Detection, Estimation and Data Analysis*, John Wiley and Sons, New York, 1988.
[16] *Digital Signal Processing Applications Using the ADSP-2100 Family*, Analog Devices, Englewood Cliffs, NJ, 1990.
[17] Orfanidis, S.J., *Optimum Signal Processing*, 2nd ed., McGraw-Hill, New York, 1988.
[18] Daniell, P.J., "On the Theoretical Specification and Sampling Properties of Autocorrelated Time-Series," *J. R. Stat. Soc., Ser. B*, Vol. 8, 1946, pp. 88–90.
[19] Bartlett, M.S., "Smoothing Periodograms from Time Series with Continuous Spectra," *Nature*, London, Vol. 161, May 1948, pp. 686–687.
[20] Welch, P.D., "The Use of the Fast Fourier Transform and the Estimation of Power Spectra: A Method Based on Time Averaging Over Short Modified Periodograms," *IEEE Trans. Audio Electroacoust.*, Vol. AU-15, June 1967, pp. 70–73.
[21] Proakis, J.G., D.G. Manolakis, *Introduction to Digital Signal Processing*, Macmillan, New York, 1988.
[22] Oppenheim, A.V., R.W. Schafer, *Digital Signal Processing*, Prentice-Hall, Englewood Cliffs, NJ, 1975.
[23] Rabiner, L.R., B. Gold, *Theory and Application of Digital Signal Processing*, Prentice-Hall, Englewood Cliffs, NJ, 1975.
[24] Rutman, J., F.L. Walls, "Characterization of Frequency Stability in Precision Frequency Sources," *Proc. IEEE*, June 1991, pp. 952–960.
[25] Percival, D.B., "Characterization of Frequency Stability: Frequency-Domain Estimation of Stability Measures," *Proc. IEEE*, June 1991, pp. 961–972.

[26] Rutman, J., "Oscillator Specifications: A Review of Classical and New Ideas," *Frequency Control Symposium,* pp. 291–301.
[27] Halford, D., et al., "Spectral Density Analysis: Frequency Domain Specifications and Measurement of Signal Stability," *27th Annual Frequency Control Symposium,* 1973, pp. 421–431.
[28] Walls, F.L., D.W. Allan, "Measurements of Frequency Stability," *Proc. IEEE,* Vol. 74, No. 1, January 1986, pp. 162–168.
[29] Cohen, L., "Time-Frequency Distributions—A Review," *Proc. IEEE,* Vol. 77, No. 77, July 1989, pp. 941–981.
[30] Allan, D., et al., "Standard Terminology for Fundamental Frequency and Time Metrology," *42nd Annual Frequency Control Symposium,* 1988, pp. 419–425.
[31] "IEEE Standard Definition of Physical Quantities for Fundamental and Time Metrology," IEEE Standard No. 1139.
[32] CCIR Recommendation 686(1990), "Glossary," in Vol. 7 (Standard Frequencies and Time Signals), International Telecommunications Union, General Secretariat–Sales Section, Place des Nations, CH-1211, Geneva, Switzerland.
[33] Gilmore, R., "Specifying Local Oscillator Phase Noise Performance: How Good is Good Enough," RF Expo, 1991.
[34] Riley, W.J., "Integrate Phase Noise and Obtain Residual FM," *Microwaves,* August 1979.
[35] Cheah, J., "Analysis of Phase Noise in Oscillators," *RF Design,* November 1991.
[36] "NLO-100 Specifications," ComFocus Corporation, January 1993.
[37] Walls, F.L., S.R. Stein, "Accurate Measurements of Spectral Density in Devices," *31st Annual Frequency Control Symposium,* 1977.
[38] Kroupa, V.F., "Noise Properties of PLL Systems," *IEEE Trans. Comm.,* October 1982.
[39] Egan, W.F., "The Effects of Small Contaminating Signals in Nonlinear Elements Used in Frequency Synthesis and Conversion," *Proc. IEEE,* Vol. 69, No. 7, July 1981, pp. 797–811.
[40] Edson, W.A., "Noise in Oscillators," *Proc. IRE,* August 1960, pp. 1454–1466.
[41] Kroupa, V.F., *Frequency Stability: Fundamentals and Measurement,* IEEE Press, New York, 1983.
[42] Flannery, B.P., *Numerical Recipes in C,* Cambridge University Press, Cambridge, MA, 1988, Sec. 12.8.
[43] Bellanger, M., *Digital Processing of Signals,* 2nd ed., John Wiley, New York, 1984.
[44] Higgins, R.J., *Digital Signal Processing in VLSI,* Prentice-Hall, Englewood Cliffs, NJ, 1990.
[45] Brigham, E.O., *The Fast Fourier Transform,* Prentice-Hall, Englewood Cliffs, NJ, 1974.
[46] Stimson, G.W., *Introduction to Airborne Radar,* Hughes Aircraft Co., El Segundo, CA, 1983.
[47] McClure, M.R., "Residual Phase Noise of Digital Frequency Dividers," *Microwave J.,* March 1992, pp. 124–130.
[48] Walls, F.L., C.M. Felton, "Low Noise Frequency Synthesis," *41st Annual Frequency Control Symposium,* 1987.
[49] Scherer, D., "Design Principles and Measurement of Low Phase Noise RF and Microwave Sources," *Hewlett-Packard RF & Microwave Measurement Symposium,* April 1979.
[50] Manassewitsch, V., *Frequency Synthesizers Theory and Design,* 2nd ed., Wiley-Interscience, New York, 1980, App. B.
[51] "Silicon/GaAs Complementary, Not Competitive," *RF Lett. RF Design,* August 1986.
[52] Bomford, M., "Selection of Frequency Dividers for Microwave PLL Applications," *Microwave J.,* November 1990.
[53] Rohde, V., *Digital PLL Frequency Synthesizers Theory and Design,* Prentice-Hall, Englewood Cliffs, NJ, 1983.
[54] Egan, W.F., *Frequency Synthesis by Phase Lock,* John Wiley & Sons, New York, 1981, Chap. 5.
[55] Best, R.E., *Phase-Locked Loops Theory, Design, & Applications,* McGraw-Hill, New York, 1984.

[56] Wolaver, D.H., *Phase-Locked Loop Circuit Design*, Prentice-Hall, Englelwood Cliffs, NJ, 1991.
[57] Walls, F.L., S.R. Stein, J.E. Gray, "Design Considerations in State-of-the-Art Signal Processing and Phase Noise Measurement Systems," *30th Annual Frequency Control Symposium*, 1976.
[58] Hock, T.F., "Synthesizer Design with Detailed Noise Analysis," *RF Design*, July 1993, pp. 37–48.
[59] Walls, F.L., et al., "Extending the Range and Accuracy of Phase Noise Measurements," *42nd Annual Frequency Control Symposium*, 1989, pp. 432–441.
[60] Martin, L., "Program Optimizes PLL Phase Noise Performance," *Microwaves & RF*, April 1992, pp. 78–91.
[61] Rice, S.O., "Mathematical Analysis of Random Noise," *Bell Syst. Tech. J.*, Vol. 23, pp. 282–332, 1944, and Vol. 24, pp. 46–157, 1945, reprinted in Nelson Wax, "Noise and Stochastic Processes," Dover Publications, New York, 1954, pp. 133–294.
[62] Gray, P.R., R.G. Meyer, *Analysis and Design of Analog Integrated Circuits*, 2nd ed., John Wiley & Sons, New York, 1984, Chap. 11.
[63] Motchenbacher, C.D., F.C. Fitchen, *Low-Noise Electronic Design*, John Wiley & Sons, New York.
[64] Evans, A.D., *Designing with Field-Effect Transistors*, McGraw-Hill, New York, 1981.
[65] Gupta, M.S., *Electrical Noise: Fundamentals & Sources*, IEEE Press, New York.
[66] Van der Ziel, A., "Noise in Solid-State Devices and Lasers," *IEEE Proc.*, August 1970.
[67] Mumford, W.W., E.H. Scheibe, *Noise Performance Factors in Communication Systems*, Horizon House, Norwood, MA.
[68] Mueller, O., W.A. Edelstein, "Low-Noise Preamplifier Design for NMR," *RF Design*, January 1987.
[69] Alley, C.L., K.W. Atwood, *Electronic Engineering*, Chap. 13.
[70] Fukui, H., "The Noise Performance of Microwave Transistors," *IEEE Trans. Electron. Dev.*, March 1966.
[71] Mitama, J., H. Katoh, "An Improved Computational Method for Noise Parameter Measurement," *IEEE Trans. MTT*, June 1979.
[72] Suter, W.A., "Smith Chart Circles Aid Gain/NF Tradeoff," *Microwaves*, July 1980.
[73] Sanders, D.P., "Predict Noise Figures of Bipolar Transistors," *Microwaves*, October 1979.
[74] Nikclas, K.B., "The Exact Noise Figure of Amplifiers with Parallel Feedback and Lossy Matching Circuits," *IEEE Trans. MTT*, May 1982.
[75] Choma, J., "A Model for the Computer-Aided Noise Analysis of Broad-Banded Bipolar Circuits," *IEEE J. Solid-State Circuits*, December 1974.
[76] Suter, W.A., "Feedback and Parasitic Effects on Noise," *Microwave J.*, February 1983.
[77] Erdi, G., "Amplifier Techniques for Combining Low Noise, Precision, and High-Speed Performance," *IEEE J. Solid-State Circuits*, Vol. SC-16, No. 6, December 1981, pp. 653–661.
[78] Bordelon, J.H., D.R. Hertling, "System Noise Analysis Using Spreadsheets," *RF Design*, Nov. 1988.
[79] Temes, G.C., S.K. Mitra, *Modern Filter Theory and Design*, John Wiley & Sons, New York, 1973, p. 83.
[80] Hudec, P., *Microwave J.*, September 1992, pp. 162–170.
[81] Martin, L., "Noise-Property Analysis Enhances PLL Designs," *EDN*, September 16, 1981, pp. 91–98.
[82] Przedpelski, A.B., "Programmable Calculator Computes PLL Noise, Stability," *Electron. Design*, March 31, 1981, pp. 183–191.
[83] O'Leary, M., "Practical Approach Augurs PLL Noise in RF Synthesizers," *Microwaves & RF*, September 1987, pp. 185–194.
[84] *The Semiconductor Data Library*, Vol. I, Series A, Motorola, 1974.

[85] Mezak, J.A., G.D. Vendelin, "CAD Design of YIG Tuned Oscillators," *Microwave J.*, December 1992.
[86] Kiefer, R., L. Ford, "CAD Tool Improves SAW Stabilized Oscillator Design," *Microwaves & RF,* December 1992.
[87] Edson, W.A., "Noise in Oscillators," *Proc. IRE,* August 1960, p. 1461.
[88] Kurokawa, K., *An Introduction to the Theory of Microwave Circuits,* Academic Press, New York, 1969.
[89] Kurokawa, K., "Injection Locking of Microwave Solid-State Oscillators," *IEEE Proc.*, Vol. 61, No. 10, October 1973, pp. 1386–1410.
[90] Palladino, G.J., "Two-Port Parameter Analysis Produces Negative Resistance," *Microwaves & RF,* April 1991.
[91] Leeson, D.B., "A Simple Model of Feedback Oscillator Noise Spectrum," *IEEE Proc.*, February 1966.
[92] Sauvage, G., "Phase Noise in Oscillators: A Mathematical Analysis of Leeson's Model," *IEEE Trans. Instrumen. Measurement,* Vol. IM-26, No. 4, December 1977, pp. 408–410.
[93] Parzen, B., "Clarification and a Generalized Restatement of Leeson's Oscillator Noise Model," *42nd Annual Frequency Control Symposium,* 1988.
[94] Everard, J.K.A., "Minimum Sideband Noise in Oscillators," *40th Annual Frequency Control Symposium,* 1986, pp. 336–339.
[95] Muat, R., A. Upham, "Low Noise Oscillator Design," *Hewlett-Packard RF & Microwave Measurement Symposium.*
[96] Muat, R., "Designing Oscillators for Spectral Purity," *Microwaves & RF,* August 1984.
[97] Muat, R., "Choosing Devices for Quiet Oscillators," *Microwaves & RF,* August 1984.
[98] Muat, R., "Computer Analysis Aids Oscillator Designers," *Microwaves & RF,* September 1984.
[99] Pergal, F., "Detail a Colpitts VCO As A Tuned One-Port," *Microwaves,* April 1979.
[100] Rogers, R.G., "Theory and Design of Low Noise Microwave Oscillators," *42nd Annual Frequency Control Symposium,* 1988, pp. 301–303.
[101] Esdale, D.J., M.J. Howes, "A Reflection Coefficient Approach to the Design of One-Port Negative Impedance Oscillators," *IEEE Trans. MTT,* Vol. MTT-29, No. 8, August 1981, pp. 770–776.
[102] Janning. E.A., "A Low Noise Oscillator," Masters Thesis, June 1967.
[103] Chua, L.O., "Nonlinear Circuits," *IEEE Trans. Circuits Syst.,* Vol. CAS-29, No. 3, January 1984, pp. 150–168.
[104] Chua, L.O., "Nonlinear Oscillation Via Volterra Series," *IEEE Trans. Circuits Syst.,* Vol. CAS-31, No. 1, March 1982, pp.69–87.
[105] Clarke, K.K., D.T. Hess, *Communication Circuits,* Addison-Wesley, 1971.
[106] Hodowanec, G., "High-Power Transistor M/W Oscillators," RCA Application Note, October 1972.
[107] Strid, G., "S-Parameters Simplify Accurate VCO Design," *Microwaves,* 1975.
[108] Spence, R., "A Theory of Maximally Loaded Oscillators," *IEEE Trans. Circuit Theory,* June 1966.
[109] Vehovec, M., et al., "On Oscillator Design for Maximum Power," *IEEE Trans. Circuit Theory,* September 1968.
[110] Kotzebue, K.L., W.J. Parrish, "The Use of Large Signal S-Parameters in Microwave Oscillator Design," *Proc. IEEE Int. Symposium on Circuits and Systems,* 1975.
[111] Meyer, R.G., M.L. Stephens, "Distortion in Variable Capacitance Diodes," *IEEE J. Solid-State Circuits,* February 1975.
[112] Peterson, D.F., "Varactor Properties for Wideband Linear Tuning Microwave VCOs," *IEEE Trans. MTT,* February 1980.

[113] Healey, D.J., S.Y. Kwan, "SC-Cut Quartz Crystal Units in Low-Noise Oscillator Applications at VHF," *35th Annual Frequency Control Symposium, 1981.*
[114] Kurokawa, K., "Injection Locking of Microwave Solid-State Oscillators," *Proc. IEEE*, Vol. 61, No. 10, October 1973.
[115] Adler, R., "A Study of Locking Phenomena in Oscillators," *Proc. IRE*, June 1946.
[116] Uzunoglu, V., "Synchronization and Tracking with Synchronous Oscillators," *37th Annual Frequency Control Symposium, 1983.*
[117] Dewan, E.M., "Harmonic Entrainment of van der Pol Oscillations: Phaselocking and Asynchronous Quenching," *IEEE Trans. Auto. Control*, October 1972.
[118] Obregon, J., A.P.S. Khanna, "Exact Derivation of the Nonlinear Negative-Resistance Oscillator Pulling Figure," *IEEE Trans. MTT*, July 1982.
[119] Behagi, A., "Derive and Measure Source Pulling Figure," *Microwaves & RF*, August 1992.
[120] Yabuki, H., et al., "VCOs for Mobile Communications," *Appl. Microwave*, Winter 91/92.
[121] Proakis, J.G., *Digital Communications*, 2nd ed., McGraw-Hill, New York, 1989.
[122] Lindsey, W.C., M.K. Simon, *Telecommunication Systems Engineering*, Prentice-Hall, Englewood Cliffs, NJ, 1973, Chap. 6.
[123] Ziemer, R.E., R.L. Peterson, *Digital Communications and Spread Spectrum Systems*, Macmillan, New York, 1985, App. E.
[124] Saito, Y., Y. Nakamura, "256 QAM Modem for High Capacity Digital Radio System," *IEEE Trans. Comm.*, August 1986, pp. 799–805.
[125] Biglieri, E. D. Divsalar, P.J. McLane, M.K. Simon, *Introduction to Trellis-Coded Modulation with Applications*, Macmillan, New York, 1991.
[126] Lee, P.J., "Computation of BER of Coherent M-ary PSK with Gray Code Bit Mapping," *IEEE Trans. Comm.*, May 1986, pp. 488–491.
[127] Ha, T.T., *Digital Satellite Communications*, 2nd ed., McGraw-Hill, New York, 1990, Chap. 9.
[128] Myers, G.A., dc Bukofzer, "A Simple Two-Channel Receiver for 8-PSK," *RF Design*, February 1991.
[129] Bukofzer, D.C., "A Two-Channel Direct Bit Detection Receiver for 16-PSK Modulated Signals," *IEE Proc. Comm. Radar Signal Proc.*, Vol. 135, Part F, No. 5, October 1988.
[130] Divsalar, D., M.K. Simon, M. Shahshahani, "The Performance of Trellis-Coded MDPSK With Multiple Symbol Detection," *IEEE Trans. Comm.*, COM-38, No. 9, September 1990, pp.1391–1403.
[131] Ho, P., D. Fung, "Error Performance of Multiple-Symbol Differential Detection of PSK Signals Transmitted Over Correlated Rayleigh Fading Channels," *IEEE Trans. Comm.*, COM-40, October 1992, pp. 1566–1569.
[132] Liu, J., et al., "An Analysis of the MPSK Scheme with Differential Recursive Detection (DRD)," *IEEE*, No. 2944-7/91/0000/0741, 1991.
[133] Anderson, J.B., T. Aulin, C.E. Sundberg, *Digital Phase Modulation*, Plenum Press, New York, 1987.
[134] Omura, J.K., B.K. Levitt, "Coded Error Probability Evaluation for Antijam Communication Systems," *IEEE Trans. Comm.*, COM-30, No. 5, May 1982, pp. 896–903.
[135] Geist, J.M., "Capacity and Cutoff Rate for Dense M-ary PSK Constellations," Milcom 1990, pp. 37.1.1–37.1.3.
[136] Feher, K., *Advanced Digital Communications*, Prentice-Hall, Englewood Cliffs, NJ, 1987.
[137] Kucar, A., K. Feher, "Practical Performance Prediction Techniques for Spectrally Efficient Digital Systems," *RF Design*, February 1991, pp. 58–66.

SELECTED BIBLIOGRAPHY

Actis, R., R.A. McMorran, "Millimeter Load Pull Measurements," *Appl. Microwave,* November/December 1989.

Alexovich, J.R., R.M. Gagliardi, "The Effect of Phase Noise on Noncoherent Digital Communications," *IEEE Trans. Comm.,* COM-38, No. 9, September 1990, pp. 1539–1548.

Alexovich, J.R., R.M. Gagliardi, "Frequency Synthesizer Effects in FSK Frequency Hopped Communications," University of Southern California, August 1987.

Analog Devices, "Ultrahigh Speed Phase/Frequency Discriminator," Data Sheet AD9901, December 1990, Analog Devices, One Technology Way, Norwood, MA 02062.

Barry, J.R., E.A. Lee, "Performance of Coherent Optical Receivers," *Proc. IEEE,* Vol. 78, No. 8, August 1990, pp. 1369–1394.

Borras, J.A., "GaAs Phase Noise Characteristics in High Frequency Synthesizer Prescalers," *IEEE GaAs IC Symposium,* 1988.

Cheah, J.Y.C., "The Design of a Sampling Gate Receiver," *RF Design,* November 1992.

Cutler, L.S., C.L. Searle, "Some Aspects of the Theory and Measurement of Frequency Fluctuations in Frequency Standards," *Proc. IEEE,* Vol. 54, No. 2, February 1966, pp. 136–154.

Fantanas, C., "Introduction to Phase Noise," *RF Design,* August 1992.

Fischer, M.C., "Frequency Domain Measurement Systems," *Tenth Annual Precise Time and Time Interval Applications and Planning Meeting,* Hewlett-Packard Co., December 1978.

Gardner, F.M., "Charge-Pump Phase-Lock Loops," *IEEE Trans. Comm.,* Vol. COM-28, No. 11, November 1980, pp. 1849–1858.

Gavin, D.T., "PLL Synthesizer Phase-Locks with GaAs Comparator," *Microwaves & RF,* July 1988, pp. 79–83.

Gavin, D.T., R.M. Hickling, "A PLL Synthesizer Utilizing a New GaAs Phase Frequency Comparator," INS EEA 89–000943, pp. 457–471.

Hafner, E., "The Effects of Noise in Oscillators," *Proc. IEEE,* February 1966.

Healey III, D.J., "Flicker of Frequency and phase and White Frequency and Phase Fluctuations in Frequency Sources," *Frequency Control Symposium.*

Hill, A., J. Surber, "The PLL Dead Zone and How to Avoid It," *RF Design,* March 1992, pp. 131–134.

Holmes, J.K., *Coherent Spread Spectrum Systems,* John Wiley & Sons, New York, 1982.

Hooper, W.P., "Noise Temperature Summary," *Frequency Technol.,* March 1970, pp. 22–25.

Huffman, D., "Extremely Low Noise Frequency Dividers," *Microwave J.,* November 1985.

Kavehrad, M., "Phase-Noise Effects on QPSK Carriers in Burst Transmission," *IEE Proc.,* Vol. 129, Part F, No. 5, October 1982.

Kimbrough, L.W., "Choosing the Optimum Synthesizer Architecture for Your Receiver Application," RF Expo.

Liu, H., A. Ghafoor, P.H. Stockmann, "Time Jitter Analysis for Quadrature Sampling," *IEEE Trans. Aerospace Electron. Syst.,* July 1989.

Merigold, C., "Improved Test Methods Ease Phase-Noise Measurement," *Microwaves & RF,* February 1985.

Osafune, K., K. Ohwada, N. Kato, "High-Speed and Low-Power GaAs Phase Frequency Comparator," *IEEE Trans. MTT,* Vol. MTT-34, No. 1, January 1986, pp. 142–146.

Paczkowski, H.C., J. Whelehan, "Understanding Noise: Part II," *IEEE MTTS Newsletter.*

Pergal, F., "Detail A Colpitts VCO As A Tuned One-Port," *Microwaves,* April 1979.

Pucel, R.A., "The GaAs FET Oscillator—Its Signal and Noise Performance," *40th Annual Frequency Control Symposium,* 1986.

Riddle, A.N., R.J. Trew, "A New Method of Reducing Phase Noise in GaAs FET Oscillators," *IEEE MTT-S Digest,* 1984.

Riddle, A.N., R.J. Trew, "A Novel GaAs FET Oscillator with Low Phase Noise," *IEEE MTT-S Digest*, 1985.
Robins, W.P., "Synthesizer Phase Jitter Contributed by TTL and ECL Components," presented at Colloquium on Low Noise Oscillators and Synthesizers, 1986.
Rohdin, H., C. Su, C. Stolte, "A Study of the Relation Between Device Low-Frequency Noise and Oscillator Phase Noise for GaAs MESFETs," *IEEE MTT-S Digest*, 1984, pp. 267–269.
Sampei, S. R. Sunaga, "Rayleigh Fading Compensation Method for 16-QAM in Digital Land Mobile Radio Channels," IEEE CH2379-1/89/0000/0640, 1989.
Schunemann, K.F., K. Behm, "Nonlinear Noise Theory for Synchronized Oscillators," *IEEE Trans. MTT*, May 1979.
Shen, E.K., F.S. Lee, "Monolithic GaAs Variable Modulus Divider for Frequency Synthesis Applications," *Microwave J.*, August 1985.
Simon, M.K., D. Divsalar, "The Performance of Trellis Coded Multilevel DPSK on a Fading Mobile Satellite Channel," *IEEE Trans. Veh. Tech.*, Vol. 37, No. 2, May 1988, pp. 78–91.
Solbach, K., "Simulation Study of Harmonic Oscillators," *IEEE Trans. MTT*, August 1982.
Ungerboeck, G., "Channel Coding with Multilevel/Phase Signals," IEEE *J. of Information Theory*, Vol. 28, No. 1, January 1982, pp. 55–67.
Ward, H.R., "An Optimum Filter for Direct A/D Conversion," *IEEE Trans. Aerospace Electron. Syst.*, November 1991.
Warren, D.A., et al., "Large and Small Signal Oscillator Analysis," *Microwave J.*, May 1989.
Znojkiewicz, M.E., B. Vassilakis, "Phase Transients in Digital Radio Local Oscillators," *IEEE MTT-S Digest*, 1987, pp. 475–478.

Chapter 4
Phase-Locked Loop Analysis for Continuous Linear Systems

Phase-locked loops constitute a necessary and important element of modern frequency synthesis. The Hewlett-Packard 8662A synthesized source incorporates six phase-locked loops within its architecture for example [1]. The phase-locked loop is one of the most versatile synthesis elements available, offering a wide range of performance options in tuning bandwidth, channel spacing, spurious and phase noise performance, as well as switching speed. This same flexibility combined with the many performance measure interdependencies that arise often leads to fairly complicated design procedures. These interdependencies mandate that any top-level synthesis architecture be rigorously evaluated before launching into detailed hardware design efforts. Failure to heed this advice can lead to a plethora of problems.

The primary emphasis of this chapter is to develop transform-based methods for analyzing continuous and pseudo-continuous phase-locked loop designs by computer. *Pseudo-continuous* in this context is meant to encompass hybrid systems that contain both analog and digital elements. With suitable bandwidth constraints, continuous system methods (i.e., classical control theory) may be extended to model these systems as well. We will develop a number of pseudo-continuous models for different digital loop elements in this chapter. The emphasis on transform methods is intended to provide an intermediate-level design tool for examining specific realizability and performance questions without resorting to highly detailed studies. Results obtained at this level of detail can be used to provide information for the systems-level analysis issues that were considered in Chapter 3. The final detailed analyses, which must necessarily follow later in the design process, will be done in part using the methods developed in Chapter 8.

The mathematical methods used in this chapter are primarily based on Laplace transforms. This precludes consideration of nonlinear effects such as cycle-slipping due to frequency pull-in or a poor signal-to-noise ratio (SNR). Unlike other phase-locked loop applications (see Fig. 4.1), the SNR in frequency synthesizers is extremely large. Phase noise effects can then be analyzed by using simple perturbation

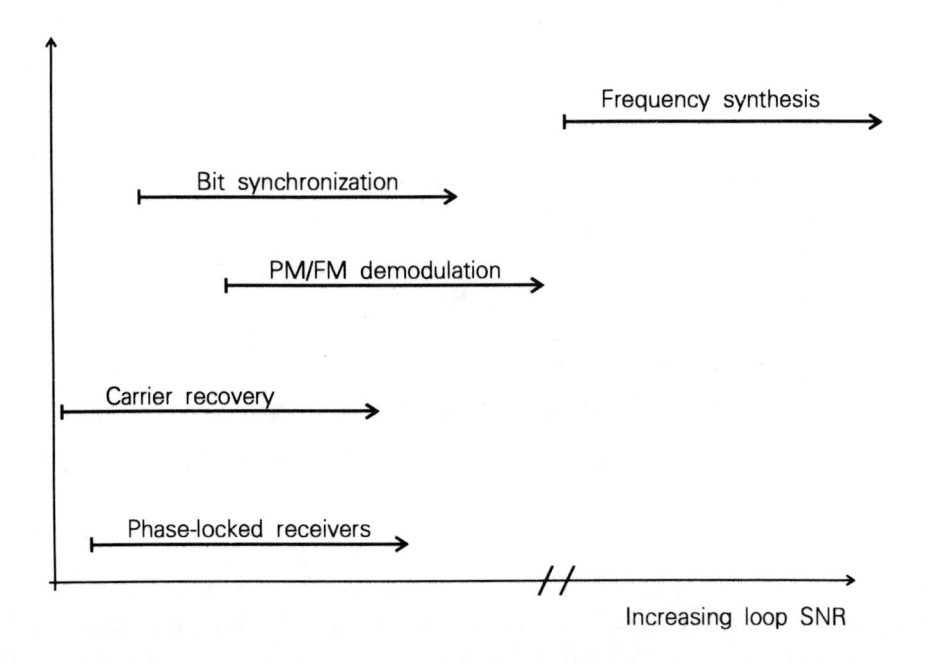

Figure 4.1 Phase-locked loops used for frequency synthesis have inherently large loop SNRs that permit phase noise effects to be analyzed by means of perturbation techniques. After: [4].

techniques as discussed in Chapter 5 rather than having to resort to the more esoteric methods of stochastic differential equations [2,3]. Therefore, both time-domain and frequency-domain performance aspects of a particular phase-locked loop can be analyzed by using Laplace transform-based methods only.

4.1 PHASE-LOCKED LOOP BASICS

The basic phase-locked loop structure consists of a phase detector, a loop filter, a voltage-controlled oscillator (VCO), and a feedback digital divider as shown in Figure 4.2. Additional complexity is often present in practice, but this simple model provides an avenue to introduce pseudo-continuous models in a straightforward manner based on classical Laplace transform techniques.

For the purposes of this introduction, an ideal type 2 phase-locked loop is considered. Throughout this text, loop *type* identifies the number of open-loop gain poles at the origin, whereas loop *order* identifies the degree of the associated characteristic equation following the standard conventions given in [4,5]. For example, a type 2 second-order phase-locked loop is physically possible whereas a type 3 second-order loop is not.

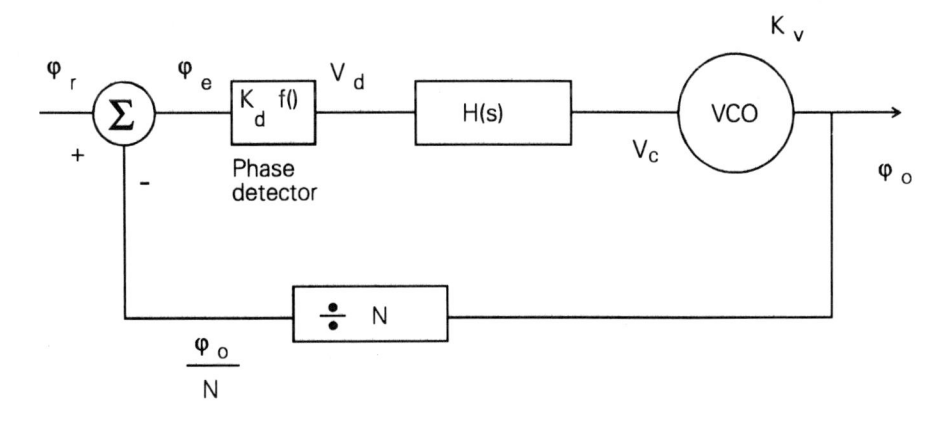

Figure 4.2 Basic frequency synthesis loop consisting of a phase detector, a loop filter, a VCO, and a feedback divider.

Integro-differential equations may be written for each of the loop elements shown in Figure 4.2. The VCO acts as an ideal integrator of phase and the output phase may be expressed as

$$\varphi_o(t) = \omega_c t + \int_0^t K_v V_c(t)\,dt + \varphi_o(0) \quad \text{for} \quad t \geq 0 \qquad (4.1)$$

where ω_c is the nominal VCO frequency in radians per second, and K_v is the VCO tuning sensitivity in radians per second per volt. The phase detector output is given by

$$V_d = K_d f(\varphi_e) \qquad (4.2)$$

where

$$\varphi_e = \varphi_r - \frac{\varphi_o}{N} \qquad (4.3)$$

which is the loop phase error, and N is the feedback divider ratio. All explicit time dependence has been omitted in these last two equations as well as those that follow. Even though the feedback divider function is discrete rather than continuous in nature, its operation will be treated as continuous, deferring further discussion until Sec. 4.2.5. The phase detector transfer function $f(\ )$ may be nonlinear in this formulation as long as it is memoryless, but in obtaining the final results, linearity will be required in order to represent the system using Laplace transforms. A constant frequency source is used to generate the reference phase function φ_r.

The loop filter transfer function, $H(s)$, for the ideal type 2 case is the lead-lag filter, which may be written in differential equation form as

$$\tau_1 \frac{dV_c}{dt} = V_d + \tau_2 \frac{dV_d}{dt} \tag{4.4}$$

where $\tau_1 = R_1 C$ and $\tau_2 = R_2 C$ in Figure 4.3. The system characteristic equation may be found by first differentiating (4.3) twice and then substituting the second derivative of (4.1) followed by substitutions of (4.4) and (4.2) to yield

$$\frac{d^2 \varphi_e}{dt^2} + \frac{K_d K_v}{N\tau_1} \tau_2 f'(\varphi_e) \frac{d\varphi_e}{dt} + \frac{K_d K_v}{N\tau_1} f(\varphi_e) = 0 \tag{4.5}$$

where f' denotes differentiation with respect to the phase error φ_e. For strictly linear phase detector operation,

$$f'(\varphi_e) = 1 \tag{4.6}$$

The result in (4.5) may then be cast in classical form by adopting the standard conventions [5] for second-order control systems where

$$\omega_n = \sqrt{\frac{K_d K_v}{N\tau_1}} \text{ (rad/s)} \tag{4.7}$$

is the natural frequency for the control loop and

$$\zeta = \frac{1}{2} \omega_n \tau_2 \tag{4.8}$$

is the control loop damping factor. The final result for the linear type 2 system characteristic equation is then

$$\frac{d^2 \varphi_e}{dt^2} + 2\zeta\omega_n \frac{d\varphi_e}{dt} + \omega_n^2 \varphi_e = 0 \tag{4.9}$$

The nonlinear system represented by (4.5) has traditionally been handled using phase-plane analysis techniques [6,7] whereas the linear system characterized by (4.9) is ideally suited for solution using Laplace transforms. Since this subject has been examined in depth by a number of authors [1,5,8], this introductory case is not addressed further.

Heaviside operator notation may be used for each of the constitutive phase-locked loop elements in this example to represent each in terms of a frequency domain transfer function as follows:

Linear phase detector

$$K_d \qquad (4.10)$$

Lead-lag loop filter

$$\frac{1 + s\tau_2}{s\tau_1} \qquad (4.11)$$

Voltage-controlled oscillator

$$\frac{K_v}{s} \qquad (4.12)$$

Feedback divider

$$\frac{1}{N} \qquad (4.13)$$

The phase-locked loop represented by these equations is shown in Figure 4.3 and the associated closed-loop transfer function in Laplace transform form is simply

$$T(s) = \frac{\varphi_e(s)}{\varphi_r(s)} = \frac{s^2}{s^2 + 2\zeta\omega_n s + \omega_n^2} \qquad (4.14)$$

Similar analysis techniques will be utilized throughout the remainder of this chapter.

In the next section, we will show that a phase-locked loop that employs a digital rather than an analog phase detector may be analyzed in much this same manner provided that the digital phase detector is suitably described by a pseudo-continuous model. The impact of using different phase detectors can then be ascertained by simply replacing the continuous phase detector transfer function [e.g., (4.10)] with the appropriate pseudo-continuous phase detector transfer function. Therefore, classical Laplace transform techniques may be retained and easily adapted to pseudo-continuous system design and analysis.

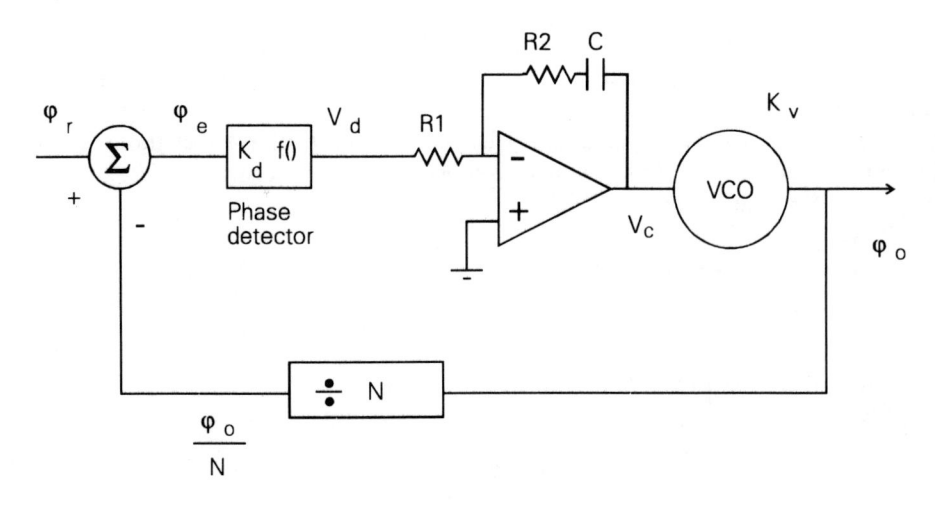

Figure 4.3 Ideal type 2 second-order phase-locked loop utilizing an active lead-lag network as the loop filter.

4.2 PSEUDO-CONTINUOUS PHASE DETECTOR MODELS

The suitability of representing discrete processes by continuous ones is determined almost exclusively by the system percentage bandwidth under consideration. If the system bandwidth is much smaller than the sampling rate of the underlying discrete process taking place, its behavior can be closely approximated by using continuous system approximations. Pseudo-continuous [9] analysis is most applicable where the phase-locked loop bandwidth is 10% or less of the reference frequency. These techniques may be used for larger percentage bandwidth systems, of course, but modeling inaccuracies will increase substantially.

Our development of pseudo-continuous models focuses on the type of digital phase detector employed. Systems that use all-digital approaches for loop filters, VCOs, etc., can be analyzed in their entirety using z transforms and, therefore are not considered further here. The hybrid systems that are considered here are completely analog aside from their discrete phase detector and optional feedback divider. Once suitable models for the digital phase detectors have been found, it will be shown that inclusion of a digital feedback divider is a trivial matter.

Many different types of phase detectors are available, particularly in the areas of bit synchronization and carrier recovery (e.g., [10]). Only several of the most common phase detectors best suited for frequency synthesizer design are examined here.

Phase-Frequency Detector

The phase-frequency detector represents one of the most widely used digital phase detectors in modern synthesizer design due to its digital simplicity and acquisition aiding characteristics. In fast frequency switching applications, initially large VCO frequency errors that exceed the loop bandwidth would normally result in cycle-slipping due to a potentially minuscule dc component at the phase detector output if a multiplicative-type phase detector is used [5]. In contrast, the frequency discriminator characteristic of this phase detector creates a much larger dc component, thereby enhancing the frequency acquisition speed dramatically. Several phase/frequency detectors were mentioned in Chapter 2.

In the next section, the tri-state phase-frequency detector will be addressed. Because of the linearity constraints imposed on these discussions, only the small phase error operating characteristics, i.e., $|\varphi_e| < 2\pi$, are considered here. Without further loss of generality, these discussions will be based on the use of this phase detector in an ideal type 2 phase-locked loop.

4.2.1 Tri-State (Output) Phase-Frequency Detector

A simplistic equivalent circuit for the tri-state phase/frequency detector is shown in Figure 4.4. This model utilizes two ideal switches, S1 and S2, whose operation is governed by the leading edges of the digital input signals that are being phase compared. Resistor R_L represents the resistive load following the phase detector. This model is closely related to work presented in [11,12]. Representative timing relationships for the detector operation are shown in Figure 4.5.

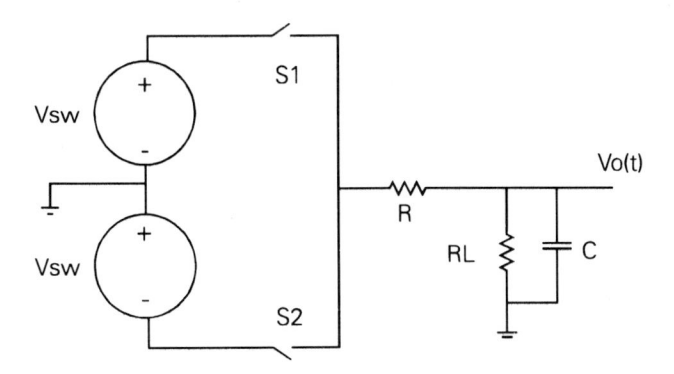

Figure 4.4 Tri-state phase-frequency detector equivalent circuit for linear operation.

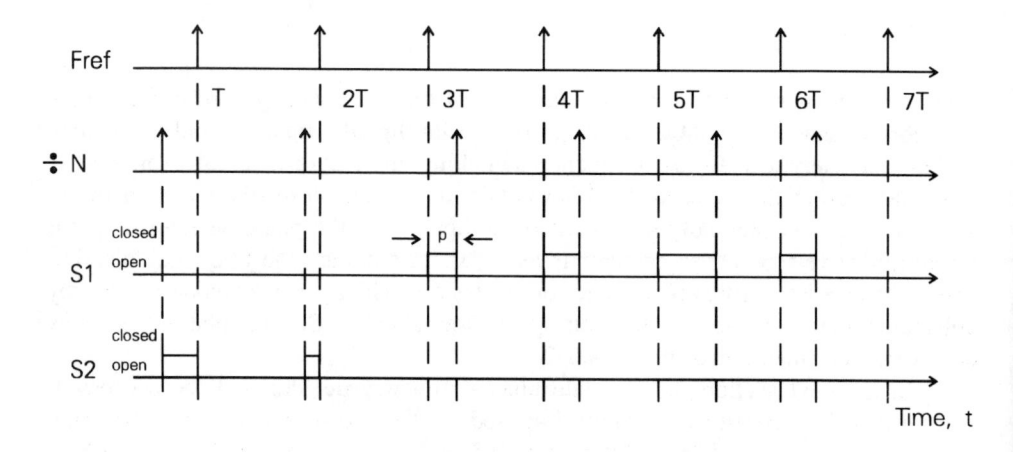

Figure 4.5 Tri-state phase-frequency detector signal timing relationships for linear operation.

Operation of the phase detector is governed solely by the signal transitions of the two phase detector input signals, F_{ref} and $\div N$, as shown in Figure 4.5. Switch S1 or S2 is closed depending on which signal transition occurs first during a specific reference period. The duration of each switch closure corresponds to the time interval between each F_{ref} transition and its paired $\div N$ transition. In ideal steady-state operation, the F_{ref} and $\div N$ transitions would be time coincident, reflecting a zero phase error.

The analysis of this phase detector is complicated by the fact that the natural modes of the circuit network in Figure 4.4 are different for a switch-closed time period versus a both-switches-open time period. In other words, a network using a tri-state phase detector is inherently time variant because the circuit time constants are a function of the phase detector gating periods. Several simplifying assumptions can nonetheless be identified that will permit linear analysis techniques to be used.

The required assumptions for linear phase detector operation may be identified by examining the operation of the phase detector in more detail. Assume that the instantaneous phase error within the phase-locked loop at a particular time instant is

$$\varphi_e = 2\pi \frac{p}{T} \qquad (4.15)$$

where p is the transition timing difference shown in Figure 4.5 and T is the reference period, both having units of seconds. The frequency-domain impulse response of the RC filter in Figure 4.4 when either one of the two switches is closed is

$$H(s) = \frac{R_L}{R_L + R} \frac{1}{1 + s\tau_1} \quad (4.16)$$

where

$$\tau_1 = C \frac{R_L R}{R_L + R} \quad (4.17)$$

The time-variant nature of this network requires that two strong assumptions be made about its use and its operation. Letting

$$\tau_2 = R_L C \quad (4.18)$$

it is necessary for the following to hold:

1. p and τ_2 be such that any stored charge on capacitor C from a previous sample must have decayed to negligible levels by the time that the next phase error sample is taken.
2. p be sufficiently small such that the approximations

$$\begin{aligned} e^{-p/\tau_1} &\approx 1 - \frac{p}{\tau_1} \\ e^{p/\tau_2} &\approx 1 \end{aligned} \quad (4.19)$$

may be made.

Both of these assumptions significantly affect the phase error dynamic range for which the final results may be assumed valid. The assumptions require that the analyses be limited to cases where the phase error is only a small fraction of 2π. The filtering time constants must also be selected to be quite small as compared to a reference period T. These assumptions are quite restrictive but necessary if we are to employ linear transform techniques.

Based on these assumptions, and arbitrarily focusing on the third output pulse in Figure 4.5, the network output voltage is given by

$3T \leq t \leq 3T + p$:

$$V_{\text{out}}(t) = \frac{V_{sw} R_L}{R_L + R} [1 - e^{-(t-3T)/\tau_1}] + V_o e^{-(t-3T)/\tau_1} \quad (4.20)$$

where V_o is the initial capacitor voltage at $t = 3T$, which by way of assumption 2 is zero, and

$3T + p \leq t < 4T$:

$$V_{out}(t) = V_{out}(3T + p)e^{-(t-3T-p)/\tau_2} \tag{4.21}$$

If the substitutions

$$t' = t - 3T \tag{4.22}$$

and (4.18) and (4.19) into (4.20) and (4.21) are made, the final results take the very simple form of

$$0 \leq t' < p: \quad V_{out}(t') \approx \frac{V_{sw}}{RC} t' \tag{4.23}$$

and

$$p \leq t' < T: \quad V_{out}(t') \approx \frac{V_{sw}}{RC} p e^{-t'/\tau_2} \tag{4.24}$$

Having been forced by linearity arguments to restrict the phase errors to small values,

$$|p| \ll T \tag{4.25}$$

the output voltage over the single reference period may be written as approximately

$$0 \leq t' < T: \quad V_{out}(t') \approx \frac{V_{sw}}{RC} p e^{-t'/\tau_2} \tag{4.26}$$

Using this result combined with (4.15), the output due to a single phase error sample may be expressed as

$$V_{out}(t') \approx K_d' \varphi_e e^{-t'/\tau_2} \tag{4.27}$$

where

$$K_d' = \frac{V_{sw} T}{2\pi RC} \quad \text{(V/rad)}$$

is the phase detector gain, and φ_e is the phase error sample at $t' = 0$. This result may be expressed in the frequency domain using Laplace transforms as

$$V_{out}(s) = \tau_2 K'_d \frac{\varphi_e}{1 + s\tau_2} \qquad (4.28)$$

which is nothing more than the transform of the normal RC filter impulse response weighted by a constant and the instantaneous phase error at $t' = 0$.

The result given in (4.28) only includes the effects of the single phase error sample at $t' = 0$. If the future phase error samples are now also included, the sole sample weight φ_e must be replaced by

$$\sum_{n=0}^{\infty} \varphi_e(nT) e^{-snT} \qquad (4.29)$$

which is incidently equivalent to the z transform of the phase error sample sequence. The instantaneous phase error sample stream can be represented in the time domain as

$$\varphi_e^*(t) = \sum_{n=0}^{\infty} \varphi_e(t) \delta(t - nT) \qquad (4.30)$$

where $\delta(\ )$ is the Dirac delta function. Its Laplace transform is given by

$$\mathcal{L}[\varphi_e^*(t)] = \sum_{n=0}^{\infty} \int_0^{\infty} \varphi_e(t) \delta(t - nT) e^{-st} dt \qquad (4.31)$$

On simplification,

$$\mathcal{L}[\varphi_e^*(t)] = \sum_{n=0}^{\infty} \varphi_e(nT) e^{-snT} = \varphi_e^*(s) \qquad (4.32)$$

which is precisely the same as (4.29). Therefore using this result, all future phase error samples may be included by rewriting (4.28) as

$$V_{out}(s) = \tau_2 K'_d \frac{\varphi_e^*(s)}{1 + s\tau_2} \qquad (4.33)$$

As will be shown in Chapter 5, the Laplace transform of a continuous phase error function of time, which undergoes ideal impulse sampling in the time domain every T seconds as described by (4.30), is related to the Laplace transform of the continuous signal by

$$\varphi_e^*(s) = \frac{1}{T} \sum_{n=-\infty}^{\infty} \varphi_e\left(s + j\frac{2\pi n}{T}\right) \qquad (4.34)$$

where

$$\varphi_e(s) = \mathcal{L}[\varphi_e(t)]$$

and

$$\varphi_e^*(s) = \mathcal{L}[\varphi^*{}_e(t)]$$

If it can be assumed that all of the high-frequency alias terms (i.e., $n \neq 0$) fall well outside the loop bandwidth and therefore have negligible effect in (4.34), the effect of the sampling process can be approximated by

$$\varphi_e^*(s) \approx \frac{1}{T} \varphi_e(s) \qquad (4.35)$$

Using this result in (4.33), the pseudo-continuous transfer function for the tri-state phase-frequency detector followed by the RC postdetection filter as shown in Figure 4.4 may be represented by

$$H_{pf}(s) = \frac{V_{out}(s)}{\varphi_e(s)} = \frac{K_d}{1 + s\tau_2} \qquad (4.36)$$

where the phase detector gain is given by

$$K_d = \frac{\tau_2}{T} K_d' = \frac{R_L V_{sw}}{R 2\pi} \quad (\text{V/rad})$$

In concluding this section, it is necessary to reiterate that the tri-state phase-frequency detector result is clearly subject to a number of restrictive requirements on network parameters as well as the dynamic range of applicability. For these reasons, the simplified phase-frequency detector discussed next is preferred from an analysis perspective.

4.2.2 Simplified Phase-Frequency (Tri-Logic State) Detector

The simplified phase-frequency detector was first introduced in Chapter 2 as the "three-state" detector. This detector exhibits three logic states rather than a tri-state output. To avoid confusion with the detector discussed in the previous section, the tri-(logic)state detector will be referred to here as the simplified phase-frequency detector. In this section, the pseudo-continuous transfer function model is derived for this phase detector. As will be seen shortly, a number of the restrictions that were necessary for dealing with the tri-state phase detector of Sec. 4.2.1 may be lifted for the simplified detector. The phase error dynamic range must, however, be once again limited to $\pm 2\pi$ range for linear analysis to be valid.

The simplified phase detector does not exhibit the erratic gain behavior due to internal race conditions or finite slew-rate as in other phase-frequency detector architectures and this fact makes it very well suited for low noise designs. Even so, the input-output relationships for the device closely parallel those of the tri-(impedance) state detector shown in Figure 4.5. The most significant modeling difference of this phase detector compared to the tri-state detector is that its output is a tri-valued ideal voltage source, which preserves the circuit network linearity. Network time constants are no longer a function of the phase detector output state.

A representative timing diagram for the simplified detector is shown in Figure 4.6. Each transition of the detector output coincides with a leading transition of the F_{ref} or the $\div N$ signal inputs. The relative position of each leading $\div N$ signal transition with respect to the reference signal, F_{ref}, has been converted to an equivalent phase error function, $\varphi_e(t)$, in Figure 4.6.

The phase error signal shown in Figure 4.6 is continuous. It therefore must be viewed as a scaled version of the loop phase error at the VCO rather than at the phase detector. This distinction is necessary since the phase detector phase error is only known at discrete time instants given by the $\div N$ signal transitions.

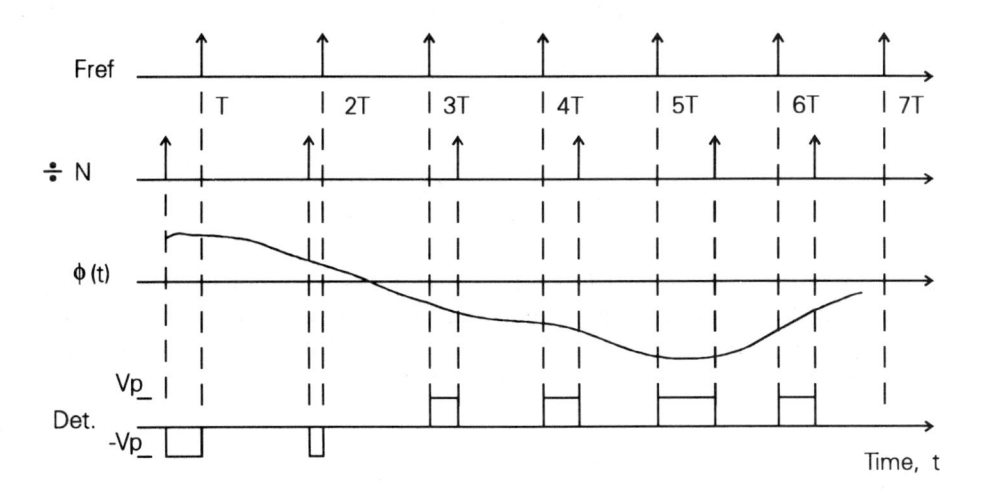

Figure 4.6 Simplified phase-frequency detector timing.

The phase detector output pulse widths are proportional to the instantaneous phase error at each sampling instant as shown in Figure 4.6. Because the source impedance of the phase detector output is always 0 Ω, this permits the effects of adjacent phase error samples to be superimposed without the restrictions encountered with the tri-state detector.

The pseudo-continuous transfer function for this phase detector may be derived by focusing on the dynamics shown in Figure 4.6. The Laplace transform of the phase detector output pulse train may be written as

$$V_d(s) = V_p \sum_{n=0}^{\infty} \left(\frac{1 - e^{-sp_n}}{s} e^{-snT} \right) \quad (4.37)$$

where p_n is the difference in time occurrence between the $\div N$ signal zero-crossing and the reference signal zero-crossing for the nth reference period, V_p is the voltage pulse magnitude in volts, and T is the constant sampling period in seconds. Each p_n is related to the nth instantaneous phase error as described earlier by (4.15). If the loop bandwidth of interest is adequately small, the complex exponential in (4.37) may be approximated by

$$e^{-sp_n} \approx 1 - sp_n \quad (4.38)$$

and on substitution into (4.37), the phase detector output may be written as

$$V_d(s) \approx K_d T \sum_{n=0}^{\infty} \varphi_e(nT) e^{-snT} \quad (4.39)$$

where

$$K_d = \frac{V_p}{2\pi} \quad (\text{V/rad})$$

The derivation may be completed by relating the instantaneous phase error samples to their equivalent transform domain representation as done earlier for the tri-state detector case. By using the results of (4.32) and (4.35) in (4.39), the pseudo-continuous transfer function for this phase detector is given by

$$H_{pfi}(s) = \frac{V_d(s)}{\varphi_e(s)} = K_d \quad (4.40)$$

This result is identical to the tri-state phase detector result aside from the cascaded transfer function for the postdetection filter in (4.36). Even though the transfer function models for the two phase detectors are identical, it is important to note that the constraints under which the two models are valid are quite different.

The accuracy of this pseudo-continuous model for the simplified phase-frequency detector is primarily dependent on the approximation made in (4.38) and its applicability can be easily examined by considering the ratio of the ideal exponential factor to the approximation as given by

$$\left(\frac{e^{-s\tau}}{1 - s\tau}\right)_{s=j\omega} = A(\omega) e^{j\theta(\omega)} \tag{4.41}$$

where the magnitude and phase of this ratio are given, respectively, by

$$A(\omega) = \frac{1}{\sqrt{1 + (\omega\tau)^2}}$$

and

$$\theta(\omega) = -\omega\tau + \arctan(\omega\tau)$$

As long as the magnitude of this ratio given by $A(\omega)$ and the phase given by $\theta(\omega)$ over the system bandwidth of interest are close to unity and zero, respectively, the approximation will be valid. The magnitude and phase are plotted versus frequency in Figure 4.7. Assuming that a 1-dB amplitude error is admissible (with a corresponding phase error of roughly 2.1 deg), then the maximum loop bandwidth that should be used with this model is $[F_{\text{ref}}/(4\pi)]$ Hz. These models may of course be used to analyze systems with larger loop bandwidths but at the expense of additional errors, which become significant due to the approximation made in (4.35). Loop bandwidths that exceed this limit should be analyzed by means of the methods in Chapter 5.

In general, some caution must be exercised in using the pseudo-continuous models for the phase-frequency detectors because they are developed on the premise that the open-loop gain function is suitably narrowband. This restriction is necessary in order to validate the approximation made in (4.35) (see Sec. 4A, the appendix at the end of this chapter).

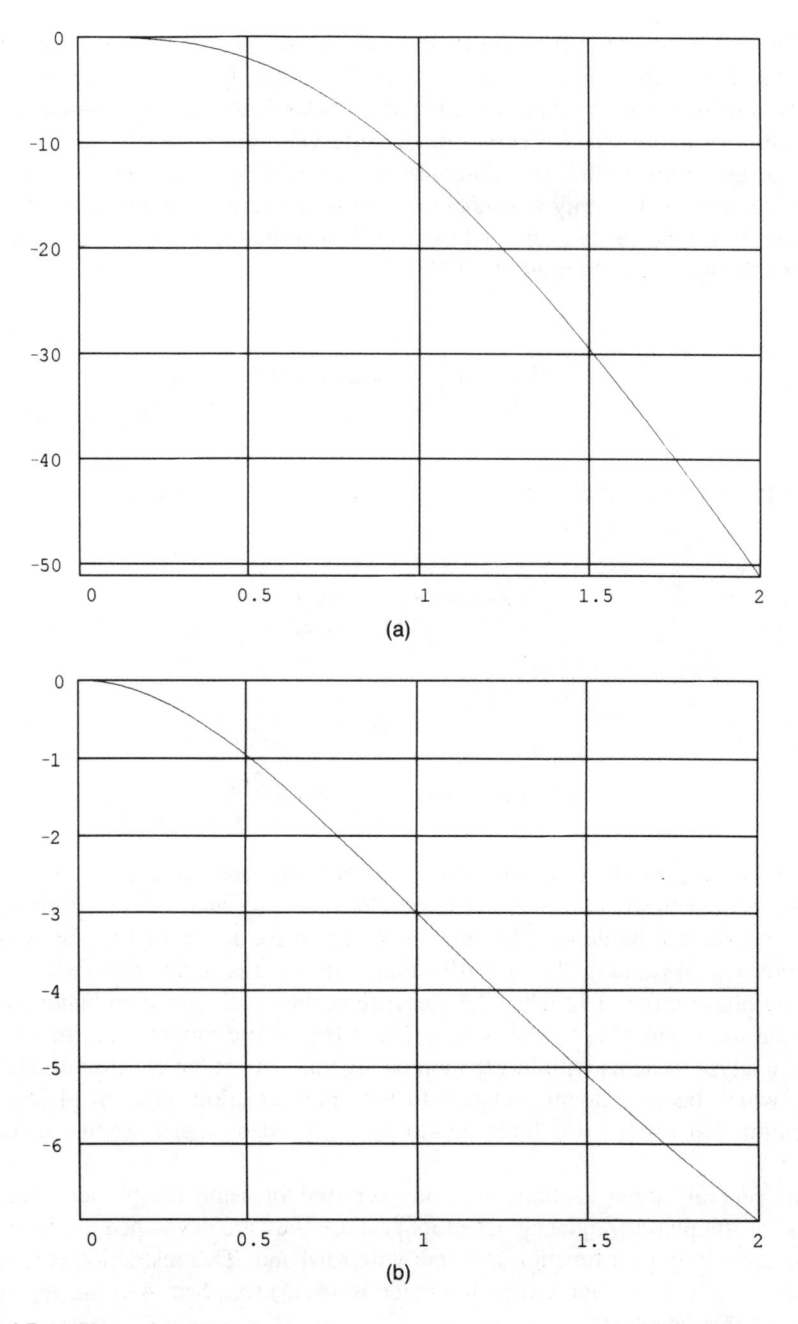

Figure 4.7 (a) Phase of (4.41) in degrees versus normalized frequency $\omega\tau$. (b) Magnitude of (4.41) in decibels versus normalized frequency $\omega\tau$.

4.2.3 Zero-Order Sample-and-Hold Phase Detector

The sample-and-hold phase detector, although having substantially more circuit complexity than the phase-frequency detector, is capable of extremely good spurious performance and optimal transient response. As shown in Chapter 5, phase-locked loops that employ this type of phase detector are theoretically capable of achieving phase lock in as little as one reference period (i.e., F_{ref}^{-1} sec). The pseudo-continuous transfer function model for this phase detector can be developed in much the same way as done for the phase-frequency detectors.

A continuous-time function, $f(t)$, which is undergoing an ideal sample-and-hold operation with output $f_{SH}(t)$ is shown in Figure 4.8. We assume that $f(t) = 0$ for $t \le 0$. The input signal is sampled once every T $(=1/F_{ref})$ sec. The analysis is somewhat simplified if the auxiliary function, $f'_{SH}(t)$, which is the time derivative of the sample-and-hold output $f_{SH}(t)$ is considered first. This auxiliary function may be written as

$$f'_{SH}(t) = \sum_{n=0}^{\infty} [f(t) - f(t - T)]\delta(t - nT) \qquad (4.42)$$

The Laplace transform of the auxiliary function may be written as

$$\mathcal{L}[f'_{SH}(t)] = F_p(s) = F(s)(1 - e^{-sT}) \otimes \frac{1}{1 - e^{-sT}} \qquad (4.43)$$

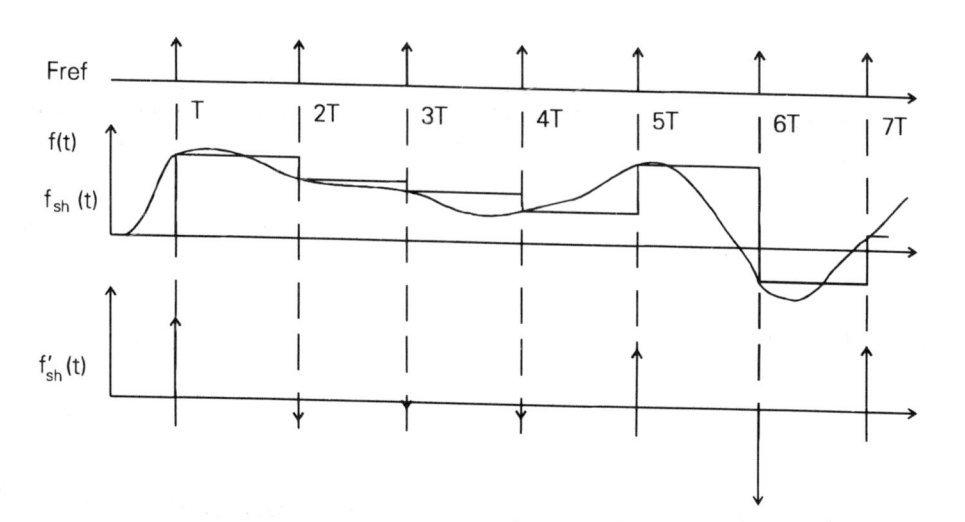

Figure 4.8 Sample-and-hold detector operation for a continuous input waveform. The time derivative of the sample-and-hold output is a series of weighted Dirac delta functions.

where ⊗ denotes convolution in the frequency domain. From [13], this convolution may be written as

$$F_p(s) = \frac{1}{j2\pi} \int_{c-j\infty}^{c+j\infty} F(\zeta)(1 - e^{-\zeta T}) \frac{1}{1 - e^{-(s-\zeta)T}} d\zeta \qquad (4.44)$$

where the normal conditions for existence apply for constant c. For all cases of interest here, the poles of $F(s)$ lie only in the left-half plane. It is therefore advantageous to evaluate (4.44) using contour integration about the right-half plane since all of the right-half plane poles are solutions of

$$1 - e^{-(s-\zeta)T} = 0 \qquad (4.45)$$

and take the particularly simple form of

$$\zeta = s + j\frac{2\pi n}{T} \qquad (4.46)$$

Employing the Cauchy residue theorem [14], the residue factors due to the poles within the denominator term in (4.44) evaluate to

$$\left\{ \frac{d}{d\zeta}[1 - e^{-(s-\zeta)T}] \right\}^{-1}_{\zeta = s + j2\pi n/T} = -\frac{1}{T} \qquad (4.47)$$

The negative sign is cancelled due to the sense of the contour integration dictated by (4.44) about the right-half plane. The final result is realized by including the residue factors resulting from the remainder of the integrand in (4.44) leading to

$$F_p(s) = \frac{1}{T} \sum_{n=-\infty}^{+\infty} F\left(s + j\frac{2\pi n}{T}\right)[1 - e^{-(s + j2\pi n/T)T}] \qquad (4.48)$$

On simplification,

$$F_p(s) = (1 - e^{-sT}) \frac{1}{T} \sum_{n=-\infty}^{+\infty} F\left(s + j\frac{2\pi n}{T}\right) \qquad (4.49)$$

Recalling that (4.49) is the Laplace transform for the auxiliary function, this result must be finally multiplied by s^{-1} in order to obtain the Laplace transform for the sample-and-hold output. The transform of the output is therefore given by

$$\mathcal{L}[f_{SH}(t)] = \frac{1 - e^{-sT}}{s} \left[\frac{1}{T} \sum_{n=-\infty}^{+\infty} F\left(s + j\frac{2\pi n}{T}\right) \right] \quad (4.50)$$

or

$$\mathcal{L}[f_{SH}(t)] = \frac{1 - e^{-sT}}{s} F^*(s) \quad (4.51)$$

where once again

$$F^*(s) = \mathcal{L}[f^*(t)]$$

From this, the pseudo-continuous transfer function model for the sample-and-hold phase detector is obtained by assuming suitable system bandwidth constraints and retaining only the $n = 0$ term in (4.50) leaving the familiar function

$$H_{SH}(s) = \frac{K_d}{T} \frac{1 - e^{-sT}}{s} \quad (4.52)$$

where K_d is the appropriate phase detector gain in volts per radian.

4.2.4 Inefficient Sample-and-Hold Phase Detector

High-speed sample-and-hold circuits are prone to exhibit poor sampling efficiency due to the finite voltage slew rate in the associated electronics. This undesirable phenomenon leads to additional gain and phase correction terms that must augment the ideal result represented by (4.52). Inefficient sampling in the case of an input unit step function is portrayed in Figure 4.9. The inefficiency is assumed to have a geometric nature, which in the context of Figure 4.9 means that the output voltage error with respect to ideal is reduced at each sample by η where $0 < \eta < 1$. The sample-and-hold aperture width is assumed to be w seconds at each sampling instant during which the hold capacitor is assumed to exponentially charge with time constant τ. For a unit-step input, the sample-and-hold output after the first sampling period is

$$f_{SH}(T) = 1 - e^{-w/\tau} \quad (4.53)$$

and, in general, the output after the nth sample is

$$f_{SH}(nT) = 1 - e^{-nw/\tau} \quad (4.54)$$

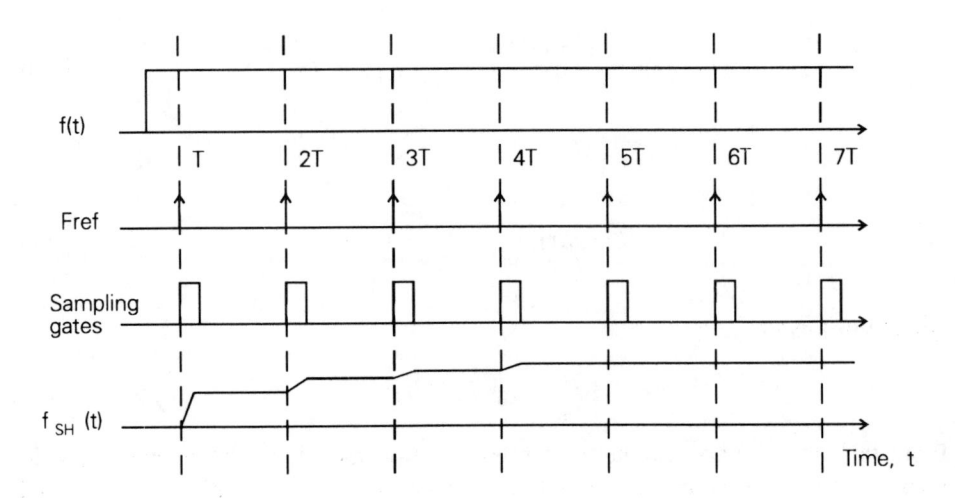

Figure 4.9 Inefficient sample-and-hold with unit-step input.

This sample sequence may be expressed in terms of its z transform as

$$F_{\text{SH}}(z) = (1 - e^{-w/\tau}) + (1 - e^{-2w/\tau})z^{-1} + \ldots \qquad (4.55)$$

which on further investigation may be rewritten as

$$F_{\text{SH}}(z) = \frac{z(z - e^{-w/\tau}) - z(z - 1)e^{-w/\tau}}{(z - 1)(z - e^{-w/\tau})} \qquad (4.56)$$

Since the applied input signal here is the unit step with z transform

$$\frac{z}{z - 1}$$

the transfer function representing the inefficient sampling operation is simply

$$T(z) = \frac{z - 1}{z} F_{\text{SH}}(z) = \frac{z(1 - e^{-w/\tau})}{z - e^{-w/\tau}} \qquad (4.57)$$

Noting that z is mathematically equivalent to e^{sT}, this may be rewritten as

$$T(s) = \frac{\eta}{1 - (1 - \eta)e^{-sT}} \qquad (4.58)$$

where

$$\eta = 1 - e^{-w/\tau}$$

This result is also given in Eq. (5.28) of [11]. The resulting pseudo-continuous transfer function for the inefficient zero-order sample-and-hold is therefore

$$H_{\text{SH}\eta}(s) = \frac{K_d}{T} \frac{1 - e^{-sT}}{s} \frac{\eta}{1 - (1 - \eta)e^{-sT}} \qquad (4.59)$$

The magnitude and phase of the correction factor (4.58) have been evaluated in [11] and are presented here in Figures 4.10 and 4.11. In practice, the most detrimental aspect of inefficient sampling is the decreased loop phase margin, which results due to the additional phase factor. In very high-speed systems, sampling efficiency can ultimately set the limit on attainable transient performance. This matter is explored in Chapter 5.

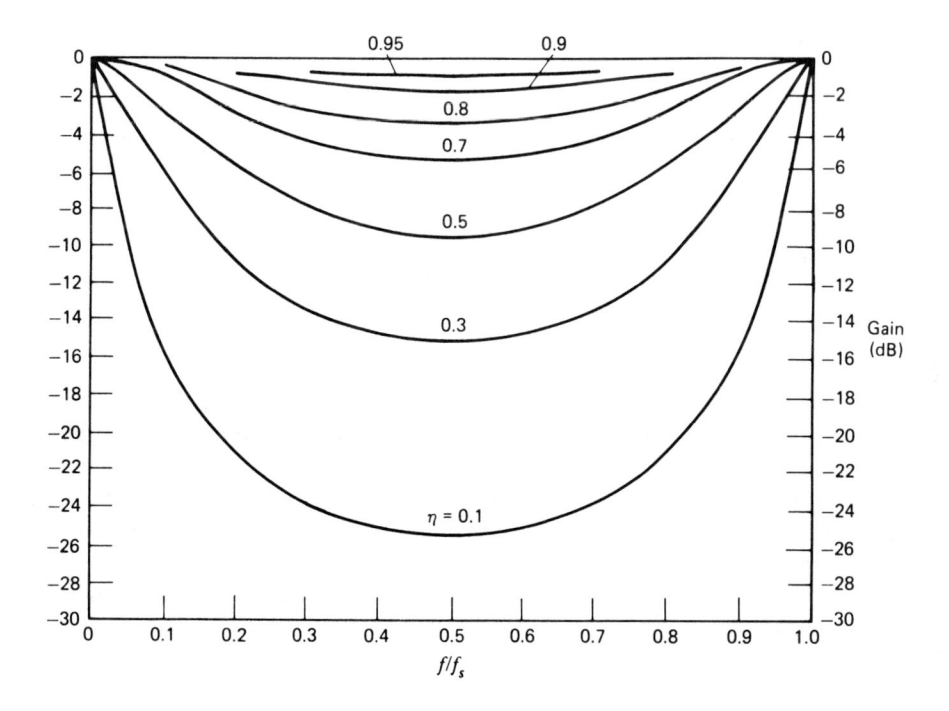

Figure 4.10 Attenuation resulting from finite sample-and-hold efficiency versus normalized frequency f/F_s. (Copyright © 1981 John Wiley & Sons, Inc. Reprinted by permission [11].)

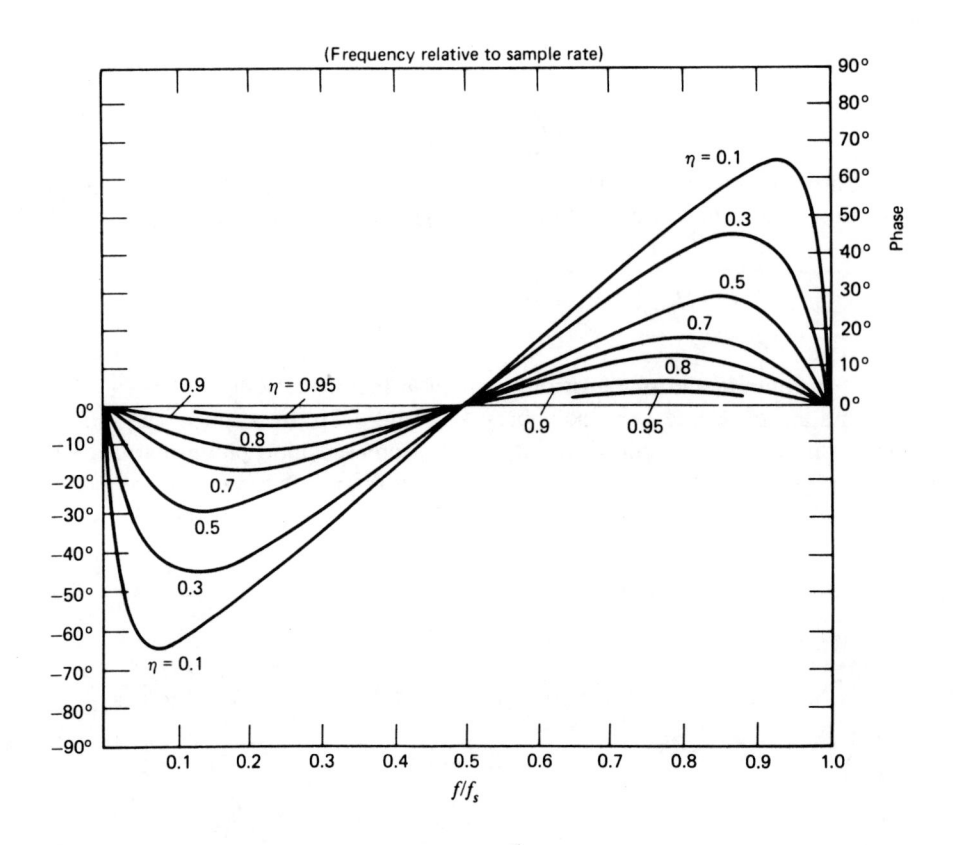

Figure 4.11 Phase argument due to inefficient sampling versus normalized frequency f/F_s. (Copyright © 1981 John Wiley & Sons, Inc. Reprinted by permission [11].)

4.2.5 Digital Feedback Dividers

Feedback dividers strictly operate on only the input signal zero-crossings as shown for a division factor of 3 in Figure 4.12. Any phase error or phase modulation on the input signal is manifested as time perturbations of the input zero-crossings when compared to a separate stable frequency source. The ideal divider in effect transports every N'th input positive-going zero-crossing to its output as shown in Figure 4.12. The input signal shown here, $S_{in}(t)$, has a frequency of precisely three times the reference frequency, F_{ref}, but with a slight phase lag. The phase error between the ideal third harmonic of F_{ref} and $S_{in}(t)$ is

$$\theta_{ei} = 2\pi \frac{3p}{T}$$

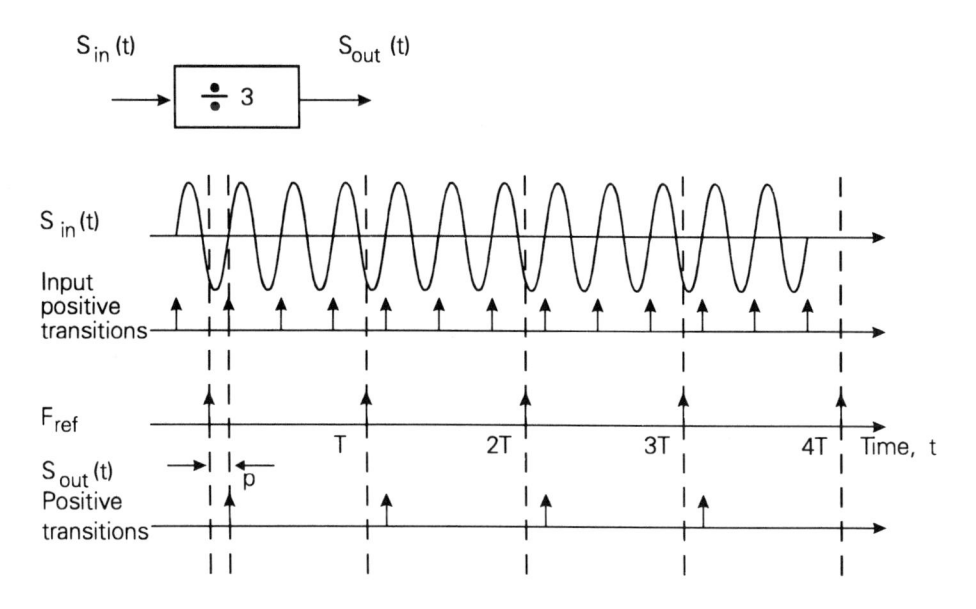

Figure 4.12 Divide-by-N input and output transition relationships.

whereas the phase error between F_{ref} and $S_{out}(t)$ is clearly

$$\theta_{eo} = 2\pi \frac{p}{T}$$

which is of course three times smaller. The important point to be made is that the feedback divider only functions to select or window which divider input transitions are conveyed to the divider output. The sampling action due to the effective transition detector at the divider input and the sampling action of a digital phase detector may be viewed as a cascade of ideal samplers as shown in Figure 4.13. Since coherent resampling of an already sampled waveform has no additional effects on the once-sampled signal [15], the cascade of the feedback divider and ideal impulse sampling phase detector may be replaced by a single sampler-phase detector and a factor of N^{-1} as shown in Figure 4.14.

Hence, inclusion of a feedback divider with the digital phase detectors only requires that a multiplicative factor of N^{-1} be included to account for the feedback divider in the open-loop gain function. The model in Figure 4.14 will be drawn on heavily in Chapter 5 where the phase-locked loop concepts for sampled systems are developed.

$X_{in}(t)$ → /‾ → /‾ → $X_{out}(t)$

         $S_N(t)$ $S(t)$

Feedback divider sampling function $\quad S_N(t) = \sum_{n=-\infty}^{\infty} \delta(t - nT/N)$

Phase detector sampling function $\quad S(t) = \sum_{n=-\infty}^{\infty} \delta(t - nT)$

Figure 4.13 Sampling action due to the feedback divider and phase detector results in a cascade of ideal impulse sampling functions.

Phase detector P(s)

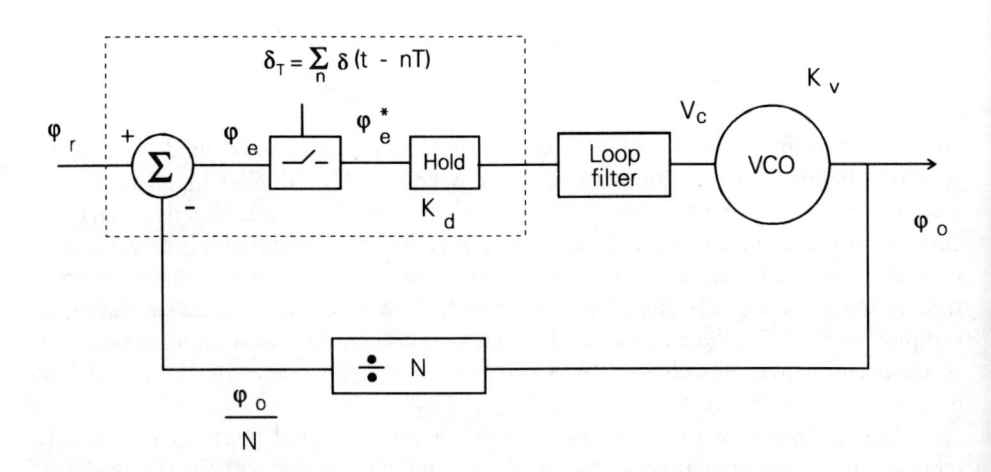

$\delta_T = \sum_n \delta(t - nT)$

φ_r → Σ → φ_e → /‾ → φ_e^* → Hold K_d → Loop filter → V_c → VCO (K_v) → φ_o

÷ N

$\dfrac{\varphi_o}{N}$

Figure 4.14 Equivalent model for a phase-locked loop incorporating a digital phase detector and digital feedback divider. The hold function may be as simple as an analog RC filter or as sophisticated as a true sample-and-hold.

4.3 MODELING TIME DELAYS IN CONTINUOUS SYSTEMS

Inclusion of ideal time delay elements within a continuous system is particularly difficult to analyze because it results in a closed-loop characteristic function that is no longer algebraic having infinite order [16]. The necessity to model time delays accurately may arise due to the use of pseudo-continuous models [e.g., (4.52)], the desire to model actual transport delays within the loop, or the need to approximate the group delay of additional loop filter elements. The following discussion is derived largely from [16].

Consider a type 1 phase-locked loop that utilizes a zero-order sample-and-hold phase detector. The sample-and-hold phase detector can be approximated by

$$H_{SH}(s) = \frac{K_d}{T} \frac{1 - e^{-sT}}{s} \approx K_d e^{-sT/2} \qquad (4.60)$$

The system open-loop gain function is then given by

$$G_{OL}(s) = \frac{K_d K_v}{N} \frac{e^{-sT/2}}{s} \qquad (4.61)$$

The associated characteristic function for this system is simply

$$s + K e^{-sT/2} = 0 \qquad (4.62)$$

where $K = K_d K_v N^{-1}$. The equation is clearly nonalgebraic, requiring transcendental equation methods for solution. The root-locus for this characteristic equation is given in [17].

The closed-loop transfer function for the example cited here is

$$T(s) = \frac{\varphi_e(s)}{\varphi_r(s)} = \frac{s}{s + K e^{-s\tau}} \qquad (4.63)$$

where $\tau = 0.5T$. Time-domain transient response calculations are no less involved due to the infinite system order, but as shown in [16], this may be solved as a sequence of open-loop analyses over appropriate time intervals. For calculation of the system impulse response, this is equivalent to repeated long-hand division as shown in Figure 4.15. The long division form may be used to represent the impulse response in an infinite summation form as

$$\varphi_e(s) = \varphi_r(s) + \sum_{n=1}^{\infty} \frac{(-K)^n}{s^n} e^{-sn\tau} \varphi_r(s) \qquad (4.64)$$

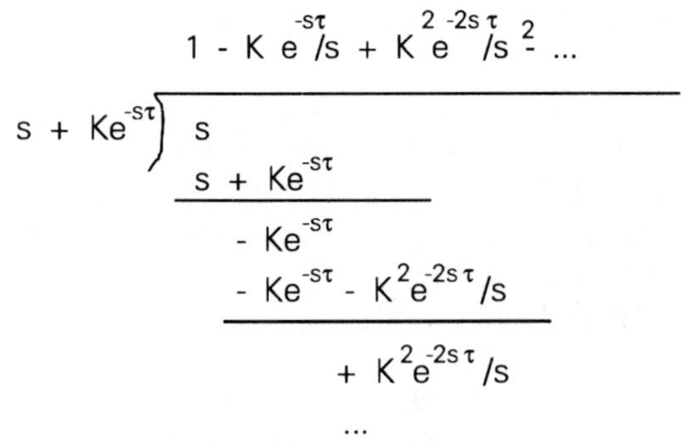

$$\begin{array}{r}1 - Ke^{-s\tau}/s + K^2e^{-2s\tau}/s^2 - \ldots\\[4pt]\hline s + Ke^{-s\tau}\overline{\smash{\big)}\,s}\\ \underline{s + Ke^{-s\tau}}\\ -Ke^{-s\tau}\\ \underline{-Ke^{-s\tau} - K^2e^{-2s\tau}/s}\\ +K^2e^{-2s\tau}/s\\ \ldots\end{array}$$

Figure 4.15 Partial long-hand division showing development of the time-domain system impulse response.

which has a corresponding time-domain impulse response [i.e., $\varphi_r(s) = 1$] of

$$\varphi(t) = \delta(t) + \sum_{n=1}^{\infty} (-K)^n \frac{(t - n\tau)^{n-1}}{(n-1)!} \mu(t - n\tau) \qquad (4.65)$$

where $u(\ )$ is the unit-step function. This system is stable for $\tau = 1$ as long as K is less than or equal to 0.5π [16].

Returning to (4.62), it is tempting to approximate the troublesome exponential by a Taylor series expansion of sufficient order, thereby making the characteristic equation algebraic once more. This approach is flawed, however, because it leads to characteristic equation root locations that are in the right-half plane, thereby implying system instability [16]. It is nonetheless possible to use a first-order approximation for the exponential

$$e^{-s\tau} \approx \frac{1}{1 + s\tau} \qquad (4.66)$$

because this transfer function corresponds to a simple one-pole RC low-pass filter whose only pole is clearly in the left-half plane.

A technique that, unlike the approximation in (4.66), preserves the proper modulus of the exponential approximation is that of Padé approximates [16]. This approach is still unsuitable for stability considerations for the same reasons cited however. The first several Padé approximates for $e^{-s\tau}$ follow [16]:

$$n = 1$$
$$\frac{1 - s\tau/2}{1 + s\tau/2}$$

$n = 2$

$$\frac{1 - s\tau/2 + (s\tau)^2/12}{1 + s\tau/2 + (s\tau)^2/12}$$

$n = 3$

$$\frac{1 - s\tau/2 + (s\tau)^2/10 - (s\tau)^3/120}{1 + s\tau/2 + (s\tau)^2/10 + (s\tau)^3/120}$$

It is interesting to note that if the nth-order Padé approximate denominator polynomial is divided into the associated numerator polynomial, the first $2n$ terms agree with the terms that would be found in a $2n$-degree Taylor series expansion of the complex exponential. Further information on Padé approximates is available in [18,19].

A much more useful approximation for the ideal time delay, which does not result in erroneous right-half plane poles, is

$$e^{-s\tau} = \lim_{n \to \infty} \left(\frac{1 - \frac{s\tau}{2n}}{1 + \frac{s\tau}{2n}} \right)^n \qquad (4.67)$$

because it is equivalent to a cascade of first-order all-pass sections. One possible circuit realization of a single section all-pass network is shown in Figure 4.16. The modeling errors that result from using this approximation for different order n may be gauged based on the data provided in Table 4.1 for a time delay of $T/2$. This

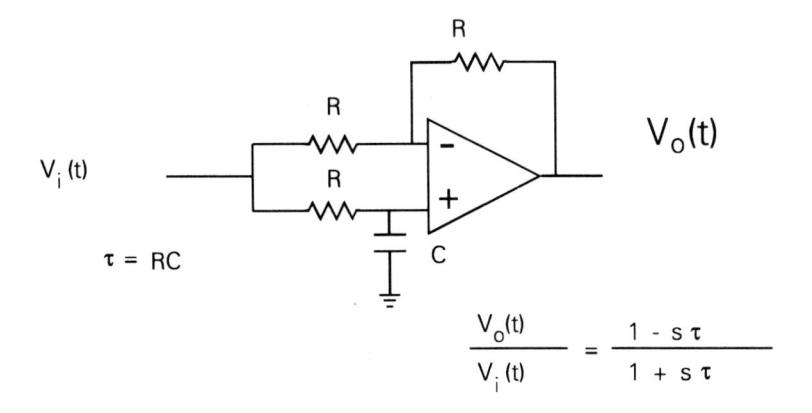

$$\frac{V_o(t)}{V_i(t)} = \frac{1 - s\tau}{1 + s\tau}$$

Figure 4.16 First-order active all-pass network.

time delay case will appear quite often in subsequent analyses. The tabular data indicate that only an $n = 3$ approximation is required if modeling phase errors of less than 2 deg are desired at the critical frequency of $0.5F_{\text{ref}}$. This level of complexity is very acceptable since all calculations will be performed by computer.

Table 4.1
All-Pass Network Delay Approximations

$$e^{-s\tau/2} \approx \left(\frac{1 - \dfrac{s\tau}{4n}}{1 + \dfrac{s\tau}{4n}}\right)^n$$

Angular error from ideal, degrees.

n	$0.01\ F_{\text{ref}}$	$0.1\ F_{\text{ref}}$	$0.2\ F_{\text{ref}}$	$0.5\ F_{\text{ref}}$
1	0.000148	0.14589	1.1188	13.708
2	0.000037	0.03687	0.2918	4.24
3	—	0.01642	0.1307	1.976
4	—	0.00924	0.0737	1.1306
5	—	0.00592	0.0473	0.7295

4.4 FREQUENCY-DOMAIN ANALYSIS

The transfer function models developed in the previous three sections provide the groundwork from which it is possible to describe any continuous or pseudo-continuous phase-locked loop in terms of its open-loop gain function $G_{\text{OL}}(s)$. The two principal closed-loop gain functions are

$$T_1(s) = \frac{\varphi_o(s)}{\varphi_r(s)} = \frac{NG_{\text{OL}}(s)}{1 + G_{\text{OL}}(s)} \tag{4.68}$$

and

$$T_2(s) = \frac{\varphi_o(s)}{\varphi_v(s)} = \frac{1}{1 + G_{\text{OL}}(s)} \tag{4.69}$$

These are used extensively in other chapters to predict the phase noise performance of candidate phase-locked loop designs. Transfer function $T_1(s)$ is used to calculate the impact of reference node phase noise at the loop output whereas $T_2(s)$ is used to assess the suppression of the VCO self-noise at the loop output.

Formulation of the open-loop gain function for a purely continuous system is straightforward so this case is not considered further. Instead, it is more instructive to look at two sample calculations for pseudo-continuous systems before proceeding.

Consider first the fairly simple type 2 system shown in Figure 4.17, which utilizes a zero-order sample-and-hold phase detector and one additional low-pass filter for improved spurious performance. The loop natural frequency is assumed to be 5% of the reference frequency and the low-pass filter pole is taken to be at 10% of the reference frequency, thereby providing a 20-dB improvement in the spurious levels of the first reference spurs. The loop damping factor is assumed to be 0.707. Following the guidelines given in the previous sections, the open-loop gain function can be formulated as

$$G_{OL}(s) = \frac{K_d}{N}\frac{1}{T}\frac{1-e^{-sT}}{s}\frac{1+s\tau_2}{s\tau_1}\frac{K_v}{s(1+s\tau_3)} \tag{4.70}$$

where $T = F_{ref}^{-1}$ sec. On substituting $j\omega$ for s in (4.70), this may be written as

$$G_{OL}(s) = -\left(\frac{\omega_n}{\omega}\right)^2 e^{-j\omega T/2} \operatorname{sinc}\left(\frac{\omega T}{2}\right)\frac{1+j\omega\tau_2}{1+j\omega\tau_3} \tag{4.71}$$

where (4.7) has been used for ω_n and $\operatorname{sinc}(x)$ is equivalent to $\sin(x)/x$.

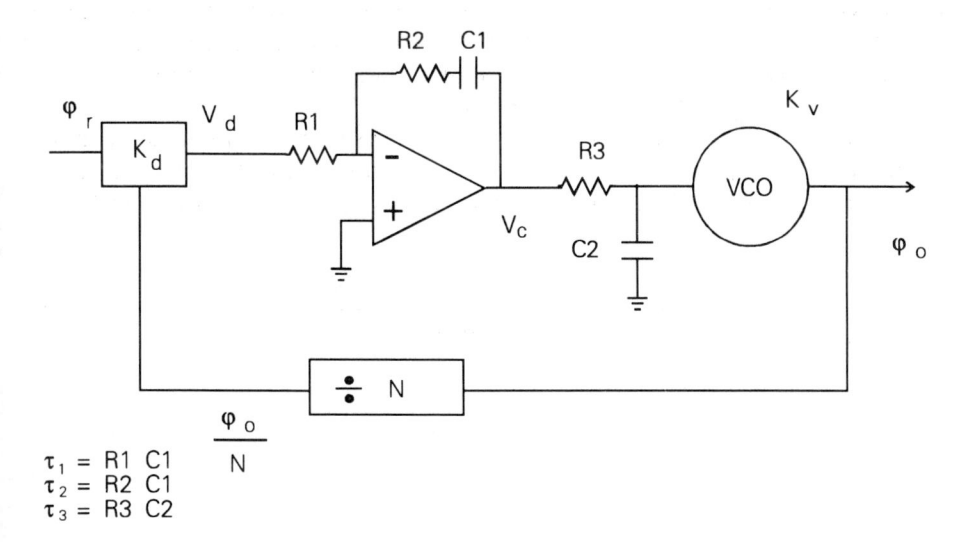

$\tau_1 = R1\ C1$
$\tau_2 = R2\ C1$
$\tau_3 = R3\ C2$

Figure 4.17 Basic type 2 third-order phase-locked loop with sample-and-hold phase detector.

The techniques developed in Chapter 5 may be used to calculate the exact open-loop gain function for this example in terms of z transforms. The final result, which is simply stated here, is

$$G_{OL}(\omega) = (\omega_n T)^2 \frac{C_2 z^2 + C_1 z + C_0}{(z-1)^2 (z-A)} \quad (4.72)$$

where

$$A = e^{-T/\tau_3}$$

$$C_2 = \left(\frac{\tau_2}{T} - \frac{\tau_3}{T}\right)\left[1 + (A-1)\frac{\tau_3}{T}\right] + \frac{1}{2}$$

$$C_1 = \left(\frac{\tau_3}{T} - \frac{\tau_2}{T}\right)\left[(2A+1)\frac{\tau_3}{T} + 1 + A - 3\frac{\tau_3}{T}\right] + \frac{1-A}{2}$$

$$C_0 = \left(\frac{\tau_2}{T} - \frac{\tau_3}{T}\right)\left[\frac{\tau_3}{T}(A-1) + A\right] - \frac{A}{2}$$

$$z = e^{j\omega T}$$

The pseudo-continuous and exact open-loop gain functions are tabulated in Table 4.2 for comparison and show excellent agreement well beyond the unity gain frequency. The close agreement between pseudo-continuous and exact calculations is primarily due to the combined reduction in high-frequency gain due to the sample-and-hold, the low-pass filter, and the VCO.

The second example we consider is quite similar but utilizes the simplified phase-frequency detector rather than a sample-and-hold phase detector. This circuit is shown in Figure 4.18. The pseudo-continuous open-loop gain function for this loop configuration is

$$G_{OL}(s) = \frac{K_d}{N} \frac{1}{1 + s\tau_3/4} \frac{1 + s\tau_2}{s\tau_1} \frac{K_v}{s} \quad (4.73)$$

This may be further reduced to

$$G_{OL}(s) = \left(\frac{\omega_n}{s}\right)^2 \frac{1 + 2\zeta s/\omega_n}{1 + s\tau_3/4} \quad (4.74)$$

and on substituting $j\omega$ for s and τ_0 for $\tau_3/4$, the final result is

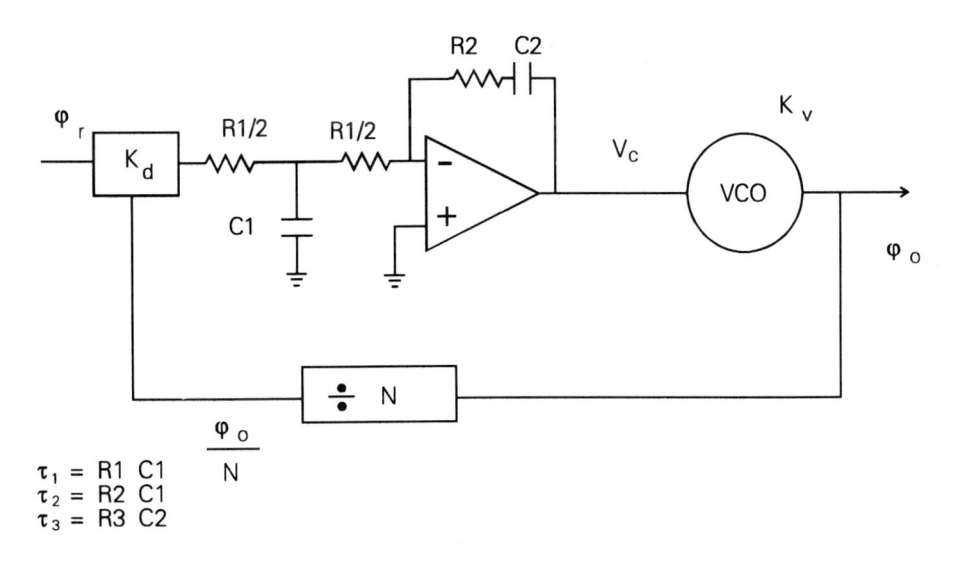

Figure 4.18 Basic type 2 third-order phase-locked loop with simplified phase-frequency detector.

$\tau_1 = R1\ C1$
$\tau_2 = R2\ C1$
$\tau_3 = R3\ C2$

Table 4.2
Comparison of Pseudo-Continuous and Exact Open-Loop Gain Functions for Type 2 Third-Order Phase-Locked Loop with Sample-and-Hold Phase Detector

Frequency f/F_{ref}	$G_{OL}(s)$ Gain (dB)	$G_{OL}(s)$ Phase (deg)	$G_{OL}(z)$ Gain (dB)	$G_{OL}(z)$ Phase (deg)
0.000001	187.959	−179.999	187.959	−179.999
0.01	28.247	−171.719	28.247	−171.719
0.02	16.946	−165.417	16.946	−165.417
0.05	3.765	−160.834	3.765	−160.830
0.07	−0.729	−164.393	−0.730	−164.382
0.10	−5.654	−172.474	−5.660	−172.442
0.15	−11.741	−186.575	−11.773	−186.456
0.20	−16.467	−199.461	−16.571	−199.152

$\omega_n T = 2\pi 0.05$
$\zeta = 0.707$
Low-pass filter corner frequency $0.1 F_{ref}$

$$G_{OL}(\omega) = -\left(\frac{\omega_n}{\omega}\right)^2 \frac{1 + j2\omega\zeta/\omega_n}{1 + j\omega\tau_0} \qquad (4.75)$$

Employing the methods of Chapter 5 once more, the exact open-loop gain expressed in z transforms for this example is

$$G_{OL}(\omega) = (\omega_n T)^2 z \frac{K_1 z + K_0}{(z-1)^2(z-A)} \qquad (4.76)$$

where

$$K_1 = \left(\frac{T_2}{T} - \frac{T_0}{T}\right)(1 - A) + 1$$

$$K_0 = \left(\frac{T_2}{T} - \frac{T_0}{T}\right)(A - 1) - A$$

$$A = e^{-T/\tau_0}$$

$$z = e^{j\omega T}$$

The pseudo-continuous and exact open-loop gain functions for this case are tabulated in Table 4.3 assuming a damping factor of 0.707, ω_n corresponding to 5% of the

Table 4.3
Comparison of Pseudo-Continuous and Exact Open-Loop Gain Functions for Type 2 Third-Order Phase-Locked Loop with Phase-Frequency Detector

Frequency f/F_{ref}	$G_{OL}(s)$		$G_{OL}(z)$	
	Gain (dB)	Phase (deg)	Gain (dB)	Phase (deg)
0.000001	187.959	−179.999	187.959	−179.999
0.01	28.185	−173.136	28.190	−173.140
0.02	16.714	−167.948	16.732	−167.974
0.05	2.683	−163.415	2.774	−163.593
0.07	−2.369	−164.516	−2.206	−164.814
0.10	−7.900	−166.992	−7.589	−167.470
0.15	−14.462	−170.268	−13.789	−171.037
0.20	−19.261	−172.370	−18.078	−173.432

$\omega_n T = 2\pi 0.05$
$\zeta = 0.707$
$R_1 C_1 = 10T$

reference frequency, and $\tau_3 = 10T$ corresponding to a simple pole at $0.0159F_{\text{ref}}$. Once again, the two gain functions track closely well beyond the unity gain frequency as desired, but not as well as in the first example. The open-loop gain function in this case does not attenuate the high-frequency gain terms as sharply and this leads to increased approximation error with the pseudo-continuous approach. Larger percentage bandwidths would, of course, lead to more significant difference.

Stability

Pseudo-continuous modeling techniques provide a close approximation to the true open-loop gain function as long as the system bandwidth is a small fraction of the system reference frequency. Although stability is considered at length in Chapter 5, a brief review of the underlying concepts is presented in this section.

Many methods are available for assessing the stability of linear time-invariant systems, including the Routh-Hurwitz criterion, the Nyquist criterion, root-locus methods, and Bode diagrams [17]. Stability studies of nonlinear systems are considerably more complex and often involve Lyapunov methods. A fine example of this technique as applied to a third-order phase-locked loop may be found in [20].

The Nyquist stability criterion is well suited to these present applications because (1) it utilizes the same open-loop gain function, which must be evaluated in (4.68) and (4.69); (2) it permits very easy inclusion of ideal time delay elements, which would otherwise have to be approximated by cascaded all-pass sections; and (3) inclusion of exact sampling effects may be achieved using a simple modification of traditional continuous control system techniques as will be shown in Chapter 5. The underlying principles of this criterion are briefly discussed in the following few paragraphs.

The Nyquist stability criterion is based on the principle of the argument theorem, which relates the encirclement of the poles and zeros of a function $P(s)$ by a contour Γ_s in the s-plane to the number of encirclements of the origin in the $P(s)$-plane by the mapped contour Γ_p. If Γ_s is selected to enclose the right-half portion of the s-plane with the restrictions that (1) Γ_s not pass through any pole or zero of $P(s)$, and (2) Γ_s is traversed in the counterclockwise direction, then this principle states that the contour Γ_p in the $P(s)$-plane will encircle the origin $N = Z - P$ times counterclockwise where Z and P are the number of zeros and poles of $P(s)$ enclosed by Γ_s, respectively.

In applying this theorem to phase-locked loops, stability studies focus on the function

$$P(s) = 1 + G_{\text{OL}}(s) \qquad (4.77)$$

In general, $G_{\text{OL}}(s)$ has one or more poles at the origin but no poles in the right-half plane. Therefore, if Γ_s is chosen as shown in Figure 4.19 with $R \to \infty$ and $\epsilon \to 0$,

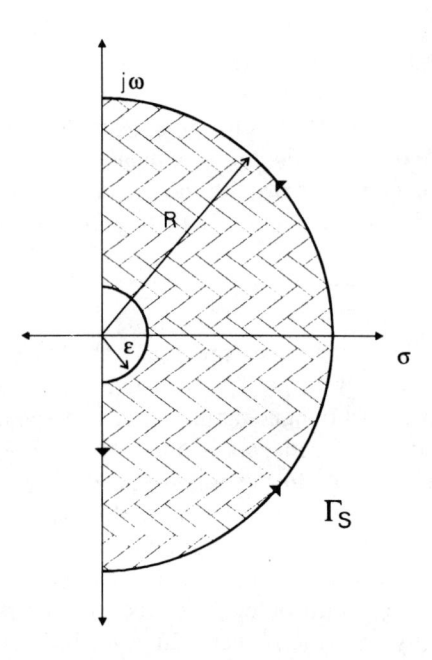

Figure 4.19 Complex path of integration for Nyquist criterion.

the system will be unstable for any encirclements of the $P(s)$-plane origin. Furthermore, since $G_{OL}(s)$ is more readily available than $1 + G_{OL}(s)$ in our calculations, it suffices to consider encirclements of the point $(-1, j0)$ in the $P(s)$-plane with $P(s) = G_{OL}(s)$.

As an example of stability assessment using the Nyquist criterion, the two phase-locked loop examples that were examined in Sec. 4.4 can be revisited. For the sample-and-hold phase detector case, the open-loop gain function (4.70) was

$$G_{OL}(s) = \frac{K_d}{N} \frac{1}{T} \frac{1 - e^{-sT}}{s} \frac{1 + s\tau_2}{s\tau_1} \frac{K_v}{s(1 + s\tau_3)}$$

$$= \frac{\omega_n^2}{Ts^3}(1 - e^{-sT})\frac{1 + s\tau_2}{1 + s\tau_3} \qquad (4.78)$$

Evaluating $G_{OL}(s)$ over the Γ_s contour in Figure 4.19, the Γ_p contour is as shown in Figure 4.20 where the same loop parameters were chosen as in Sec. 4.4. Since there are no encirclements of the 0-dB $\angle$180-deg point, the system is stable.

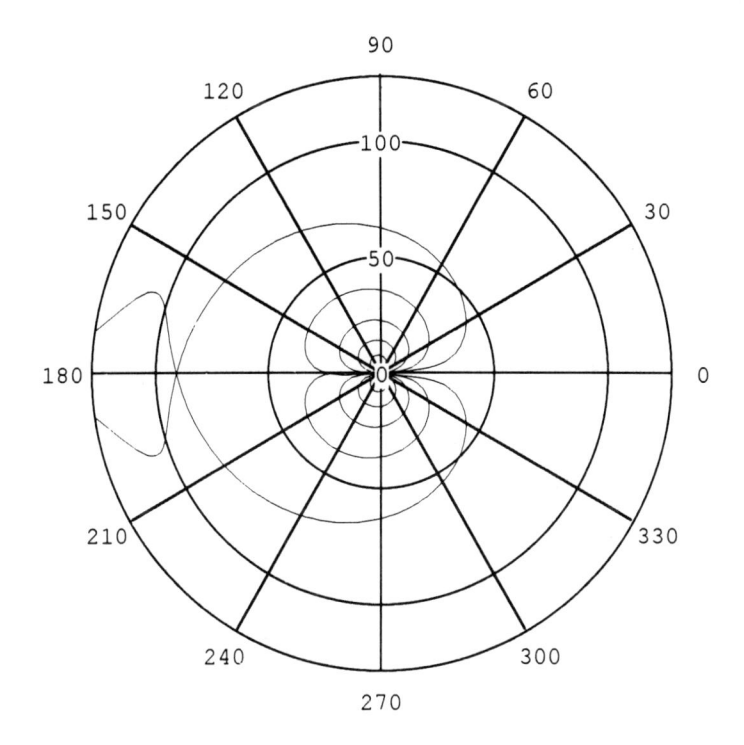

[Polar Nyquist plot with angles 0–330° labeled around circumference and radial decibel values 50 and 100 shown.]

Figure 4.20 Portion of the Nyquist stability plot for example 1 with radial units of decibels. The unity gain circle is given by the circle marked 100 dB.

For the second phase-locked loop case, the open-loop gain function was (4.74)

$$G_{\text{OL}}(s) = \left(\frac{\omega_n}{s}\right)^2 \frac{1 + 2\zeta s/\omega_n}{1 + s\tau_3/4} \tag{4.79}$$

The $P(s)$ contour that results, again evaluating (4.79) over the Γ_s contour in Figure 4.19, is shown in Figure 4.21 and, once again, the system is stable.

The points at which the gain and phase margins are calculated are clearly visible in Figures 4.20 and 4.21 using the polar representation. The gain and phase margins for case 1 are 9 dB and 16 deg, respectively, and for case 2, 27 dB and 16 deg.

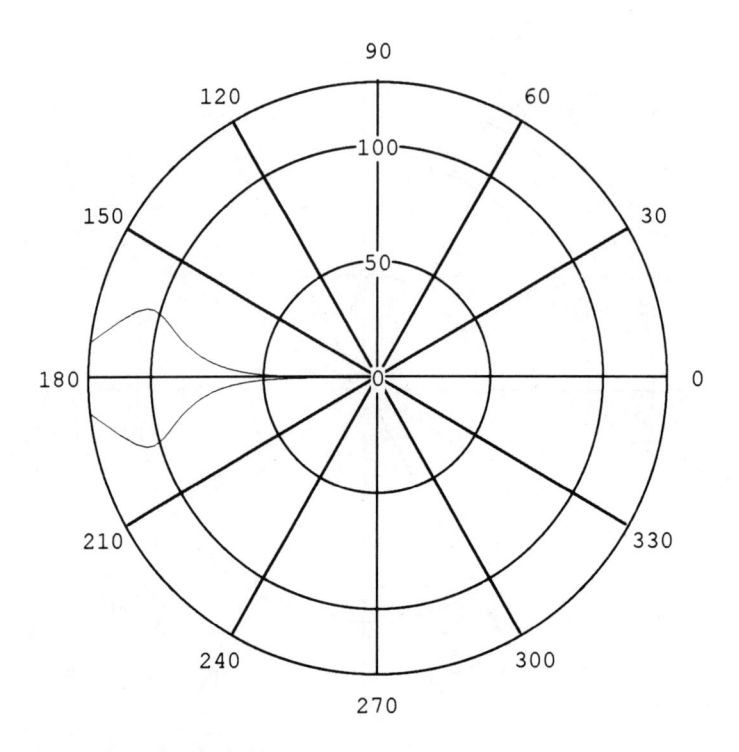

Figure 4.21 Portion of the Nyquist stability plot for example 2 with radial units of decibels. The unity gain circle is given by the circle marked 100 dB.

4.5 TRANSIENT RESPONSE EVALUATION FOR CONTINUOUS SYSTEMS

Time-domain transient response calculations for a phase-locked loop subjected to a specified input are fairly intensive and prone to numerical complications for high-order systems. Many transient response methods (e.g., [21–27]) have been proposed over time. The methods examined in this chapter are the Tustin method [9] and a circuit equivalence method.

Assume for the time being that the open-loop gain function under consideration, $G_{OL}(s)$, may be expressed as a cascade of quadratic rational factors in s as represented by

$$G_{OL}(s) = \frac{N(s)}{D(s)} = \prod_{i=1}^{k}\left[\frac{P_i(s)}{Q_i(s)}\right]$$

$$= \prod_{i=1}^{k} F_i(s) \qquad (4.80)$$

This statement automatically implies that any time delay elements within the loop have already been approximated by a suitable cascade of first-order all-pass sections. The loop transient phase error response to a system input excitation, $\varphi_x(s)$, may be expressed using Laplace transforms as

$$\varphi_e(s) = \frac{\varphi_x(s)}{1 + G_{OL}(s)}$$

$$= \varphi_x(s) \frac{D(s)}{N(s) + D(s)} \qquad (4.81)$$

Formulation of the final portion of (4.81) can lead to numerical problems in high-order systems due to inadequate precision in calculating the denominator polynomial, $N(s) + D(s)$. In situations where the polynomial coefficients of $N(s)$ and $D(s)$ differ substantially in magnitude, finite precision in the resulting sum can shift some of the roots into the right-half plane. In general, it is preferable to manipulate poles and zeros rather than polynomial coefficients.

In contrast, the form of (4.80) is very attractive for a block diagram level analysis because each block is clearly identifiable. As shown momentarily, both methods considered allow this clarity to be retained throughout the solution process.

The Tustin method provides a straightforward technique for converting an arbitrary rational Laplace transform in pole-zero form into a representative z transform utilizing a simple substitution for s^{-1}. This well-known substitution method leads to fairly stable, although not necessarily accurate, difference equations for simulating a continuous system [28]. The accuracy of the technique is primarily driven by the ratio of sampling rate to the highest frequency pole or zero in the system. The method is based on approximating z as

$$z = e^{sT} \approx \frac{1 + sT/2}{1 - sT/2} \qquad (4.82)$$

which is the first-order Padé approximant. This result may be rearranged to give

$$s^{-1} = \frac{T}{2} \frac{z + 1}{z - 1} \qquad (4.83)$$

It may be shown [28] that the transfer function s^{-1} in the Tustin method is equivalent to trapezoidal integration, which is an implicit numerical integration form. This is a highly desirable attribute [24,28] as discussed in Chapter 8 and is largely responsible for the numerical stability virtues of this technique.

As mentioned previously, the accuracy of this technique is primarily driven by the sampling rate as compared to the highest frequency pole or zero in the system.

Guidelines for sampling rate selection may be found by examining the accuracy of (4.82) for dominant poles as is done in [9], or by accepting the recommendation developed in [28], which suggests a sampling rate of at least 15 times the maximum pole or zero frequency in the system. The latter approach is adopted here for the sake of brevity.

Based on the problem formulation described by (4.80), it is convenient to think of transforming the continuous analysis problem using the Tustin method into an equivalent finite difference equation formulation as shown in Figure 4.22. Because all of the internal block diagram nodes remain accessible after the transformation, the time-domain response at each internal node is a natural by-product of this approach. It only remains to consider the details of transforming each $F_i(s)$ in (4.80) into a representative $F_i(z)$.

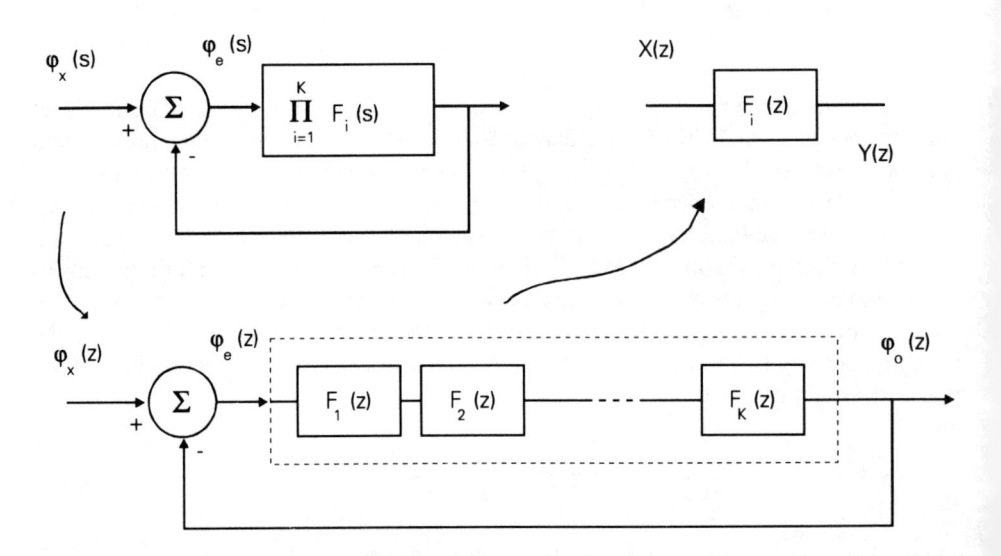

Figure 4.22 Transformation of the Laplace transform description into a cascade of second-order sections by means of the Tustin transformation.

The numerator and denominator factors of each $F_i(s)$ in (4.80) were restricted to be quadratic factors for improved precision, and to permit an easy transformation of each $F_i(s)$ into $F_i(z)$ to be made. Although this approach simplifies the computations considerably, it also introduces an additional unit delay between each $F_i(z)$ block. The delays added in the case of a large system can affect the closed-loop stability significantly if this effect is not anticipated. In the case of a single pole $F(s)$,

$$F(s) = \frac{1}{1+s\tau} = \frac{\frac{1}{s}}{\frac{1}{s}+\tau} \qquad (4.84)$$

Substituting (4.83) into (4.84) and simplifying,

$$F(z) = \frac{(z+1)}{\frac{2\tau}{T}(z-1)+(z+1)} \qquad (4.85)$$

If $F(z)$ is thought of as a transfer function relating the z transform of the input $X(z)$ to the z transform of the output $Y(z)$ as shown in Figure 4.22, then (4.85) may be rewritten as

$$Y(z) = \frac{1+z^{-1}}{1+2\tau/T}X(z) - \frac{1-2\tau/T}{1+2\tau/T}z^{-1}Y(z) \qquad (4.86)$$

Recognizing that z^{-1} is a unit-sample delay operator, (4.86) may be cast in its final difference equation form as

$$y(n) = \frac{x(n)+x(n-1)}{1+2\tau/T} - \frac{1-2\tau/T}{1+2\tau/T}y(n-1) \qquad (4.87)$$

This result is represented schematically using unit delay elements in Figure 4.23.

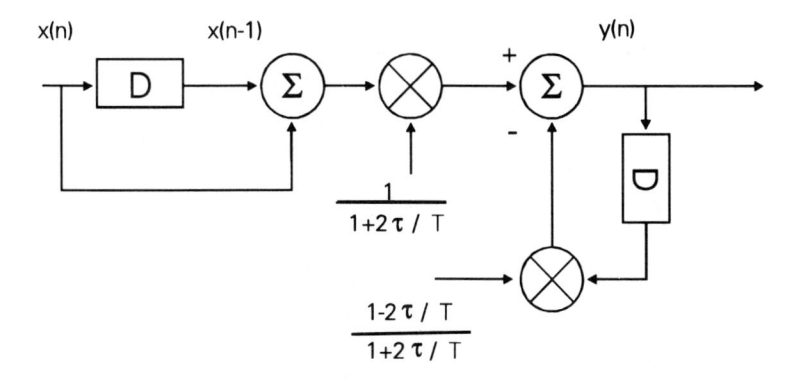

Figure 4.23 Unit delay implementation for a single-pole low-pass filter.

In the case of a generalized rational quadratic transfer function, the same substitution technique may be used. If the general quadratic rational Laplace transform is represented as

$$F(s) = \frac{a_2 s^2 + a_1 s + a_0}{b_2 s^2 + b_1 s + b_0} \qquad (4.88)$$

and (4.83) is substituted into (4.88), the resulting second-order difference equation is

$$y(n) = \frac{1}{K_0}[K_1 x(n) + K_2 x(n-1) + K_3 x(n-2)] \\ - \frac{1}{K_0}[M_1 y(n-1) + M_2 y(n-2)] \qquad (4.89)$$

where

$$K_0 = b_2 + \frac{b_1 T}{2} + \frac{b_0 T^2}{4}$$

$$K_1 = a_2 + \frac{a_1 T}{2} + \frac{a_0 T^2}{4}$$

$$K_2 = -2a_2 + \frac{a_0 T^2}{2}$$

$$K_3 = a_2 - \frac{a_1 T}{2} + \frac{a_0 T^2}{4}$$

$$M_1 = -2b_2 + \frac{b_0 T^2}{2}$$

$$M_2 = b_2 - \frac{b_1 T}{2} + \frac{b_0 T^2}{4}$$

This result is shown schematically in Figure 4.24.

A particularly insightful method for performing the bilinear s to z transformation for an arbitrary rational Laplace transform may be found in [29]. A Turbo

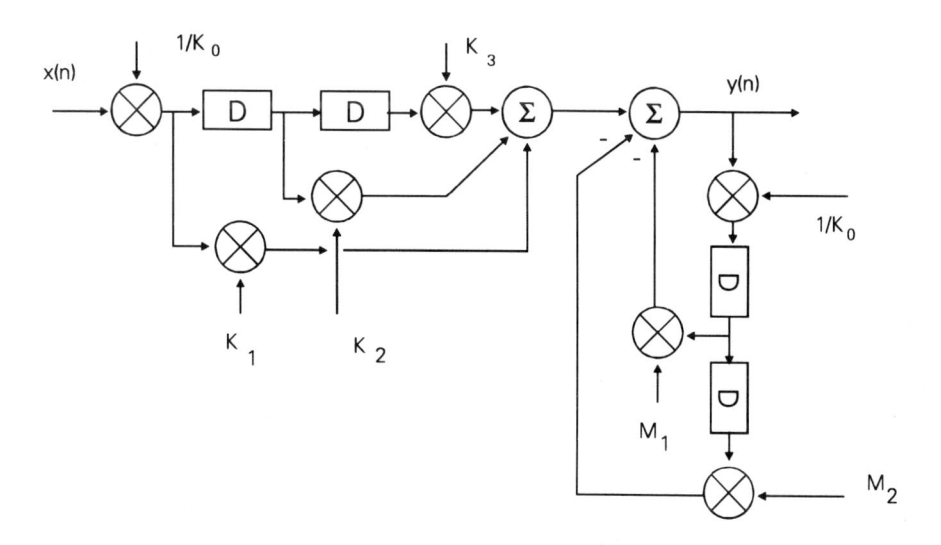

Figure 4.24 Unit delay implementation for the generalized second-order system.

Pascal program based on this method can be found on disk with the companion computer program to this text.

As pointed out in [28], extension of this substitution method to other numerical integration formulas is very simple. The procedure involves expressing the desired integration formula as a z transform, and then substituting it for s^{-1} as done earlier. This is further highlighted in [29].

Since each node of the original block diagram remains assessable in the solution process, inclusion of nonlinear transfer functions between the $F_i(z)$ in Figure 4.22 is also possible. This may be used to simulate the impact of finite voltage supply rails in an actual circuit implementation for example.

Solution of the transient response then reduces to representing the open-loop gain function as a cascade of at most second-order finite difference modules as discussed earlier and effectively closing the feedback loop by subtracting $\varphi_o(n)$ from $\varphi_x(n)$ at each time step to form $\varphi_e(n)$ as shown in Figure 4.22. Closing the feedback path in this manner precludes the algebraic manipulations done in (4.81) prior to employing the Tustin method.

Although the Tustin method is very attractive due to its simplicity, the added unit delay elements that are introduced can be troublesome since we are primarily concerned with feedback systems. Second, we will find that a variable time-step capability will be required in order to deal with the transient response of sampled phase-locked loops, and the unit delay formulations of Figures 4.23 and 4.24 are difficult to modify for this purpose.

The second method, which is only briefly mentioned here, is based on equivalencing each principal block diagram element (e.g., a fourth-order low-pass Chebyshev filter) or the quadratic factors in Figure 4.22 with an RLC circuit that has the desired voltage transfer function. Expressed in circuit form, a number of excellent analysis techniques are available for calculation of the transient response that do not have the aforementioned shortcomings of the Tustin method. This method will be developed in substantial detail in Chapter 8.

4.6 SUMMARY

Through the developments in this chapter, the concept of pseudo-continuous models was introduced. Considerable time was spent on developing models for several discrete loop elements. Although the model results were not mathematically new, the underlying assumptions and their connections with sampled control systems were explicitly discussed so as to justify the design approximations that have traditionally been made in the past without regard to sampling effects.

Techniques for assessing control loop stability and calculating the transient response for a phase-locked loop were also considered. As shown in Chapter 5, complete sampling effects will be analyzed with only minor modifications to the techniques that have been developed in this chapter.

Appendix 4A
Loop Bandwidth Considerations Using the Phase-Frequency Detector

The phase-frequency detector, by virtue of its pulse-like output, is only intended for use in "suitably" narrowband phase-locked loop applications as discussed in the text. The focal point in determining this suitability is the approximation made in (4.35), specifically, the exclusion of the high-frequency alias terms in (4.34). This is as much a function of the pole excess (i.e., number of open-loop gain poles minus number of open-loop gain zeros) in the open-loop gain function as it is of the percentage closed-loop bandwidth. Having a phase-locked loop that has a small $\omega_n T$ factor does not automatically imply that the loop is suitably narrowband in the present context. Equation (4.34) is, in fact, only valid for pole-zero excesses greater than or equal to two.

If the open-loop gain function exhibits only a single attenuation pole for frequencies beyond the unity gain frequency, in general, the $n = \pm 1, \pm 2, \pm 3, \ldots$ terms in (4.34) cannot be neglected, regardless of the loop $\omega_n T$ value. Also, as shown in [12], the phase-frequency detector transfer function must be appended with an additional delay-like term, $e^{-sT/2}$, in the limit as all filtering within the control loop is removed. The appearance of this term has often been blamed on divider delay, but it is solely a result of sampling.

In proper use, the phase-frequency detector should be followed by some form of analog hold function (e.g., a single-pole RC low-pass filter with corner frequency less than F_{ref}/π Hz) that adequately attenuates the high-frequency components present. In this context, no phase correction to the phase detector transfer function [i.e., (4.40)] is needed, and inclusion of any $e^{-sT/2}$ factor is in fact erroneous. This point is worth developing further.

In [12], the appearance of the $e^{-sT/2}$ term was shown based on a simplistic argument that considered a type 1 phase-locked loop with a single-pole low-pass loop filter. The open-loop gain function for this case was given as

$$G_{\text{OL}}(s) = \frac{K_d K_v}{N} \frac{1}{s(1 + s\tau)}$$

where τ was the time constant of the low-pass filter. Sampling effects were then included by transforming this open-loop gain function into z transforms giving

$$G_{OL}(z) = \frac{K_d K_v T}{N} \frac{z(1-A)}{(z-1)(z-A)}$$

where $A = e^{-T/\tau}$. In the next step, the low-pass filter was effectively eliminated by taking $\tau \to 0$, which caused $A \to 0$ and resulted in an open-loop gain function of

$$G_{OL}(z) = \frac{K_d K_v T}{N} \frac{1}{(z-1)}$$

The so-called "small" percentage bandwidth open-loop gain function was then obtained by substituting $e^{j\omega T}$ for z and assuming $\omega T \ll 1$, which gave

$$G_{OL}(\omega) \approx \frac{K_d K_v}{N} \frac{e^{-j\omega T/2}}{j\omega}$$

This simple argument has often been construed as advocating inclusion of the delay-like term wherever the phase-frequency detector is employed. However, its inclusion is only accurate in the limiting case where $\tau/T \to 0$. Otherwise, the "corrective term" when the loop filter is a simple one-pole low-pass filter is instead given by

$$\frac{e^{j\omega T}(1-A)}{e^{j\omega T} - A}$$

which is of nontrivial magnitude and phase. In most if not all cases of practical interest, no phase or gain corrective term is required at all.

Before considering an example, it is interesting to consider the open-loop gain function for the type 1 system in the absence of any loop filter. For this case, the open-loop gain function is

$$G_{OL}(s) = \frac{K_d K_v}{N} \frac{1}{s}$$

which on transforming to z transforms to include sampling effects becomes

$$G_{OL}(z) = \frac{K_d K_v T}{N} \frac{z}{z-1}$$

Carrying out the same approximations as done before, the phase term that appears is $e^{+sT/2}$ rather than $e^{-sT/2}$! Hence, the inclusion of the low-pass filter in the earlier argument really was necessary in order to obtain the exponential delay term we desired. Reiterating, this erroneous result is caused by excluding the low-pass filter all together, which invalidates (4.34) because of insufficient pole excess as discussed earlier.

The important role that this aforementioned analog hold plays in validating the approximation made in (4.35) is best illustrated using a numerical example. The type 2 case that was studied earlier (Table 4.3) clearly shows excellent agreement in the open-loop gain phase argument for $G_{OL}(s)$ and $G_{OL}(z)$. The analog hold low-pass filter in this case had an effective corner frequency of roughly $0.064F_{ref}$. In sharp contrast, large phase angle discrepancies between $G_{OL}(s)$ and $G_{OL}(z)$ occur if the low-pass corner frequency is increased to $5F_{ref}$ as shown in Table 4A.1. Inclusion of the additional $e^{-sT/2}$ factor with $G_{OL}(s)$ is required to eliminate this disparity. In actual practice, the low-pass corner frequency is always made a fraction of the reference frequency in order to realize some suppression of the reference spurs. This example only serves to demonstrate the underlying concepts.

In conclusion, the phase-frequency detector should be used with an analog hold low-pass filter (corner frequency $< F_{ref}/\pi$ Hz) in order to obtain valid results when using the pseudo-continuous models. Inclusion of the additional exponential phase factor for cases that do not obey this guideline is strongly discouraged in favor of the techniques developed in Chapter 5 to deal with sampling effects exactly.

Table 4A.1

Comparison of Pseudo-Continuous and Exact Open-Loop Gain Functions for a Type 2 Third-Order Phase-Locked Loop with Phase-Frequency Detector and Insufficient Analog Hold

Frequency f/F_{ref}	$G_{OL}(s)$			$G_{OL}(z)$	
	Gain (dB)	Phase (deg)	Phase*	Gain (dB)	Phase (deg)
0.000001	187.959	−179.998	−179.998	187.959	−179.998
0.01	28.291	−164.322	−166.122	28.360	−164.454
0.02	17.122	−150.736	−154.336	17.344	−151.583
0.05	4.770	−125.841	−134.841	5.377	−131.428
0.07	1.073	−117.602	−130.202	1.817	−126.835
0.10	−2.502	−110.619	−128.619	−1.611	−125.207
0.15	−6.302	−104.983	−131.983	−5.184	−128.171
0.20	−8.905	−102.317	−138.317	−7.511	−133.885

$\omega_n T = 2\pi 0.05$
$\zeta = 0.707$
$R_1 C_1 = 0.1273T$

*The $e^{-sT/2}$ term is included with $G_{OL}(s)$.

Transient Response Calculations for Continuous Systems

Ross Method

Throughout the remainder of this appendix we will assume that the Laplace transform of the transient error response has been calculated and is available in a rational polynomial form as described by (4.81),

$$\varphi_e(s) = \frac{a_{m-1}s^{m-1} + a_{m-2}s^{m-2} + \cdots + a_1 s + a_0}{b_m s^m + b_{m-1}s^{m-1} + \cdots + b_1 s + b_0} \quad (4A.1)$$

The first transform inversion technique we will examine is that of Ross [21].

The rational Laplace transform in (4A.1) is equivalent to a differential equation with specific initial conditions. To support this claim, consider the linear m'th-order differential equation

$$b_m \frac{d^m}{dt^m}\varphi_e = -b_0 \varphi_e - b_1 \frac{d\varphi_e}{dt} - b_2 \frac{d^2 \varphi_e}{dt^2} - \cdots$$

$$- b_{m-1} \frac{d^{m-1}\varphi_e}{dt^{m-1}} \quad (4A.2)$$

Transforming (4A.2) into Laplace transforms,

$$\varphi_e(b_m s^m + b_{m-1}s^{m-1} + \cdots + b_1 s + b_0) = b_1 \varphi_0$$
$$+ b_2[s\varphi_e(0) + \varphi_e'(0)] \quad (4A.3)$$
$$+ b_3[s^2 \varphi_e(0) + s\varphi_e'(0) + \varphi_e''(0)] + \cdots$$
$$+ b_m[s^{m-1}\varphi_e(0) + s^{m-2}\varphi_e'(0) + \cdots + \varphi_e^{m-1}(0)]$$

where $\varphi_e^k(t)$ denotes the k'th-order time derivative evaluated at t. Collecting like terms in (4A.3) and solving for the a_i coefficients in (4A.1),

$$a_i = \sum_{j=i+1}^{m} b_j \varphi_e^{j-i-1}(0) \quad \text{for } 0 \le i \le m-1 \quad (4A.4)$$

which may be inverted, thereby allowing a recursive solution for the value of each time derivative as

$$\varphi_e^i(0) = \frac{1}{b_m}\left[a_{m-i-1} - \sum_{j=0}^{i-1} b_{m-i+j}\varphi_e^j(0)\right] \quad \text{for } 1 \leq i \leq m-1 \quad (4A.5)$$

and $\varphi_e(0) = a_{m-1}/b_m$. Given the characteristic equation from (4A.3) and the initial conditions from (4A.5), any of a large number of techniques could be used to integrate this differential equation. Ross employed a particularly simple method, which we develop now.

The $m-1$ derivatives at time zero are available from (4A.5) and may be used to form a Taylor series expansion. This makes it possible to calculate φ_e at a new time instant $t' = t + T$ as

$$\varphi_e(t') = \varphi_e(t) + \sum_{j=1}^{m-1}\left[\varphi_e^j(t)\frac{T^j}{j!}\right] \quad (4A.6)$$

This process may be iterated for additional time steps, thereby evaluating the complete system transient response. Following each time iteration, the derivative values are recalculated in preparation for the next time-step iteration. Additional derivatives may be included in (4A.6) for improved accuracy by further differentiating (4A.2) giving

$$\varphi_e^{m+i}(t) = \frac{-1}{b_m}\sum_{j=0}^{m-1} b_j \varphi_e^{i+j}(t) \quad \text{for } i = 0, 1, \ldots, v \quad (4A.7)$$

where v is the number of additional derivatives desired beyond the m'th order. The derivatives are easily recalculated at each time increment using

$$\varphi_e^i(t) = \sum_{j=0}^{m+v-i} \varphi_e^{i+j}(t)\frac{T^j}{j!} \quad \text{for } 1 \leq i \leq m+v-1 \quad (4A.8)$$

and using (4A.7) for the $(m+v)$'th derivative.

A Turbo Pascal computer program is provided on disk for calculating the time-domain transient response using the Ross method. Since the parameter T is raised to a large power in this algorithm, the Laplace transform that is being inverted should first be frequency-scaled such that T is in the range 0.01 to 0.5 in order to avoid numerical difficulties. An example calculation using the Ross method is shown in Table 4A.2. Finally, note that a more recent publication dealing with the use of time moments for calculating the system response is available [30].

Table 4A.2
Ross Method of Laplace Transform Inversion

$$\text{Example: } \varphi_e(S) = \frac{s^2 + 3}{s^4 + s^3 + 2s^2 + 5s + 10}$$

Time	φ_e
0.00000	0.00000000
0.01000	0.00995033
0.02000	0.01980263
0.03000	0.02955883
0.04000	0.03922080
0.05000	0.04879035
0.06000	0.05826927
0.07000	0.06765928
0.08000	0.07696203
0.09000	0.08617914
0.10000	0.09531219
0.11000	0.10436266
0.12000	0.11333203
0.13000	0.12222169
0.14000	0.13103299
0.15000	0.13976722

$v = 10, T = 0.01$

Corrington Method

The second inversion method we will examine is that of Corrington [22]. It is interesting from the perspective that it clearly relates the time-sampled linear system transient response with a finite difference equation and an associated z transform.

Starting again from (4A.1), the denominator polynomial may be repeatedly divided into the numerator polynomial resulting in

$$\varphi_e(s) = \sum_{r=0}^{\infty} \frac{C_r}{s^{r+1}} \qquad (4A.9)$$

and on taking the inverse Laplace transform,

$$\varphi_e(t) = \sum_{r=0}^{\infty} C_r \frac{t^r}{r!} \quad \text{for } t \geq 0 \qquad (4A.10)$$

Once again, frequency scaling of the original Laplace transform is necessary in order to avoid numerical problems in evaluating (4A.10) for even a short time period.

Corrington showed that $\varphi_e(t)$ can be expressed as a linear homogeneous difference equation, which greatly simplifies the calculation of $\varphi_e(t)$ for $t \geq (2m - 1)T$. More specifically,
Corrington showed that

$$\varphi_e(t) = \sum_{i=1}^{n} (-1)^{i+1} F_{n,i} \varphi_e(t - iT) \quad \text{for } t = kT, k > 2m - 1 \quad (4A.11)$$

To calculate the $F_{n,i}$ coefficients, (4A.10) must be calculated at $2m - 1$ equally spaced time intervals of T and the simultaneous equations which result from (4A.11) must be solved. For a third-order system, the problem statement would take the form of

$$\begin{vmatrix} \varphi_e(3) & \varphi_e(2) & \varphi_e(1) \\ \varphi_e(4) & \varphi_e(3) & \varphi_e(2) \\ \varphi_e(5) & \varphi_e(4) & \varphi_e(3) \end{vmatrix} \begin{vmatrix} F_{3,1} \\ -F_{3,2} \\ F_{3,3} \end{vmatrix} = \begin{vmatrix} \varphi_e(4) \\ \varphi_e(5) \\ \varphi_e(6) \end{vmatrix} \quad (4A.12)$$

where $T = 1$ has been assumed.

Given the $2m - 1$ initial transient error sample values from (4A.10), and the linear homogeneous difference equation (4A.11), the z transform of the transient response may be expanded using the basic definition

$$\varphi_e(z) = \sum_{n=0}^{\infty} \varphi_e(nT) z^{-n} \quad (4A.13)$$

Equivalently,

$$\varphi_e(z) = \frac{N(z)}{D(z)} \quad (4A.14)$$

where $N(z)$ and $D(z)$ are real coefficient polynomials in z. Corrington showed that the coefficients of $D(z)$ are given by the $F_{n,i}$ taken with alternating signs. Using this fact, the coefficients of $N(z)$ may be obtained from (4A.13) and (4A.14) by equating the first m polynomial coefficients of

$$N(z) = D(z) \sum_{n=0}^{\infty} \varphi_e(nT) z^{-n} \quad (4A.15)$$

Corrington develops these final points in much greater detail in [22].

A sample computation using the Corrington method is shown in Table 4A.3. A computer program written in Turbo Pascal that utilizes the Corrington method for transform inversion is provided on disk.

The z transform representation of the transient response is a natural by-product of the Corrington method just discussed. An alternative method for obtaining a z transform representation for the response (close approximation) is by direct substitution of the bilinear transformation relating s to z (4.83) into the Laplace transform representing the continuous system response. The Turbo Pascal program provided on disk performs this computation based on the method described in [29]. It is interesting to compare the z transform obtained using the Corrington method to that obtained using the bilinear transform method. Results from the bilinear transformation method are provided in Table 4A.4 in order to facilitate such a comparison for the same example case used in creating Table 4A.3. Note that although the denominator polynomials in z match very closely between the Corrington and bilinear methods, the numerator polynomials are quite different. Even so, the time-domain responses are almost identical since the simulation time increment selected is small compared to the system dynamics.

Table 4A.3
Corrington Method of Laplace Transform Inversion

$$\text{Example: } \varphi_e(s) = \frac{s^2 + 3}{s^4 + s^3 + 2s^2 + 5s + 10}$$

$$\varphi_e(z) = \frac{\sum_{i=0}^{3} n_i z^i}{\sum_{i=0}^{4} d_i z^i}$$

Time	φ_e		
0.00000	0.0000000		
0.01000	0.0095033124669		
0.02000	0.0198026332279	$n_0 =$	0.0
0.03000	0.0295588304542	$n_1 =$	0.0099502910
0.04000	0.0392207966663	$n_2 =$	-0.0198976395
0.05000	0.0487903544704	$n_3 =$	0.0099503312
0.06000	0.0582692749868		
0.07000	0.0676592775133	$d_0 =$	0.990045665
0.08000	0.0769620292015	$d_1 =$	-3.969940637
0.09000	0.0861791447486	$d_2 =$	5.969739405
0.10000	0.0953121861017	$d_3 =$	-3.989844334
0.11000	0.1043626621774	$d_4 =$	1.0
0.12000	0.1133320285944		
0.13000	0.1222216874214		

Table 4A.4
Bilinear Transform Method of Laplace Transform Inversion

$$\text{Example: } \varphi_e(s) = \frac{s^2 + 3}{s^4 + s^3 + 2s^2 + 5s + 10}$$

$$\varphi_e(z) = \frac{\sum_{i=0}^{3} n_i z^i}{\sum_{i=0}^{4} d_i z^i}$$

Time	φ_e		
0.00000	0.00248762		
0.01000	0.00992598		
0.02000	0.01977892	$n_0 =$	0.002487623431
0.03000	0.02953573	$n_1 =$	0.000000746231
0.04000	0.03919829	$n_2 =$	−0.004973754401
0.05000	0.04876842	$n_3 =$	0.000000746231
0.06000	0.05824789	$n_4 =$	0.002487623431
0.07000	0.06763842		
0.08000	0.07694168	$d_0 =$	0.990049008787
0.09000	0.08615928	$d_1 =$	−3.969950469100
0.10000	0.09529278	$d_2 =$	5.969749036462
0.11000	0.10434369	$d_3 =$	−3.989847470000
0.12000	0.11331347	$d_4 =$	1.000000000000
0.13000	0.12220353		

Classical Residue Method for Transform Inversion

The final transient response method we will examine is the classical residue method. This method basically consists of three steps:

1. Calculate the characteristic equation roots.
2. Expand the rational Laplace transform in partial fractions.
3. Compute the inverse Laplace transform of each partial fraction term.

Only the first two steps of this process are examined here.

Calculating Characteristic Equation Roots

Assume that the characteristic equation of interest is given by the denominator of (4A.1),

$$f(s) = b_m s^m + b_{m-1} s^{m-1} + \cdots + b_1 s + b_0 \tag{4A.16}$$

It will be assumed that the polynomial has been suitably scaled in frequency in order to prevent numerical problems. In the opinion of the author, the Moore root finding method [25] is best suited to this present application. This method is also known as the complex Newton-Raphson method by some.

Assume now that the initial estimate for a root of (4A.16) is $x + jy$ and that $f(x + jy) = u + jv$. The solution approach is based on minimizing the error function

$$F(x + jy) = |f(x + jy)|^2 \qquad (4A.17)$$

Taking the partial derivatives of (4A.17) with respect to x and y respectively,

$$\frac{\partial F}{\partial x} = 2\left(u\frac{\partial u}{\partial x} + v\frac{\partial v}{\partial x}\right) \qquad (4A.18)$$

and

$$\frac{\partial F}{\partial y} = 2\left(u\frac{\partial u}{\partial y} + v\frac{\partial v}{\partial y}\right) \qquad (4A.19)$$

Since $f(z)$ is an analytic function of $z = x + jy$, its real and imaginary parts, u and v, respectively, obey the Cauchy-Riemann differential equations, which state that

$$\frac{\partial u}{\partial x} = \frac{\partial v}{\partial y} \qquad (4A.20)$$

and

$$\frac{\partial u}{\partial y} = -\frac{\partial v}{\partial x} \qquad (4A.21)$$

Using this fact, (4A.20) and (4A.21) may be substituted into (4A.19) to obtain

$$\frac{\partial F}{\partial y} = 2\left(-u\frac{\partial v}{\partial x} + v\frac{\partial u}{\partial x}\right) \qquad (4A.22)$$

which means that only two partial derivatives must be evaluated rather than four. The improved estimate of the true root location following [25] is then given by

$$x' = x - \frac{1}{2}\frac{\frac{\partial F}{\partial x}}{|f'|^2} \qquad (4A.23)$$

$$y' = y - \frac{1}{2}\frac{\frac{\partial F}{\partial y}}{|f'|^2} \qquad (4A.24)$$

where $f(z)$ and $f'(z)$ are most easily evaluated using synthetic division. Once a root location has been identified that meets a user-specified error criterion for F, the root is factored out of the polynomial using synthetic division.

A Turbo Pascal computer program listing is provided on disk that utilizes the Moore method for computing the roots of a given characteristic equation. The program was initially based on a translation from [25], but since that time, enhancements have been made to the initial root estimate for each root extraction iteration utilizing several concepts from [31].

Partial Fraction Expansion

Once the characteristic roots have been found, the rational polynomial may be expanded into a series of partial fractions that may be easily inverse-transformed into the time domain. A number of authors have provided program listings that perform this task in the case of distinct roots. Equation (A4.25) from [25] demonstrates a method that is also effective for cases involving repeated roots. The interested reader is referred to [29] for further details.

$$\begin{aligned}
P(x) &= \frac{2x^5 + 9x^4 - x^3 - 26x^2 + 5x - 1}{(x+1)^2(x-1)^3(x-2)} \\
&= \frac{1}{(x+1)(x-1)^3(x+2)}\left(2x^4 + 7x^3 - 8x^2 - 18x + 23 - \frac{24}{x+1}\right) \\
&= \frac{1}{(x-1)^3(x+2)}\left\{2x^3 + 5x^2 - 13x - 5 + \frac{1}{x+1}\left[28 - \frac{24}{(x+1)^2}\right]\right\} \\
&= \frac{1}{(x-1)^2(x+2)}\left\{2x^2 + 7x - 6 + \frac{1}{x-1}\left[-11 + \frac{28}{x+1} - \frac{24}{(x+1)^2}\right]\right\} \\
&= \frac{1}{(x-1)(x+2)}\left\{2x + 9 + \frac{1}{x-1}\left[2 + \frac{-11+8}{x-1} + \frac{-8}{x+1} + \frac{12}{(x+1)^2}\right]\right\} \\
&= \frac{1}{x+2}\left\{2 + \frac{1}{x-1}\left[11 + \frac{3-1}{x-1} + \frac{-3}{(x-1)^2} + \frac{1}{x+1} + \frac{-6}{(x+1)^2}\right]\right\} \\
&= \frac{1}{x+2}\left[2 + \frac{11-1}{x-1} + \frac{2}{(x-1)^2} + \frac{-3}{(x-1)^3} + \frac{1}{x+1} + \frac{3}{(x+1)^2}\right] \\
&= \frac{1}{x+2} + \frac{3}{x-1} + \frac{1}{(x-1)^2} + \frac{-1}{(x-1)^3} + \frac{-2}{x+1} + \frac{3}{(x+1)^2} \qquad (4A.25)
\end{aligned}$$

REFERENCES

[1] Manassewitsch, V., *Frequency Synthesizers Theory and Design*, 2nd ed., John Wiley and Sons, New York, 1980, pp. 502–503.
[2] Snyder, D.L., "The State Variable Approach to Analog Communication Theory," *IEEE Trans. Info. Theory*, Vol. IT-14, No.1, January 1968.
[3] Lindsey, W.C., *Synchronization Systems in Communication and Control*, Prentice-Hall, Englewood Cliffs, NJ, 1972.
[4] Crawford, J.A., "The Phase-Locked Loop Concept for Frequency Synthesis," *M/ACOM Linkabit*, February 1987.
[5] Gardner, F.M., *Phaselock Techniques*, 2nd ed., John Wiley and Sons, New York, 1979, pp. 11, 75.
[6] Viterbi, A.J., *Principles of Coherent Communication*, McGraw-Hill, New York, 1966, pp. 45–75.
[7] Clarke, K.K., D.T. Hess, *Communication Circuits*, Addison-Wesley, Reading, MA, 1971, Chap. 6.
[8] Blanchard, A., *Phased-Locked Loops Application to Coherent Receiver Design*, John Wiley and Sons, New York, 1976.
[9] Houpis, C.H., et al., "Refined Design Method for Sampled-Data Control Systems: The Pseudo-Continuous Time (PCT) Control System Design," *IEE Proc.*, Vol. 132, Part D, No. 2, March 1985, pp. 69–74.
[10] Halgren, R.C., et al., "Improved Acquisition in Phase-Locked Loops with Sawtooth Phase Detectors," *IEEE Trans. Comm.*, Vol. COM-30, No. 10, October 1982, pp. 2364–2375.
[11] Egan, W.F., *Frequency Synthesis by Phase Lock*, John Wiley and Sons, New York, 1981, pp. 115–123, 126–129.
[12] Crawford, J., "The Phase/Frequency Detector," *RF Design*, February 1985, pp. 46–57.
[13] Kuo, B.C., *Digital Control Systems*, Holt, Rinehart, and Winston, New York, 1980, p. 48.
[14] Kreyszig, E., *Advanced Engineering Mathematics*, 3rd ed., John Wiley and Sons, New York, 1972, p. 604.
[15] Franklin, G.F., J.D. Powell, *Digital Control of Dynamic Systems*, Addison-Wesley, Reading, MA, 1980, p. 86.
[16] Marshall, J.E., *Control of Time-Delay Systems*, Peter Peregrinus, London, Chap. 3.
[17] Kuo, B.C. *Automatic Control Systems*, 3rd ed., Prentice-Hall, Englewood Cliffs, NJ, 1975, pp. 316–374, 434–444.
[18] Gibson, J.A., et al., "Transfer Function Models of Sampled Systems," *IEE Proc.*, Vol. 130, Part G, No. 2, April 1983, pp. 37–44.
[19] Horiguchi, K., N. Hamada, "System Theoretical Considerations on N-Point Padé Approximation," *Electron. Comm. Japan*, Vol. 69, Part 1, No. 4, 1986, pp. 10–20.
[20] Abramovitch, D.Y., "Analysis and Design of a Third-Order Phase-Locked Loop," Milcom 1988, pp. 455–459.
[21] Ross, R.I., "Evaluating the Transient Response of a Network Function," *Proc. IEEE*, May 1967, pp. 693–694.
[22] Corrington, M.S.,"Simplified Calculation of Transient Response," *Proc. IEEE*, March 1965, pp. 287–292.
[23] Vlach, J., *Computerized Approximation and Synthesis of Linear Networks*, pp. 106–112.
[24] Chua, L.O., P. Lin, *Computer-Aided Analysis of Electronic Circuits: Algorithms & Computational Techniques*, Prentice-Hall, Englewood Cliffs, NJ, 1975, Chap. 12.
[25] Cuthbert, T.R., *Circuit Design Using Personal Computers*, John Wiley and Sons, New York, 1983, Chap. 3.

[26] Henderson, K.W., W.H. Kautz, "Transient Response of Conventional Filters," *IRE Trans. Circuit Theory*, December 1958, pp. 333–347.
[27] Jagerman, D.L., "An Inversion Technique for the Laplace Transform," *Bell Syst. Tech. J.*, Vol. 61, No. 8, October 1982, pp. 1995–2003.
[28] Smith, J.M., *Mathematical Modeling and Digital Simulation for Engineers and Scientists*, 2nd ed., John Wiley and Sons, New York, 1987, App. A.
[29] Malvar, H.S., "Transform Analog Filters into Digital Equivalents," *Electron. Design*, April 30, 1981, pp. 145–148.
[30] Holder, M.E., V.A. Thomason, "Using Time Moments to Determine System Response," *IEEE Trans. Circuits Syst.*, Vol. CAS-35, No. 9, September 1988, pp. 1193–1195.
[31] Waldhauer, F.D., *Feedback*, John Wiley and Sons, New York, 1982.

SELECTED BIBLIOGRAPHY

Blake, J., "Design of Practical Wideband Type 2 Second-Order Phase-Locked Loop Frequency Synthesizers," RF Expo, 1989, INS EEA 89–000897, pp. 181–193.

Chin, F.Y., K. Steiglitz, *IEEE Trans. Circuits Syst.*, Vol. CAS-24, No. 1, January 1979.

Chinichiaan, M., C.T. Fulton, "Software for Closed-Form Solution of Linear Time-Invariant Differential Systems," *IEEE Circuits Dev.*, May 1986, pp. 25–32.

Egan, W.F., "Sampling Delay—Is It Real?," *RF Design*, February 1991, pp. 114–116.

El-Hawary, M.E., *Control System Engineering*, Reston Publishing, Reston, VA, 1984, Chaps. 4, 8.

Gardner, F.M., "A Transformation for Digital Simulation of Analog Filters," *IEEE Trans. Comm.*, Vol. COM-34, No. 7, July 1986, pp. 676–680.

Gardner, F.M., "Charge-Pump Phase-Locked Loops," *IEEE Trans. Comm.*, Vol. COM-28, No. 11, November 1980, pp. 1849–1858.

Hein, J.P., "z-Domain Model for Discrete-Time PLLs," *IEEE Trans. Circuits Syst.*, Vol. 35, No. 11, November 1988, pp. 1393–1400.

Kovacs, J., "Analyze PLLs With Discrete Time Modeling," *Microwaves & RF*, May 1991, pp. 224–229.

Maisel, J.E., "Bilinear Transform Quickly Converts Analog Systems Into Digital Domain," *Personal Eng. Instrum. News*, February 1991, pp. 63–65.

Orchard, H.J., "The Laguerre Method for Finding the Zeros of Polynomials," *IEEE Trans. Circuits Syst.*, Vol. 36, No. 11, November 1989, pp. 1377–1381.

Schneider, A.M., et al., "Higher Order s-to-z Mapping Functions and Their Application to Digitizing Continuous-Time Filters," *Proc. IEEE*, Vol. 79, No. 11, November 1991, pp. 1661–1674.

Schreiber, H.H., "Digital Simulation of Analog Systems for a Restricted Class of Inputs," *IEEE Circuits Syst.*, December 1982, pp. 4–9.

Chapter 5
Frequency Synthesis Using Sampled-Data Control Systems

Sampled-data phase-locked loops have in principle played a significant role in frequency synthesis ever since digital elements (e.g., digital dividers, phase detectors) began supplementing the analog design process [1]. Nevertheless, classical control theory design methods have remained the most prevalent even today. The continued success of these methods for the relatively small percentage bandwidth systems was demonstrated in Chapter 4 where these techniques were developed for pseudo-continuous system modeling. In this chapter, the system bandwidth constraint is removed and the sampling effects that result are considered in some detail. Once the theoretical groundwork for sampled-data phase-locked loops has been discussed, several basic design cases of general interest are examined.

5.1 SAMPLED-DATA CONTROL SYSTEM BASICS

A block diagram for a general phase-locked loop that utilizes a digital feedback divider and discrete phase detector is shown in Figure 5.1. The power spectral density $S_r(\omega)$ represents a baseband reference noise source, whereas $S_v(\omega)$ represents an oscillator bandpass phase noise source, both of which are considered in Sec. 5.6.

For the time being, only the deterministic sources in Figure 5.1 will be considered. Following the discussions in Chapter 4, the sampled phase error process $\varphi_e^*(t)$ may be written in terms of Laplace transforms as

$$\varphi_e^*(s) = \left[\varphi_r(s) - \frac{\varphi_o(s)}{N} \right]^* \quad (5.1)$$

where the asterisk denotes ideal impulse sampling in the time domain. In many circumstances, $\varphi_r(t)$ will be a discrete digital input itself, in which case it would already

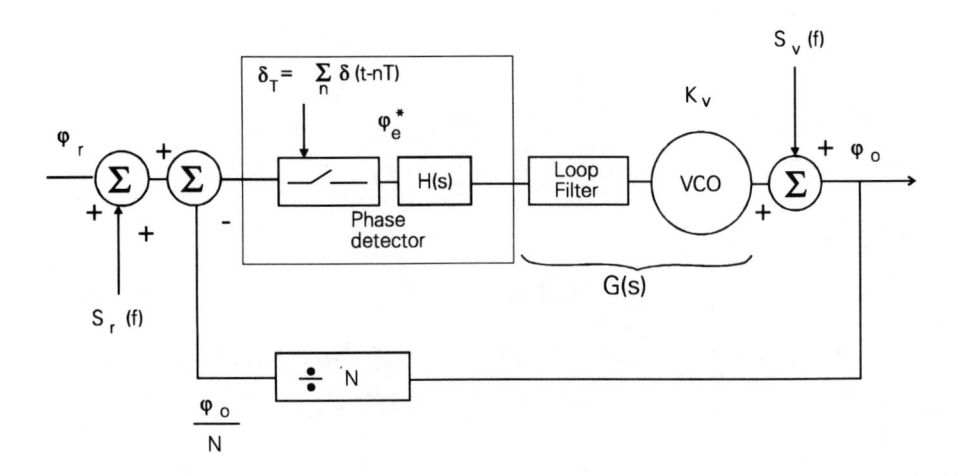

Figure 5.1 Sampled-data phase-locked loop including principal noise sources at the reference and voltage-controlled oscillator (VCO) nodes.

be in a sampled-data form [i.e., $\varphi_r^*(s)$]. As justified momentarily, this will not affect the final results of this derivation. The feedback divider is assumed to contribute only a factor of N^{-1} to the open-loop gain function based on the discussions in Chapter 4.

Similarly, the Laplace transform of the loop output phase may be written as

$$\varphi_o(s) = \varphi_e^*(s)H(s)G(s) \qquad (5.2)$$

On substitution into (5.1), the sampled phase error transform is given by

$$\varphi_e^*(s) = \left[\varphi_r(s) - \frac{\varphi_e^*(s)H(s)G(s)}{N}\right]^* \qquad (5.3)$$

Applying a basic property of sampled-data systems, Eq. (4.17) of [2], the sampling operation may be moved within the parentheses and brackets in (5.3) to give

$$\varphi_e^*(s) = \left[\varphi_r^*(s) - \frac{\varphi_e^*(s)HG^*(s)}{N}\right] \qquad (5.4)$$

This is equivalent to stating that resampling an already sampled waveform with the same clock is a transparent operation. Therefore, since

$$[\varphi_e^*(s)]^* = \varphi_e^*(s)$$

the fact that $\varphi_r(t)$ may have initially been discrete in nature rather than continuous is of no consequence. Finally, with minor rearrangements to (5.4), the final transfer function results are obtained as

$$\varphi_e^*(s) = \frac{\varphi_r^*(s)}{1 + \frac{HG^*(s)}{N}} \tag{5.5}$$

$$\varphi_o(s) = \frac{H(s)G(s)}{1 + \frac{HG^*(s)}{N}} \varphi_r^*(s) \tag{5.6}$$

where

$$HG^*(s) = [H(s)G(s)]^* \tag{5.7}$$

These last three equations will be utilized throughout the remainder of this chapter. It is significant to recognize that the sampling operation primarily affects the denominator factors in (5.5) and (5.6). As such, sampling unquestionably enters into matters concerning system stability and transient response behavior.

5.1.1 Frequency-Domain Representation for Sampled Signals

Before continuing, it is advantageous to develop the relationship between the frequency-domain representation for $f(t)$ and its time-sampled counterpart $f^*(t)$. This relationship will ultimately allow these last results to be expressed concisely by means of z transforms.

Assume that $F(s)$ is the Laplace transform of $f(t)$ and that $f(t) = 0$ for $t \leq 0$. The ideal impulse sample sequence $\delta_T(t)$ in Fig. 5.1 may be written as

$$\delta_T(t) = \sum_{n=-\infty}^{\infty} \delta(t - nT) \tag{5.8}$$

where T is the time interval between impulses ($T^{-1} = F_s$) and $\delta(t)$ is the Dirac delta function. If the signal $f(t)$ is now ideally impulse sampled, this is equivalent to multiplying $f(t)$ by $\delta_T(t)$. This product may be mathematically written as

$$f^*(t) = f(t)\delta_T(t) = \sum_{n=-\infty}^{\infty} f(t)\delta(t - nT) \tag{5.9}$$

Adopting the approach taken in [2], the impulse sampling sequence may be represented equivalently as a Fourier series

$$\delta_T(t) = \sum_{n=-\infty}^{\infty} C_n e^{j\omega_s nt} \tag{5.10}$$

where $\omega_s = 2\pi/T$ and

$$C_n = \frac{1}{T} \int_{-T/2}^{T/2} \delta(t) e^{-j\omega t} dt$$

$$= \frac{1}{T} \tag{5.11}$$

If (5.10) is now substituted for $\delta_T(t)$ in (5.9) and the Laplace transform taken,

$$F^*(s) = \mathscr{L}[f^*(t)]$$

$$= \int_0^\infty f(t) \sum_{n=-\infty}^{\infty} \frac{1}{T} e^{jn\omega_s t} e^{-st} dt$$

$$= \frac{1}{T} \sum_{n=-\infty}^{\infty} \int_0^\infty f(t) e^{-(s-jn\omega_s)t} dt$$

$$= \frac{1}{T} \sum_{n=-\infty}^{\infty} F(s - jn\omega_s) \tag{5.12}$$

This fundamentally important relationship reveals how the frequency-domain description $F(s)$ is altered by the introduction of ideal sampling. An alternative derivation of this result [3] reveals that $F(s)$ must have a pole-zero excess greater than or equal to 2 in order for (5.12) to be valid. This criterion, which will be met in any practical system we may wish to study, is equivalent to disallowing any step discontinuities in $f(t)$ at $t = 0$ as was initially assumed, which is very nonrestrictive for practical phase-locked loop situations.

This section concludes by making use of (5.12) to express the earlier results (5.5) and (5.6), respectively, as

$$\frac{\varphi_e^*(s)}{\varphi_r^*(s)} = \frac{1}{1 + \dfrac{1}{NT} \sum_{n=-\infty}^{\infty} H(s - jn\omega_s) G(s - jn\omega_s)} \tag{5.13}$$

$$\frac{\varphi_o(s)}{\varphi_r^*(s)} = \frac{H(s)G(s)}{1 + \dfrac{1}{NT} \sum_{n=-\infty}^{\infty} H(s - jn\omega_s)G(s - jn\omega_s)} \tag{5.14}$$

where

$$\frac{1}{N} HG^*(s) = G_{\text{OL}}^*(s)$$

$$= \frac{1}{NT} \sum_{n=-\infty}^{\infty} H(s - jn\omega_s)G(s - jn\omega_s) \tag{5.15}$$

These relationships provide a means for calculating the frequency-domain sampled-data transfer functions with only knowledge of the classical continuous transfer functions and also, as shown in Sec. 5.3, a convenient means for assessing system stability as well.

5.1.2 $G_{\text{OL}}^*(s)$ and Its Relationship to z Transforms

The results obtained thus far may be expressed more concisely if z transforms are introduced. Taking Laplace transforms of both sides of (5.9),

$$\mathcal{L}[f^*(t)] = \int_0^\infty \sum_{n=-\infty}^{\infty} f(t)\delta(t - nT)e^{-st}\,dt$$

$$= \sum_{n=0}^{\infty} \int_0^\infty f(t)\delta(t - nT)e^{-st}\,dt$$

$$= \sum_{n=0}^{\infty} f(nT)e^{-snT}$$

$$= F(z) \tag{5.16}$$

which is of course the z transform of $f(t)$ where $z = e^{sT}$. This outcome may be used to replace the infinite summations in (5.13) and (5.14) with their equivalent z transforms, thereby simplifying these important results further.

As a matter for future reference, the forward and inverse z transform operations are given, respectively, by

$$F(z) = \sum_{n=0}^{\infty} f(nT)z^{-n} \tag{5.17a}$$

$$f(kT) = \frac{1}{2\pi j} \oint F(z) z^{k-1} dz \quad \text{for} \quad k \geq 0 \tag{5.17b}$$

Systems that have nonintegral internal time delays may be modeled using modified z transforms [3], which have a slightly altered form compared to (5.17a) and (5.17b). A short table of transforms is provided in Table 5.1.

Table 5.1
Table of Commonly Used z Transforms

Laplace Transform	Time Function $f(t)\ t>0$	z transform $F(z)$	Modified z transform $F(z, m)$
1	$\delta(t)$	1	0
e^{-kTs}	$\delta(t - kT)$	z^{-k}	z^{-k-1+m}
$\dfrac{1}{s}$	$u_s(t)$	$\dfrac{z}{z-1}$	$\dfrac{1}{z-1}$
$\dfrac{1}{s^2}$	t	$\dfrac{Tz}{(z-1)^2}$	$\dfrac{mT}{z-1} + \dfrac{T}{(z-1)^2}$
$\dfrac{2}{s^3}$	t^2	$\dfrac{T^2 z(z+1)}{(z-1)^3}$	$T^2 \dfrac{m^2 z^2 + (2m - 2m^2 + 1)z + (m-1)^2}{(z-1)^3}$
$\dfrac{1}{s+a}$	e^{-at}	$\dfrac{z}{z - e^{-aT}}$	$\dfrac{e^{-amT}}{z - e^{-aT}}$
$\dfrac{a}{s(s+a)}$	$1 - e^{-at}$	$\dfrac{z(1 - e^{-aT})}{(z-1)(z - e^{-aT})}$	$\dfrac{(1 - e^{-aT})z + (e^{-amT} - e^{-aT})}{(z-1)(z - e^{-aT})}$
$\dfrac{1}{(s+a)(s+b)}$	$\dfrac{1}{(b-a)}(e^{-at} - e^{-bt})$	$\dfrac{1}{(b-a)}\left[\dfrac{z}{z - e^{-aT}} - \dfrac{z}{z - e^{-bT}}\right]$	$\dfrac{1}{(b-a)}\left[\dfrac{e^{-amT}}{z - e^{-aT}} - \dfrac{e^{-bmT}}{z - e^{-bT}}\right]$

("Appendix A" from *Digital Control Systems*, Second Edition, by Benjamin Kuo, copyright © 1992 by Saunders College Publishing, reprinted by permission of the publisher [4].)

5.2 FREQUENCY-DOMAIN ANALYSIS

The infinite summation form for the open-loop gain (5.15) permits sampling effects to be included with no additional information beyond that needed for pseudo-continuous analysis methods. Since $G_{OL}(s) = H(s)G(s)$ must be inherently band-limited, generally the infinite summation in (5.15) may be truncated to a sum over $-k \leq n \leq k$ where k is typically < 5 and ω is restricted to the primary range $|\omega| \leq \pi/T$. As shown in Figure 5.2, ideal sampling causes $G^*_{OL}(\omega)$ to be composed of an infinite number of $G_{OL}(\omega)$ aliases each centered about an integer multiple of the sampling frequency. The quantity $G^*_{OL}(s)$ will then normally be approximated in numerical analysis using a truncated sum as

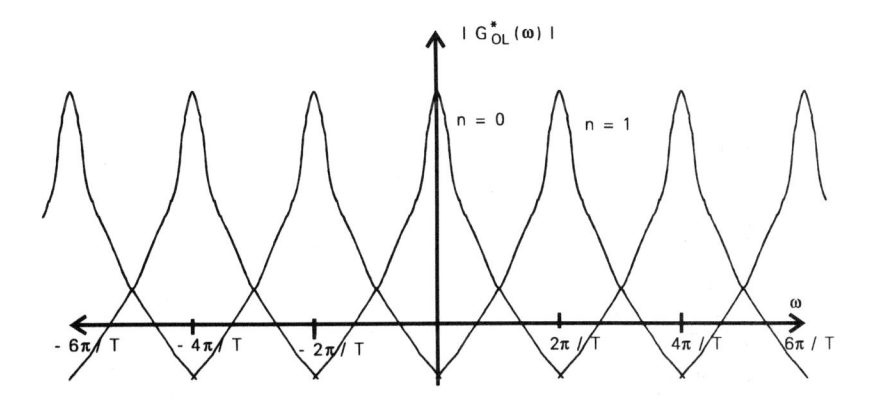

Figure 5.2 The sampled open-loop gain function consists of an infinite number of aliased copies of $G_{OL}(\ )$ aliased about multiples of the sampling frequency F_s. Source: [1] reprinted with permission.

$$G_{OL}^*(s) \approx \frac{1}{T} \sum_{n=-k}^{k} G_{OL}(s + jn\omega_s) \qquad (5.18)$$

where k is chosen to meet a user-specified error criterion [3].

5.3 STABILITY

The stability margins predicted by means of sampled-data control theory (i.e., gain and phase margin) are, in general, smaller than the stability margins calculated using pseudo-continuous control theory. As shown in Figure 5.2, the effect of aliasing increases the effective transfer function gain over the normal continuous system gain (i.e., $n = 0$ summation term only) thereby reducing the stability margins. This is most pronounced near $\omega = \pi/T$ in Figure 5.2 where the $n = 0$ and $n = 1$ summation terms both have the same amplitude and, as it turns out, equal phase arguments.

Due to the periodic nature of $G_{OL}^*(\omega)$ as evidenced by (5.12) and (5.16), the open-loop frequency response may be completely characterized by evaluating (5.18) over the principal frequency range $|\omega| < \pi T^{-1}$ where

$$G_{OL}^*(\omega) = \frac{1}{T} \sum_{n=-k}^{k} G_{OL}(\omega + n\omega_s) \qquad (5.19)$$

and k is chosen to meet a user-specified error criterion. From this information, the stability of the system may be assessed by using standard Bode plot methods. This

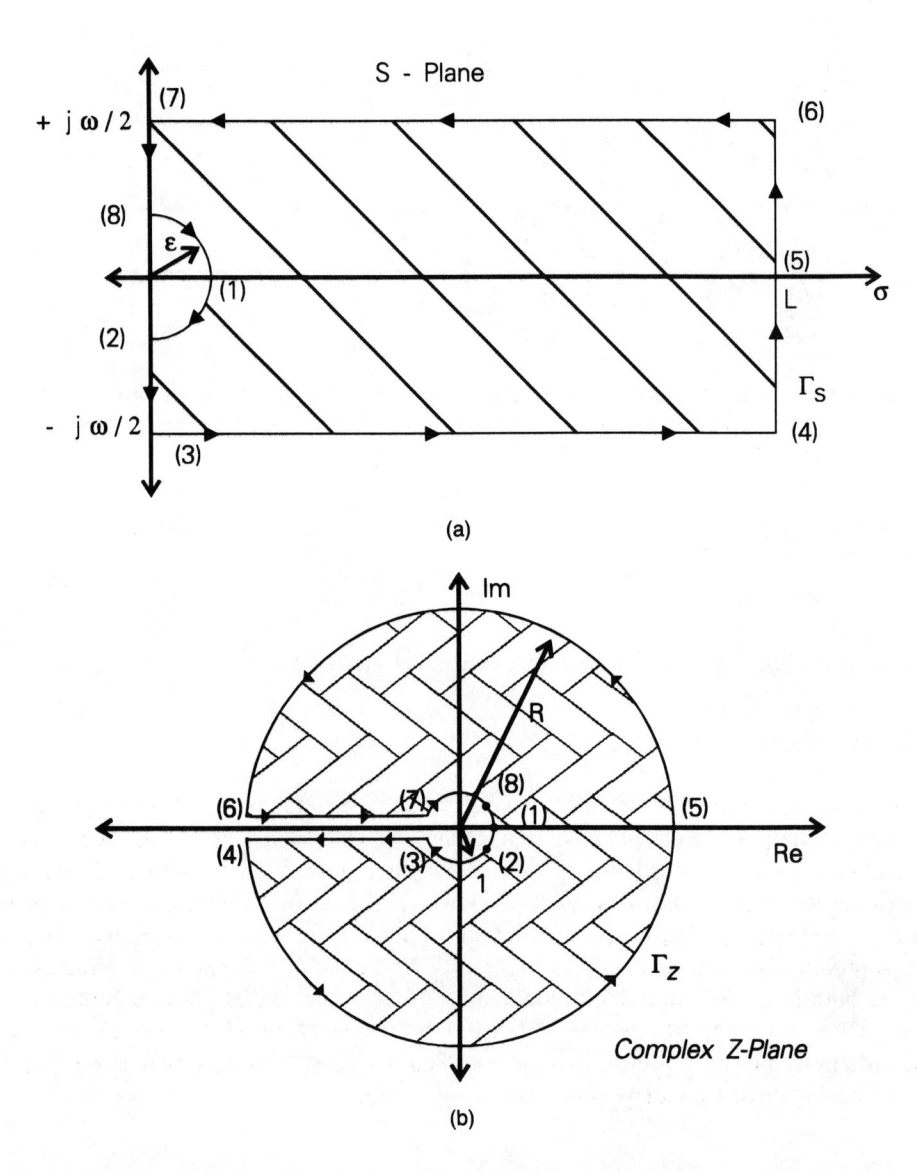

Figure 5.3 (a) Complex path of integration for sampled-data system stability assessment using the Nyquist criterion in the complex s-plane. (b) Equivalent path of integration for Nyquist stability considerations in the complex z plane.

fact precludes the need to compute the equivalent z transform before stability issues can be considered.

Alternatively, system stability may be assessed by employing the Nyquist criterion, which was discussed in Sec. 4.4. However, since $G_{\text{OL}}^*(s)$ is a periodic function of ω, the integration contour Γ_s in Figure 4.19 may be modified as shown in Figure 5.3(a) where in the limit $L \to \infty$ and $\varepsilon \to 0$ are taken. The equivalent integration path in the complex z-plane may be constructed by mapping the Γ_s contour into Γ_z by means of the mapping function $z = e^{sT}$ as shown in Figure 5.3(b). In the limit, Γ_z encompasses the complete exterior of the unit circle. If the assumption is made that $G_{\text{OL}}(s)$ has no poles in the right-half plane as done in Chapter 4, the subject sampled-data control system will be stable so long as the $G_{\text{OL}}^*(s)$-plane contour [or $G_{\text{OL}}(z)$ contour] does not encircle the point $(-1, j0)$.

5.4 TIME-DOMAIN ANALYSIS

The time-domain analysis of greatest interest is concerned with the transient response of a sampled-data phase-locked loop to a step change in frequency and or phase. The time-domain transient response for sampled-data systems is most often discussed in the context of z transforms. Given that the system transient response has been found in terms of z transforms, the time-domain response may be found by calculating the inverse z transform using (5.17b), or repeated long division may be performed, thereby expanding the z transform into an infinite power series whose coefficients are the actual time-domain sample values of the transient response. Both of these approaches necessitate that the system transient response be calculated in terms of z transforms, an operation that has thus far been avoided. A major portion of this reluctance to use z transforms directly is that calculation of the z transform is prone to numerical problems for high-order systems. Their use would further preclude inclusion of even the most basic memoryless nonlinearities. It is therefore preferable to use the circuit equivalence transient response method, which was only briefly mentioned in the final section of Chapter 4.

Operation of the simplified phase-frequency detector and the sample-and-hold phase detector is completely governed by the zero-crossings of the input reference signal and the feedback signal as discussed in Chapter 4. Between successive zero-crossings, the phase-locked loop may be modeled essentially as a continuous control system in which the phase detector output is constant and the feedback path has been broken. The primary computational task is then to precisely determine the temporal occurrence of each zero-crossing such that each time simulation increment occurs over periods of time where the phase detector output is constant (i.e., between zero-crossings).

To have a stable phase-locked loop, the reference phase $\varphi_r(t)$ and the feedback phase $\varphi_f(t)$ must be strictly increasing functions of time modulo 2π at the phase detector. Assuming that the simulation time increment is given by the quantity h, and that $\varphi_r(t)$ and $\varphi_f(t)$ are the reference and feedback phase values immediately following the previous time iteration, respectively, then the new phase values after the next time increment h are

$$\varphi_r(t + h) = \varphi_r(t) + \int_t^{t+h} \omega_r(u)\,du \tag{5.20}$$

and

$$\varphi_f(t + h) = \varphi_f(t) + \int_t^{t+h} \omega_f(u)\,du \tag{5.21}$$

where $\omega_r(t)$ and $\omega_f(t)$ are the instantaneous reference and feedback radian frequencies, respectively. If either $\varphi_r(t + h)$ or $\varphi_f(t + h)$ equals or exceeds 2π radians, a zero-crossing has occurred and a smaller time increment h' must be found such that the time iteration extends only as far as the zero-crossing occurrence. Following the iteration over h', the phase detector output is adjusted, the phase values φ_r and φ_f are reduced modulo 2π, and the iterative process repeated.

In principle, calculation of h' at a zero-crossing is a difficult problem. However, since h is selected to be roughly $1/15$ of the smallest system time constant or smaller per the simple guideline given in Chapter 4, $\varphi_r(t)$ and $\varphi_f(t)$ increase smoothly with each new time increment and h' can be closely approximated with the aid of a second-order interpolation polynomial as shown in Figure 5.4.

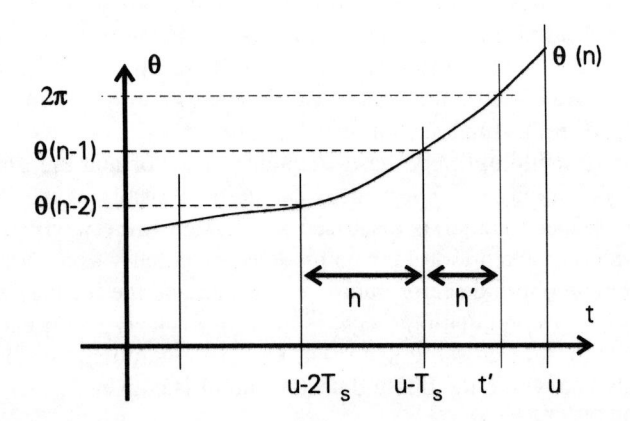

Figure 5.4 Calculation of an interpolated zero-crossing occurrence, h', using a second-order interpolation formula.

In this figure, the phase function $\theta(t)$ exceeds 2π radians after the nth time iteration implying that a zero-crossing occurred somewhere in the range $u - h < t' < u$. A valid second-order interpolation polynomial for the three phase samples shown in Figure 5.4 is

$$\theta(n - r) = ar^2 + br + c \tag{5.22}$$

where r is the time offset of interest from the nth sample in h units. The undetermined coefficients a, b, and c may be found by equating the polynomial at each time step to the respective calculated phase values and solving the three resulting equations

$$\theta(n) = c$$

$$\theta(n - 1) = a - b + c \tag{5.23}$$

$$\theta(n - 2) = 4a - 2b + c$$

simultaneously to give

$$c = \theta(n)$$

$$b = \frac{\theta(n - 2)}{2} - 2\theta(n - 1) + \frac{3}{2}\theta(n) \tag{5.24}$$

$$a = \frac{\theta(n - 2)}{2} - \theta(n - 1) + \frac{\theta(n)}{2}$$

The unknown h' may then be found by solving the quadratic equation in α given by

$$a\alpha^2 + b\alpha + c = 2\pi \tag{5.25}$$

where only the root α_0 is admitted such that $-1 < \alpha_0 < 0$, thereby giving $h' = (1 - \alpha_0)h$.

The preceding methodology clearly requires the use of a variable time-step numerical integration algorithm in order to accommodate the periodically nonstandard integration time interval h'. This subject is discussed at length in Chapter 8.

In summary, the circuit equivalence approach introduced in Chapter 4 is equally applicable for sampled-data systems provided that the preceding provisions are included for properly adjusting the iteration time increment at each phase detector input zero-crossing in order to handle the discontinuous nature of the phase detector output. Since the additional calculations are fairly simple and have a low duty factor, the additional computational load compared to pseudo-continuous analysis will be minimal.

5.5 CLOSED-FORM SAMPLED-DATA SYSTEM RESULTS

Sampled-data phase-locked loops offer unequaled performance in several categories compared to their continuous system counterparts as has been alluded to several times in earlier chapters. Only phase-locked loops that utilize a sample-and-hold type phase detector are considered in this section. In particular, we show that the ideal type 1 and type 2 sampled-data loops can theoretically attain steady-state phase-lock in only one and two reference periods, respectively [1,4]. A number of concise results for predicting the stability margins for these sampled-data systems are also derived. The closed-form solutions that follow are derived using the Nyquist stability criterion and further confirmed using root locus techniques.

5.5.1 Case 1: Ideal Type 1

The open-loop gain for an ideal type 1 sampled-data phase-locked loop is given by

$$G_{OL}(s) = \frac{K_d}{N} \frac{1 - e^{-sT}}{s} \frac{K_v}{s} \qquad (5.26)$$

The T^{-1} factor is missing because no approximation for sampling effects [e.g., (4.35)] has been made in this representation. Rather than deal with the infinite summation form for $G_{OL}^*(s)$, the z transform formulation is used with (5.26), which is

$$\begin{aligned} G_{OL}(z) &= \mathbf{Z}\left(\frac{K_d}{N} \frac{1 - e^{-sT}}{s} \frac{K_v}{s}\right) \\ &= \frac{K_d K_v}{N}(1 - z^{-1})\mathbf{Z}\left(\frac{1}{s^2}\right) \\ &= \frac{K_d K_v}{N}(1 - z^{-1})\frac{Tz}{(z-1)^2} \\ &= \frac{K}{z - 1} \end{aligned} \qquad (5.27)$$

where $K = K_d K_v T N^{-1}$. From this result, the open-loop gain may be rewritten as

$$G_{OL}^*(\omega) = \frac{K e^{-j\omega T/2}}{j2\sin(\omega T/2)} \qquad (5.28)$$

where the substitution $e^{j\omega T}$ has been made for z. Since the phase of $G_{OL}^*(\omega)$ passes through -180 deg at precisely $\omega = \pi/T$, the gain margin for this case is

$$G_M = -20 \log_{10}\left(\frac{K}{2}\right)$$

$$= -20 \log_{10}\left(\pi \frac{\omega_n}{\omega_s}\right) \quad (5.29)$$

where $\omega_n = K_d K_v N^{-1}$ and $\omega_s = 2\pi T^{-1}$. Utilizing (5.28) once more, the phase margin is easily computed as

$$\varphi_M = \frac{\pi}{2} - \arcsin\left(\frac{K}{2}\right) = \frac{\pi}{2} - \arcsin\left(\pi \frac{\omega_n}{\omega_s}\right)$$

$$= \arccos\left(\frac{\pi \omega_n}{\omega_s}\right) \quad (5.30)$$

The gain and phase margins for the ideal type 1 loop are plotted in Figure 5.5 using (5.29) and (5.30), respectively, where $K = \omega_n T$.

The transient phase error response to a step frequency change may be evaluated by first expressing the error response as a z transform. Utilizing (5.13) and (5.16),

$$\varphi_e(z) = \frac{\varphi_r(z)}{1 + G_{\text{OL}}(z)} \quad (5.31)$$

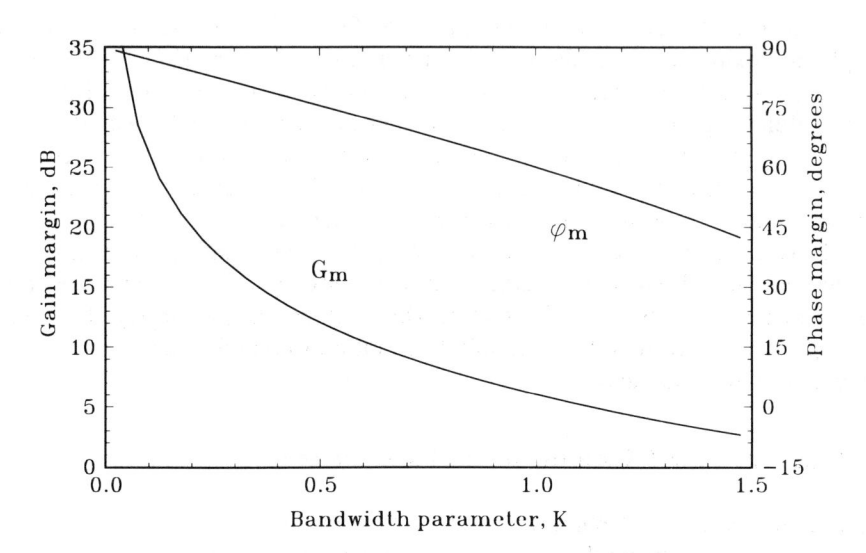

Figure 5.5 Gain and phase margins for the ideal type 1 loop with sample-and-hold phase detector.

where $\varphi_r(z)$ represents the system input excitation. For a step change of Δf Hz in the loop output frequency, the system input excitation is a phase ramp whose z transform is

$$\varphi_r(z) = \frac{2\pi\Delta f}{N} \frac{Tz}{(z-1)^2} \qquad (5.32)$$

Substituting (5.32) and (5.27) into (5.31) yields

$$\varphi_e(z) = \frac{2\pi\Delta f}{N} \frac{Tz}{(z-1)^2} \frac{1}{1 + K/(z-1)}$$

$$= \frac{2\pi\Delta f T}{N} \frac{z}{(z-1)(z-1+K)} \qquad (5.33)$$

and on taking the inverse z transform,

$$\varphi_e(nT) = \frac{1}{2\pi j} \oint \frac{2\pi\Delta f T}{N} \frac{z}{(z-1)(z-1+K)} z^{n-1} dz$$

$$= \frac{2\pi\Delta f T}{NK} [1 - (1-K)^n] \quad \text{for} \quad n \geq 0 \qquad (5.34)$$

This result reveals that the transient phase error decreases geometrically with each sample and exhibits no oscillatory behavior as long as K ($= K_d K_v T/N$) ≤ 1. The optimal phase settling speed (i.e., one reference period) is achieved if $K = 1$ is chosen. Referring to Figure 5.5, the gain and phase margins for the speed-optimized case (i.e., $K = 1$) are 6 dB and 60 deg, respectively. It is worth noting here that if pseudo-continuous analysis had been used for the $K = 1$ case, the predicted gain margin would have been erroneously about 7 dB larger.

Type 1 sampled-data phase-locked loops have not been used extensively in practice primarily because the output phase is susceptible to oscillator post-tuning drift problems as discussed in Chapter 6. This problem can be largely eliminated if the output can be followed by a suitably large ratio external divider. Often however, this option is not available.

5.5.2 Case 2: Type 1 With Inefficient Phase Detector

The analysis for a type 1 phase-locked loop that has an inefficient sample-and-hold phase detector is quite similar to the preceding case 1. The open-loop gain function for this case is given by

$$G_{\text{OL}}(s) = \frac{K_d}{N}\frac{1-e^{-sT}}{s}\frac{\eta}{1-(1-\eta)e^{-sT}}\frac{K_v}{s} \qquad (5.35)$$

where η is the sampling efficiency defined in (4.58). The sampled system open-loop gain may be expressed in z transforms as

$$G_{\text{OL}}(z) = \frac{K_d K_v}{N}\eta\frac{1-z^{-1}}{1-(1-\eta)z^{-1}}\mathcal{Z}\left(\frac{1}{s^2}\right)$$

$$= \frac{K_d K_v T}{N}\eta\frac{z}{(z-1)(z-1+\eta)} \qquad (5.36)$$

The gain margin for this case is found by evaluating (5.36) at the frequency for which its phase argument is $-\pi$, which occurs for $\omega = \pi T^{-1}$ resulting in

$$G_{\text{OL}}(z)\big|_{z=-1} = 20\log_{10}\left[\frac{K_d K_v T}{N}\eta\left|\frac{1}{-2(-2+\eta)}\right|\right]$$

$$= 20\log_{10}\left(\frac{K}{2}\frac{\eta}{2-\eta}\right) \qquad (5.37)$$

and in expanded form, the system gain margin may be written as

$$G_M = -20\log_{10}\left(\frac{K}{2}\right) + 20\log_{10}\left(\frac{2}{\eta}-1\right) \quad \text{dB} \qquad (5.38)$$

The first term in (5.38) is the gain margin for the ideal type 1 case (5.29) whereas the second term is due to the inefficient sampling. The increase in system gain margin predicted by (5.38) can be misleading because it really results from a decrease in the system bandwidth rather than improved system stability. The detrimental aspects of nonideal sampling efficiency would be much more clear if it were possible to examine the system gain margin as a function of η while maintaining a constant closed-loop bandwidth.

To calculate the phase margin for this system, it is necessary to first determine the radian frequency ω_u, at which the open-loop gain magnitude is unity. By equating the magnitude of (5.35) to unity, the solution is given by

$$\omega_u T = \arccos(x) \qquad (5.39)$$

where x is the solution of

$$x^2 - \frac{4(1-\eta) + 2 + 2(1-\eta)^2}{4(1-\eta)} x + \frac{2 + 2(1-\eta)^2 - \gamma^2}{4(1-\eta)} = 0 \qquad (5.40)$$

and only the solution $|x| < 1$ is retained. The parameter $\gamma = K\eta = K_d K_v T \eta N^{-1}$. Given the solution x, it is a simple matter to calculate the phase of $G^*_{OL}(\omega)$ using (5.36) from which the system phase margin is given by

$$\varphi_M = \frac{\pi}{2} + \frac{1}{2}\arccos(x) - \arctan\left(\frac{\sqrt{1-x^2}}{x+\eta-1}\right) \qquad (5.41)$$

The phase margin for this case is shown graphically in Figure 5.6.

The transient response of this system to a step change in output frequency may be found using (5.31) and (5.32) once more giving

$$\varphi_e(z) = \frac{2\pi\Delta f}{N} \frac{Tz}{(z-1)^2} \frac{1}{1 + \dfrac{\gamma z}{(z-1)(z+\eta-1)}}$$

$$= \frac{2\pi\Delta f T}{N} \frac{z(z+\eta-1)}{z^3 + z^2(\gamma+\eta-3) + z(3-\gamma-2\eta) + \eta - 1} \qquad (5.42)$$

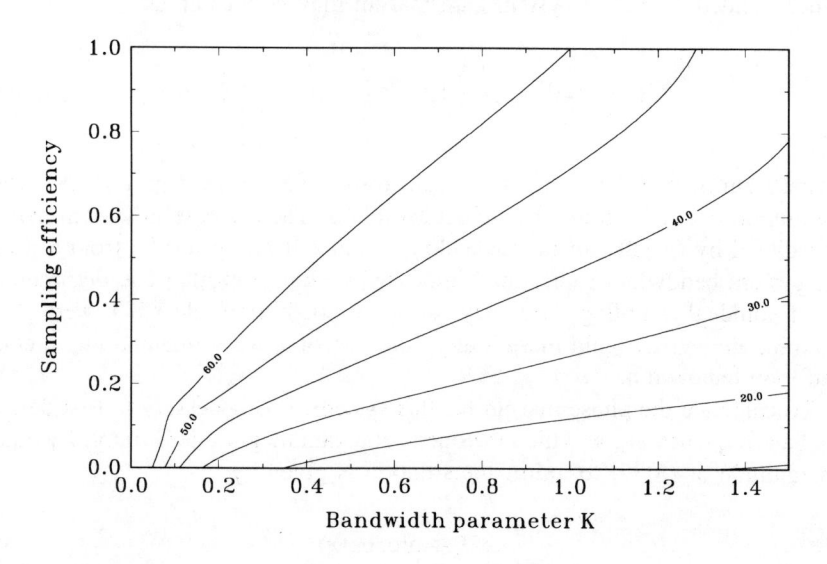

Figure 5.6 Phase margin contours for the ideal type 1 loop with inefficient sample-and-hold phase detector.

This result may be used to quickly calculate the transient phase error response as shown in Figure 5.7.

Inefficient sampling affects the choice of the loop bandwidth parameter K for optimal phase-locking speed. The phase-locking speed can be roughly optimized by minimizing the sum of the phase error sample variances with respect to the final steady-state phase error of the system. In other words, using the sample error sequence available from (5.42), minimization of

$$E(K) = \sum_{i=1}^{\infty} W(i)[\varphi_e(iT) - \varphi_e(\infty)]^2 \qquad (5.43)$$

with respect to K given an arbitrary sampling efficiency η and a fixed frequency step Δf must be performed. The function $W(i)$ in (5.43) is a weighting factor that can be used to accentuate different performance objectives. For instance, if it is important to weight error samples that occur later in the transient response more than the overshoot behavior of the system, $W(i)$ can be chosen to be an increasing function of the error sample index such as $W(i) = i^2$.

The steady-state phase error of the system can be calculated from (5.42) by means of the final value theorem of z transforms, which gives

$$\varphi_e(\infty) = \lim_{z \to 1} [(1 - z^{-1})\varphi_e(z)]$$

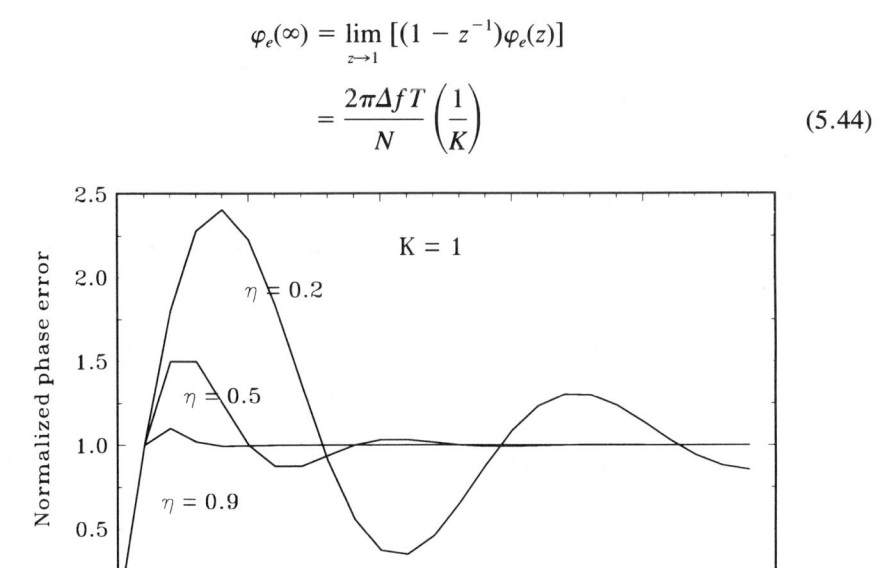

$$= \frac{2\pi \Delta f T}{N}\left(\frac{1}{K}\right) \qquad (5.44)$$

Figure 5.7 Transient error responses for the inefficient type 1 loop subjected to a step change in frequency.

and without loss of generality, the normalization $2\pi\Delta fTN^{-1} = 1$ can be made from this point on.

A Turbo Pascal program that calculates the optimum K value given a user-specified sampling efficiency factor is provided on disk in the companion software program to this text. This program utilizes repeated long division to calculate the time-domain sample error sequence from (5.42) combined with a simple linear search technique to iteratively calculate the optimum value for K. This program was used to create Figure 5.8, which shows the optimum loop K value as a function of sampling efficiency. Three different curves are shown, one for each different weighting function used. It is interesting to note that regardless of the weighting function selected, the optimum K value is essentially unity as long as the sampling efficiency is greater than roughly 85%. Although a larger K factor generally results in faster phase-locking speed when the sampling process is inefficient, this is of course accompanied with a decrease in system phase margin.

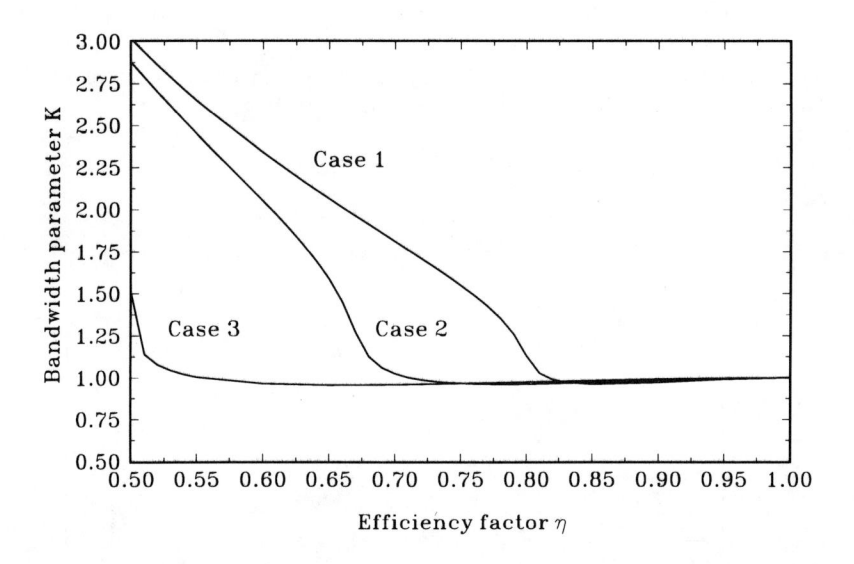

Figure 5.8 Type 1 loop optimum K factor as a function of sampling efficiency. Case 1: $W(i) = i^2$; Case 2: $W(i) = i$; Case 3: $W(i) = 1$.

5.5.3 Case 3: Type 1 With Internal Delay

The preceding considerations ignore the presence of possible transport delay, which is always present in any physically realizable system. Transportation delay can limit transient response performance in very high speed systems as discussed in this section. The delay often arises from the group delay of additional filters that may have been added within the phase-locked loop for improved spurious rejection. The analysis used here is very similar to that employed in case 1 except that modified z transforms must be used in order to include the time delay. In this case, the open-loop gain is given by

$$G_{\text{OL}}(s) = \frac{K_d}{N} \frac{1 - e^{-sT}}{s} \frac{K_v}{s} e^{-s\tau_d} \tag{5.45}$$

where τ_d is the transportation delay in seconds. Converting this to z transforms, the open-loop gain function is given by

$$\begin{aligned} G_{\text{OL}}(z) &= \frac{K_d K_v}{N} (1 - z^{-1}) \mathbf{Z}_m\left(\frac{1}{s^2}\right) \\ &= \frac{K_d K_v T}{N} z^{-1} \frac{mz + (1 - m)}{z - 1} \end{aligned} \tag{5.46}$$

where $m = 1 - \tau_d/T$ and $\tau_d < T$. Larger time delays may be accommodated by appending integral powers of z^{-1} to (5.46) as required. Evaluation of the gain margin for this case is more involved than for the inefficient phase detector case because the radian frequency at which the phase of $G_{\text{OL}}(e^{j\omega T})$ is $-\pi$, ω_π, is a function of m. From (5.46), it is possible to show that

$$\omega_\pi T = \begin{cases} \pi & \text{for } 0 \leq \dfrac{\tau_d}{T} \leq \dfrac{1}{4} \\ 2 \arccos \sqrt{1 - \dfrac{T}{4\tau_d}} & \text{for } \dfrac{1}{4} \leq \dfrac{\tau_d}{T} < 1 \end{cases} \tag{5.47}$$

With this result, the gain margin for this system can be written as

$$G_m = -20 \log_{10}\left(\frac{K}{2}\right) - 20 \log_{10}\left(1 - \frac{2\tau_d}{T}\right) \tag{5.48}$$

for $0 \leq \tau_d/T \leq 1/4$ and

$$G_m = -20 \log_{10}\left[\frac{K}{2 \sin(\omega_\pi T/2)}\right]$$
$$- 10 \log_{10}\{[m \cos(\omega_\pi T) + 1 - m]^2 + m^2 \sin^2(\omega_\pi T)\} \tag{5.49}$$

for $1/4 \leq \tau_d/T < 1$. The result in (5.48) is of particular interest because the first term is the gain margin for an ideal type 1 loop and the second term is due solely to the presence of internal loop transport delay. Gain margin contours for this case are shown in Figure 5.9.

The system phase margin may be computed by first calculating the unity gain radian frequency ω_u for which the open-loop gain from (5.46) satisfies

$$1 = \left| K \frac{mz + (1 - m)}{z - 1} \right| \tag{5.50}$$

where $z = e^{j\omega T}$. After some algebraic manipulations,

$$\omega_u T = \arccos\left[\frac{2 - K^2(2m^2 - 2m + 1)}{2 + 2K^2 m(1 - m)} \right] \tag{5.51}$$

from which the system phase margin is given by

$$\varphi_M = \frac{\pi}{2} - \frac{3}{2}\omega_u T + \arctan\left[\frac{m \sin(\omega_u T)}{m \cos(\omega_u T) + 1 - m} \right] \tag{5.52}$$

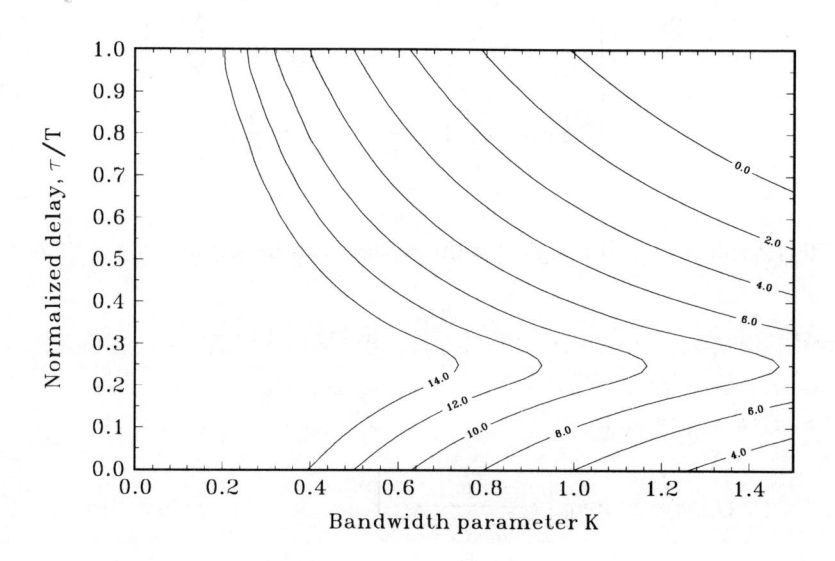

Figure 5.9 Constant gain margin contours for the ideal type 1 loop with internal time delay.

The system phase margin is only given by (5.52) in instances where the system gain margin is greater than 0 dB; otherwise, it is undefined. Phase margin contours for this system are shown in Figure 5.10. This figure clearly shows the impact of transport delay on the system stability as well as the region in the $(K, \tau_d/T)$ plane for which the system is unstable.

The transient response of this system to a step change in output frequency is given by

$$\varphi_e(z) = \frac{2\pi \Delta f T}{N} \frac{z^2}{z^3 + z^2(Km - 2) + z[K(1 - 2m) + 1] - K(1 - m)} \quad (5.53)$$

This result may be used to quickly calculate the transient phase error response as shown in Figure 5.11.

With minor computer program modifications, and again using $W(i) = i^2$, it is possible to optimize the phase-locking speed for a given delay factor τ_d/T by again minimizing (5.43) with respect to K but with $\varphi_e(z)$ now given by (5.53). The results are shown in Figure 5.12. If the phase margin for each optimized parameter pair $(\tau_d/T, K_{opt})$ is also evaluated using (5.51) and (5.52), it can be easily shown that the K_{opt} locus results in a nearly constant 60-deg phase margin for $\tau_d < T$. The topic of constant phase margin system design is explored in Chapter 7 in the context of constant phase loop filter networks for achieving a nearly constant damping factor over large changes in closed-loop bandwidth.

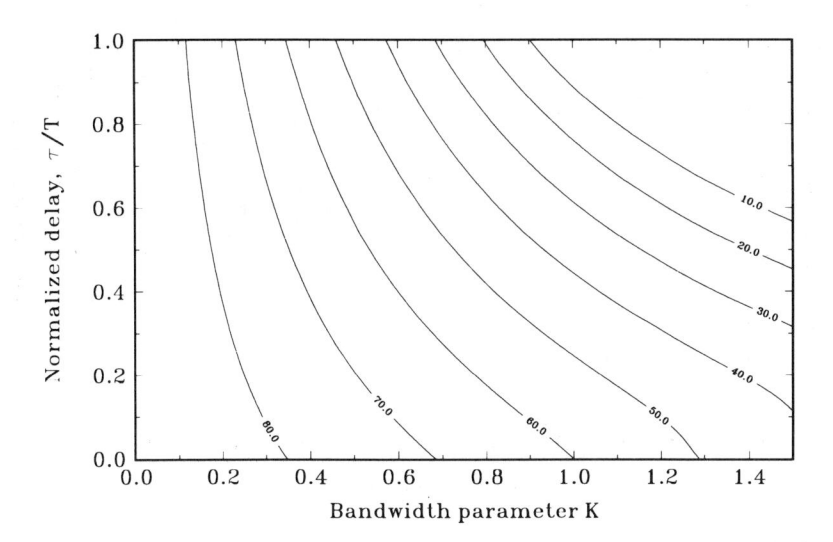

Figure 5.10 Constant phase margin contours for the ideal type 1 loop with internal time delay.

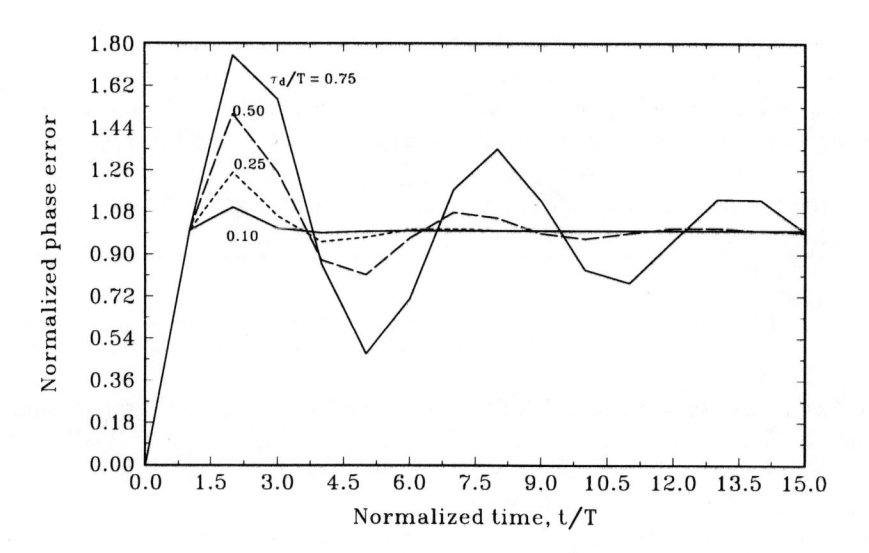

Figure 5.11 Transient error responses for the type 1 loop with internal time delay subjected to a step change in frequency.

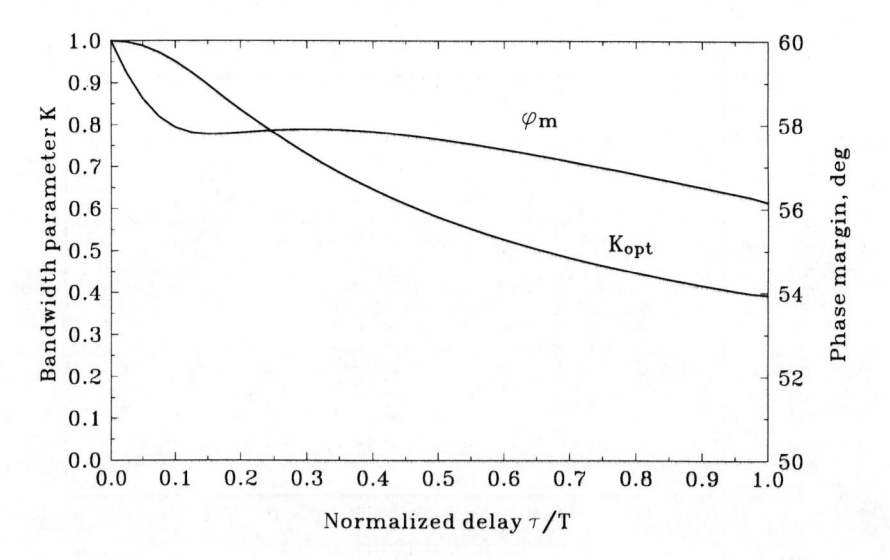

Figure 5.12 Optimization of the loop bandwidth parameter K for a type 1 loop with internal time delay results in a nearly constant phase margin design criterion.

5.5.4 Case 4: Ideal Type 2

The open-loop gain function for the ideal type 2 loop is given by

$$G_{OL}(s) = \frac{K_d}{N} \frac{1 - e^{-sT}}{s} \frac{1 + s\tau_2}{s\tau_1} \frac{K_v}{s} \qquad (5.54)$$

In this case, the sampled system open-loop gain function can be expressed in terms of z transforms as

$$G_{OL}(z) = \omega_n^2(1 - z^{-1})Z\left(\frac{1 + s\tau_2}{s^3}\right)$$

$$= (\omega_n T)^2 \frac{z(0.5 + \tau_2/T) + (0.5 - \tau_2/T)}{(z - 1)^2} \qquad (5.55)$$

The gain margin for this system is given by

$$G_M = -20 \log_{10}(\zeta \omega_n T) \quad \text{dB} \qquad (5.56)$$

for $\omega_n T < 4\zeta$. For $\omega_n T \geq 4\zeta$, the gain margin is undefined. As before, the loop damping factor is given by $\zeta = \omega_n \tau_2/2$. Constant gain margin contours for this system are shown graphically in Figure 5.13.

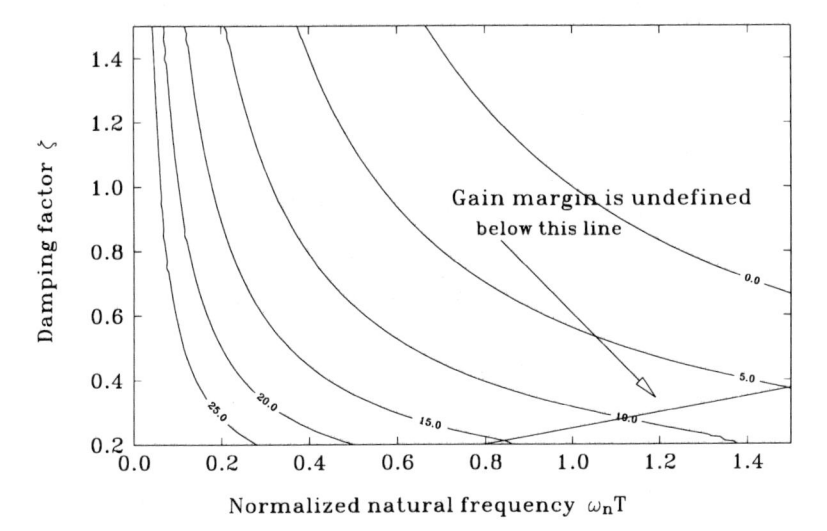

Figure 5.13 Constant gain margin contours for the ideal type 2 loop as a function of normalized natural frequency $\omega_n T$.

The phase margin for this system is given by

$$\varphi_m = -\omega_u T + \text{arctan4}[a\sin(\omega_u T), a\cos(\omega_u T) + b] \qquad (5.57)$$

where the normalized unity-gain frequency $\omega_u T$ is given by

$$\omega_u T = \arccos\left[1 + abc - \sqrt{a^2 b^2 c^2 + c}\right] \qquad (5.58)$$

and

$$a = 0.5 + \frac{2\zeta}{\omega_n T}$$

$$b = 0.5 - \frac{2\zeta}{\omega_n T}$$

$$c = \frac{(\omega_n T)^4}{4}$$

The arccosine must exist in order for the phase margin to be defined, but in addition, it is also necessary that $\omega_n T < 4\zeta$ if this result is to be valid. The phase margin contours for this system as a function of $\omega_n T$ and ζ are shown in Figure 5.14. The system preference for a damping factor close to 0.707 is very apparent.

The transient phase error response of this system to a step change in output frequency Δf may be derived in terms of z transforms as

$$\varphi_e(z) = \frac{\varphi_r(z)}{1 + G_{\text{OL}}(z)}$$

$$= \frac{2\pi \Delta f T}{N} \frac{z}{z^2 + z[a(\omega_n T)^2 - 2] + [1 + b(\omega_n T)^2]} \qquad (5.59)$$

and formulating the time-domain error sequence using the residue method,

$$\varphi_e(nT) = \frac{1}{2\pi j} \oint \frac{2\pi \Delta f T}{N} \frac{z^n}{(z - A)(z - B)} dz \qquad (5.60)$$

where A and B are the characteristic equation roots in (5.59). Carrying out the indicated integration, the final result for the transient phase error sample sequence is

$$\varphi_e(nT) = \frac{2\pi \Delta f T}{N} \frac{A^n - B^n}{A - B} \qquad (5.61)$$

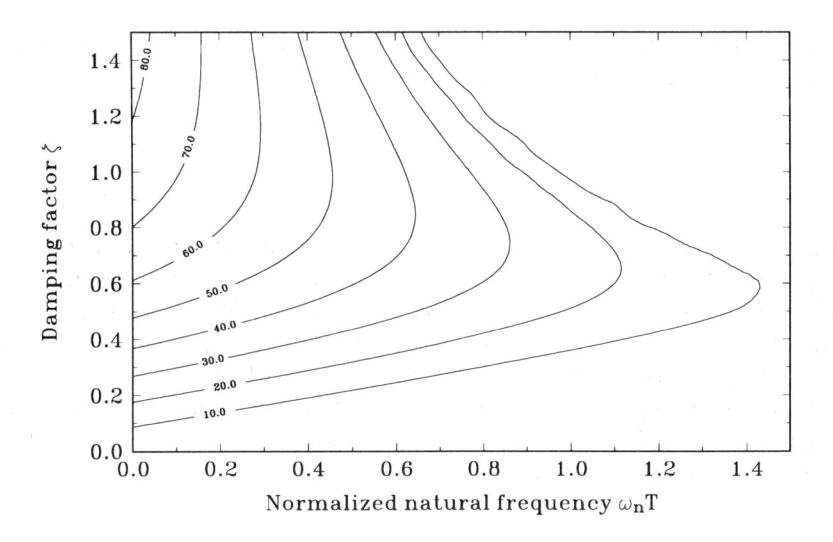

Figure 5.14 Constant phase margin contours for the ideal type 2 loop as a function of normalized natural frequency $\omega_n T$.

Just as in the continuous-time case, the transient phase error response may be classified into three different categories depending on the roots of the characteristic equation [5]. For the underdamped case, $A = \alpha + j\beta$ and $B = \alpha - j\beta$ in which case

$$\varphi_e(nT) = \frac{2\pi \Delta f T}{N} \frac{R^n}{\beta} \sin(n\theta) \tag{5.62}$$

where

$$R = \sqrt{\alpha^2 + \beta^2}$$

$$\theta = \arctan\left(\frac{\beta}{\alpha}\right)$$

For the critically damped situation, $A = B = \alpha$, which leads to

$$\varphi_e(nT) = \frac{2\pi \Delta f T}{N} n\alpha^{n-1} \tag{5.63}$$

Finally for the overdamped case, $A = \alpha$, $B = \mu$, resulting in

$$\varphi_e(nT) = \frac{2\pi \Delta f T}{N} \frac{\alpha^n - \mu^n}{\alpha - \mu} \tag{5.64}$$

The particular case of interest is the critically damped case given by (5.63). In this case, if $\alpha = 0$, the transient error response is given by

$$\varphi_e(nT) = \frac{2\pi\Delta fT}{N} \delta(t - T)$$

and the phase settling theoretically is achieved in only two sample periods. This value of α is obtained by selecting $\omega_n T = 1$ and $\zeta = 0.75$. In practice, this ideal dead-beat response is impossible to obtain due to other nonideal factors such as phase detector sampling efficiency and internal loop delays. The gain margin for this ideal dead-beat response parameter choice is only 2.5 dB, which makes precise control over all loop gain parameters mandatory to ensure system stability. Had this analysis been done with pseudo-continuous models, the computed gain margin would have appeared to be roughly 7 dB larger.

5.5.5 Case 5: Type 2 With Inefficient Phase Detector

The open-loop gain for a type 2 loop that has an inefficient sample-and-hold phase detector is given by

$$G_{OL}(s) = \frac{K_d}{N} \frac{1 - e^{-sT}}{s} \frac{\eta}{1 - (1 - \eta)e^{-sT}} \frac{1 + s\tau_2}{s\tau_1} \frac{K_v}{s} \qquad (5.65)$$

This case is representative of a phase-locked loop (PLL) intended for maximum switching speed. This case has no additional filtering to reduce sampling spurs in the output spectrum, which, from a spurious performance perspective, is primarily limited by nonideal aspects of the phase detector. For the sampled open-loop gain function in terms of z transforms, the open-loop gain function is given by

$$G_{OL}(z) = (\omega_n T)^2 \frac{\eta z}{z - (1 - \eta)} \frac{z\left(\frac{1}{2} + \frac{2\zeta}{\omega_n T}\right) + \left(\frac{1}{2} - \frac{2\zeta}{\omega_n T}\right)}{(z - 1)^2} \qquad (5.66)$$

The system gain margin is given by

$$G_M = -20 \log_{10}(\zeta\omega_n T) + 20 \log_{10}\left(\frac{2}{\eta} - 1\right) \qquad (5.67)$$

for

$$\zeta > \frac{\omega_n T}{4}\left(\frac{2}{\eta} - 1\right)$$

but is otherwise undefined. This result is quite similar to the result obtained for the inefficient type 1 loop case in that the first term is the gain margin for an ideal type 2 loop, whereas the second term results from the sampling inefficiency. Since the finite efficiency simply adds a corrective factor to the gain margin of the ideal type 2 case, the gain margin contours that were developed in Figure 5.13 are equally applicable for the inefficient type 2 case providing that the range of valid design parameters is observed.

The phase margin for this case is difficult to express in a closed form because of the third-order characteristic equation that is involved. The constant phase margin contours can nonetheless be calculated numerically and these results are shown in Figures 5.15a through 5.15c for a number of sampling efficiency cases.

The transient response for the inefficient type-2 loop may be expressed in terms of z transforms as

$$\varphi_e(z) = \frac{2\pi \Delta f T}{N}$$
$$\times \frac{z^2 + (\eta - 1)z}{z^3 + z^2[\eta - 3 + (\omega_n T)^2 \eta a] + z[3 - 2\eta + (\omega_n T)^2 \eta b] + \eta - 1} \quad (5.68)$$

Figure 5.15 Constant phase margin contours for the type 2 loop having a sample-and-hold phase detector efficiency factor of (a) $\eta = 0.90$, (b), $\eta = 0.75$, and (c) $\eta = 0.50$.

(b)

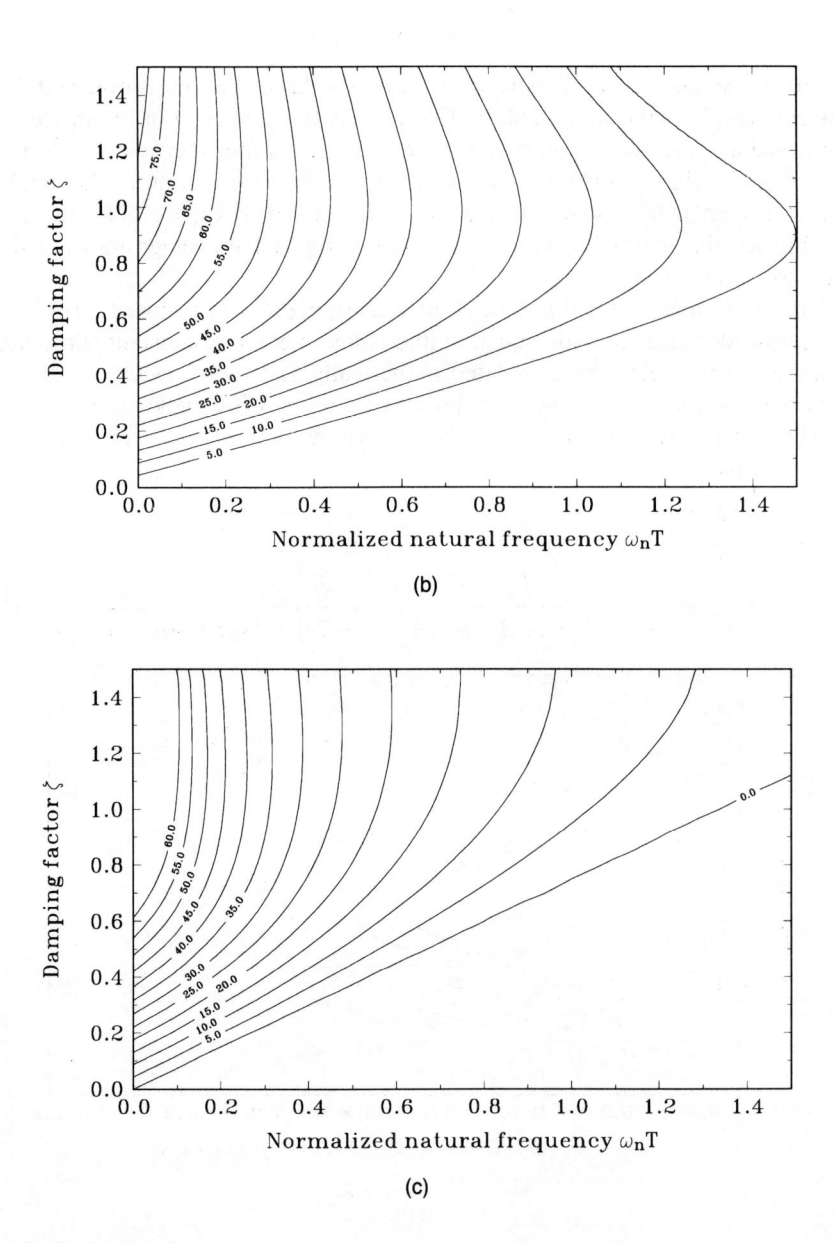

(c)

Figure 5.15 (Continued)

The phase-locking speed for this case can be optimized by minimizing (5.43) given the transient error response, which is available from (5.68) and again taking $W(i) = i^2$. The resulting parameter optimization as a function of sampling efficiency is shown in Figure 5.16.

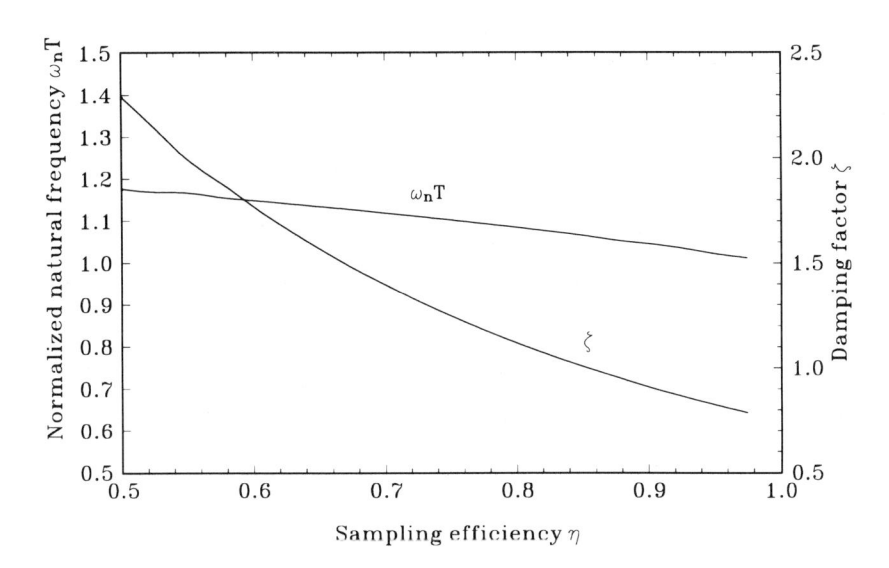

Figure 5.16 Optimum loop parameters for maximizing the phase-locking speed of a type 2 loop having an inefficient sample-and-hold phase detector.

5.5.6 Case 6: Type 2 With Internal Delay

The open-loop gain for a type 2 loop that utilizes an ideal sample-and-hold phase detector but has a nonzero internal time delay is given by

$$G_{\text{OL}}(s) = \frac{K_d}{N} \frac{1 - e^{-sT}}{s} \frac{1 + s\tau_2}{s\tau_1} \frac{K_v}{s} e^{-s\tau_d} \tag{5.69}$$

This case is representative of the phase-locked loop that includes additional low-pass filtering in the transfer function to reduce the spurious sampling components in the output spectrum. The delay term results from the group delay introduced by the additional filtering. The sampled control system open-loop gain function in z transforms is given in this case as

$$G_{\text{OL}}(z) = (\omega_n T)^2 \frac{az^2 + bz + c}{z(z-1)^2} \tag{5.70}$$

where

$$a = m\left(\frac{m}{2} + \frac{2\zeta}{\omega_n T}\right)$$

$$b = \frac{2\zeta}{\omega_n T}(1 - 2m) + m - m^2 + \frac{1}{2}$$

$$c = \frac{2\zeta}{\omega_n T}(m - 1) + \frac{(m-1)^2}{2}$$

and $m = 1 - \tau_d/T$. Larger time delays may be dealt with by appending additional factors of z^{-1} as mentioned for case 3.

The Nyquist criterion may be used once again to express the system gain margin by

$$G_M = -20 \log_{10}\left[\left(\frac{\omega_n T}{2}\right)^2\right] - 20 \log_{10}\left[\frac{\sqrt{(aC_2 + bC_1 + c)^2 + (aS_2 + bS_1)^2}}{\sin^2(\omega_\pi T/2)}\right] \quad (5.71)$$

where

$$C_1 = \cos(\omega_\pi T)$$

$$C_2 = \cos(2\omega_\pi T)$$

$$S_1 = \sin(\omega_\pi T)$$

$$S_2 = \sin(2\omega_\pi T)$$

and

$$\omega_\pi T = \left[\pi, \arccos\left(-\frac{b}{2c}\right)\right] \quad (5.72)$$

In cases where the arccosine does not exist, the solution $\omega_\pi T = \pi$ must be automatically selected in (5.72). Even though these equations result in defined quantities for any parameter choice, the system gain margin is in fact only defined for parameter selections that also satisfy the inequality

$$\frac{b + 2c}{a + b + c} < 0$$

These results were used to construct the gain margin contours shown in Figures 5.16(a) through (c). Similarly, (5.70) may be used to numerically calculate the phase margin contours shown in Figures 5.17(a) through (c) for this system.

The transient error response to a step change in output frequency Δf for this case is given by

$$\varphi_e(z) = \frac{2\pi \Delta f T}{N} \frac{z}{(z-1)^2} \frac{1}{1 + (\omega_n T)^2 \dfrac{az^2 + bz + c}{z(z-1)^2}}$$

$$= \frac{2\pi \Delta f T}{N}$$

$$\times \frac{z^2}{z^3 + z^2[-2 + a(\omega_n T)^2] + z[1 + (\omega_n T)^2 b] + (\omega_n T)^2 c} \quad (5.73)$$

The parameter locus that minimizes the error function given by (5.43) for this case is shown as a function of τ_d/T in Figure 5.19. Here again, $W(i) = i^2$ has been used.

The foregoing case studies are valuable for the initial design process where block-level performance budgets must be estimated. The transport delay cases are particularly useful for estimating the amount of additional reference spur attenuation

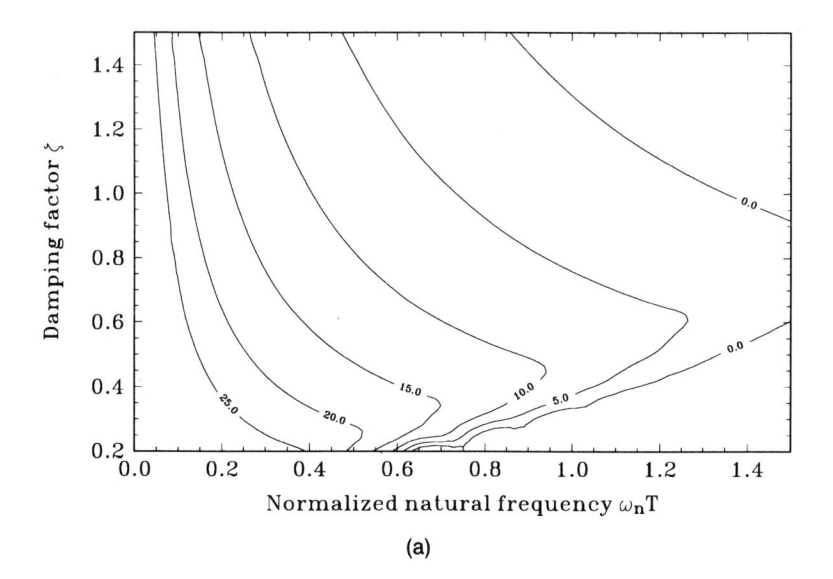

Figure 5.17 Constant gain margin contours for a type 2 loop having internal time delay (a) $\tau_d/T = 0.10$, (b) $\tau_d/T = 0.50$, and (c) $\tau_d/T = 0.75$.

(b)

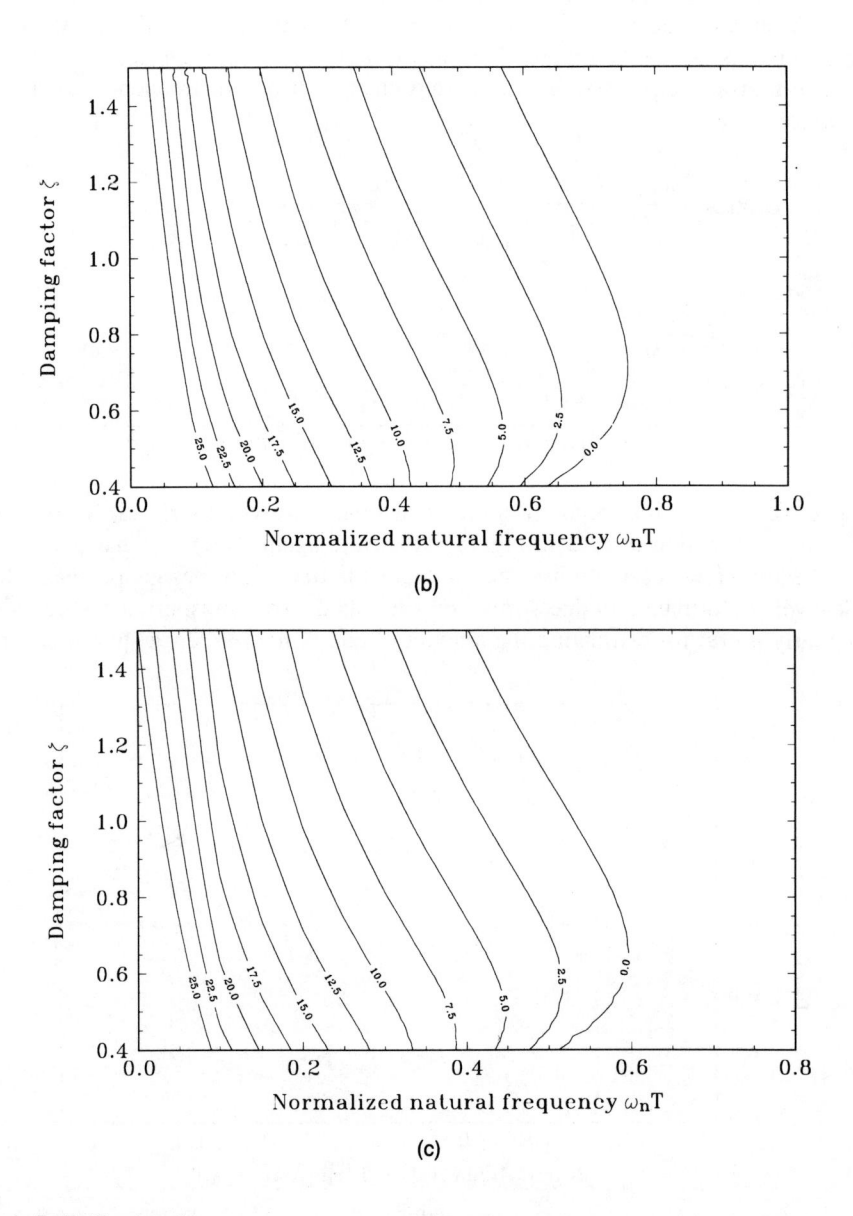

(c)

Figure 5.17 (Continued)

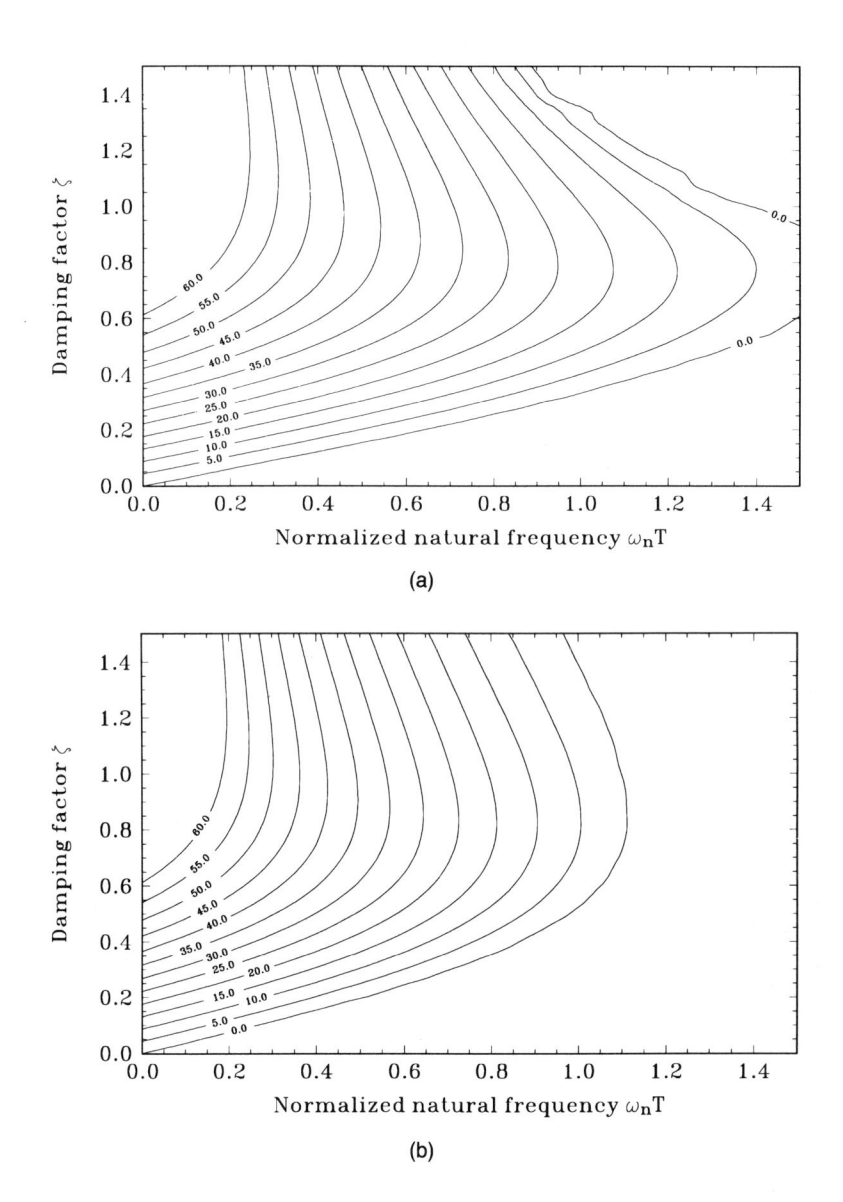

Figure 5.18 Constant phase margin contours for a type 2 loop having internal time delay (a) $\tau_d/T = 0.10$, (b) $\tau_d/T = 0.25$, and (c) $\tau_d/T = 0.50$.

(c)

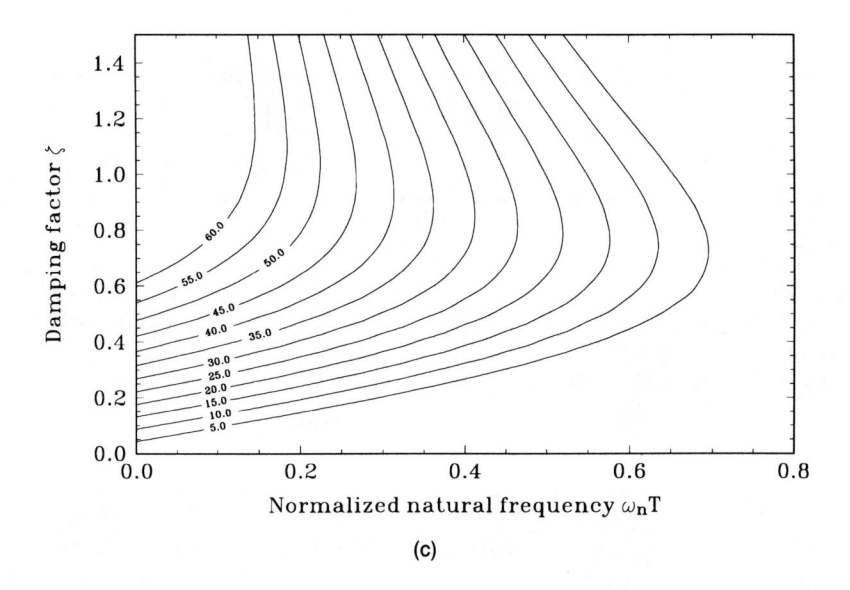

Figure 5.18 (Continued)

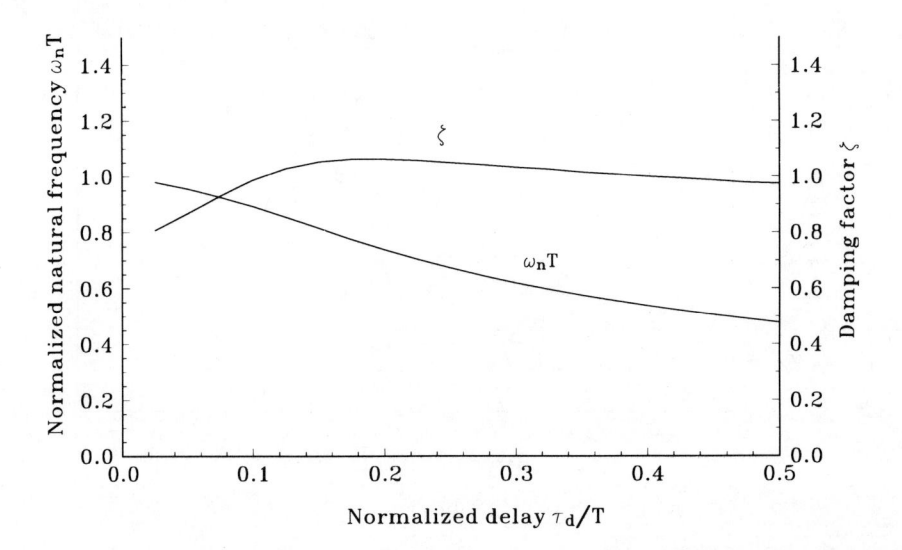

Figure 5.19 Optimum phase-locking speed loop parameters for a type 2 loop with internal time delay.

that may be obtained through additional filtering within the phase-locked loop. Given that the stability margins are suitable for a given loop delay τ_d, the designer may select a low-pass filter that maximizes the reference spur attenuation while contributing a group delay less than or equal to roughly τ_d. The sampling efficiency cases may be used in a similar fashion to estimate the needed performance level for the sample-and-hold phase detector. Clearly, case studies such as these are at best laborious algebraic exercises; as such, they provide a strong motivation for the computer-aided analysis methods discussed in Chapter 8. A similar case study is presented in Chapter 6, which addresses the simplified phase/frequency detector. Additional design guidelines for a popular type 2 phase-locked loop that utilizes a simplified phase-frequency detector along with an embedded $N = 3$ elliptic filter for added spurious rejection are provided in Chapter 7.

5.6 PHASE NOISE

Sampling action within a phase-locked loop used for frequency synthesis can significantly alter the output phase noise spectrum compared to that predicted using traditional pseudo-continuous analysis methods. These sampling effects are considered in some detail in this section.

The qualitative impact of sampling on phase noise spectra has been recognized for a long time [9–11]. Because most PLL designs have historically used a small percentage bandwidth, however, sampling effects have largely gone unnoticed except in cases where far-removed discrete spurs were aliased within the loop bandwidth, causing close-in spurious problems [9].

Our discussions here focus on the model presented in Figure 5.1 where only two phase noise sources are shown, one corresponding to reference-related phase noise, and the second corresponding to the VCO self-noise. Any other noise sources within the system may be represented by transforming them into an equivalent noise source at one of these two points, thereby making the following discussions generally applicable to any case that might be encountered.

Any study of phase noise must address the random nature of the noise. Generally, this leads to discussions involving stochastic random processes and ensemble averaging. However, because most laboratory spectrum measurements are made using time averages rather than ensemble (probabilistic) averages, as discussed in Chapter 3, the phase noise analysis problem is addressed here using primarily time-domain measures. This approach is further supported by the fact that the theory of random processes based on time averages is substantially more solid than the theory based on ensemble averages [12]. In replacing ensemble averages with time averages, the underlying random noise is assumed to be an ergodic process. More extensive treatment of this subject may be found in [12–14].

5.6.1 Reference Phase Noise

Since the reference and oscillator noise are in general statistically independent, each noise source in Figure 5.1 will be considered separately. Reference noise and its effect on the output phase noise spectrum of the synthesizer are considered in this section. The derivation begins by noting that the phase error impulse response in terms of z transforms may be written as

$$\theta_e(z) = \frac{1}{1 + G_{OL}(z)} = M(z) \tag{5.74}$$

from (5.5) and (5.16). This impulse response may be expanded into a power series as

$$\theta_e(z) = \sum_{i=0}^{\infty} a_i z^{-i} \tag{5.75}$$

Referring now to Figure 5.1, it will be assumed that the analog hold function $H(s)$ has an impulse response given by $h(t)$. The analog voltage at the output of the phase detector in Figure 5.1 therefore has the impulse response

$$V_i(t) = \sum_{m=0}^{\infty} a_m h(t - mT) \tag{5.76}$$

Thus far, these discussions have been strictly deterministic. Assume now that the reference phase noise source after impulse sampling may be represented by the random sample sequence $\theta_r(m)$, which is actually given by $\theta_r(mT)$. For simplicity, the time dependence T will not be carried in the subsequent discussions. The phase detector analog output voltage is then given by

$$V(t) = \sum_{k=-\infty}^{\infty} \theta_r(k) \left[\sum_{m=0}^{\infty} a_m h(t - mT - kT) \right] \tag{5.77}$$

In order to compute the noise power spectral density at the phase-locked loop output, it is first necessary to compute the power spectral density of $V(t)$. This power spectral density is given by

$$S_v(f) = \lim_{T_w \to \infty} \frac{1}{T_w} \mathbf{E}[|V_{T_w}(f)|^2] \tag{5.78}$$

where $\mathbf{E}$ represents a statistical expectation, which will be replaced here with a time average based on earlier comments [14–16]. Letting $T_w = 2NT$,

$$V_{T_w}(f) = \int_{-T_w/2}^{T_w/2} V(t)\, e^{-j2\pi fu}\, du$$

$$\approx \int_{-\infty}^{\infty} \sum_{k=-N}^{N-1} \theta_r(k) \left[\sum_{m=0}^{\infty} a_m h(t - mT - kT) \right] e^{-j2\pi ft}\, dt \qquad (5.79)$$

Performing the integration,

$$V_{T_w}(f) = \sum_{k=-N}^{N-1} \theta_r(k) \sum_{m=0}^{\infty} a_m H(f)\, e^{-j2\pi fT(m+k)} \qquad (5.80)$$

Computing the magnitude-squared value of (5.80),

$$|V_T(f)|^2 = |H(f)|^2 \left[\sum_{u=0}^{\infty} \sum_{m=0}^{\infty} a_m a_u\, e^{-j2\pi fT(m-u)} \right]$$

$$\times \left[\sum_{k=-N}^{N-1} \sum_{p=-N}^{N-1} \theta_n(k)\theta_n(p)\, e^{-j2\pi fT(k-p)} \right] \qquad (5.81)$$

The first term in brackets is completely deterministic, resulting from the impulse error response of the phase-locked loop. The random features of the problem are contained within the second bracketed term alone. Only this term is affected by the expectation and limit operations that remain to be included from (5.78).

Focusing on the second bracketed term in (5.81), and assuming that the phase noise process is ergodic, the averaging step results in

$$\mathbf{E}\left[\sum_{k=-N}^{N-1} \sum_{p=-N}^{N-1} \theta_r(k)\theta_r(p) e^{-j2\pi fT(k-p)} \right] = \sum_{\nu=-2N+1}^{2N-1} (2N - |\nu|) R_{\theta_r}(\nu)\, e^{-j2\pi fT\nu} \qquad (5.82)$$

where the normal substitution $\nu = k - p$ has been made. Rigorously speaking, the kernel in $R_{\theta_r}(\ )$ is actually νT but the time factor T has been dropped to conform with the notation used with the sample sequence θ_r. Due to the assumed ergodicity, the correlation function for the random phase error process is given by

$$R_{\theta_r}(\nu) = \mathbf{E}[\theta_r(k)\theta_r(k - \nu)]$$

$$= \lim_{N \to \infty} \frac{1}{2N + 1} \sum_{k=-N}^{N} \theta_r(k)\theta_r(k - \nu) \qquad (5.83)$$

Finally, taking the limit $T_w \to \infty$ in (5.78) is equivalent to taking $N \to \infty$ in (5.82), which finally transforms the second bracketed term to

$$\lim_{N\to\infty} \frac{1}{2NT} \sum_{\nu=-2N+1}^{2N-1} [2N-|\nu|] R_{\theta_r}(\nu) e^{-j2\pi fT\nu} = \frac{1}{T} \sum_{\nu=-\infty}^{\infty} R_{\theta_r}(\nu) e^{-j2\pi fT\nu} \quad (5.84)$$

From this result, along with the Poisson sum formula given by

$$\sum_{\nu=-\infty}^{\infty} R_{\theta_r}(\nu) e^{-j2\pi fT\nu} = \frac{1}{T} \sum_{m=-\infty}^{\infty} S_{\theta_r}\left(f - \frac{m}{T}\right) \quad (5.85)$$

the expectation in (5.82) may be finally represented as

$$\frac{1}{T^2} \sum_{m=-\infty}^{\infty} S_{\theta_r}\left(f - \frac{m}{T}\right) \quad (5.86)$$

where

$$S_{\theta_r}(f) = \int_{-\infty}^{\infty} R_{\theta_r}(\tau) e^{-j2\pi f\tau} d\tau \quad (5.87)$$

Concentrating now on the first bracketed term in (5.81), this simplifies to

$$\sum_{u=0}^{\infty} \sum_{m=0}^{\infty} a_m a_u e^{-j2\pi fT(m-u)} = \left|\sum_{m=0}^{\infty} a_m e^{-j2\pi mfT}\right|^2 = |M(z)|^2_{z=e^{j2\pi fT}} \quad (5.88)$$

Finally, combining the results given in (5.86) and (5.88), the power spectral density of $V(t)$ is given by

$$S_v(f) = |H(f)|^2 |M(e^{j2\pi fT})|^2 \frac{1}{T^2} \sum_{m=-\infty}^{\infty} S_{\theta_r}\left(f - \frac{m}{T}\right) \quad (5.89)$$

Since the open-loop gain function is linear and time invariant, the phase-locked loop output phase noise power spectral density due to the reference phase noise is given by

$$S_o(f) = S_v(f)|G(f)|^2$$

$$= |H(f)|^2 |M(e^{j2\pi fT})|^2 \frac{|G(f)|^2}{T^2} \sum_{m=-\infty}^{\infty} S_{\theta_r}\left(f - \frac{m}{T}\right) \quad (5.90)$$

On closer examination, this may be rewritten in the final form

$$S_o(f) = \left| \frac{H(f)G(f)}{1 + G_{OL}^*(f)} \right|^2 S_{\theta_r}^*(f) \tag{5.91}$$

where

$$S_{\theta_r}^*(f) = \frac{1}{T^2} \sum_{m=-\infty}^{\infty} S_{\theta_r}\left(f - \frac{m}{T}\right) \tag{5.92}$$

The magnitude-squared quantity should be recognized as nothing more than the transfer function derived earlier in (5.6). Although this result could have been written down largely by inspection by recognizing that (5.6) is a linear time-invariant function from the beginning, the approach taken here clearly illustrates the mechanics involved in a more rigorous manner.

This result may be used to develop a closed-form expression for the reference-related phase noise at the output of a sampled type 1 phase-locked loop as follows. For the ideal sample-and-hold case,

$$\frac{H(f)G(f)}{1 + G_{OL}^*(f)} = NT \frac{\sin(\pi fT)}{(\pi fT)} \frac{\omega_n T}{e^{j2\pi fT} - 1 + \omega_n T} \tag{5.93}$$

where $\omega_n = K_d K_v N^{-1}$. Letting $K = \omega_n T$, the magnitude-squared value of (5.93) is [17] given by

$$T^2 N^2 \frac{K^2}{K^2 - 2K + 2 + 2(K-1)\cos(2\pi fT)} \left[\frac{\sin(\pi fT)}{\pi fT}\right]^4 \tag{5.94}$$

In the speed-optimized case where $K = 1$, the output power spectral density is then given by

$$S_o(f) = N^2 \left[\frac{\sin(\pi fT)}{\pi fT}\right]^4 \sum_{m=-\infty}^{\infty} S_{\theta_r}\left(f - \frac{m}{T}\right) \tag{5.95}$$

As would be expected based on classical phase-locked loop theory, the phase noise close to the reference carrier ($m = 0$) is enhanced by 20 $\log_{10}(N)$ dB. A further ramification of the sampling action shown in (5.95) is that phase noise near any harmonic of the reference frequency is aliased in frequency such that it appears within the closed-loop bandwidth of the system, from that point on taking on the same character as the $m = 0$ phase noise spectrum term. Therefore, the system designer cannot neglect the presence of possible reference phase noise contributions that appear far removed from the carrier frequency. This statement in fact applies for any circuitry that utilizes digital (i.e., sampling) components.

5.6.2 VCO-Related Phase Noise

The output phase noise contribution that arises from the self-noise of the VCO is considered in this section. In general, this analysis is more complicated than the reference phase noise case and it will be necessary to use additional approximations in order to reach usable closed-form results. Because the phase noise spectrum of the VCO does not, in general, satisfy the Nyquist sampling criterion with respect to the sampling rate used within the phase-locked loop, spectral components are aliased to baseband within the control loop itself.

Returning to the deterministic closed-loop analysis with which we began this chapter, the impulse phase error response of the system to a deterministic disturbance that is injected at the VCO sum port in Figure 5.1 is given by

$$\theta_o(s) = \theta_v(s) - \frac{H(s)G(s)/N}{1 + HG^*(s)/N} \theta_v^*(s) \quad (5.96)$$

This may be represented in the time domain as

$$\theta_o(t) = \theta_v(t) - f(t) \otimes \theta_v^*(t) \quad (5.97)$$

where

$$f(t) = \mathcal{L}^{-1}\left[\frac{H(s)G(s)/N}{1 + HG^*(s)/N}\right] \quad (5.98)$$

Since $HG^*(s)/N = G_{OL}^*(s)$ may be equivalenced to a z transform from (5.16),

$$\frac{1}{1 + HG^*(s)/N} = \sum_{i=0}^{\infty} b_i z^{-i} \quad \text{for} \quad z = e^{sT} \quad (5.99)$$

which leads finally to

$$f(t) = \sum_{i=0}^{\infty} b_i G_{OL}(t - iT) \quad (5.100)$$

provided that the b_i meet certain convergence criteria (which are, in practice, almost always met). Given that it is possible to compute $f(t)$, the output phase function may be written in terms of the (deterministic) VCO phase disturbance as

$$\theta_o(t) = \theta_v(t) - \sum_{n=-\infty}^{\infty} \theta_v(t-nT)f(t-nT) \qquad (5.101)$$

If it is now assumed that $\theta_v(t)$ is band-limited such that the system sampling rate $(T)^{-1}$ is at least the Nyquist rate, then $\theta_v(t)$ may be expressed as

$$\theta_v(t) = \sum_{n=-\infty}^{\infty} \theta_v(nT) \frac{\sin[\pi(t/T-n)]}{\pi(t/T-n)} \qquad (5.102)$$

Combining (5.101) and (5.102),

$$\theta_o(t) = \sum_{n=-\infty}^{\infty} \theta_v(nT) \left[\frac{\sin[\pi(t/T-n)]}{\pi(t/T-n)} - f(t) \right]$$

$$= \sum_{n=-\infty}^{\infty} \theta_v(nT) q(t-nT) \qquad (5.103)$$

where

$$q(t) = \frac{\sin\left(\frac{\pi}{T}t\right)}{\frac{\pi}{T}t} - f(t) \qquad (5.104)$$

In this context, $q(t)$ may be viewed as a difference function between the two sampled impulse responses.

Thus far, $\theta_v(t)$ has been viewed as a deterministic phase disturbance. If instead it is now assumed to be a band-limited widesense stationary random phase noise process, these results can be used to calculate the resulting output phase noise spectrum. Calculating the power spectral density using (5.78), it can be shown that the output power spectral density due to the VCO noise is given by

$$S_{\theta_o}(f) = |Q(f)|^2 \frac{1}{T^2} \sum_{k=-\infty}^{\infty} S_{\theta_v}\left(f - \frac{k}{T}\right) \qquad (5.105)$$

where $Q(f)$ is the Fourier transform of $q(t)$.

Although this is the final result in the case where the VCO power spectral density is itself band-limited, it provides little insight into the system phase noise behavior since it requires that $f(t)$ first be computed and then the function $q(t)$; these are generally fairly complex computations. There is also no insight into dealing with situations where the VCO phase noise is not band-limited (with respect to the sampling frequency), which is generally the case. To proceed, it is necessary to examine the impact of sampling on a general nonband-limited random phase noise spectrum.

In addressing the more general case, it is helpful to return to Figure 5.1. Assume now that the control loop is broken immediately following the analog hold function $H(s)$, and examine the noise spectrum due to the VCO phase noise at that point. The equivalent phase error process that will be witnessed at this point due to the VCO phase noise may be represented by

$$\theta(t) = \frac{1}{N} \sum_{n=-\infty}^{\infty} \theta_v(nT) h(t - nT) \tag{5.106}$$

where $h(t)$ is again the impulse response of the analog hold function. The power spectral density of this phase error process is given by

$$S_\theta(f) = \frac{|H(f)|^2}{N^2} \left[\frac{1}{T^2} \sum_{m=-\infty}^{\infty} S_{\theta_v}\left(f - \frac{m}{T}\right) \right] \tag{5.107}$$

The $m = 0$ term is the spectral component that would be expected from continuous system theory. It is not difficult to show that the $m = 0$ spectrum term is reduced at the phase-locked loop output by roughly the open-loop gain whereas the $m \neq 0$ terms appear as an equivalent reference phase noise source, which is only attenuated outside the closed-loop bandwidth. The output phase noise spectrum due to the VCO noise spectrum may therefore be closely approximated by

$$S_{\theta_o}(f) = \frac{S_{\theta_v}(f)}{|1 + HG^*(f)/N|^2}$$

$$+ \left| \frac{HG(f)}{1 + HG^*(f)/N} \right|^2 \frac{1}{T^2} \sum_{\substack{m=-\infty \\ m \neq 0}}^{\infty} S_{\theta_v}\left(f - \frac{m}{T}\right) \tag{5.108}$$

This result reveals that the VCO phase noise and spurious components that are far removed from the carrier can nonetheless lead to close-in spectrum degradation due to aliasing if they are not adequately attenuated prior to the sampling process.

REFERENCES

[1] Crawford, J.A., "Understanding the Specifics of Sampling in Synthesis," *Microwaves & RF*, Vol. 23, No. 8, August 1984, pp. 120–126, 144.
[2] Franklin, G.F., J.D. Powell, *Digital Control of Dynamic Systems*, Addison-Wesley, Reading, MA, 1980, pp.79–80.
[3] Kuo, B.C., *Digital Control Systems*, Holt, Rinehart, Winston, New York, 1980, pp. 52–55, 114–120.
[4] Kolumbán, G., "Transient Properties of High Speed Frequency Synthesizers Based Upon Sampled PLL," EIM 87–12, pp. 314–317.
[5] Blake, J., "Design of Wideband Frequency Synthesizers," *RF Design*, May 1988, pp. 26–34
[6] Egan, W.F., "The Effects of Small Contaminating Signals in Nonlinear Elements Used in Frequency Synthesis and Conversion," *Proc. IEEE*, Vol. 69, No. 7, July 1981, pp. 797–811.
[7] Peregrino, L., D.W. Ricci, "Phase Noise Measurement Using a High Resolution Counter with On-Line Data Processing," *Proc. 30th Annual Symp. Frequency Control*, Atlantic City, NJ, June 1976, pp. 309–317.
[8] Bennett, "Methods of Solving Noise Problems," *Proc. IRE*, May, 1956, pp. 609–638.
[9] Gardner, W.A., *Introduction to Random Processes*, 2nd ed., McGraw-Hill, New York, 1990.
[10] Papoulis, A., *Probability, Random Variable, and Stochastic Processes*, McGraw-Hill, New York, 1965.
[11] Marple, S.L., *Digital Spectral Analysis*, Prentice-Hall, Englewood Cliffs, NJ, 1987.
[12] Wheatley, C.E., D.E. Phillips, "Spurious Suppression in Direct Digital Synthesizers," *Proc. 35th Annual Frequency Control Symp.*, Ft. Monmouth, NJ, May 1981, pp. 428–435.
[13] Blachman, N., *Noise and Its Effect on Communication*, 2nd ed., Robert E. Kreiger Publishing, Malabar, FL, 1982.
[14] Crawford,J.A., "Extending Sampling to Type 2 Phase-Locked Loops," *Microwaves & RF*, September 1984, pp. 171–174.

SELECTED BIBLIOGRAPHY

Barab, S.A.L. McBride, "Uniform Sampling Analysis of a Hybrid Phase-Locked Loop with a Sample-and-Hold Phase Detector," IEEE Trans. AES, Vol. AES-11, No. 2, March 1975, pp. 210–216.

Chua, L.O., P.Lin, Computer-Aided Analysis of Electronic Circuits: Algorithms and Computational Techniques, Englewood Cliffs, NJ: Prentice-Hall, 1975, Ch. 13.

Underhill, M.J., R.I.H. Scott, "The Effect of the Sampling Action of Phase Comparators on Frequency Synthesizer Performance," Proc. 33rd Annual Frequency Control Symposium, 1979, pp. 449–457.

Chapter 6
Fast-Switching Frequency Synthesizer Design Considerations

The term *fast switching* has often been misused in the frequency synthesizer community largely because its definition is quite nebulous. In this chapter, as throughout most of this text, only indirect synthesis methods are considered; specifically, the design of individual phase-locked loops. This chapter primarily expands on some of the concepts presented in Chapter 5, relating these concepts more directly to hardware implementation.

In the strict sense, frequency switching speed can only be measured in the context of a (1) given magnitude for the frequency change performed and (2) the degree to which final phase-lock must be obtained before the system can, in fact, be considered locked. In phase modulation systems such as BPSK, it is common to define the locking speed as that time required for a given worst case frequency change to be reduced by the loop such that the voltage-controlled oscillator (VCO) phase remains within 0.1 rad of final steady-state phase error. In FM systems, particularly analog voice systems, the requirement is often given in terms of the instantaneous frequency error, defining the time at which lock is achieved as that point beyond which the instantaneous frequency error with respect to steady state remains less than, say, 100 Hz. Any meaningful discussion of switching speed must address the issue in this more formalistic manner.

To add two more issues, it is not adequate to judge switching speed based solely on the absolute time required to achieve lock because (3) channel spacing and (4) spurious performance also play a significant role in any true comparison. In comparing the switching speed between loops that might have different reference frequencies (i.e., channel spacing), the time to achieve phase-lock should first be normalized with respect to the respective reference frequency being used. Although more difficult to evaluate, differences in spurious performance dramatically alter the overall performance assessment as well.

Although this discussion provides motivation for defining a fast-switching speed *figure of merit,* such a quantity would be driven by so many conflicting factors that its use would be at best questionable. Add to the four itemized issues given earlier the issues of (5) phase noise performance and (6) cost, and the figure-of-merit approach is doomed. Fractional-N synthesis, which is discussed in Chapters 7 and 9, further obscures identifying such a figure of merit.

As shown in Chapter 5, the optimum frequency switching speed for the type 1 and type 2 architectures are one and two sample periods, respectively, for ideal systems. As alluded to at that time, such performance is generally unattainable in analog systems for a number of reasons. In this chapter, several of the finer points involved in actually implementing a high-speed phase-locked loop such as that theorized in Chapter 5 are discussed. Later in the chapter, some comparisons will address choosing between the sample-and-hold phase detector and the simplified phase-frequency detector for different applications. In performing this comparison, attention will be focused on switching speed and sampling spur performance for the basic type 2 architecture shown in Figure 4.18. This figure was the subject of case 4 in Chapter 5 but incorporated a sample-and-hold phase detector type. This comparison should be helpful in deciding between phase detector types when such a choice is available.

6.1 TYPE 1 CASE STUDY

The departures from ideal type 1 loop performance that were considered in Chapter 5 only addressed (1) finite sample-and-hold phase detector sampling efficiency and (2) internal loop time delays resulting from device propagation delays and light additional filtering for reducing spectrum sampling spurs at the loop output. Other high-level issues that should be considered if near-optimal switching speed is required include (3) the degradation resulting from nonlinear VCO tuning characteristics and (4) sampling interval modulation, which occurs during the transient portion of the system response.

In this section, the design approach for a high-speed type 1 loop, which was reported in [1,2], is examined. Through the remainder of this chapter, high-speed switching will be taken to be equivalent to requiring that no cycle-slipping at the phase detector occurs during the transient response. Any inclusion of cycle-slipping effects (i.e., nonlinear effects) would otherwise result in a very difficult nonlinear problem that is most expediently investigated using simulation methods such as those discussed in Chapter 8.

A top-level block diagram for this loop is shown in Figure 6.1. A number of important design considerations are worth elaborating at this level of detail. First of all, the 1.8-GHz offset local oscillator was used primarily to reduce the range of input frequencies as seen at the $\div N$ input. At the time this circuitry was designed, dual-modulus counters (i.e., a 10/11 divider in this case) that operated above 550 MHz were a rarity, let alone those that operated over the entire military temperature range.

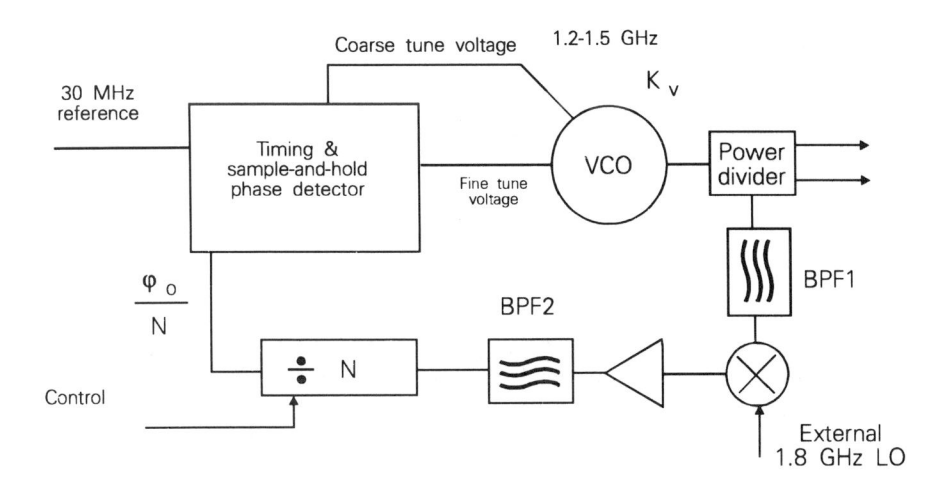

Figure 6.1. Type 1 phase-locked loop design example under consideration. After: [1].

Two other benefits are also achieved by this high-side local oscillator approach. Because the maximum $\div N$ is reduced, phase noise performance is also improved. In addition, because most VCO tuning characteristics flatten out near the upper end of their tuning range (i.e., K_v becomes smaller), this is compensated for to some degree by the corresponding decrease in the $\div N$ ratio as the desired synthesizer output frequency is increased. In this particular design, gain compensation was used across the full 1.2- to 1.5-GHz tuning range so this factor was less of an issue than it might otherwise have been.

The actual reference frequency used within the loop is 30 MHz divided by 10 or 3 MHz. It is very important to note that the 1.8-GHz local oscillator was also phase-locked to the same 30-MHz reference and that 1800 MHz is divisible by the reference frequency used within the loop. Otherwise, the aliasing effects of sampling [e.g., (5.12)] would have dramatically increased the filtering and isolation requirements of the frequency offset circuitry. Because all of the frequencies involved have integral relationships with 3-MHz reference, bandpass filter BPF2 was only required to reduce the unwanted mixer products to the point where false-clocking of the feedback divider input was prevented. As supported by discussions in Chapter 7 of [3], spurious product suppression for the frequency plan here by way of BPF2 need be quite minimal.

The reference frequency distributed to the phase-locked loop in Figure 6.1 was purposely made as high as possible to help alleviate possible problems with outside signal contamination. Since the 30-MHz source was divided down by a factor of 10 within the timing-sample-hold (TS&H) hybrid, in effect, the 3-MHz reference used by the loop had any phase noise contaminants reduced by 20 dB compared to bringing in the 3-MHz reference directly.

6.1.1 Internal Design Details

The type 1 control loop was designed in the spirit of an ideal type 1 loop where adjustment of the phase detector linear ramp slope was used to keep the normalized design parameter K [from (5.27)]

$$K = \frac{K_d K_v}{N} T \qquad (6.1)$$

as close to the optimal value of unity as possible. Even with coarse VCO pretuning errors as large as 40 MHz, full-band frequency hops to within 7 deg of a steady-state phase were achieved typically within less than eight samples or equivalently 2.5 μs while maintaining reference-related spurs below -55 dBc. Further speed enhancement would have been easily achieved by using more elaborate VCO frequency pretuning techniques.

The primary factors that limited the loop switching speed were (1) sampling efficiency and (2) internal transport delay. Hence, that was the rational for examining the impact of these two factors as was done in Chapter 5.

The heart of the design is unquestionably the sample-and-hold phase detector portion of the hardware, for which many of its finer points come under an issued patent [4]. The key elements of this phase detector are shown in Figure 6.2. In principle, the sample-and-hold phase detector was designed to operate very much like those discussed in other references, for example, [5]. Many refinements were made, however, in order to achieve the needed performance at the fairly high reference frequency. Since time delays and noise performance were crucial, discrete custom designs were used throughout the TS&H hybrid rather than utilizing more highly integrated devices with lower performance.

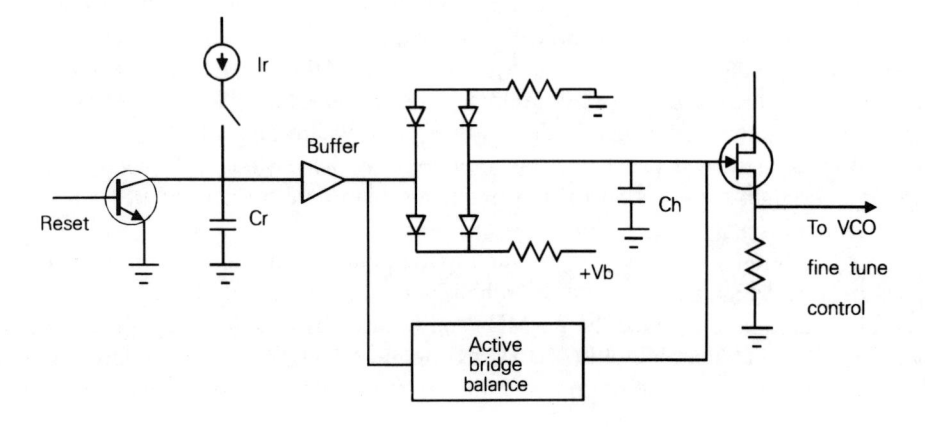

Figure 6.2. Important features of the sample-and-hold phase detector design.

A timing diagram for the principal signals within the TS&H is shown in Figure 6.3. The timing details have largely been duplicated within the Motorola MC145159 monolithic device where a linear ramp is formed, halted for sampling, and then quickly reset.

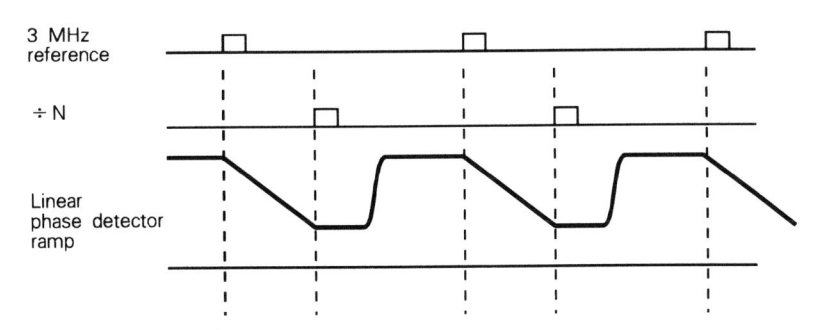

Figure 6.3. Sample-and-hold phase detector internal timing relationships.

Several factors forced the allocation of loop gain between the VCO tuning sensitivity K_v and the phase detector gain K_d. Since very high speed devices are usually limited to operation with 5 to 10V maximum, the total linear ramp excursion was limited to 0 to 6V here. It was also verified experimentally that very high slew rates in the phase detector linear ramp area led to degraded phase noise performance. Combining these constraints with (6.1) and with the desire that any frequency hop still result in linear phase detector operation (i.e., the first and all succeeding error samples must land on the linear portion of the phase detector ramp voltage characteristic in Figure 6.3), the range of linear ramp characteristics that was possible are as shown in Figure 6.4.

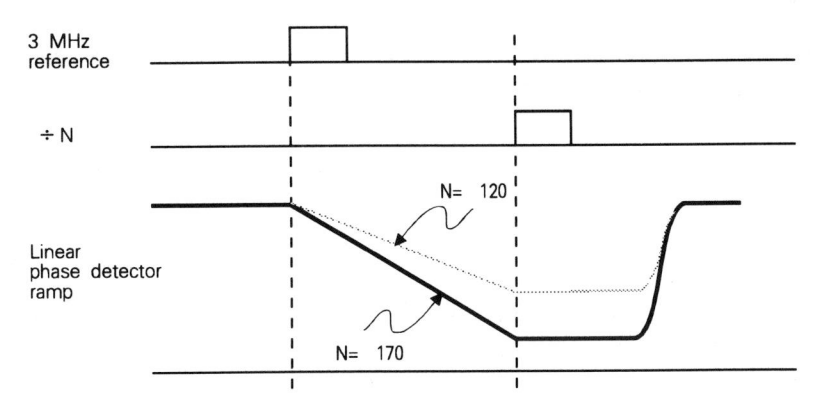

Figure 6.4. Gain compensation for changing feedback divider ratio $\div N$ was achieved by varying the slope of the linear phase detector ramp.

The latter objective of keeping all phase error samples on the linear ramp portion of the phase detector characteristic required additional special-purpose circuitry within the TS&H. Following each new VCO presetting and divide-by-N load operation, the linear ramp was allowed to slew its full range during one reference period, and a high-speed discrete comparator was used to detect when the ramp was at its mean-center value. At this instant, the feedback divider was enabled to begin accumulating counts. Given no VCO presetting error, no time delay through the comparator circuitry, and no initial counts in the dual-modulus prescaler within the divide-by-N, the next feedback divider output pulse would occur precisely at the nominal mean-center value of the next reference period as shown in Figure 6.5. Between old and new frequency word load commands, the linear ramp slope was changed to effect the needed gain compensation per (6.1). From Figure 6.4, it should be clear that such a change in slope necessarily altered the desired ramp sampling point within each reference period if the same sample-and-hold output voltage (nominal midramp value) was to result. This method of gain compensation mandated that some form of time base realignment be done similar to that shown in Figure 6.5. For this purpose, the multiphase clock (10 clock phases) possible from utilizing the input 30-MHz reference signal proved invaluable.

As shown in the upcoming discussion of a high-speed type 2 system, this issue of initial phase error following a command to change frequency is a substantial factor in attaining faster switching speed. Although adaptive VCO pretuning schemes can be created to deal with the initial VCO frequency error, the initial VCO phase is at least equally as important as faster and faster switching speed is sought. This is simply true because the loop electronics can only interpret any initial phase error as proportional to an underlying initial frequency error.

In the context of this previous discussion with the type 1 system, one conceptually simple method for dealing with this problem is to shift the reference phase at the beginning of a new load command sequence [6], which, in principle, is very similar to the reasoning used in Figure 6.5. During the first complete reference period following a new frequency word load command, the first phase error sample observed at the phase detector is given by

$$\theta_{e1} = \frac{2\pi \Delta f T + \theta_{\text{VCO}}}{N} \qquad (6.2)$$

where Δf is the residual VCO pretuning error remaining in hertz, and θ_{VCO} is the residual VCO phase error due to preexisting counts in the prescaler, circuit delays, etc. Since it is fairly impractical to reset any high-speed frequency prescaler precisely, the θ_{VCO} phase term in (6.2) can be substantial. Left uncorrected, from (6.2) the loop electronics can only interpret θ_{e1} as proportional to the residual frequency error even though the θ_{VCO} term is present. If, however, the loop electronics delay

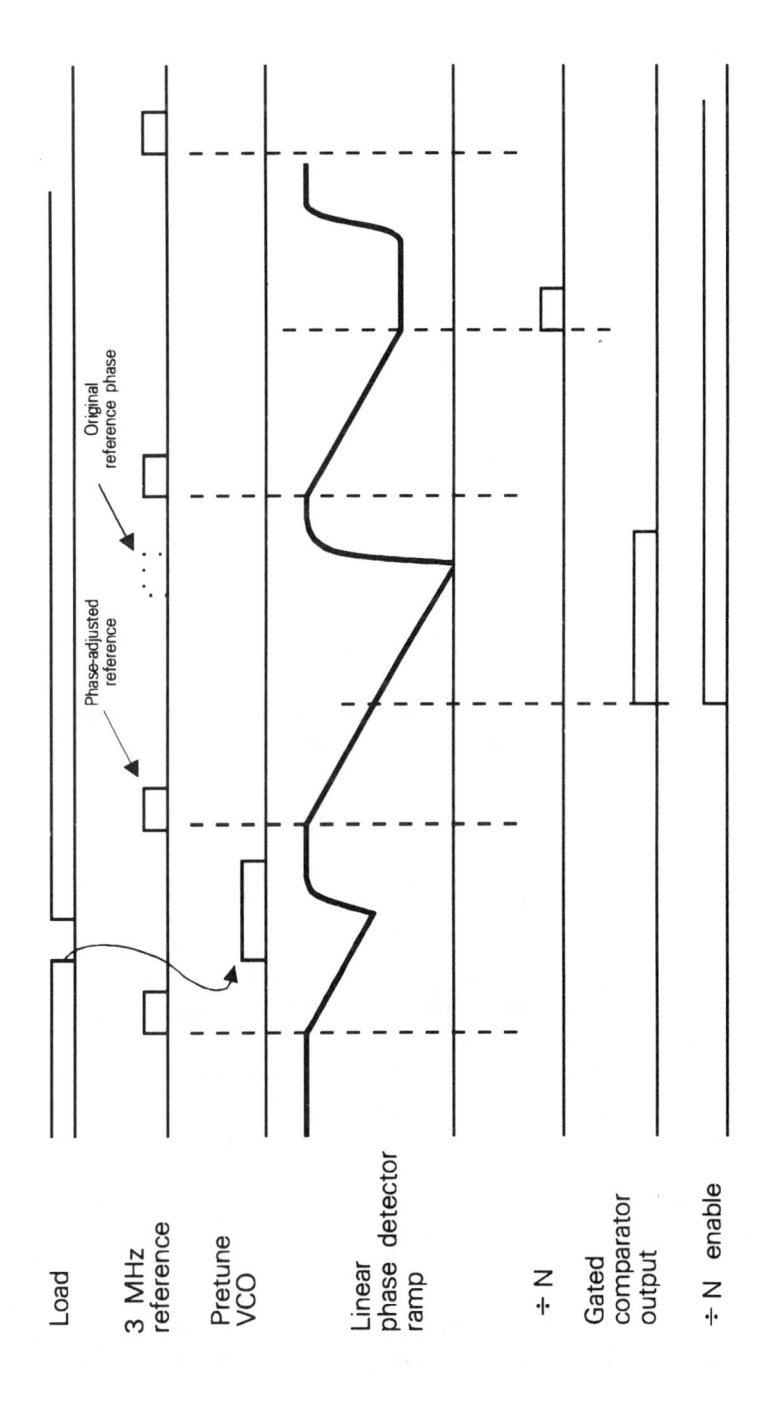

Figure 6.5. Loop timing initialization on receipt of a new frequency word load command. Following a new load command, (1) the VCO is pretuned, (2) the next available clock of the 30-MHz reference is used to initiate a new 3-MHz clock edge, (3) the nominal ramp center point is located, at which time the ÷N is enabled, and (4) normal loop operation commences.

any action on θ_{e1} and calculate the phase error at the end of a second reference period θ_{e2},

$$\theta_{e2} = \frac{2\pi\Delta f(2T) + \theta_{VCO}}{N} \tag{6.3}$$

From (6.2) and (6.3) then,

$$\theta_{bias} = \frac{\theta_{VCO}}{N} = 2\theta_{e1} - \theta_{e2} \tag{6.4}$$

and the accurate phase error (which is proportional to Δf alone) is given by

$$\theta'_{e2} = \theta_{e2} - \theta_{bias} = \frac{2\pi\Delta f T}{N} \tag{6.5}$$

and the real frequency error related to reference frequency phase error is properly visible. Therefore, if the reference phase can be adjusted by θ_{bias}, theoretically the negative impact of a nonzero θ_{VCO} can be eliminated. Although this reasoning is conceptually simple, hardware realizations are considerably more difficult and often impractical to attempt.

Sampling Bridge Design Considerations

Computer-aided analysis was used throughout the hardware design of this synthesizer in order to obtain the best type 1 switching speed and spurious performance possible. The trade-offs between switching speed and spurious performance were particularly acute as described in the following paragraphs.

The sampling gate used in the type 1 system was a quad Schottky diode bridge, which was balanced with respect to the linear phase detector ramp and the sampling pulses used to actually take the ramp voltage sample [4]. Since the linear ramp duty cycle can be nearly 100%, blow-through or leakage of the linear ramp through the stray bridge capacitance was a major spurious problem that had to be solved.

As described in [4], the bridge balancing with respect to the linear ramp waveform was dramatically improved by a simple cancellation technique. One such implementation is shown in Figure 6.6. In this method, the feedback signal is fed back 180 deg out of phase with respect to the leakage signal, thereby creating an effective negative capacitance, which ideally just cancels the leakage through the bridge.

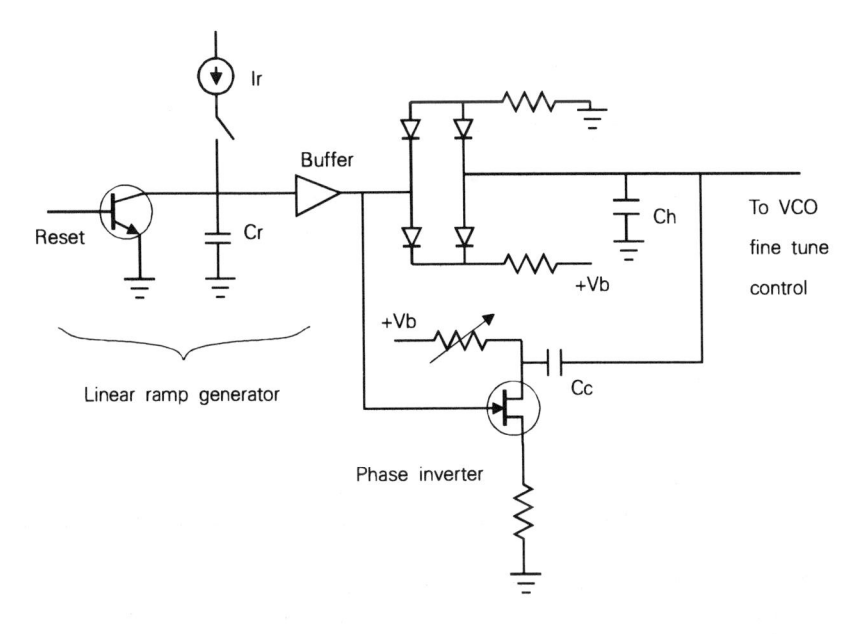

Figure 6.6. Balancing of the sampling bridge with respect to the ramp input signal may be accomplished by means of feedforward cancellation.

The blow-through problem may be described as presented in Figure 6.7 where capacitance C_s represents the stray capacitance of each of the four Schottky bridge diodes. The stray capacitances combined with the hold capacitor C_H form a simple voltage divider at high frequencies, with the resulting leakage serving to modulate the VCO directly. Since no additional filtering can be inserted between the sample-and-hold output and the VCO tuning port input without introducing disastrous group delay, the leakage term must be kept very small if low reference spurs are to be obtained. In the present system, if a 100% duty cycle triangular ramp voltage is assumed with a peak value of 6V, taking the VCO tuning sensitivity as 20 MHz per volt with a 3-MHz reference frequency, the leakage-induced reference spur levels at ±3 MHz from the carrier are approximately

$$L = 20 \log_{10} \left(\frac{6 \text{ V}}{\pi} \frac{20 \text{ MHz/V}}{2 \cdot 3 \text{ MHz}} \alpha \right) \quad \text{dBc} \qquad (6.6)$$

where α is the capacitive voltage divider ratio through the sampling bridge. To realize -60-dBc sampling spurs, α must be ≤ 0.000157, which is extremely small given that C_s is on the order of 0.2 pf and C_H can be made no larger than roughly

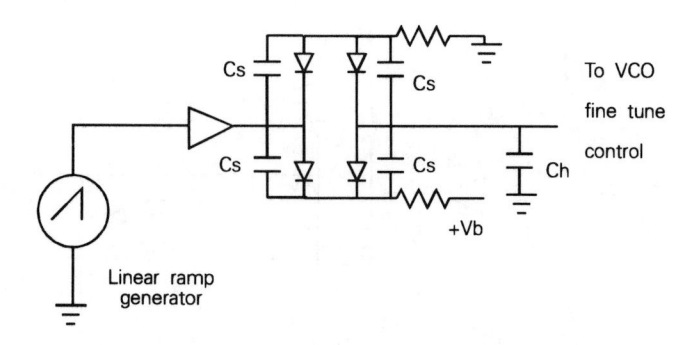

Figure 6.7. Diode sampling bridge stray capacitance is one of the primary ramp leakage mechanisms.

56 to 75 pf in order to have adequate sampling efficiency to attain the desired switching speed. From this initial discussion, it should be clear that the output impedance of the ramp buffer driving the diode bridge input, the on-resistance of the Schottky diodes, bridge balance effects, and the sampling pulse width (which directly affects sampling efficiency) plus other details were all important issues affecting the overall performance of this design.

Although modulation of the Schottky diode reverse bias capacitance by the ramp input voltage and other secondary effects may have to be considered in balancing this sampling bridge with respect to the linear ramp signal, balancing the bridge with respect to the sampling pulses also had to be done. This can be considerably more complex. A fairly simple yet effective sampling pulse generation scheme is shown in Figure 6.8 [7], which is suitable for reference frequencies of less than approximately 1 MHz. A considerably more involved method had to be utilized in the subject design, which exploited step-recovery diodes in order to obtain the desired sampling pulse width of approximately 3 ns at the 3-MHz sampling rate, but the following discussion is still applicable even for that case. In Figure 6.8, transformer $T1$ forms ideally equal but opposite voltage pulses at its two outputs. Transformer $T2$ operates as a balun to balance the bridge driving current waveforms effectively, thereby equalizing the forward and reverse sample pulse currents. Any imbalance in the pulse driving currents or the Schottky bridge diodes creates small pulse-like contamination during each sampling interval, which creates sampling spurs at the phase-locked loop output. In high-speed systems, analysis and experimental verification showed that the reverse bias voltage V_B in Figure 6.8 should ideally be maintained at a value equal to twice the sample-and-hold output voltage in order to keep the bridge symmetry intact [4].

Focusing on the bridge portion of Figure 6.8 and assuming quasi-static dc operating conditions during the sampling pulse interval, some insight into circuit balance requirements can be gained by considering Figure 6.9. The quasi-static bridge

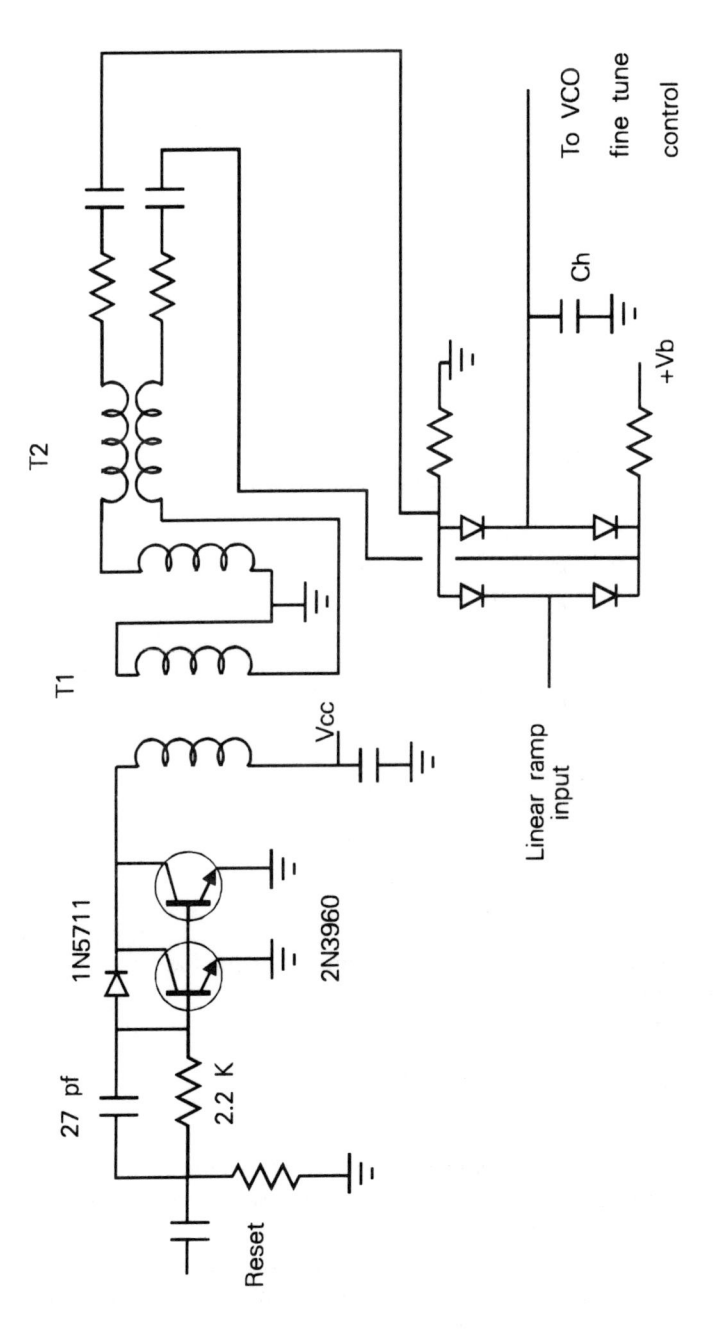

Figure 6.8. Sampling pulse generator and quad-diode bridge for a low reference frequency (i.e., <1 MHz) sample-and-hold phase detector.

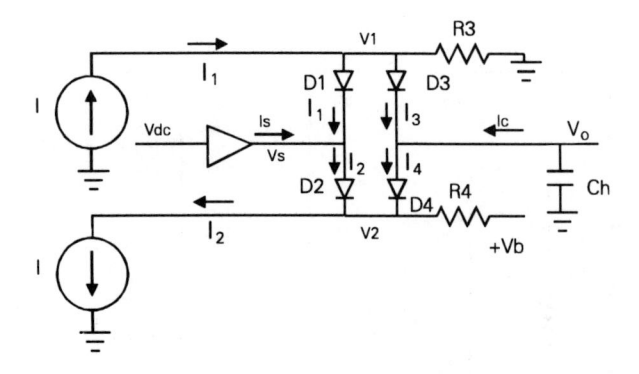

Figure 6.9. Simplified quasi-static diode bridge balance problem formulation.

is shown as driven by equal but opposite sampling currents I_1 and I_2. Summing the currents at voltage nodes V_1 and V_2,

$$I_1 = \frac{V_1}{R_3} + I_{d1} + I_{d3}$$

$$= \frac{V_1}{R_3} + f_{d1}(V_1 - V_s) + f_{d3}(V_1 - V_0) \tag{6.7}$$

$$I_2 = \frac{V_B - V_2}{R_4} + I_{d2} + I_{d4}$$

$$= \frac{V_B - V_2}{R_4} + f_{d2}(V_s - V_2) + f_{d4}(V_0 - V_2) \tag{6.8}$$

where $f_i(\ )$ represents the current-voltage relationship for the ith diode. This diode relationship is taken to be

$$f_i(v) = I_i R_i + \frac{kT}{q} \ln\left(\frac{I_i}{I_s}\right) \tag{6.9}$$

which is the conventional diode equation with a series bulk resistance R_i included. These equations may be solved numerically in an iterative manner for the hold capacitor current I_c, which initially flows each reference period during the sampling interval while the hold capacitor C_H is charging. Any nonzero I_c is effectively integrated by C_H, thereby causing small perturbations to the VCO tuning voltage every reference period. Ideally, for a constant input $V_s = V_0$, $I_c = 0$. To further quantify the impact of any imbalance, assume that the undesired bridge imbalance causes I_c to flow for τ sec of each reference period T where $\tau \ll T$. The output voltage Fourier component at the reference frequency then results in a (first sideband) spurious level of approximately

$$L = 20 \log_{10}\left(\frac{I_c \tau^2}{8\pi C_H} K_v\right) \text{ dBc} \qquad (6.10)$$

This result is only applicable where τ is small compared to the true charging time for C_H. In the present context with $C_H = 75$ pf, $K_v = 2\pi\ 20$ MHz/V, and taking $\tau = 5$ ns, I_c must be ≤ 0.337 mA in order to realize ≤ -65-dBc reference spurs. Normally, the sampling currents must be on the order of 10 mA to realize adequately low diode on resistance during the sampling action. Although it is true that better sampling pulse balance is obtainable for $R_3 = R_4$ large given a fixed voltage V_B, increasing R_3 in low reference frequency cases can exacerbate the aforementioned ramp blow-through problem. Using (6.7) through (6.10), the predicted spurious output levels as a function of different bridge imbalance are shown in Figure 6.10. In this figure, the resulting spurious performance was calculated for (1) imbalance in the bridge driving currents I_1 and I_2, and for (2) small differences in the bulk resistances for diodes D_1 and D_2. For (1), $R = I_1/I_2$ whereas for (2), $R = R\text{bulk}_1/R\text{bulk}_2$.

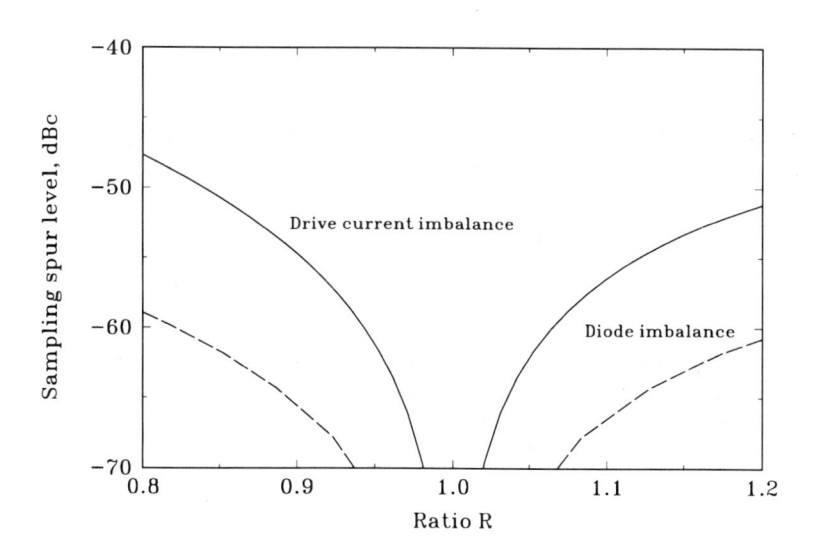

Figure 6.10. Low reference (sampling) spurs require very good matching between all of the sampling bridge diodes. Imbalances in the bridge driving currents and diode bulk resistances were considered separately.

Other analyses, particularly those concerning the magnetic components in Figure 6.8, are generally also required in order to guarantee that the drive currents are maintained in a sufficiently balanced fashion during each sampling pulse [7]. These and other detailed considerations that are not addressed here should underscore why the vast majority of modern synthesizer phase-locked loop designs utilize digital rather than analog (sample-and-hold) phase detectors wherever possible.

Monte Carlo Loop Analysis

As mentioned earlier, considerable computer design work was performed during the hardware design effort in order to identify potential circuit parameter problems as well as make prudent initial design choices. The system linearity assumptions adopted for the acquisition process made z transform analysis like that presented in Chapter 5 very valuable. Once loop parameter ranges were obtained as expected in production quantities, the z transform description also permitted fast Monte Carlo analysis of the end-product loop transient performance to be assessed.

Starting as we did in Chapter 5, the open-loop gain function considered here may be written as

$$G_{\text{OL}}(s) = \frac{1}{N} K_d \frac{1 - e^{-sT}}{s} \frac{\eta}{1 - (1 - \eta)e^{-sT}} \frac{K_v}{s} e^{-s\tau} \qquad (6.11)$$

where inefficient sampling effects (i.e., η) and internal time delay effects (i.e., τ) have been included. This may be converted to modified z transforms as

$$G_{\text{OL}}(z) = \frac{K_d K_v T \eta m}{N} \frac{z + (1 - m)/m}{z + (\eta - 1)} \frac{1}{(z - 1)} \qquad (6.12)$$

Given that the primary quantity of interest is the loop transient response to a step frequency change Δf Hz plus a step phase change $\Delta \theta$ effectively at the VCO, the reference port phase error excitation to the loop is given by

$$\theta_r(z) = \frac{1}{N} \frac{\Delta \theta z^2 + z(2\pi \Delta f T - \Delta \theta)}{(z - 1)^2} \qquad (6.13)$$

The resulting closed-loop transient phase error response at the VCO output is then

$$\theta_o(z) = \frac{n_3 z^3 + n_2 z^2 + n_1 z + n_0}{d_3 z^3 + d_2 z^2 + d_1 z + d_0} \tag{6.14}$$

where

$$n_3 = \Delta\theta$$

$$n_2 = \Delta\theta(\eta - 1) + 2\pi\Delta fT - \Delta\theta$$

$$n_1 = (2\pi\Delta fT - \Delta\theta)(\eta - 1)$$

$$n_0 = 0$$

$$d_3 = 1$$

$$d_2 = (\eta - 1) - 2 + K\eta m \tag{6.15}$$

$$d_1 = 2(1 - \eta) + 1 - 2\alpha K\eta m$$

$$d_0 = \eta - 1 - \alpha K\eta m$$

$$\alpha = \frac{(1 - m)}{m}$$

$$K = \frac{K_d K_v T}{N}$$

In this form, evaluation of the transient error response can be easily performed by using repeated long-hand division like that employed in Chapter 5.

A typical Monte Carlo analysis that utilizes this approach for a type 1 system is shown in Figure 6.11. A computer run examining 100 different parameter sets falling within user predefined limits only requires a few seconds on a 10-MHz 80286 machine equipped with a math coprocessor. The analysis shown includes a first-order low-pass filter as an additional degree of freedom beyond the formula presented here. With proper circuit parameter extraction, this design methodology provides tangible evidence for deciding whether a design is suitable for release into production. A similar Monte Carlo analysis is shown in Figure 6.12 for a type 2 system.

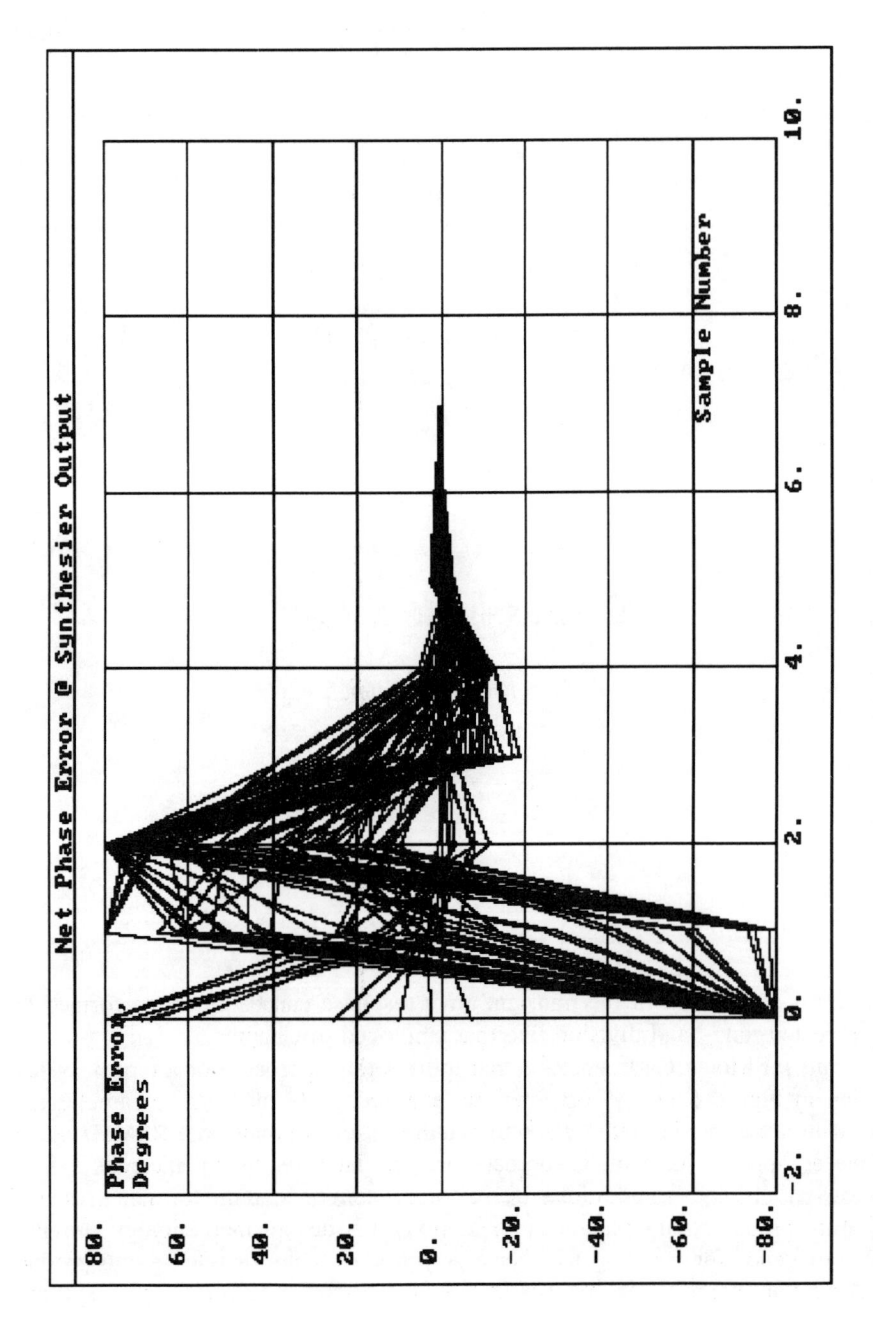

Figure 6.11. Monte Carlo transient response analysis for a typical high-performance type 1 system.

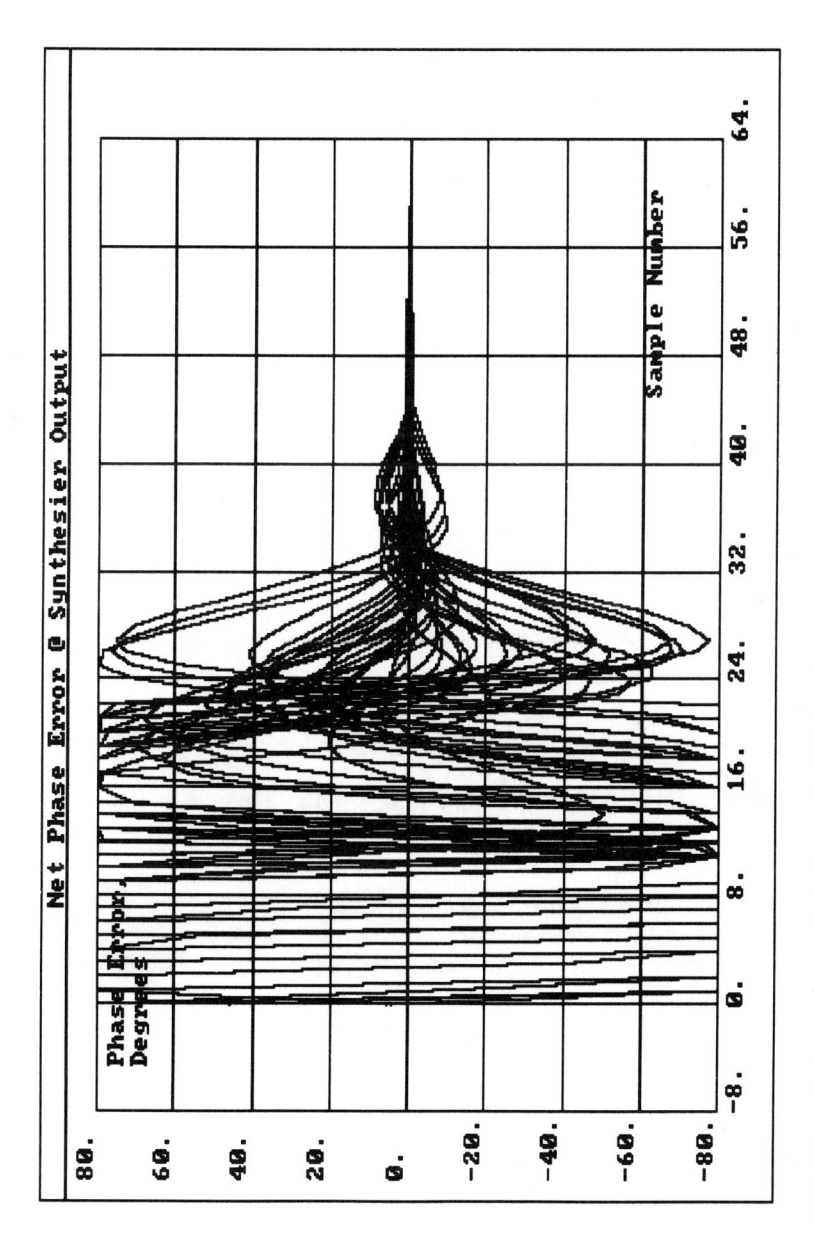

Figure 6.12. Monte Carlo transient analysis for a typical high-performance type 2 system.

6.1.2 Post-Tuning Drift (Nemesis of Type 1 Designs)

Before moving on to other material, it is important to emphasize one of the primary drawbacks of a type 1 system: post-tuning drift. Although the steady-state frequency error is zero for a type 1 system, this is not true of the steady-state phase error. Whether the cause is a thermal time constant in the VCO or vibration, any change in the VCO tuning characteristics is manifested as a phase change at the loop output; in other words, post-tuning drift. Generally, thermal time constant related post-tuning drift is not a problem for oscillators below 1.5 GHz, but the designer must thoroughly evaluate this issue within his or her own design constraints. In the multi-gigahertz range where severe VCO post-tuning drift is commonplace, even type 2 systems must consider drift effects because mitigation of these effects by a type 2 architecture is a strong function of closed-loop bandwidth. Therefore, type 2 systems with inadequate closed-loop bandwidth can also fall prey to this drift problem. More is said concerning this important subject in Sec. 6.5.

6.2 INITIAL PHASE ERROR IMPACT ON SWITCHING SPEED

The presence of a combined phase and frequency error can lead to substantially longer switching times in a phase-locked loop than that predicted by only a frequency error analysis. This is only noticeable in systems intended for near-optimal switching speed where (6.2) can lead to fairly large initial loop phase error contamination. The underlying analysis is quite straightforward and similar to that undertaken in [6].

The case considered here utilizes a 10-MHz reference frequency and a type 2 sample-and-hold architecture. Aside from an internal loop transport time delay of 20 ns, the loop is otherwise assumed to be ideal. The criterion for determination of phase-lock is that the output phase error had to remain within 15 deg of steady state from that point onward. The analysis and following figures were based on a damping factor grid spacing of 0.05 and a normalized natural frequency grid spacing of 0.03. The study assumed that a fairly sophisticated VCO presetting capability existed that made the initial frequency errors quite small. A brief examination of Figures 6.13 through 6.16 clearly shows the dramatic change in settling speed performance as a function of initial phase error. The phase error impact becomes increasingly small as the initial frequency error becomes large. This viewpoint is supported by the information presented in Figure 6.17. The time required for each parameter pair ($\omega_n T$, ζ) to reach the phase-lock criteria is given in terms of sample count for each figure. A value of 5 implies that the locking process requires five reference sample periods or 0.5 μs, for example, at 10 mHz.

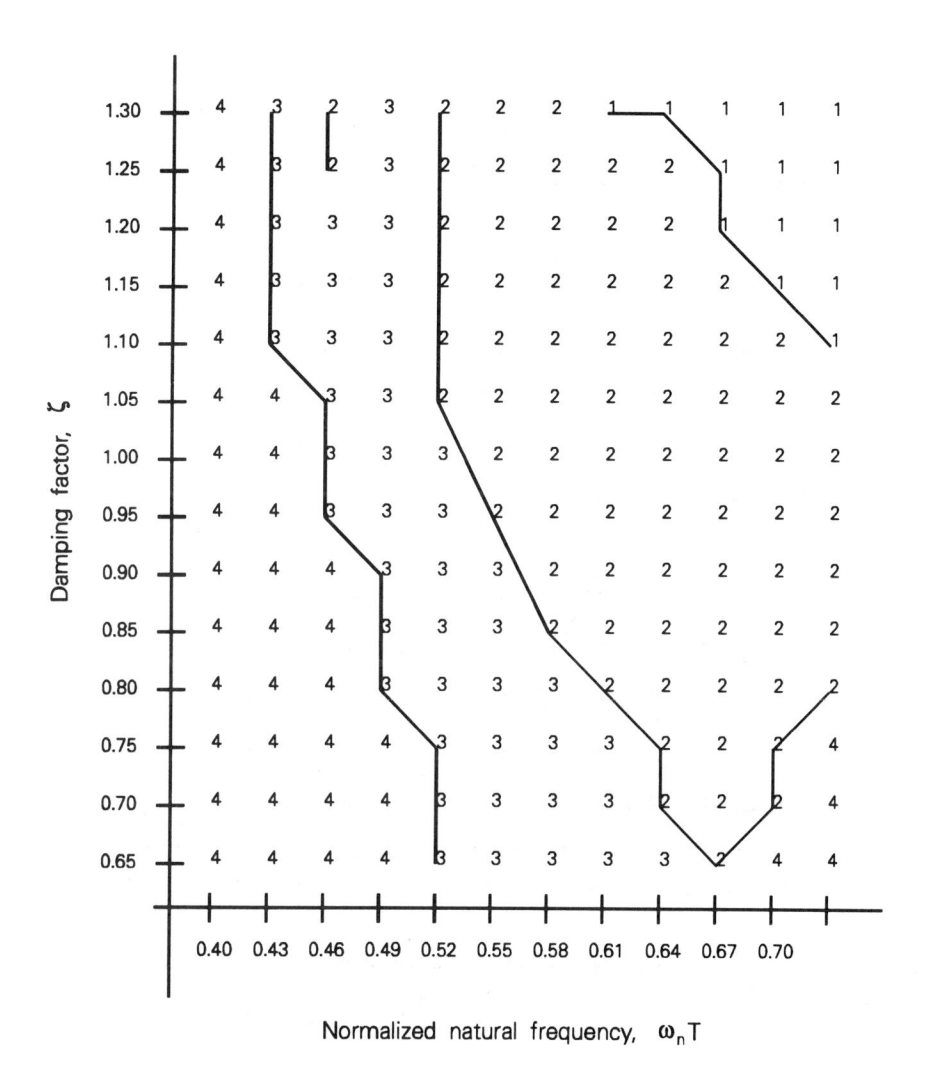

Damping factor, ζ												
1.30	4	3	2	3	2	2	2	1	1	1	1	1
1.25	4	3	2	3	2	2	2	2	2	1	1	1
1.20	4	3	3	3	2	2	2	2	2	1	1	1
1.15	4	3	3	3	2	2	2	2	2	2	1	1
1.10	4	3	3	3	2	2	2	2	2	2	2	1
1.05	4	4	3	3	2	2	2	2	2	2	2	2
1.00	4	4	3	3	3	2	2	2	2	2	2	2
0.95	4	4	3	3	3	2	2	2	2	2	2	2
0.90	4	4	4	3	3	3	2	2	2	2	2	2
0.85	4	4	4	3	3	3	2	2	2	2	2	2
0.80	4	4	4	3	3	3	3	2	2	2	2	2
0.75	4	4	4	4	3	3	3	3	2	2	2	4
0.70	4	4	4	4	3	3	3	3	2	2	2	4
0.65	4	4	4	4	3	3	3	3	3	2	4	4

Normalized natural frequency, $\omega_n T$: 0.40 0.43 0.46 0.49 0.52 0.55 0.58 0.61 0.64 0.67 0.70

Figure 6.13. Settling speed for an initial frequency error of 1 MHz, initial phase error of 50 deg.

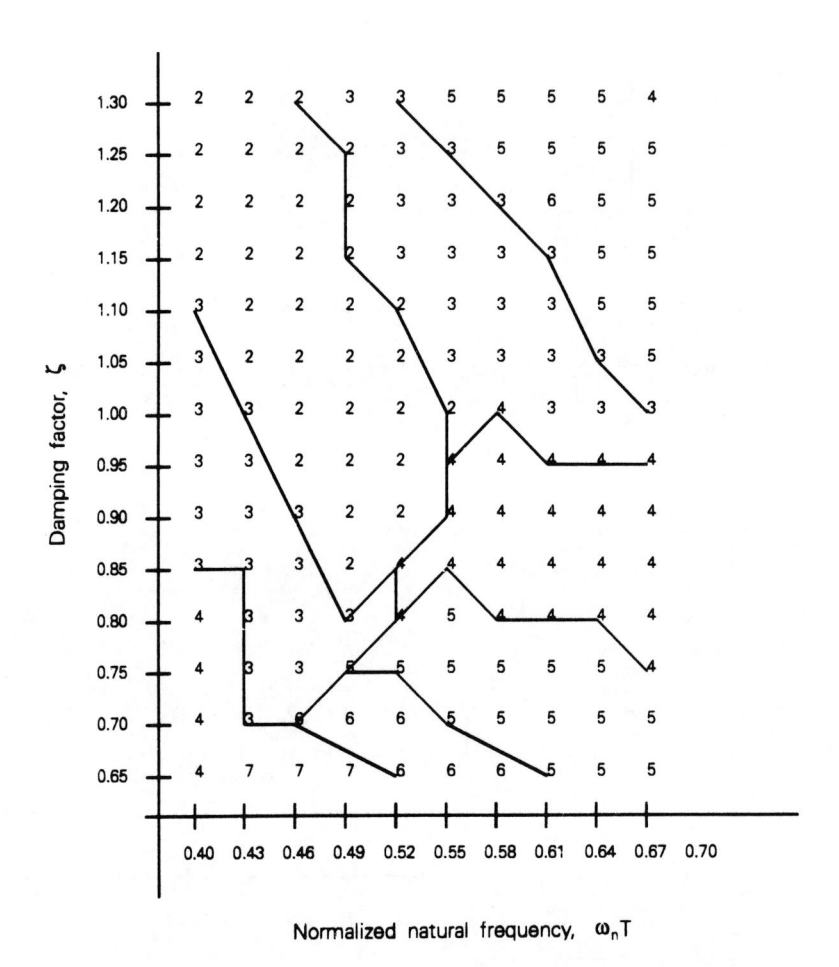

Figure 6.14. Settling speed for an initial frequency error of 1 MHz, initial phase error of 100 deg.

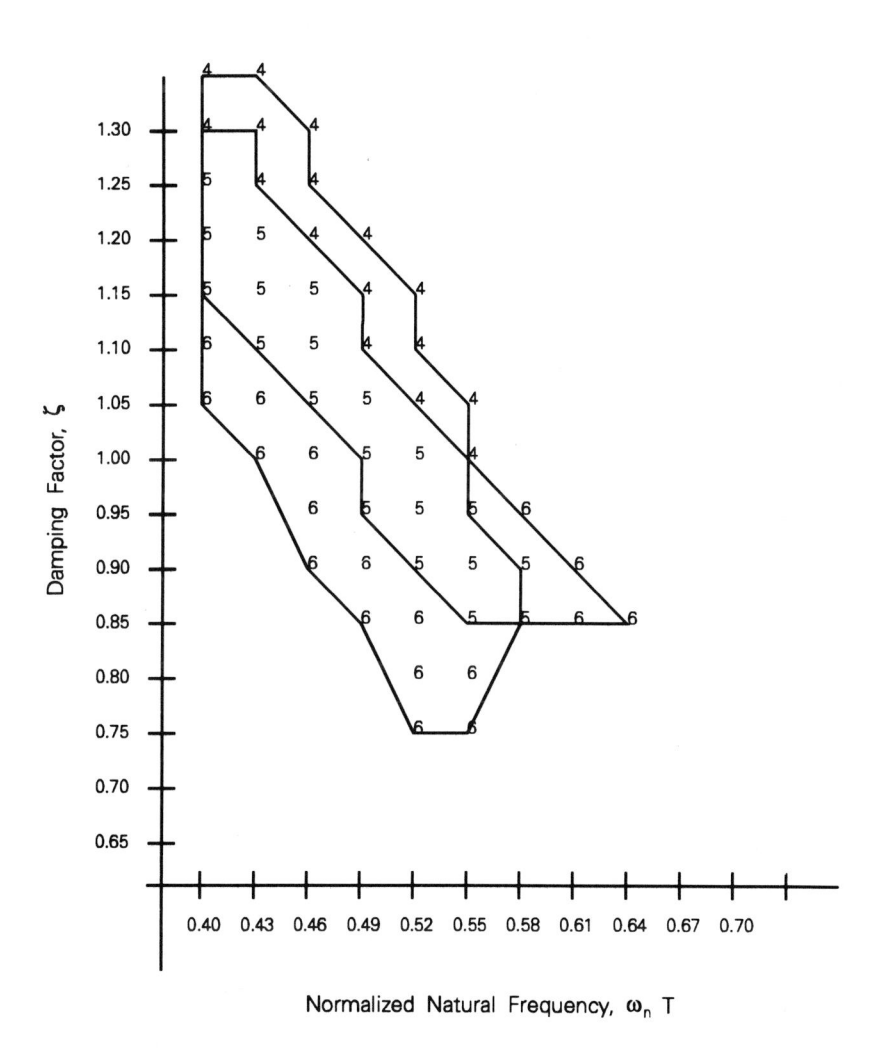

Figure 6.15. Settling speed for an initial frequency error of 1 MHz, initial phase error of 360 deg.

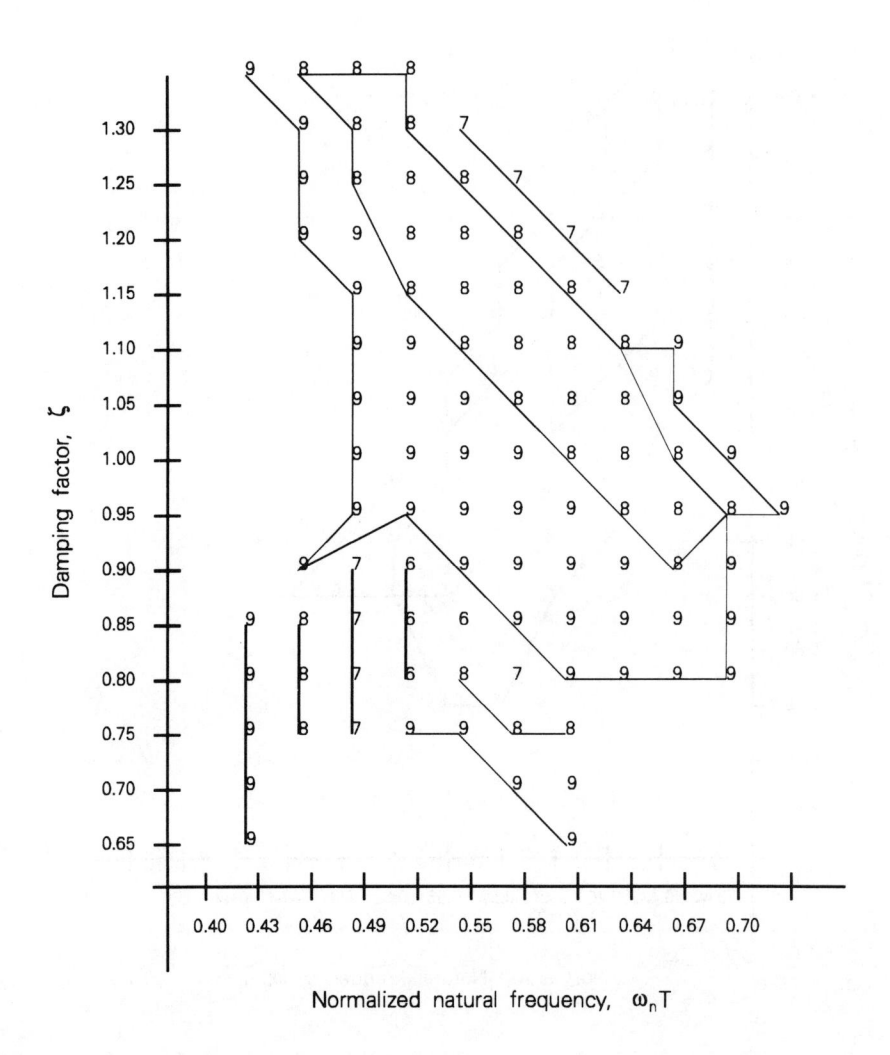

Figure 6.16. Settling speed for an initial frequency error of 1 MHz, intial phase error of 720 deg.

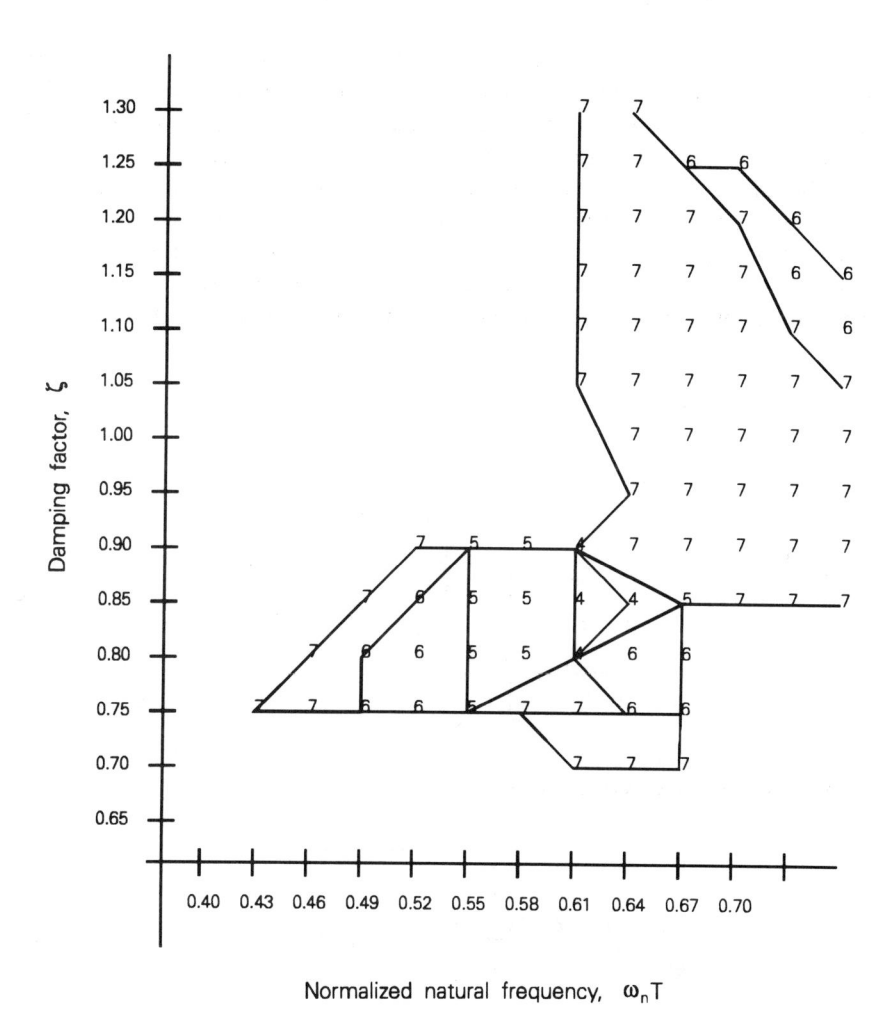

Figure 6.17. Settling speed for an initial frequency error of 5 MHz, initial phase error of 0 deg.

6.3 SAMPLE-AND-HOLD PHASE DETECTOR NOISE

The circuit complexity of the sample-and-hold phase detector can lead to noise problems if careful design precautions are not taken. In this section, the noise associated with the linear ramp portion of the phase detector is considered. In contrast, this noise source is not present in the simplified phase-frequency detector.

Ideally, the linear ramp current generator I_r in Figure 6.6 should be free of noise. Assume for the sake of this discussion that the I_r current source is that shown in Figure 6.18 where resistor R determines the current magnitude. Although the PNP transistor will contribute noise to I_r, only the noise associated with resistor R will be examined at this time. The noise equivalent representation for R is shown in Figure 6.19.

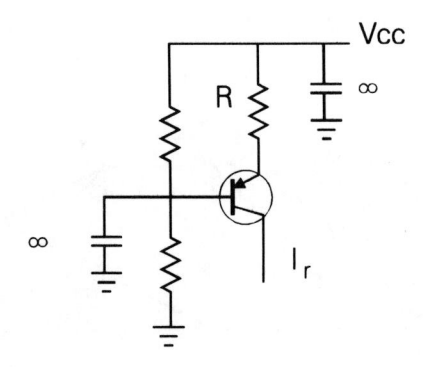

Figure 6.18. Representative current source for ramp current I_r.

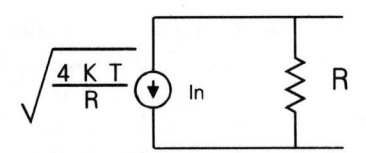

Figure 6.19. Equivalent circuit for a noisy resistor (i.e., Johnson noise).

During the linear ramping portion of each reference period, the desired ramp current I_r and the accompanying ramp noise current i_n are integrated in capacitor C_r in Figure 6.6. The capacitor is assumed to be noiseless. After allowing the ramp current to be integrated for T_r sec, the ramp voltage is given by

$$V_r = \frac{I_r T_r}{C_r} + \int_0^{T_r} \frac{i_n(t)}{C_r} dt \qquad (6.16)$$

The mean value of V_r at the sampling instant is simply

$$\bar{V}_r = \frac{I_r T_r}{C_r} \tag{6.17}$$

and the variance is given by

$$\sigma_r^2 = \mathbf{E}\left\{\int_0^{T_r} \frac{i_n(t)}{C_r} dt \int_0^{T_r} \frac{i_n(t')}{C_r} dt'\right\}$$

$$= \frac{1}{C_r^2}\int_0^{T_r} \frac{4kT}{R} dt$$

$$= \frac{4kT}{C_r^2 R} T_r \tag{6.18}$$

These results represent the mean and variance for a single sample-and-hold output sample only. To calculate the output noise power spectral density, which is really of interest from a phase noise point of view, some additional calculations must be performed also.

The sampled noise process occurring at the sample-and-hold output may be represented by

$$n_M(t) = \lim_{M \to \infty} \sum_{k=-M}^{M} d_k \, \text{rect}(t - kT_s) \tag{6.19}$$

where the d_k are the random noise sample values having variance σ_r^2, T_s is the sampling time interval, and rect(t) is unity for $-T_s/2 \leq t \leq T_s/2$ and zero otherwise.

The power spectral density for $n(t)$ is given by

$$S_n(f) = \lim_{M \to \infty} \frac{\mathbf{E}\{|\mathbf{F}[n_M(t)]|^2\}}{2MT_s} \tag{6.20}$$

where $\mathbf{F}$ represents the Fourier transform operation and $\mathbf{E}$ represents statistical expectation. Straightforward calculation shows that

$$\mathbf{F}\{n_M(t)\} = \sum_{k=-M}^{M} d_k \frac{\sin(\pi f T_s)}{\pi f} e^{-j2\pi f k T_s} \tag{6.21}$$

Performing the modulus-squared and the statistical expectation operations,

$$\mathbf{E}\{\cdot\} = \sum_{k=-M}^{M} \sum_{p=-M}^{M} R_d(|k-p|) \frac{\sin(2\pi f T_s)}{(\pi f)^2} e^{-j2\pi f(k-p)T_s} \qquad (6.22)$$

where it has been assumed that the d_k noise samples exhibit widesense stationarity. Taking the limit on M, the power spectral density for the noise-corrupted sample sequence is given by

$$S_n(f) = 2T_s \left[\frac{\sin(\pi f T_s)}{\pi f T_s}\right]^2$$

$$\times \left[\frac{R_d(0)}{2} + \sum_{k=1}^{\infty} R_d(k)\cos(2\pi f k T_s)\right] \qquad (6.23)$$

where $R_d(\ )$ is the autocorrelation function for the random d_k noise samples. In the present case, the noise samples may be assumed to be statistically independent, thereby simplifying this result to

$$S_n(f) = T_s \sigma_r^2 \left[\frac{\sin(\pi f T_s)}{\pi f T_s}\right]^2$$

$$\approx T_s \sigma_r^2 \ \frac{\mathrm{V}^2}{\mathrm{Hz}} \qquad (6.24)$$

for small frequency offsets. Converting this to an equivalent phase noise floor at the phase detector output,

$$S_\theta(0) = 10 \log\left(\frac{T_s \sigma_r^2}{K_d^2}\right) \ \frac{\mathrm{dBc}}{\mathrm{Hz}} \qquad (6.25)$$

where K_d is the phase detector gain in volts per radian and σ_r^2 is given by (6.18). The final noise floor result due to this single noise source is therefore

$$S_\theta(0) = 10 \log\left(\frac{T_s}{K_d^2} \frac{4kT}{C_r^2 R} T_r\right) \qquad (6.26)$$

These results may be used to illustrate one design example. Assume that a sample-and-hold phase detector with the following characteristics is to be examined:

K_d 2 V/rad
R 3.3 kΩ
T_s 10 μs (100-kHz reference)
T_r 5 μs ($T_s/2$)
k Boltzmann constant
T absolute temperature, 290 K
C_r 8000 pf

From these assumptions and (6.26), the resulting noise floor is $S_\theta(0) = 9.45 \times 10^{-19}$ rad^2/Hz, which is very good. Normally, other factors result in a substantially higher noise floor that masks this contribution.

In conclusion, this discussion provides some insight into the level of details that must be considered in high-performance design situations.

6.4 TYPE 2 PHASE-LOCKED LOOP WITH SIMPLIFIED PHASE-FREQUENCY DETECTOR

With so much attention in the design community focused on the fairly classic type 2 architecture utilizing the simplified phase-frequency detector, it is important that additional material be devoted to this case. This loop architecture was introduced in Chapter 4, Figure 4.18, and is repeated here as Figure 6.20. From Chapter 4, the open-loop gain function for this case was found as

$$G_{OL}(z) = (\omega_n T)^2 z \frac{K_1 z + K_0}{(z - 1)^2 (z - A)} \tag{6.27}$$

where

$$K_1 = \left(\frac{\tau_2}{T} - \frac{\tau_0}{T}\right)(1 - A) + 1$$

$$K_0 = \left(\frac{\tau_2}{T} - \frac{\tau_0}{T}\right)(A - 1) + A$$

$$A = e^{-T/\tau_0}$$

$$z = e^{j\omega T}$$

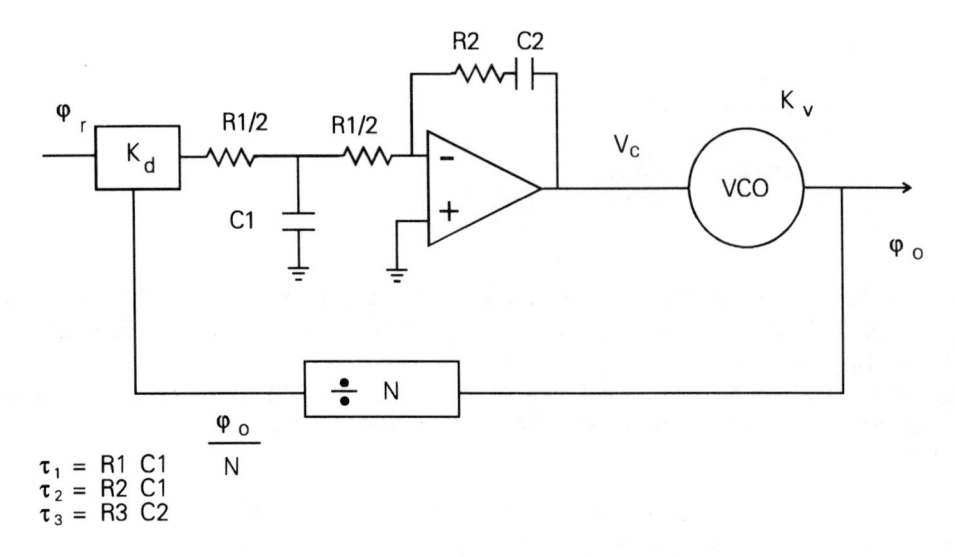

Figure 6.20. Basic type 2 third-order phase-locked loop with simplified phase-frequency detector.

$\tau_1 = R1\ C1$
$\tau_2 = R2\ C1$
$\tau_3 = R3\ C2$

While it is possible to derive algebraic results for the gain and phase margins as was done in Chapter 5, it is much more expedient to calculate these quantities numerically instead. The gain margin may be calculated by computing the maximum value of K such that the roots of

$$1 + KG_{OL}(z) = 0 \tag{6.28}$$

lie within the unit circle. Constant gain margin contours are shown versus (ζ, ω_n) for several different normalized low-pass filter corner frequency values $[f_c T = T/(2\pi\tau_0)]$ in Figures 6.21 through 6.24.

Similarly, constant phase margin contours may be found by calculating the maximum value of φ such that the roots of

$$1 + e^{j\varphi} G_{OL}(z) = 0 \tag{6.29}$$

lie within the unit circle. These contours are shown versus (ζ, ω_n) in Figures 6.25 through 6.28 for the same normalized low-pass filter cases. The curves of these figures are helpful, particularly when additional filtering for sampling spur reduction must be taken into account. The normalized filter corner frequencies $f_c T$ of 0.5, 0.25, 0.1, and 0.05 represent additional attenuation of the first reference frequency spurs of 7.0, 12.3, 20.0, and 26 dB, respectively.

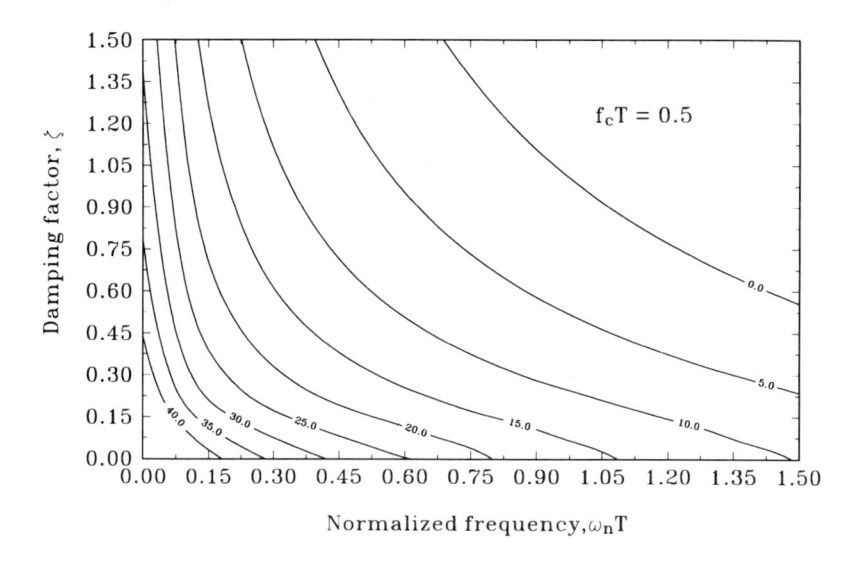

Figure 6.21. Constant gain margin contours for low-pass filter parameter $f_c T = 0.50$ in Figure 6.20.

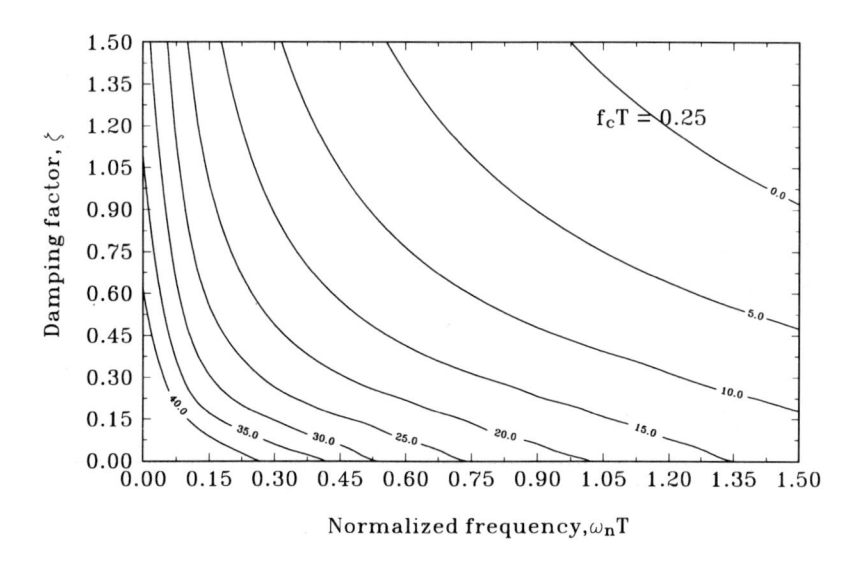

Figure 6.22. Constant gain margin contours for low-pass filter parameter $f_c T = 0.25$ in Figure 6.20.

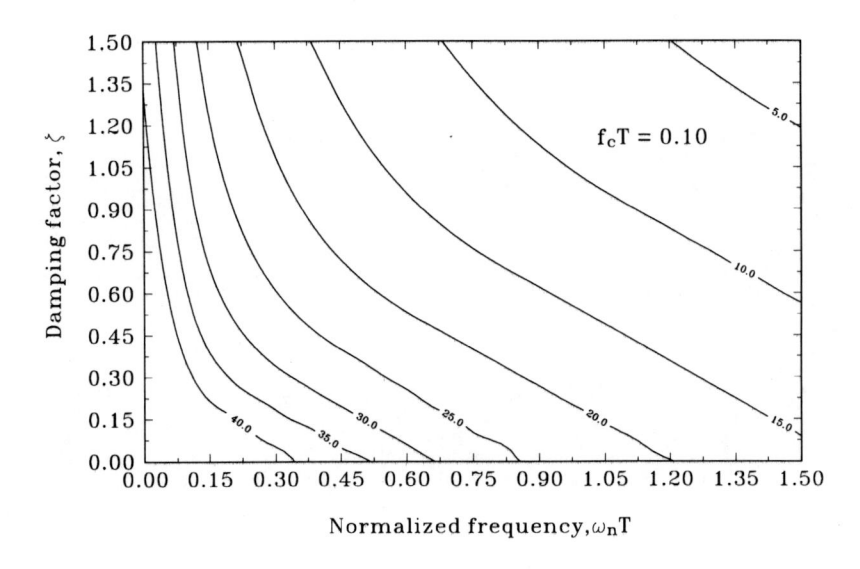

Figure 6.23. Constant gain margin contours for low-pass filter parameter $f_cT = 0.10$ in Figure 6.20.

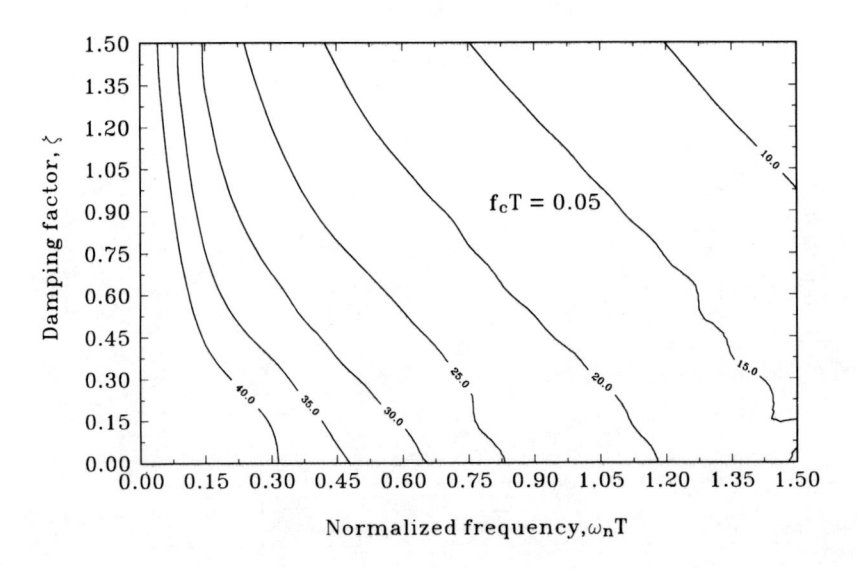

Figure 6.24. Constant gain margin contours for low-pass filter parameter $f_cT = 0.05$ in Figure 6.20.

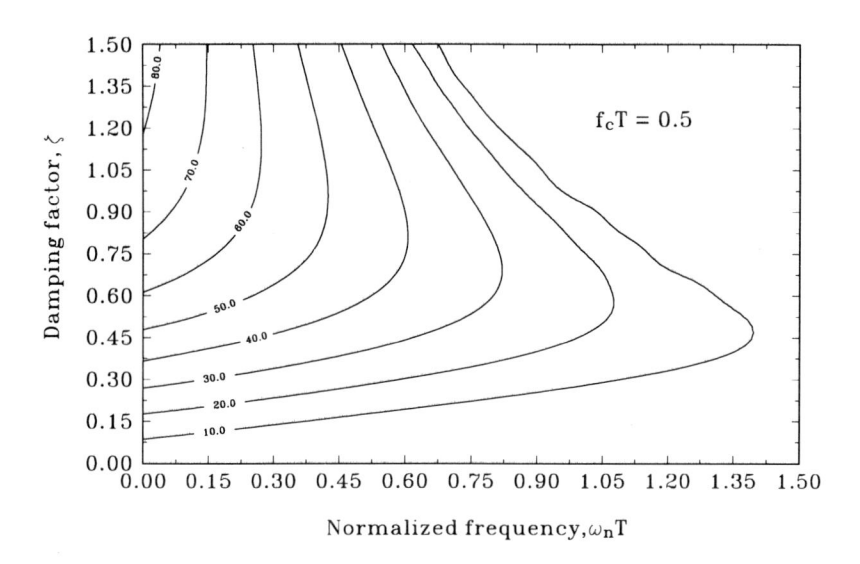

Figure 6.25. Constant phase margin contours for low-pass filter parameter $f_cT = 0.50$ in Figure 6.20.

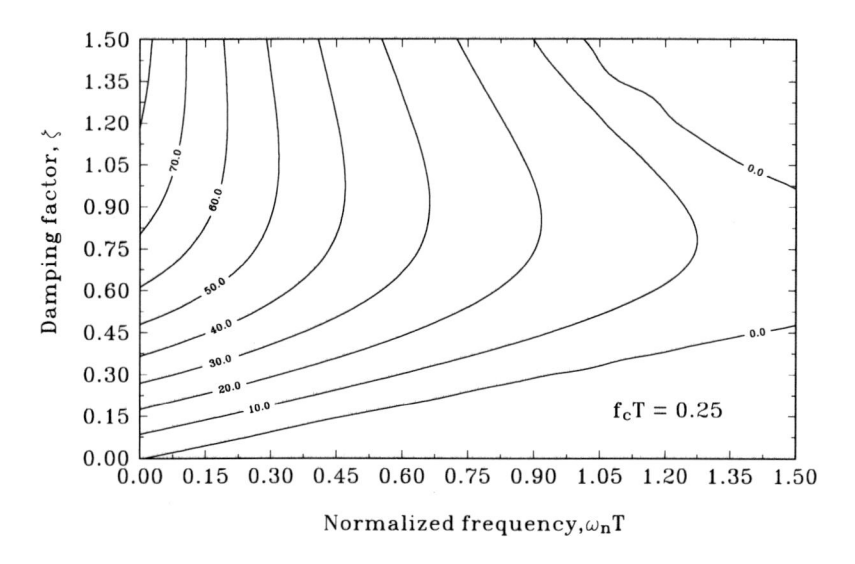

Figure 6.26. Constant phase margin contours for low-pass filter parameter $f_cT = 0.25$ in Figure 6.20.

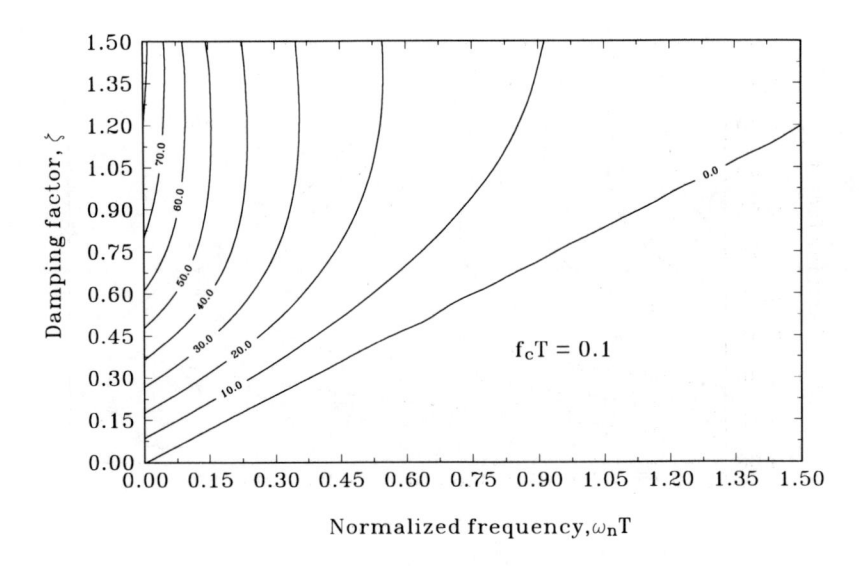

Figure 6.27. Constant phase margin contours for low-pass filter parameter $f_cT = 0.10$ in Figure 6.20.

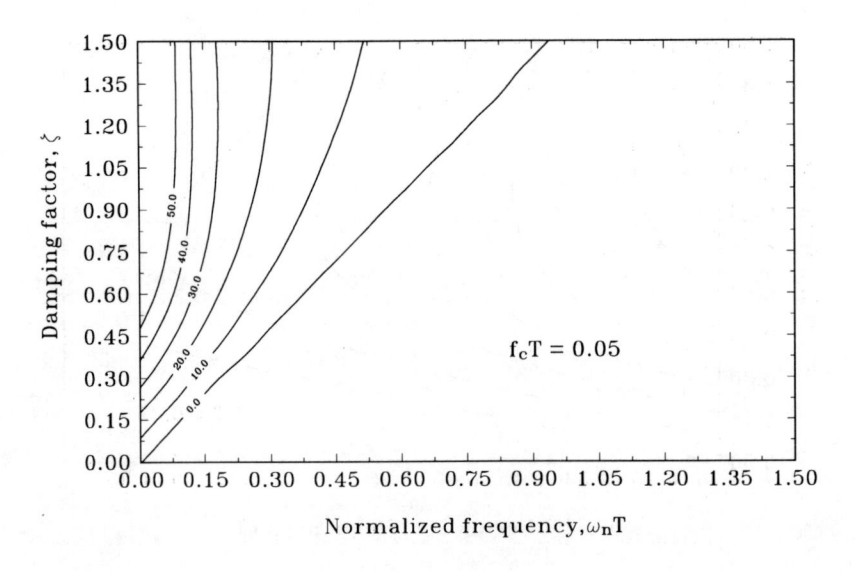

Figure 6.28. Constant phase margin contours for low-pass filter parameter $f_cT = 0.05$ in Figure 6.20.

An overlay of the gain and phase margin contours for the $f_c T = 0.50$ case is shown in Figure 6.29. This figure accentuates the fact that parameter choices must result in both adequate gain as well as phase margin in order to have a stable control system. Similar design guidelines are presented in Chapter 7 where the additional reference spur filtering is provided by a third-order elliptic filter rather than the simple RC low-pass filter assumed here.

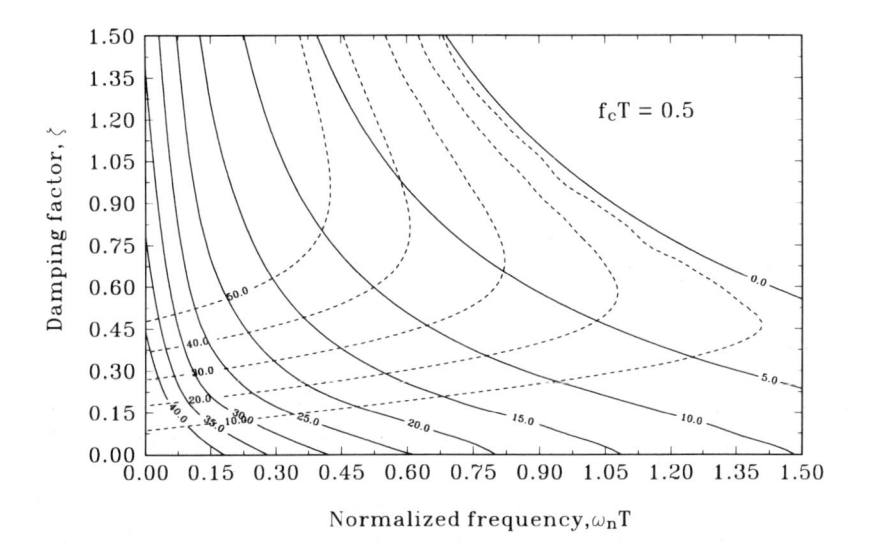

Figure 6.29. Overlay of constant gain and phase margin contours for the phase-locked loop shown in Figure 6.20 with low-pass filter parameter $f_c T = 0.50$.

An investigation into the loop transient response with respect to the additional filtering represented by $f_c T$ is shown in Figure 6.30. This diagram reveals the optimal choice of normalized loop natural frequency $\omega_n T$ and loop damping factor ζ as a function of the additional loop filtering $f_c T$ as dictated by the mean-square error criterion discussed in Chapter 5, specifically (5.43) where $W(i) = i^2$. This figure bears much resemblance to Figure 5.20 where a similar analysis was performed with respect to internal loop time delay. This resemblance should not be surprising because, for reasonably small $|s\tau|$, $e^{-s\tau} \approx (1 + s\tau)^{-1}$.

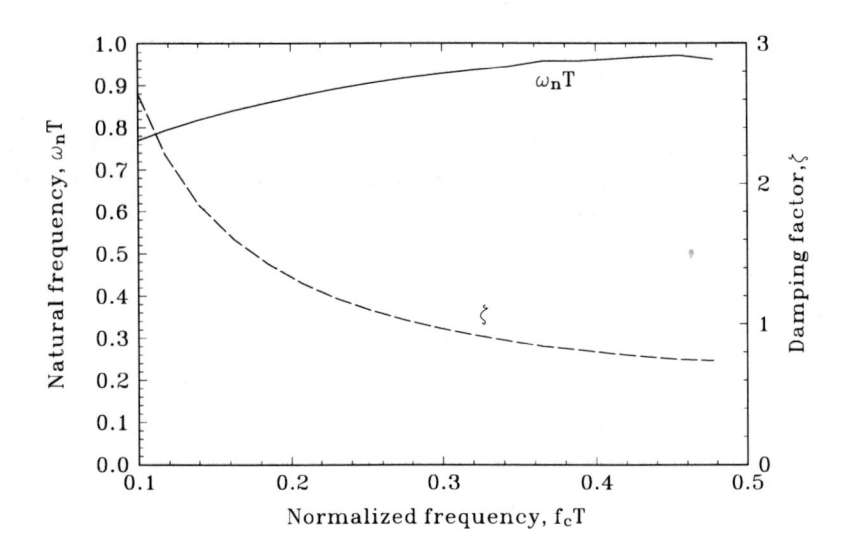

Figure 6.30. Optimization of the transient error reponse with respect to loop parameters ($\omega_n T$, ζ) for Figure 6.20 as a function of low-pass filter parameter $f_c T$.

6.4.1 Phase-Frequency Detector Refinements

Although the phase-frequency detector is much more simple than the sample-and-hold phase detector, its use also requires some special considerations of its own. Two popular configurations are shown in Figure 6.31 for this phase detector type.

The key point to realize about the phase detector portions of Figure 6.31 is that in steady-state operation, the outputs to the lead-lag filter will be almost continuously a logic high ($x - x'$ connections used) or a logic low ($y - y'$ connections used). This may appear to be a fairly mute point—that is, until power supply noise is considered in the design. Quite often, the supply voltage for the digital circuitry can be very noisy, and this noise would appear almost directly at the phase detector outputs if the $x - x'$ connections are used. On the other hand, if the $y - y'$ connections are used, the logic outputs will be pulled down to near ground potential during most of each reference period, which should make for an effectively lower output noise level to the lead-lag filter.

Special attention must also be paid to the device technology being used in the phase detector because the noise output levels for the high- and low-logic state levels are often quite different. Robins [8] reports, for instance, that for a Schottky TTL gate, the output noise levels for high and low states are, respectively, 6 and 50 nV/$\sqrt{\text{Hz}}$. For ECL, the extremes are even more substantial—3.1 and 77.5 nV/$\sqrt{\text{Hz}}$ for high and low levels, respectively. Although only speculation, it is anticipated that ambient output noise levels from CMOS devices are probably not state dependent.

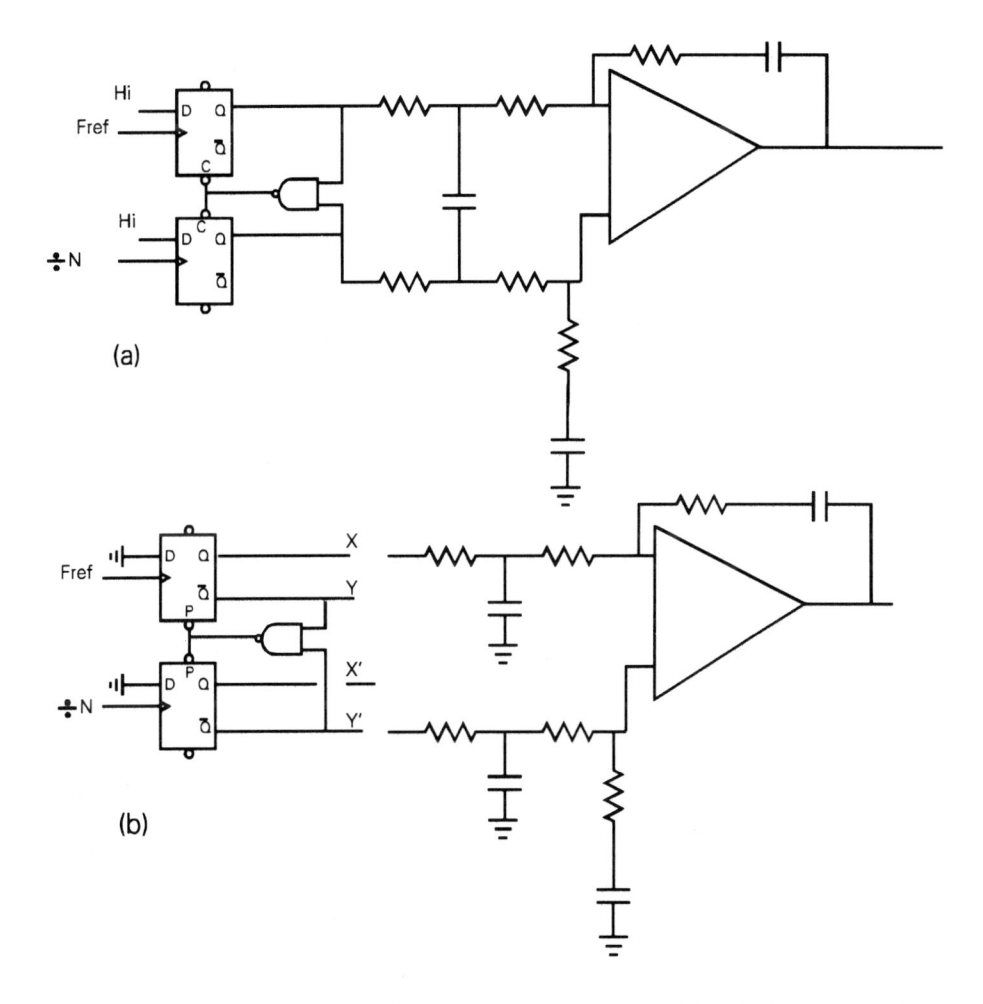

Figure 6.31. Two of the most popular phase-frequency topologies found in widespread use. Aside from one less capacitor in (a), the achieved noise performance may vary significantly between these two topologies as described in the text.

Additional consideration must be given to the flip-flop phase detector outputs if the best sampling spur performance is to be obtained. Even if both flip-flops are on the same monolithic device, it is very common for the two Q outputs to have different high- and low-voltage state values. Generally, this imbalance must be removed by proper adjustment of the circuit RC values immediately following the phase detector.

Compared to the sample-and-hold phase detector, the phase detector gain of the phase-frequency detector is limited to $V_{\text{supply}}/2\pi$ or nominally 0.8 V/rad for standard 5V HCMOS. This makes resistor and lead-lag operational amplifier noise potentially more difficult to suppress than with the sample-and-hold detector.

6.5 POST-TUNING DRIFT WITH TYPE 2 SYSTEMS

As alluded to earlier, even type 2 control loops will have difficulty with oscillator post-tuning drift if it is severe. To investigate this issue of severity further, a classical second-order continuous phase-locked loop will be considered and its behavior analyzed when subjected to VCO post-tuning drift.

Normally, the oscillator post-tuning drift can be modeled as a decaying exponential frequency error that has an initial frequency error roughly proportional to the magnitude of the step change in VCO frequency being made. In Figure 6.32, the post-tuning drift is therefore modeled as an additive external voltage affecting an ideal VCO where ΔF is the magnitude of the initial frequency error in hertz and τ is the time constant associated with the post-tuning drift decay.

Standard Laplace transform techniques may be used now to examine the loop transient response behavior to this undesired post-tuning drift component. The loop phase error response at the output is given by

$$\theta_o(s) = 2\pi\Delta F \frac{s}{(s + 1/\tau)(s^2 + 2\zeta\omega_n s + \omega_n^2)} \quad (6.30)$$

Assume also that the system-level switching speed requirement is given in terms of instantaneous frequency. Ideally then, the quantity of interest is the derivative of the output phase with respect to time or in terms of Laplace transforms

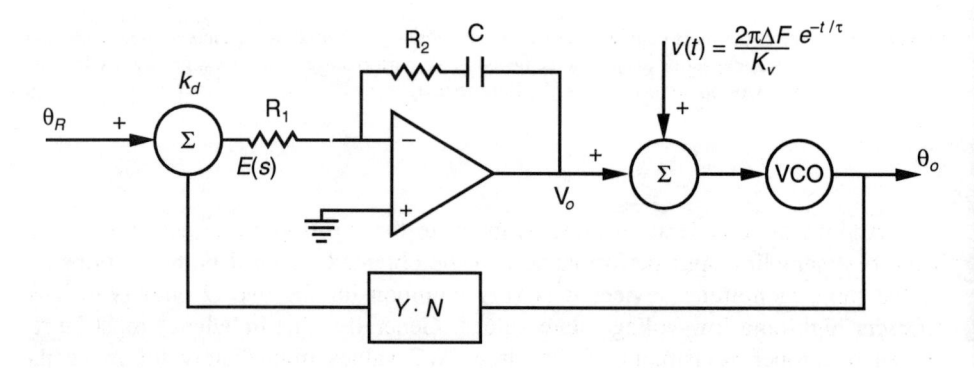

Figure 6.32. Model for analyzing VCO post-tuning drive impact on a type 2 phase-locked loop.

$$\theta'_o(s) = 2\pi\Delta F \frac{s^2}{(s + 1/\tau)(s^2 + 2\zeta\omega_n s + \omega_n^2)} \tag{6.31}$$

Expanding this result into partial fractions and taking the inverse Laplace transforms, the time-domain response to the exponentially decaying frequency term is

$$\frac{d}{dt}\theta_o(t) = \left[A e^{-t/\tau} + \frac{C e^{-\zeta\omega_n t}}{\omega_n\sqrt{1-\zeta^2}} \sin(\omega_n\sqrt{1-\zeta^2}\, t) \right] 2\pi\Delta F$$

$$+ \left\{ B \frac{d}{dt}\left[\frac{e^{-\zeta\omega_n t}}{\omega_n\sqrt{1-\zeta^2}} \sin(\omega_n\sqrt{1-\zeta^2}\, t) \right] \right\} 2\pi\Delta F \tag{6.32}$$

where

$$A = \frac{1/\tau}{\tau\omega_n^2 + (1/\tau) - 2\zeta\omega_n}$$

$$B = 1 - A$$

$$C = -A\tau\omega_n^2$$

Of all the terms in (6.32) only the $A e^{-t/\tau}$ term displays the potentially long time constant associated with the post-tuning drift phenomenon; the other terms decay with a time constant given by the loop parameters alone. Hence, the error term of interest here is given by

$$\Delta f(t) = \frac{\Delta F}{(\omega_n\tau)^2 + 1 - 2\zeta\omega_n\tau} e^{-t/\tau} \tag{6.33}$$

To illustrate this result further, assume that based on phase noise and other considerations, the loop parameters are given as $\omega_n = 2\pi\ 100$ kHz and $\zeta = 0.9$. Further, assume that the instantaneous frequency error must be less than 10 Hz 300 μs following the frequency change. Substituting these quantities into (6.33), the locus of permissible VCO post-tuning drift parameters (ΔF, τ) is as shown in Figure 6.33. This perspective is well advised in very high frequency design situations where post-tuning drift is anticipated to be a problem and fast switching speed is a requirement. It clearly dictates parameter guidelines for the VCO post-tuning drift requirements.

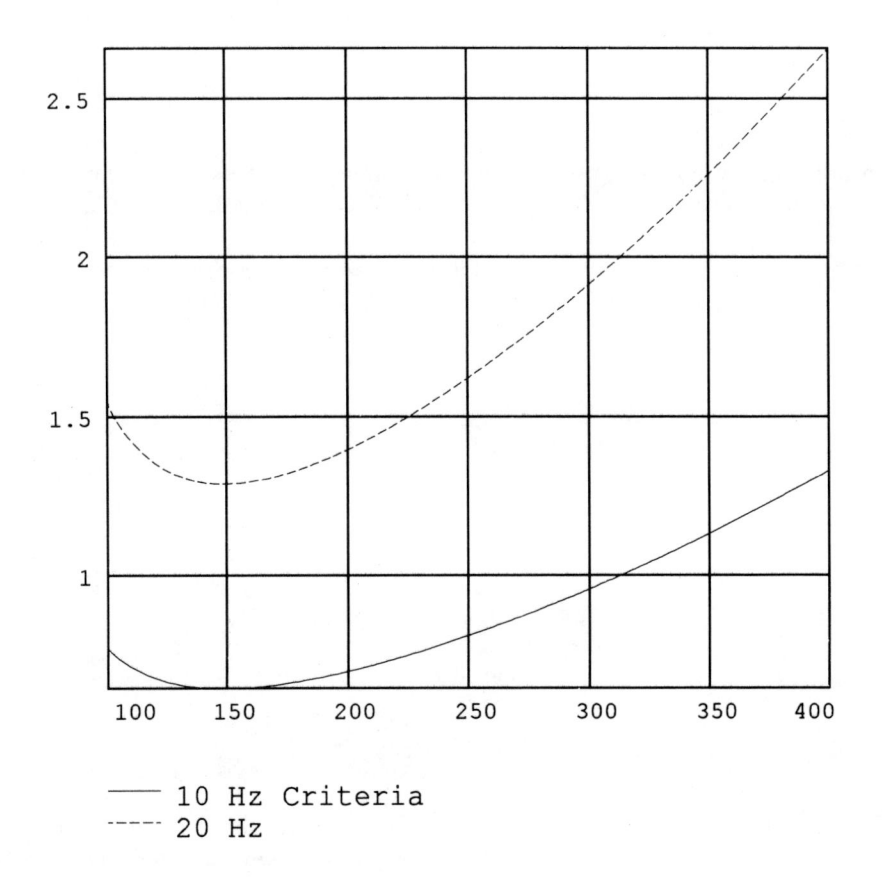

Figure 6.33. Evaluation of permissible VCO post-tuning drift is an important design consideration for high-speed high-frequency designs.

6.6 CHOOSING BETWEEN SAMPLE-AND-HOLD AND PHASE-FREQUENCY DETECTORS

A number of monolithic devices now permit designers to choose between using a sample-and-hold phase detector for their application or a phase-frequency detector. Unfortunately, design guidelines for making this choice are rather limited primarily because the guidelines are very application dependent. A general discussion should nonetheless be helpful.

When ultimate switching speed is the primary design goal, the sample-and-hold phase detector is unquestionably the proper design choice as long as the reference frequency is not too high [9]. As shown in Chapters 4 and 5, the switching speed can theoretically be one and two sample periods for the type 1 and type 2 systems, respectively. Additionally, the type 1 architecture is only suitable for a sample-and-hold phase detector due to obvious spurious reasons. As increased filtering is added within the loop to attenuate sampling spurs, the type 1 architecture can display a sizable speed advantage compared to the type 2 architecture. If only moderate sampling spurious performance is required, the phase-frequency detector type 2 architecture can attain phase-lock in roughly 25 samples. In very low spurious cases, 50 sample periods is a better figure for design purposes.

When the primary design issue is phase noise, no clear statements concerning phase detector choice can be made without closely examining device details. At first glance, the potentially larger phase detector gain of the sample-and-hold detector would appear to make it the favorite since this would lead to reduced degradation by operational amplifier noise. Unfortunately, noise processes within the monolithic type sample-and-holds can often handicap this approach, thus nullifying this apparent advantage. This is particularly true of devices that boast extremely large phase detector gains. Both phase detector types tend to perform more poorly if on-chip buffers or charge pumps are used.

In low-sampling spur applications, the choice between detectors can be equally difficult. At high reference frequencies, the phase-frequency detector becomes increasingly more desirable as ramp and sampling signal leakage in the sample-hold become more and more a problem. Clever cancellation schemes can improve the situation as discussed earlier in this chapter but at the cost of complexity.

The phase-frequency detector, of course, offers wide frequency pull-in capability by virtue of its dual modes of operation, whereas this is not true with the sample-and-hold detector. Most of the modern sample-and-hold devices, however, now include an additional frequency steering type output to remedy this problem. Even so, these outputs are often additional sources of noise and care must be appropriately given. Since the phase-frequency detector ideally displays zero group delay whereas the sample-and-hold has a group delay of $T/2$, some additional filtering is possible with the phase-frequency detector approach compared to the latter approach for a given degree of loop stability.

In conclusion, high-performance designs still require the designer to have an intimate knowledge base for the devices being used. Since phase noise and sampling spur performance are generally not measurable except in a closed loop, a test fixture approach, in which the candidate device is first evaluated, is often highly recommended.

REFERENCES

[1] Crawford, J.A., "Understanding the Specifics of Sampling in Synthesis," *Microwaves & RF*, Vol. 23, No. 8, August 1984, pp. 120–126, 144.
[2] Crawford, J.A., "Extending Sampling to Type 2 Phase-Locked Loops," *Microwaves & RF*, September 1984, pp. 171–174.
[3] Egan, W.F., *Frequency Synthesis by Phase Lock,* John Wiley & Sons, New York, 1981.
[4] Crawford, J.A., Phase-Lock Frequency Synthesizer. U.S. Patent No. 4,668,922, May 1987.

[5] Manassewitsch, V., *Frequency Synthesizers: Theory and Design,* 2nd ed., John Wiley & Sons, New York, 1980.
[6] Crawford, J.A., "Frequency Synthesizer IR&D," TRW Interoffice Correspondence C423.25–85, TRW Electronic Systems Group, May 23, 1985.
[7] Crawford, J.A., "ASM Frequency Synthesizer," Linkabit, 1986.
[8] Robins, W.P., "Synthesizer Phase Jitter Contributed by TTL & ECL Components," *Colloquium on Low Noise Oscillators and Synthesizers,* 1986.
[9] Kolumban, G., "Transient Properties of High Speed Frequency Synthesizers Based on Sampled PLL," *EIM* 87–12.

Chapter 7
Hybrid Phase-Locked Loops

The hybrid phase-locked loop terminology is meant to address cases where conventional phase-locked loop techniques (such as those discussed in Chapter 4) are married with other techniques in order to realize an improved frequency synthesis element [1]. In the case of multi-loop architectures, the word *hybrid* implies that some interaction between individual phase-locked loops is involved beyond simply frequency combining their outputs via mixing type operations. In single-loop architectures, hybrid approaches range from the addition of VCO coarse pretuning for faster switching speed to advanced fractional-N techniques.

7.1 HYBRID ARCHITECTURES

A logical starting point for our discussions is at the architectural level. The principal architectures considered in this chapter are shown in Figure 7.1.

In Figure 7.1(a), two phase-locked loops are utilized to synthesize an output channel spacing that is much lower than the reference frequency used by either loop. A typical application might use a basic standard frequency of 10.1 MHz with $R_1 = 101$ and $R_2 = 100$. In this case, the output channel spacing would be 1 kHz even though each individual loop is using a reference frequency of roughly 100 kHz. Many design considerations must be addressed in this approach as developed momentarily.

In Figure 7.1(b), three synthesis elements are combined to form the synthesis function. Although certainly more complex than Figure 7.1(a), this approach can in principle deliver superior spurious and phase noise performance.

Finally, a simplistic representation of advanced fractional-N synthesis is shown in Figure 7.1(c). Traditional fractional-N synthesis is dealt with at length in Chapter 9. The concepts discussed in this chapter go beyond the foundational concepts presented in Chapter 9 and take advantage of relatively new concepts that are really more closely related to direct digital synthesis. This material, therefore, follows the in-depth look at modern direct digital synthesizers that is included in this chapter.

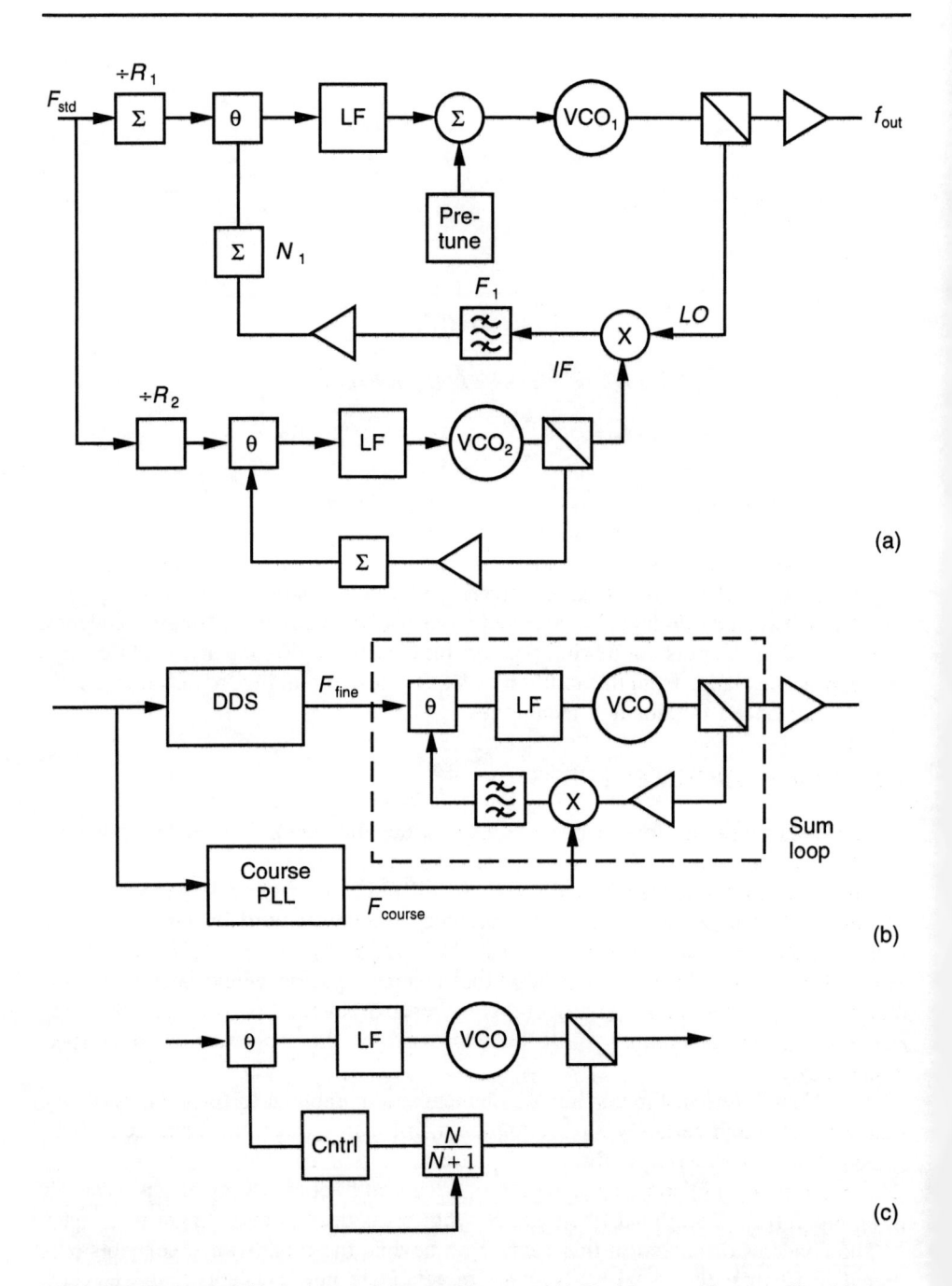

Figure 7.1 Popular hybrid phase-locked loop architectures: (a) Dual-loop architecture, (b) triple-loop architecture, (c) advanced fractional-N synthesis using a single loop. After: [1], Figure 5.

7.2 DUAL-LOOP FREQUENCY SYNTHESIS

The tight coupling between the phase-locked loop elements shown in Figure 7.1(a) results in the need to consider a wide range of design constraints. Component limitations in many situations will limit the flexibility of this approach. Even so, when performance beyond that achievable in a single phase-locked loop (PLL) is required, this is a natural extension to at least consider.

7.2.1 Frequency Channel Spacing

Assuming that reference divider ratios R_1 and R_2 are such that $R_1 > R_2$, the delivered channel spacing is simply given by

$$\Delta f = F_{\text{std}} \left(\frac{1}{R_2} - \frac{1}{R_1} \right) \tag{7.1}$$

where F_{std} is the provided frequency standard as shown in Figure 7.1(a). The use of unequal reference divider ratios R_1 and R_2 obviously implies that the two individual phase-locked loops utilize different reference frequencies. Normally, these divider ratios are chosen to be given by D and $D + 1$ where D is a fixed integer.

7.2.2 Phase Noise Performance

Traditional designs generally organize the frequency plan of the dual-loop synthesizer such that the upper loop utilizes the lowest feedback divider ratios (N_1) possible. Generally, this results in the lower loop divider ratio (N_2) being substantially larger than N_1 thereby making the lower loop the primary phase noise contributor for frequency offsets close to the carrier. Well outside the closed-loop bandwidth of the upper loop, the phase noise performance is, of course, dominated by that of VCO$_1$.

7.2.3 Offset Mixer Considerations

The frequency plan choices made in the mixer area are absolutely crucial for the dual-loop architecture based on three primary reasons. First of all, since a mixer is involved, a full spurious analysis between the local oscillator (LO) and RF frequencies applied to the mixer must be done in principal for each output frequency channel to be synthesized. Since the two phase-locked loops utilize different reference frequencies, the sampling action of divider N_1 can cause otherwise far-removed mixer spurious components to be aliased within the loop bandwidth of the upper loop. This normally dictates that the LO port frequency (f_{out}) be maintained higher than the RF port frequency for best spurious performance and that the maximum IF frequency used out of the mixer be as small as possible with respect to both LO and RF frequencies.

Secondly, the specific range of IF frequencies that can be used in the approach is dictated by several constraints. The minimum IF that can be used is normally dictated by the swallow counter prescaler, which is used to implement the divide-by-N_1 function. If the prescaler divider ratios are given by P and $P + 1$, in general the minimum IF frequency that can be used is given by $P(P - 1)F_{std}/R_1$. In the case where $P = 32$, $F_{std} = 10.1$ MHz, and $R_1 = 101$, the minimum usable IF is 99.2 MHz. Therefore, the upper loop frequency fed to its feedback divider is constrained to be 99.2 MHz or higher. Clearly, if f_{out} is needed in the 200-MHz range or lower, serious mixer spurious issues may arise. On the other hand, if f_{out} is always above several hundred megahertz, the concern will be much less.

The IF span that must be passed by filter $F1$ is dictated by the choices made for R_1, R_2, and F_{std}. In general, the minimum span is given by

$$\text{span} = \frac{F_{std}}{R_1} \frac{R_2}{R_1 - R_2} \qquad (7.2)$$

which for the present example equates to 10 MHz, thereby placing the lowest range of mixer output frequencies from 99.2 to 109.2 MHz. To avoid a complex filter for filter $F1$ in Figure 7.1(a), any significant spurious frequencies resulting from the mixing operation should be well removed from this frequency band.

Obviously, filter $F1$ cannot be made so narrow as to significantly attenuate the desired difference frequency that occurs during frequency switching since divider N_1 would interpret the absence of an input signal as a "zero" frequency signal. Therefore, this filter must be made adequately wide to pass the frequency excursions that result when changing frequencies.

Finally, in some cases where small prescaler ratios P can be used, it may at first glance appear attractive to move the center of the IF span as low in frequency as possible in order to ease the mixer spur filter issues. This normally introduces two serious problems: first, the need to provide loop gain compensation in the upper loop and, second, the difficulty of proper sideband selection in the mixing operation.

If the IF frequency span is moved very low in frequency, the ratio of the maximum and minimum N_1 divider ratios can lead to required gain compensation in the upper loop, which adds complexity and cost. If the IF is very low in frequency, it can also be troublesome to ensure that the upper PLL frequency is always maintained higher than that of the lower PLL. If the sense of these two frequencies is mistakenly interchanged, a sign change is introduced into the upper loop's transfer function, which normally leads to a hang-up condition with the upper loop VCO grossly mistuned. Pretuning of the upper VCO can alleviate this problem as long as adequate setting accuracy can be guaranteed, but other solutions are also possible. These different issues are sketched graphically in Figure 7.2.

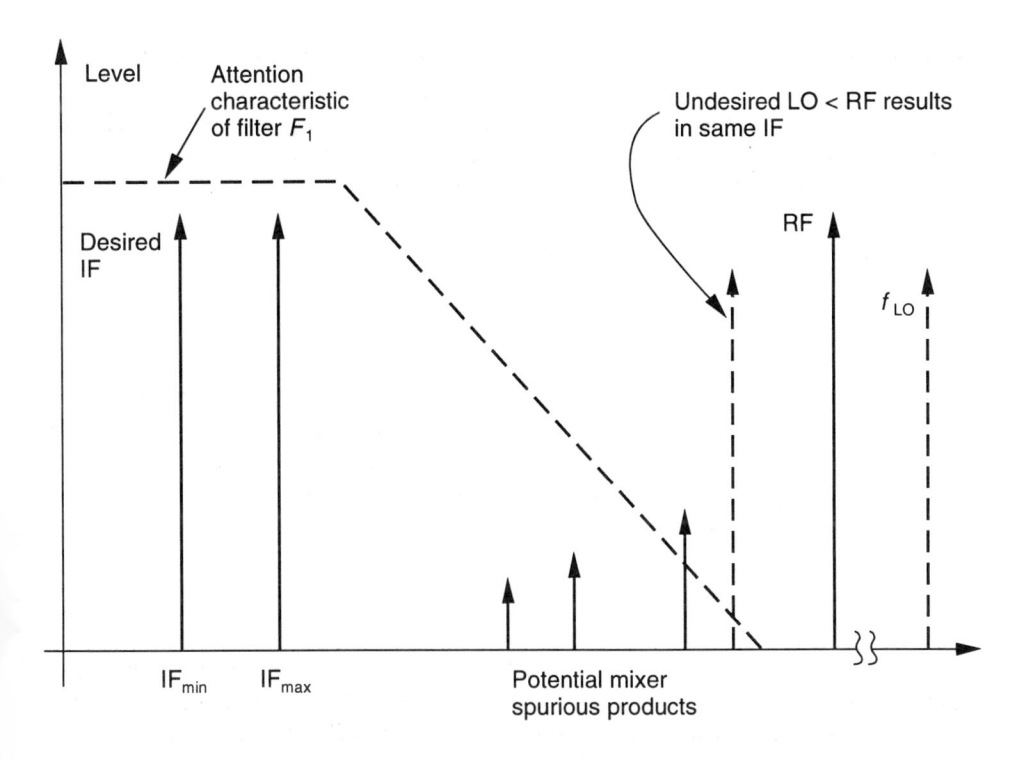

Figure 7.2 Offset mixing considerations for the dual-loop architecture.

7.2.4 Signal Isolation

Reverse isolation is an important consideration with the involvement of two voltage-controlled oscillators (VCOs) in the dual-loop architecture. The degree of isolation required between the two VCOs can be estimated using the oscillator concepts presented in Chapter 3.

Normally, the use of an active mixer helps to minimize the potential isolation issues because (1) the signal levels involved are normally lower and (2) port-to-port mixer isolation is often superior to that obtainable with passive mixers.

Since the VCO_1 output is the desired output signal from the synthesizer, it is only natural to use it for the LO input. To do otherwise would simply heighten the isolation problems that are present.

To lessen the isolation issue, an often helpful observation can be made concerning the RF signal level at the mixer input. Since the lower PLL is only important for close-in spectral control, where the output synthesizer phase noise spectrum is rarely better than, say, −120 dBc/Hz, the RF signal level at the mixer input can be made quite low (e.g., −40 dBm) as long as the output IF signal is maintained well above the thermal noise floor at that point in the system. Lowering the RF input level also reduces the spurious component production in the mixer as well. Unfortunately, this observation cannot be implemented without a price. Additional gain must be provided to the IF signal to ensure adequate signal level for feedback divider operation. Normally, the degree to which this method is effective is dictated by the mixer LO leakage level compared to the desired IF signal level appearing after filter $F1$ and the discrete spurious performance requirements.

7.2.5 Gain Compensation Strategies

In most commercial synthesizer applications such as cellular phones, gain compensation is not required because the fractional tuning bandwidths involved are normally fairly small. When gain compensation is required in more sophisticated applications, it can often add considerable complexity. Therefore, some possible approaches outside the normal literature are presented in this section.

VCO Gain Compensation

The need for gain compensation in a phase-locked loop normally results from either large changes in the feedback divider ratio or changes in the VCO tuning sensitivity. This section specifically addresses linearization of the VCO tuning curve.[1]

Unless specifically designed for linear tuning, most VCOs display a tuning curve that exhibits some form of curvature, which manifests itself as a significant change in the tuning sensitivity K_v. A typical example of such a tuning curve is shown in Figure 7.3 along with the desired linear tuning characteristic. To obtain stable loop operation, the VCO tuning curve must be strictly monotonic (i.e., have a slope versus tuning voltage that is either strictly positive or strictly negative). Large changes in open-loop gain that are not compensated for result in substantial changes in the primary closed-loop parameters ω_n and ζ. Transmission line based VCOs understandably exhibit different tuning curves than lumped element VCOs but the need for linearization remains.

In situations where the oscillator's negative resistance source can be modeled as shown in Figure 7.4, very simple compensation can be used to fit the desired linear tuning curve in a first-order Chebyshev manner. In Figure 7.4, the series resonator inductor is represented by L_s and the varactor by C_v. It can be shown that proper selection of the fixed capacitor C_o can achieve the desired first-order correction.

In steady-state oscillator operation, assume that $-R = R_s$ and $L_s + L_T = L$. Returning to Figure 7.4, the resonant frequency is obviously given by

---

[1]First presented to the author by Gary D. Frey.

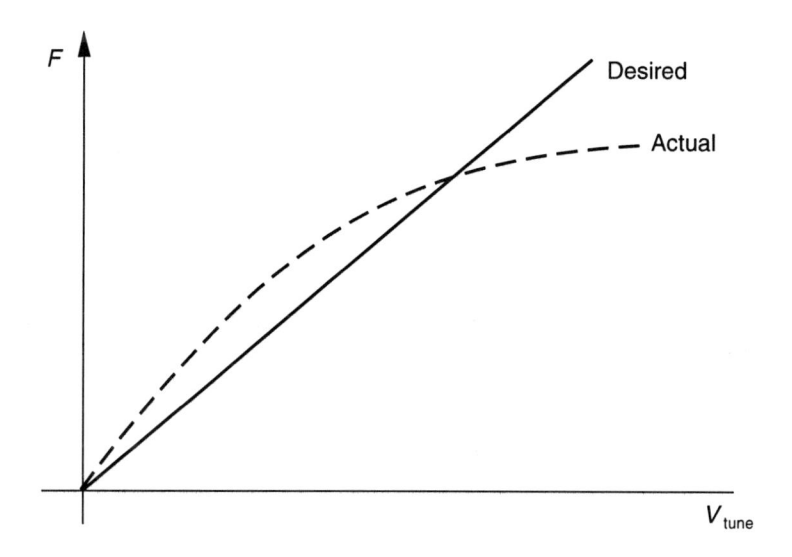

Figure 7.3 Normal nonlinear VCO tuning curve less compensation.

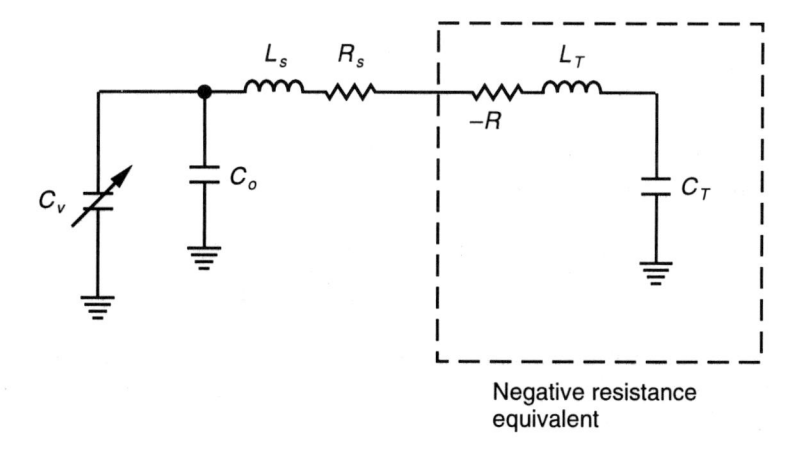

Figure 7.4 First-order negative resistance VCO model for gain compensation considerations.

$$f = \frac{1}{2\pi}\frac{1}{\sqrt{L}}(C)^{-1/2} \qquad (7.3)$$

where C is the appropriate total series capacitance of the resonator. The slope of the tuning curve is then given by

$$s = \frac{df}{dv} = \frac{1}{2\pi}\frac{1}{\sqrt{L}}\left(-\frac{1}{2}\right)C^{-3/2}\frac{dC}{dv} \qquad (7.4)$$

The second derivative of the tuning characteristic with respect to v is indicative of the curvature and is given by

$$\frac{d^2f}{dv^2} = \frac{1}{2\pi}\frac{1}{\sqrt{L}}\left(-\frac{1}{2}\right)\left[C^{-3/2}\frac{d^2C}{dv^2} - \frac{3}{2}C^{-5/2}\left(\frac{dC}{dv}\right)^2\right] \qquad (7.5)$$

and is ideally zero. Setting this quantity to zero requires that

$$\frac{d^2C}{dv^2} = \frac{3}{2}\frac{(dC/dv)^2}{C} \qquad (7.6)$$

For the case at hand,

$$C = \frac{(C_o + C_v)C_T}{C_o + C_v + C_T} \qquad (7.7)$$

Computing the first and second derivatives of C with respect to v from (7.7) and substituting into (7.6) leads finally to

$$\frac{d^2C_v}{dv^2} = \frac{(dC_v/dv)^2}{C_o + C_T + C_v}\frac{4C_v + 4C_o + 3C_T}{2(C_o + C_v)} \qquad (7.8)$$

For most hyper-abrupt varactor diodes, the capacitance versus tuning voltage characteristic can be closely modeled by an exponential for tuning voltages from roughly 3 to 9V as

$$C_v(v) = k_1 e^{-k_2 v} \qquad (7.9)$$

Computing the first and second derivatives of C_v with respect to v, on substitution into (7.8), the required capacitance relationship for linear tuning is given by

$$C_o^2 + C_o C_T - C_v^2 - \frac{C_v C_T}{2} = 0 \tag{7.10}$$

The required choice for capacitor C_o can then be placed into three categories as

$$\begin{array}{ll} C_T \to \infty & C_o \to \dfrac{C_v}{2} \\ \\ C_o \to 0 & C_T \to -2C_v \end{array} \tag{7.11}$$

$$\text{General } C_o = -\frac{C_T}{2} + \sqrt{\left(\frac{C_T}{2}\right)^2 + C_v^2 + \frac{C_v C_T}{2}}$$

In the case where the oscillator negative resistance source is broadband, $C_T \to \infty$ and the first solution applies. Since a fixed value for C_o is sought, the best choice for C_o is roughly that value which applies at the middle of the tuning range, which is 6V in this case. A representative tuning curve that uses this method is shown in Figure 7.5 under the assumption that $C_o = C_v(6\text{V})/2$. Other parameters were selected for illustrative purposes as $k_1 = 30$, $k_2 = 0.35$, and $C_T = 5.0k_1$. The normalized tuning curve is shown in Figure 7.5(a) with some slight bowing evident. The true tuning error is much more apparent once the linear tuning term is subtracted, leaving an error residue as shown in Figure 7.5(b). Finally, numerical differentiation was used to compute the normalized tuning sensitivity as shown in Figure 7.5(c).

In principle, it is possible to implement VCO gain compensation by appropriately designing the doping profile of the varactor diode [2]. In practice, however, this is rarely done because the varactor supplier and VCO supplier are rarely the same firm.

Phase Detector Gain Compensation

Overall loop gain compensation using the phase detector gain as a controlled variable was described in Chapter 6 for the sample-and-hold phase detector case. Wide range gain variations with this detector type are difficult, however, and may lead to other additional problems. In addition, most present-day integrated PLL devices use a charge pump type phase detector, which is quite different. One means for making the gain parameter controllable with charge pump type phase detectors is shown in Figure 7.6. An external charge pump is shown here for additional clarity. Without resistor R_v, all of the phase detector current is channeled through R_1. Since the lead-lag filter creates a virtual ground at its inverting input, it is a simple matter to channel a certain proportion of this current away from the integration path with the addition of resistor

R_v. Unlike the sample-and-hold phase detector case, this phase detector approach allows the gain to be dramatically reduced without affecting other system parameters such as the phase detector's linear range of operation or noise output.

Systematic Gain Compensation

If the primary purpose for employing gain compensation is one of loop stability rather than maintaining constant loop bandwidth or switching speed, an entirely different approach may be used with good success. From a stability point of view, the entire purpose of gain compensation is to maintain a system phase margin of at least some specified amount. Phase margin was considered at some length in Chapters 4 and 5.

(a)

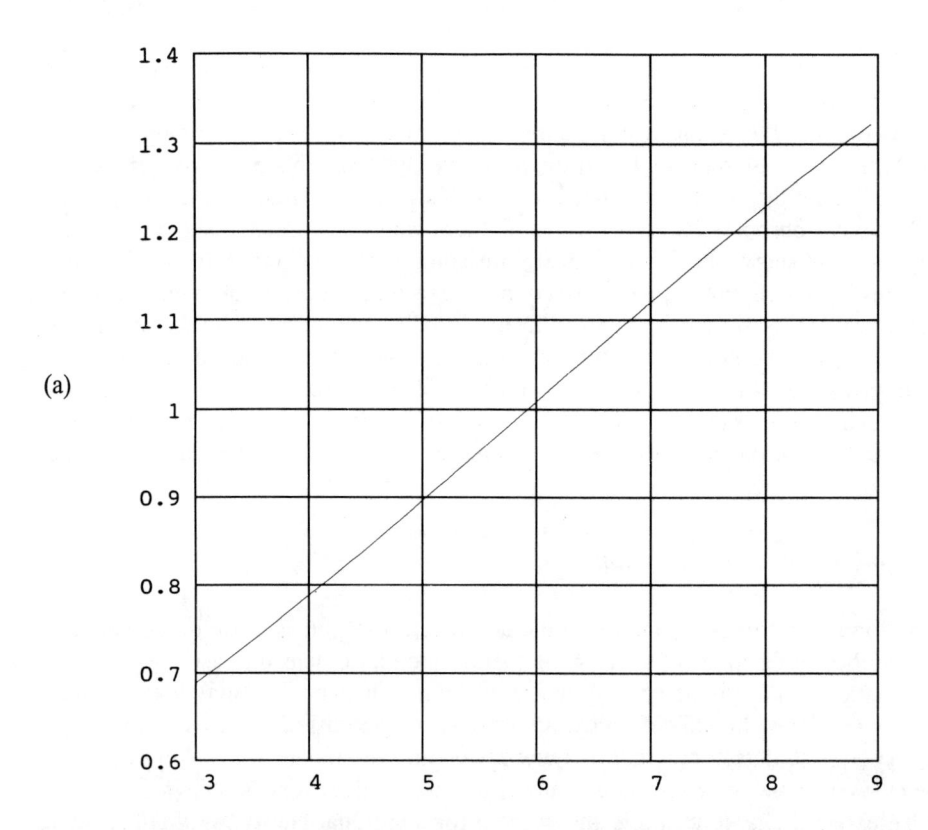

Figure 7.5 (a) Normalized VCO tuning curve including first-order linearizing gain compensation. Normalized frequency versus tuning voltage shown. (b) Normalized VCO tuning curve after removal of the linear tuning component. Normalized frequency deviation from linear versus control voltage. (c) Normalized VCO tuning sensitivity versus control voltage after linear tuning compensation.

(b)

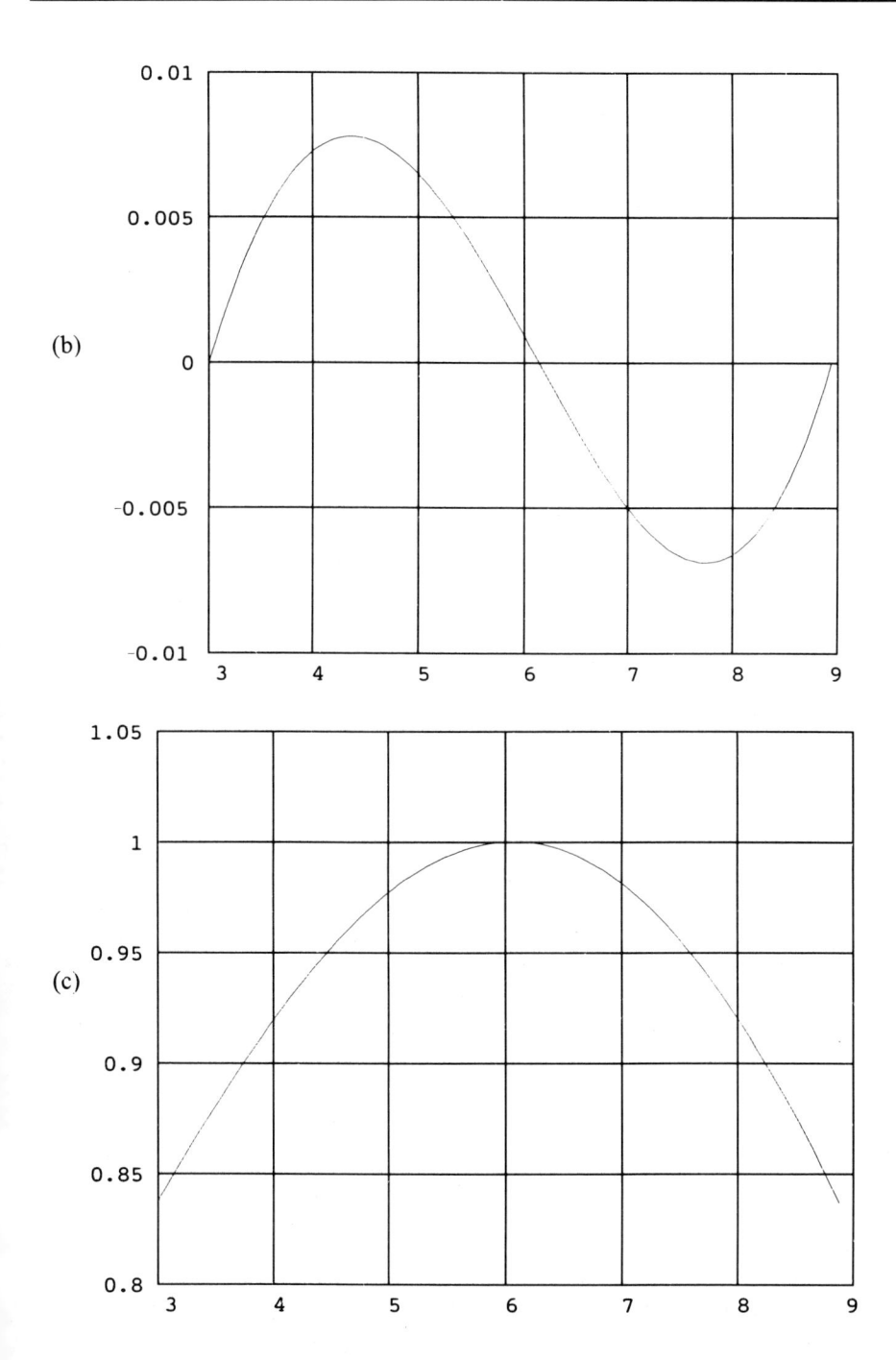

(c)

Figure 7.5 (Continued)

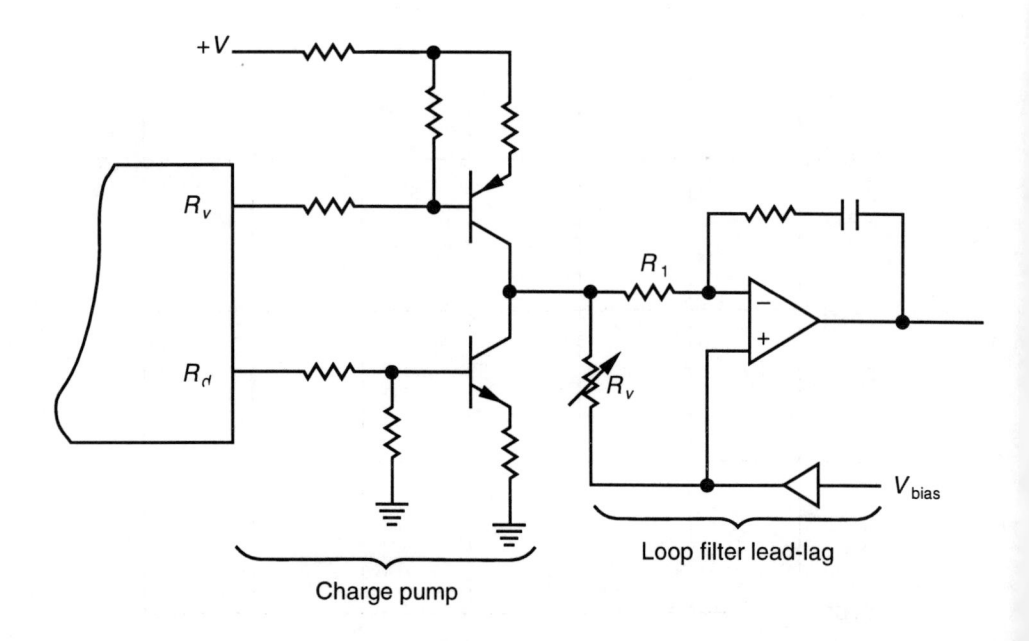

Figure 7.6 Phase detector gain control using a current steering method.

To illustrate the concept, the simplified phase-frequency based phase-locked loop circuit shown in Figure 4.18 will be considered here. The open-loop gain transfer function for this loop is given by (4.75). The gain and phase plots for this loop with $\omega_n = 1.0$, $\zeta = 1.0$, and $\tau = 0.125$ are shown in Figures 7.7 and 7.8, respectively. The key observation that should be made here is that the transfer function phase remains in the vicinity of -120 deg only briefly with respect to frequency, thereby leading to decreased phase margin if the loop gain terms are either too large or too small.

In sharp contrast, the gain and phase plots for the proposed compensation scheme are shown in Figures 7.9 and 7.10. Since the transfer function phase is nearly equiripple at -140 deg for nearly two decades in frequency, large loop gain changes can be absorbed without any loss of system phase margin. It is not difficult to obtain three decades or more for the flat phase response region if desired.

The loop filter used to accomplish this compensation only requires a modification of the lead-lag filter feedback elements as shown in Figure 7.11. For the case under consideration, the normalized component values are

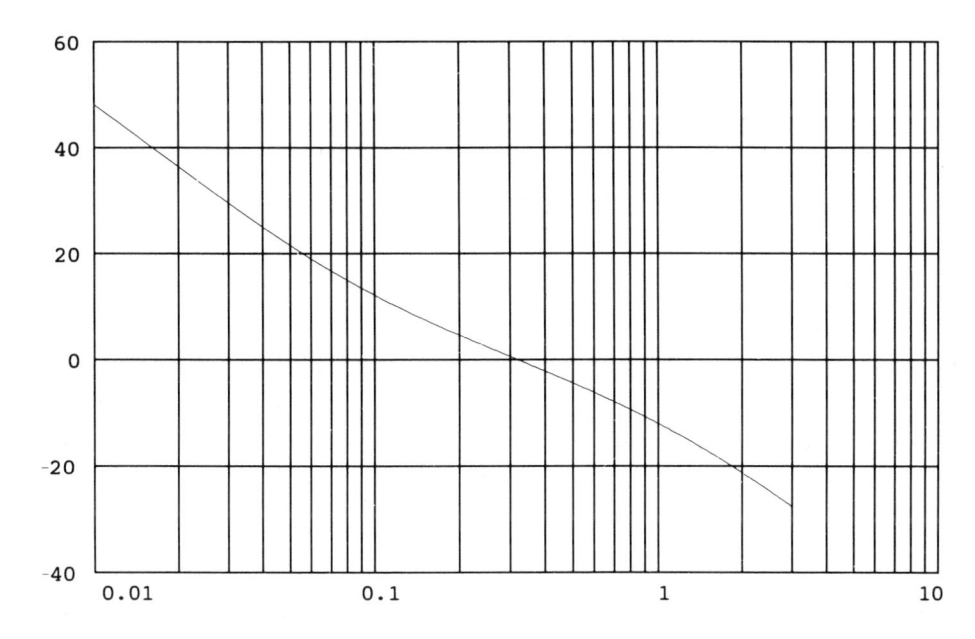

Figure 7.7 Open-loop transfer function magnitude in decibels versus frequency for a classic type 2 phase-locked loop.

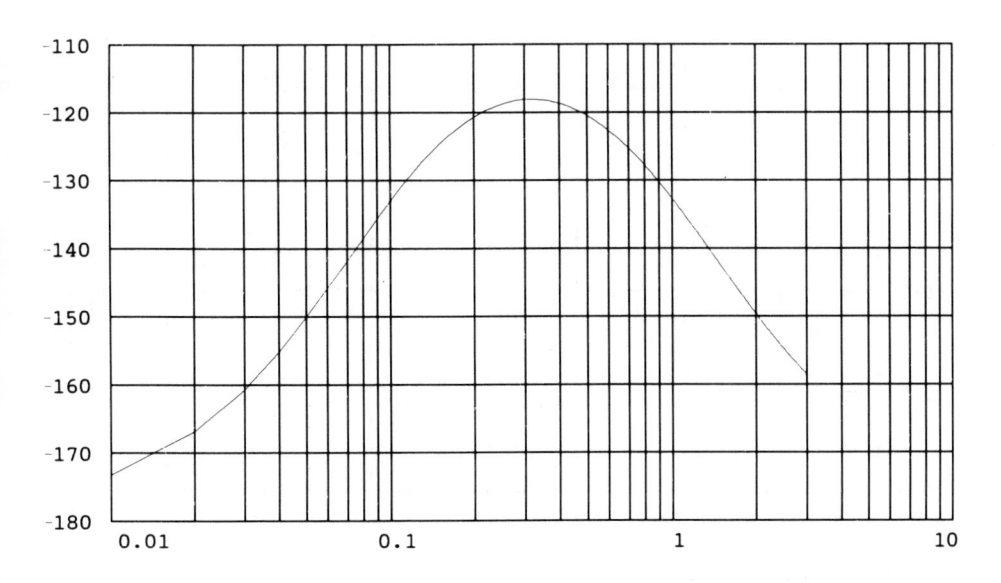

Figure 7.8 Open-loop transfer function phase versus frequency for a classic type 2 phase-locked loop.

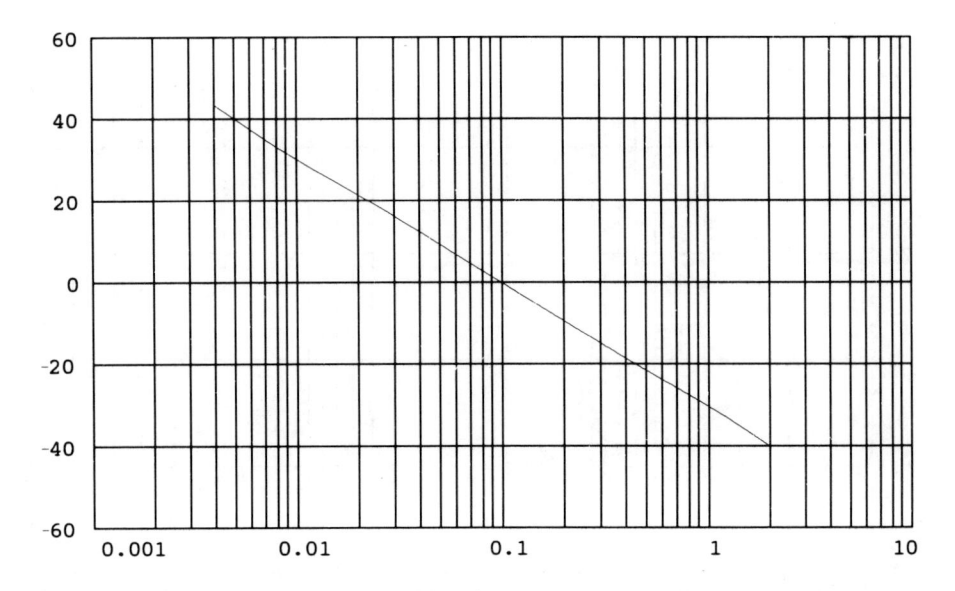

Figure 7.9 Open-loop transfer function magnitude after systematic compensation showing approximately −9 dB per octave slope.

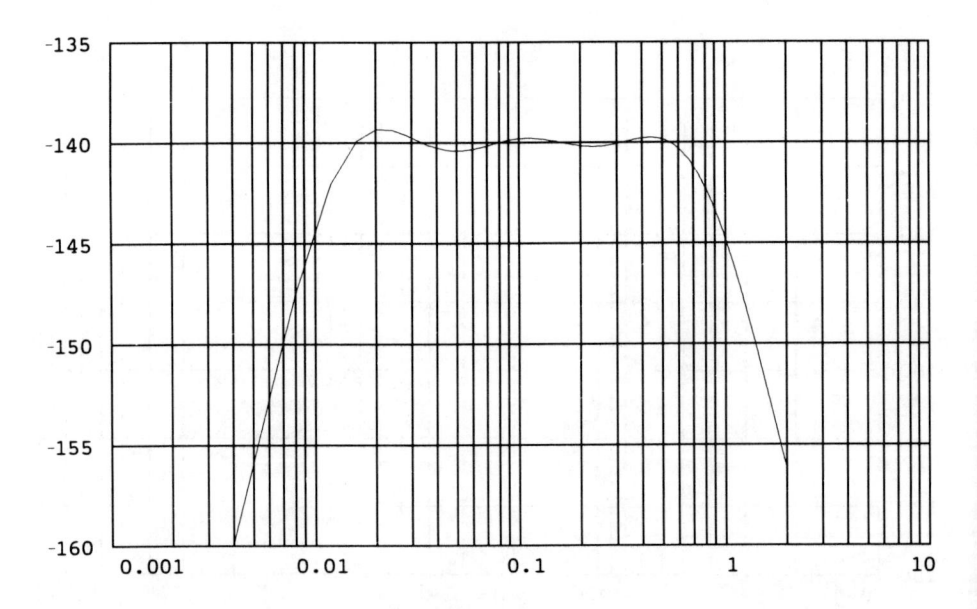

Figure 7.10 Open-loop transfer function phase after systematic compensation displaying equal-ripple phase across almost two decades in frequency.

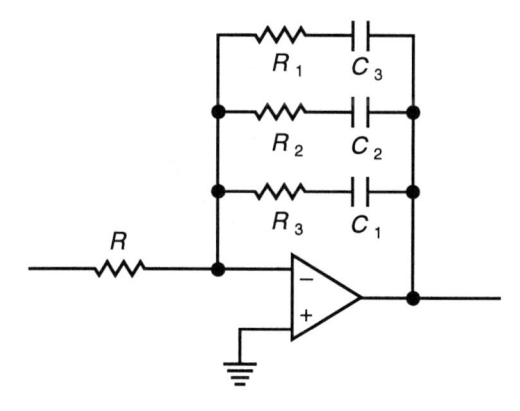

Figure 7.11 Modification of the traditional lead-lag loop filter to incorporate systematic gain compensation.

$$R_1 = 1.282$$
$$R_2 = 2.182$$
$$R_3 = 0.307$$
$$C_1 = 1.695 \quad (7.12)$$
$$C_2 = 9.028$$
$$C_3 = 1.265$$

In general, any number of feedback series RC sections may be used for widening the flat phase response width or reducing the peak-to-peak phase error ripple.

Unfortunately, no closed-form means is available for computation of the series feedback element time constants. For the computed case shown, the desired phase margin was chosen to be 40 deg, and ω_n and τ were left unchanged. Under these conditions, the time constant values were then numerically optimized for equi-ripple error performance.

Although constant phase networks were first examined long ago by Bode, the use of them in a PLL and specifically the loop filter architecture shown in Figure 7.11 were first brought to the author's attention through a patent by Haggai [3]. Although his application was completely different from that proposed here, the constant phase approach was first conceived in that work.

This technique can prove advantageous in a number of ways. For instance, it may be desirable to easily change a PLL's closed-loop bandwidth in order to optimize the output phase noise spectrum with respect to the VCO and reference phase noise with a single adjustment. This technique permits that to be done without incurring a large change in the loop damping factor ζ. In data recovery work using bit synchronizers, a wide range of transition density values generally leads to an equally wide range of effective phase detector gains in the bit synchronizer, which can also be tamed by this same approach.

Switching speed of any PLL using this technique is degraded, however, compared to a simple unmodified PLL. Although the initial frequency pull-in can be made equally rapid, the introduction of large time constants in the loop filter causes the phase tail of the transient response to decay at a rate dictated by the lowest frequency zero used in the compensation network.

7.2.6 Proper Sideband Selection

Proper sense between the LO and RF frequencies applied to the offset mixer must be maintained in order to ensure proper operation of the PLL. As stated earlier, failure to guarantee this objective normally leads to a hang-up condition in which the upper VCO is tuned to one extreme of its tuning range.

The most simple means to achieve this objective is to force the upper VCO to a known frequency extreme each time a frequency change is commanded, thereby always guaranteeing that pull-in by the upper PLL always begins with a predetermined sense with respect to the lower PLL frequency. Parameters for the upper PLL must be selected that ensure that the subsequent transient response does not result in a frequency overshoot where the LO and RF frequencies cross over. Although simple to implement, this approach delivers fairly poor switching speed because, in principle, the lower PLL must settle before releasing the upper PLL.

A second fairly obvious technique that can be used is to augment the upper and lower PLLs with some form of frequency discriminator, which can be used to ensure that the proper frequency sense is maintained. Normally, the frequency discriminator is implemented using digital dividers, which are then followed by either time interval counters or an analog circuit equivalent. Unfortunately, circuit complexity is high with this approach and, although better than the first method, switching speed is still substantially poorer than ideal.

One of the best suited acquisition aids that can be used, particularly when fast switching speed is important, is closely related to the quadricorrelator concept given in [4]. Unlike the quadricorrelator shown there, however, since it is not mandatory to know the magnitude of the frequency difference between the LO and RF frequencies—only the sign—considerable simplification can be made. For lack of an established name for this approach, the technique will be called *quadrature-based sideband selection*. The idealized circuitry needed to perform the desired operation is shown in Figure 7.12.

The fundamental principle behind the quadrature-based sideband selection concept is that if a constant phase shift is added to the otherwise unmodified LO signal as shown in Figure 7.12, on frequency translation of the shifted and nonshifted sig-

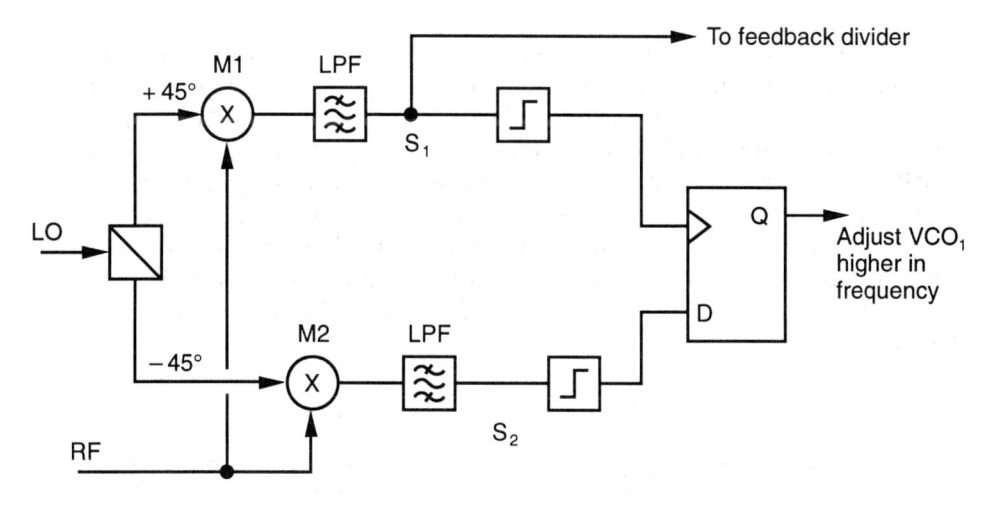

Figure 7.12 Quadrature-based sideband selection ($F_{LO} - F_{RF}$ = desired IF).

nals to a lower frequency using the provided RF signal, the phase-advanced signal should still be phase advanced compared to the other signal unless a spectral inversion has occurred, i.e., the provided RF frequency is greater than the LO frequency. A phasor diagram and a simple time-domain plot of the two (low-pass filtered) mixer output signals S_1 and S_2 are shown in Figure 7.13. As shown, if at any time signal S_2 leads signal S_1, the D flip-flop is set for the next time interval and its output can be used to slew the upper VCO higher in frequency as required. The slewing action is disengaged immediately once the proper sense of the LO and RF frequencies is restored.

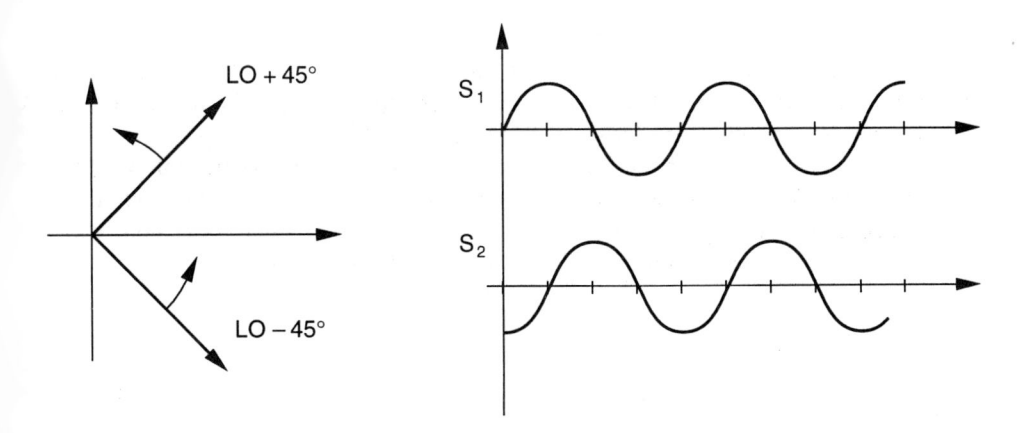

Figure 7.13 Given that $F_{RF} < F_{LO}$, then signal S_1 always leads S_2 by 90 deg.

Since one of the mix-and-filter stages in Figure 7.12 is already needed in the dual-loop approach, the cost of this acquisition aid is primarily represented by the quadrature splitter, additional mixer, and the additional low-pass filter (LPF). However, only the additional LPF represents any sizable impact given the state of modern MMIC technology. Since the time delay through the two low-pass filters must be kept equal, in principle, the two low-pass filters should be identical. This is unfortunate, however, since the filter may have to be of fairly high order in order to deliver adequate spurious performance at the feedback divider input. A simplifying modification that eases the filtering complexity of Figure 7.12 can be made as shown in Figure 7.14.

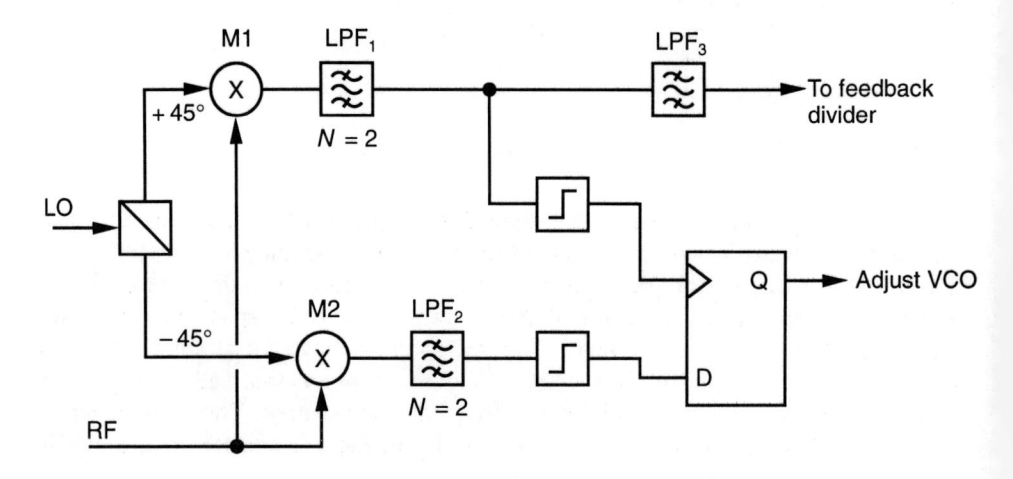

Figure 7.14 Modified quadrature-based sideband selection.

Low-order filters are used in Figure 7.14 for filters LPF_1 and LPF_2, and any additional required filtering needed to further suppress spurious signals in the feedback divider path can be provided by the remaining filter, LPF_3.

Extremely fast adjustment for improper frequency sense is achievable with this technique. In addition, no new frequency components are created due from additional divide operations. This makes the concept ideal for monolithic implementation, where signal isolation is at a premium, and for low-power consumption.

7.2.7 Achievable Switching Speed

Much has been said about switching speed performance in earlier chapters, but most of this material has addressed situations that could be handled in almost a closed-form manner for instructional purposes. To provide a better look at an actual real-world implementation and the speed degradation that accompanies increased filtering for discrete spur reduction, some consideration will be given to the phase-locked loop represented by Figure 7.15. In this figure, a normal type 2 phase-locked loop is shown with an embedded $N = 3$ elliptic low-pass filter, which is allowed to vary. Switching speed, system phase margin, and transient response overshoot are examined as a function of the reference spur attenuation level provided by the complete open-loop transfer function.

To normalize the results to some degree while eliminating parameters of secondary concern, the following assumptions pertaining to Figure 7.15 were made:

$$F_{\text{ref}} = 1 \text{ MHz}$$

$$N_{\text{stop}} = 100$$

$$\zeta = 0.75$$

$$K_v = 2\pi \, 25 \text{ MHz/V} \quad (7.13)$$

$$K_d = 0.8 \text{ V/rad}$$

$$R_1 = R_4 = 1 \text{ K}$$

$$C_1 = C_2 = 500 \text{ pF}$$

$$C_3 = C_4 = 1000 \text{ pF}$$

The study of switching speed versus elliptic filter attenuation was then conducted as described here. Since only linear loop operation (i.e., no phase detector cycle slips) was considered, a frequency hop size (change in feedback divider ratio N) was first selected for the lowest loop bandwidth case that met this criterion. For subsequently larger loop bandwidths, if the bandwidth was increased by a factor of α, the change in N (i.e., $N_{\text{start}} - N_{\text{stop}}$) was increased by the same factor. Achievement of phase-lock was defined to be the length of time from the initial hop command to the point at which the VCO output phase reached and remained within a window of ± 5 deg of its steady-state value. By simulating the PLL's response with different degrees of elliptic filtering, the impact of this filtering was assessed as shown in Figure 7.16. Circuit element values for the different normalized loop bandwidths are as given in Table 7.1. Parameters for the different elliptic filters that were used are similarly shown in Table 7.2.

Given that the elliptic filter was always an $N = 3$ LPF with a reflection coefficient $\rho = 20\%$, the total loop filtering for each case was quantified in terms of the total attenuation delivered at $0.95 \, F_{\text{ref}}$, which is essentially where the first reference

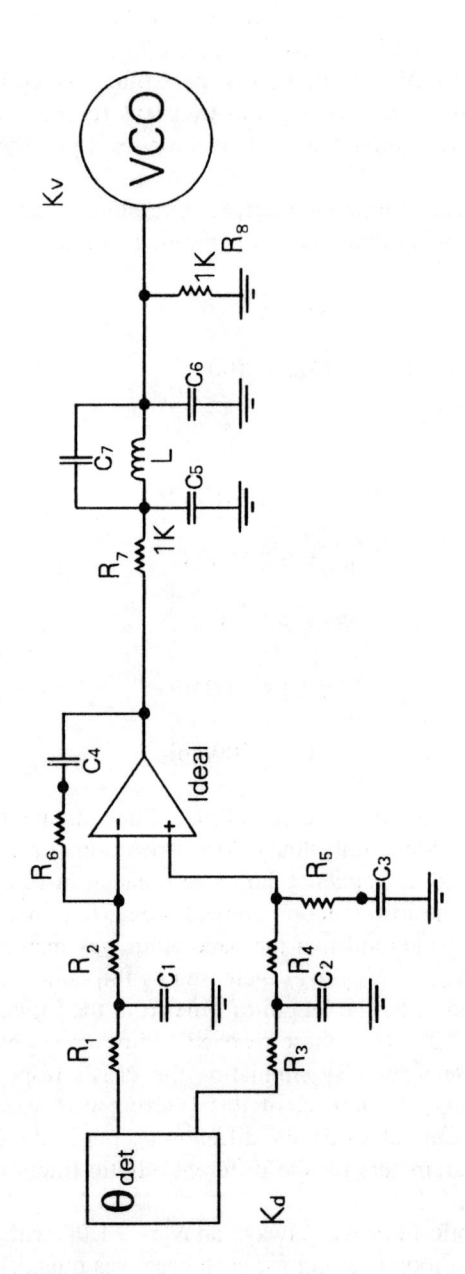

Figure 7.15 Type 2 phase-locked loop with $N = 3$ elliptic low-pass filter used for case study. Source: [1], Figure 1. Reprinted with permission.

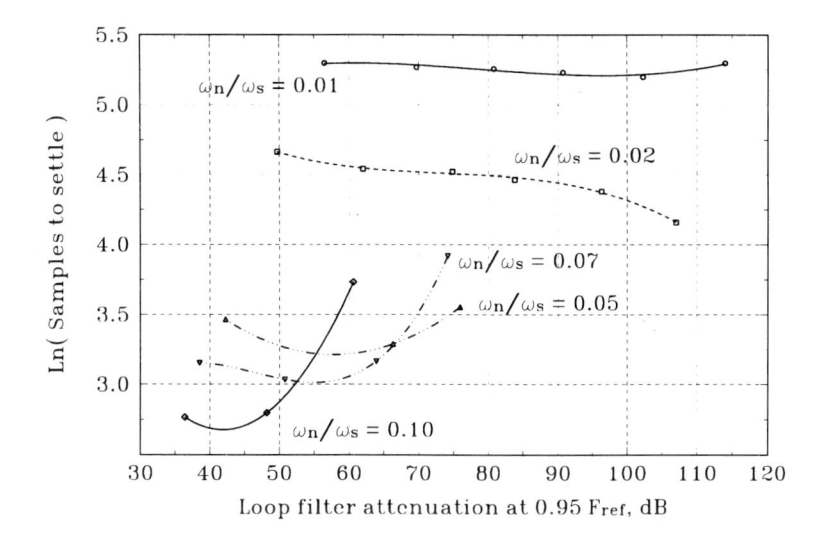

Figure 7.16 Switching speed versus open-loop transfer function attenuation for the first reference discrete spurious sideband. Source: [1], Figure 2. Reprinted with permission.

Table 7.1
Circuit Element Values for Case Study

ω_n/ω_s	τ_2 (μs)	K_v (MHz/V)	N_{start}	ΔF (MHz)	R_5, R_6
0.01	23.87	0.314	107	7	23.87 K
0.02	11.94	1.257	114	14	11.94
0.05	4.77	7.854	135	35	4.77
0.07	3.41	15.4	149	49	3.41
0.10	2.39	31.42	170	70	2.39
0.15	1.59	—	205	105	1.59

Table 7.2
Elliptic Filter Component Values ($N = 3$, $\rho = 20\%$, $R = 1$ K)

Minimum Stopband Attenuation (dB)	Filter θ (deg)	C_1, C_3 (pF)	C_2 (pF)	L (μH)
10	54	162	214	118
20	38	281	103	246
30	26	449	61	418
40	18	674	40	641
50	12	1029	26	990
60	8	1556	16.9	1504

sideband spur will be observed. As shown in Figure 7.16, the presence of some elliptic filtering actually improved the switching speed. On the other hand, system speed degradation rose quickly for the larger loop bandwidth cases as a function of elliptic filter attenuation.

To investigate the degree of VCO phase noise peaking that can be expected in each case, the system phase margin was computed and plotted in Figure 7.17. Substantial spectral peaking will be observed whenever the system phase margin is less than roughly 40 deg.

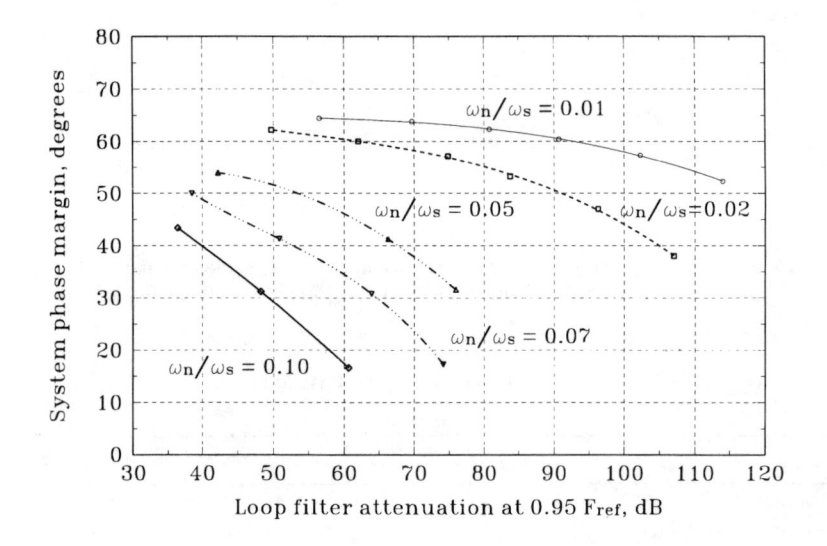

Figure 7.17 System phase margin versus open-loop transfer function attenuation for the first reference discrete spurious sideband. Source: [1], Figure 2. Reprinted with permission.

Finally, loop transient response overshoot was computed for the different cases, as shown in Figure 7.18. Per earlier discussions, substantial frequency overshoot forces the IF filtering to be wider for low IF cases, and further complicates the sideband selection process in multiloop designs.

7.3 DIRECT DIGITAL SYNTHESIZER FUNDAMENTALS

Although direct digital frequency synthesizers (DDSs) have been widely used for some time now, they have only come into prominence during roughly the past 10 years. Substantiating this claim, only about one-half page is devoted to the subject in [4] whereas no mention is made at all in [5]; both publications appeared in the early 1980s. At the same time, the rapid ascension of custom VLSI and low-cost

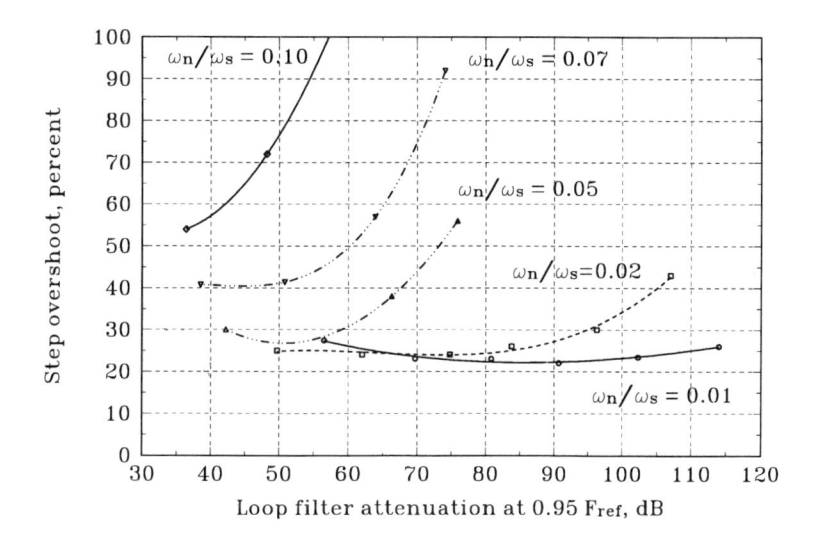

Figure 7.18 Loop overshoot versus open-loop transfer function attenuation for the first reference discrete spurious sideband. Source: [1], Figure 2. Reprinted with permission.

consumer electronics have led to a peak in standalone DDS products in the mid-1980s, which was followed by a softening of the market as more and more large-volume customers began integrating more than just the single DDS function into single devices. This trend is certain to continue as further evidenced by material presented in the final sections of this chapter.

7.3.1 General Concepts

A basic diagram for the standard DDS is shown in Figure 7.19. Almost all DDSs are composed of these same fundamental building blocks, albeit with some minor enhancements or modifications. In this figure, F_{Clk} represents the fixed high-purity reference clock being used by the DDS. At each leading edge of F_{Clk}, the phase accumulator effectively performs the computation described by

$$\theta_{k+1} = (\theta_k + \Delta\theta) \bmod 2\pi \tag{7.14}$$

where θ_k and $\Delta\theta$ are represented in actual practice by unsigned binary numbers and the modulo 2π operation is equivalent to discarding any carry-out operation from the accumulator operation. Quite simply, for a constant frequency output, θ must be advanced the same amount $\Delta\theta$ at each F_{Clk} edge. As a matter of notational convenience, when $\Delta\theta$ is represented in its binary form, it will be denoted by F_r in this material.

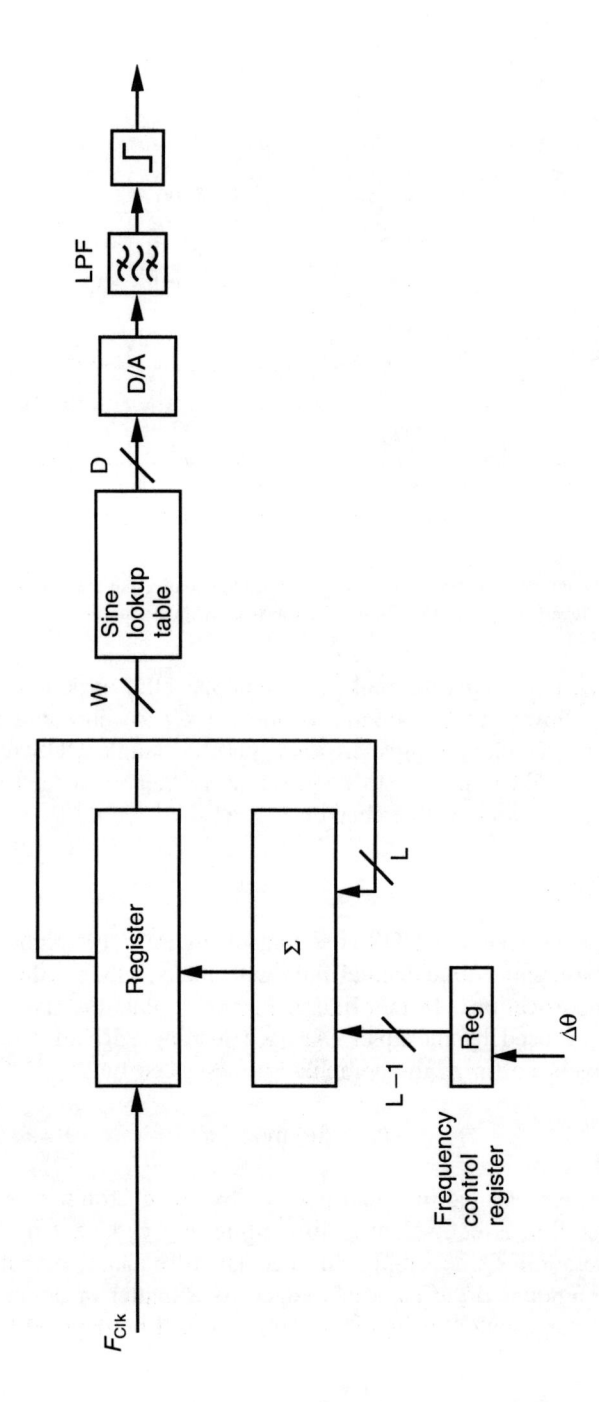

Figure 7.19 Traditional architecture for a DDS.

Given an L-bit-wide accumulator, the total number of possible phase states is simply 2^L. The lowest frequency that can be produced by the DDS corresponds to the case in which $\Delta\theta$ is represented by a binary 1. For this case, it will take 2^L clock cycles in order to pass though all 2^L phase states before the process is then repeated. The minimum output frequency achievable is then given by

$$f_{\min} = \frac{F_{\text{Clk}}}{2^L} \qquad (7.15)$$

In general, the DDS output frequency is given by

$$f_{\text{out}} = \frac{\Delta\theta F_{\text{Clk}}}{2\pi} = \frac{F_r F_{\text{Clk}}}{2^L} \qquad (7.16)$$

where $0 < F_r < 2^{L-1}$ for normal operation.

Of the L accumulator bits present, only the W MSBs of the accumulator are used to compute the $\sin(\theta)$ value. This reduction is necessary in order to keep the sine lookup table of reasonable size.

The W-bit sine table lookup value is returned then as a D-bit binary value, which is then converted to an actual analog voltage value by the digital-to-analog (D/A) converter. Both the W- and D-bit truncations lead to undesirable spectral degradations as discussed later.

Since the DDS is essentially a sampled system, the D/A output must normally be passed through a reconstruction filter, which removes unwanted alias frequencies from the output as well as any F_{Clk} leakage that would otherwise be present. Finally then, the output is passed through an ideal hard limiter in order to remove any residual AM that may be present.

Different sets of design issues arise whether one is designing an actual DDS component or whether one is simply an end-user of DDS devices available in the marketplace. Issues relevant to both perspectives are presented in the material that follows.

Other forms of digital synthesis have appeared in the literature also. In [6], synthesis is considered in terms of Walsh functions. In [7], digital oscillators based on finite difference equations are considered with the additional requirement of absolute periodicity. One of the more recent works that extends this discussion further is provided in [8]. Overall, perhaps the best introductory discussion of evolving DDS techniques is that given in [9].

7.3.2 Representation of the Ideal DDS Output Spectrum

In the ideal DDS, no phase accumulator truncation occurs ($W = L$) and the sine lookup table is assumed to have infinite precision. The DDS output can then be represented by

$$s(t) = \sum_{n=-\infty}^{\infty} \sin(2\pi f_o n T_c) h(t - nT_c) \qquad (7.17)$$

where f_o is the DDS fundamental output frequency, $T_c = F_{Clk}^{-1}$, and $h(t) = 1$ for $0 \le t < T_c$ and 0 otherwise. The function $h(t)$ represents the output sample-and-hold. In the more general case where the sampled signal is not an ideal sinusoid but rather is represented by the waveform $v(t)$,

$$s(t) = \sum_{n=-\infty}^{\infty} v(nT_c) h(t - nT_c) \qquad (7.18)$$

Schematically, the situation is as shown in Figure 7.20, which closely resembles some of the material presented in Chapter 4. Carrying this notion through further, from the sampling theorem (4.34) and (5.12), the Fourier transform of the sampled waveform is given by

$$\mathbf{F}[v^*(t)] = \frac{1}{T_c} \sum_{n=-\infty}^{\infty} V\left(f - \frac{n}{T_c}\right) \qquad (7.19)$$

Since $v(t)$ is assumed to be periodic with some frequency f_o, it may of course be represented by its Fourier series as

$$V(f) = \sum_{m=-\infty}^{\infty} c_m \delta(f - mf_o) \qquad (7.20)$$

where

$$c_m = \frac{1}{T_m} \int_{-T_o/2}^{T_o/2} v(t) \, e^{-j2\pi f_o mt} \, dt \qquad (7.21)$$

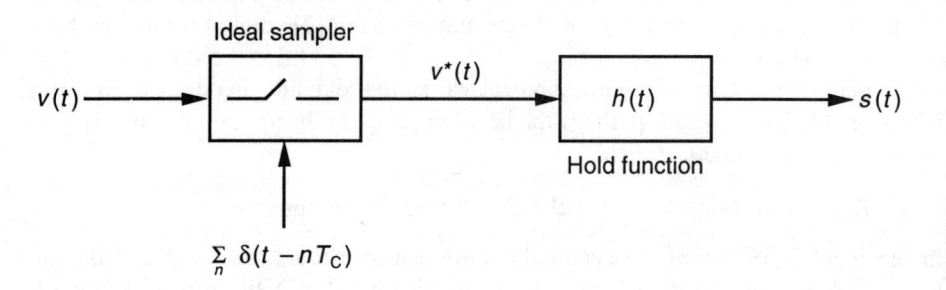

Figure 7.20 Conceptualization of the sampling process underlying direct digital frequency synthesis.

where $T_o = f_o^{-1}$. On substitution of (7.21) into (7.19), the Fourier representation is given by

$$\mathbf{F}[v^*(t)] = F_{\text{Clk}} \sum_{n=-\infty}^{\infty} \sum_{m=-\infty}^{\infty} c_m \delta(f - nF_{\text{Clk}} - mf_o) \qquad (7.22)$$

The holding function simply appends a shaping function on $\mathbf{F}[v^*(t)]$ and is given by

$$H(f) = \frac{1 - e^{-j2\pi f T_c}}{j2\pi f} \qquad (7.23)$$

Hence, the final result is given as [9]

$$S(f) = e^{-\pi f T_c} \frac{\sin(\pi f T_c)}{\pi f T_c} \sum_{n=-\infty}^{\infty} \sum_{m=-\infty}^{\infty} c_m \delta(f - nF_{\text{Clk}} - mf_o) \qquad (7.24)$$

In summary, the DDS action causes all frequency components residing in $v(t)$ to be aliased about every harmonic of the clock frequency F_{Clk}. Therefore, if $v(t)$ is a perfect sine wave, only the nearest alias frequency is normally of concern. On the other hand, if $v(t)$ were chosen to be triangular instead, the harmonics of the triangular wave would be aliased arbitrarily close to the desired fundamental output frequency and substantially different spurious performance would then be realized.

7.3.3 Phase Truncation Related Spurious Effects

Phase truncation in the DDS is caused by using only $W < L$ bits in Figure 7.19 to address the sine lookup table. In this case, the DDS output (less any additional filtering) can be written as

$$s(t) = \sum_{n=-\infty}^{\infty} \sin[2\pi f_o n T_c + e(n)] h(t - nT_c) \qquad (7.25)$$

where $e(n)$ represents the phase error at each clock instance due to truncation. The fundamental output frequency f_o is restricted to values given by

$$f_o = \frac{F_r}{2^L} F_{\text{Clk}} \qquad (7.26)$$

for some integer F_r. The phase error sample sequence is also restricted in magnitude as

$$|e(n)| < \frac{2\pi}{2^W} = \frac{\pi}{2^{W-1}} \tag{7.27}$$

and is also periodic with some period Q where $1 < Q \leq 2^{L-W}$. It is important to understand how this implementation limitation is manifested in terms of a degraded output spurious spectrum. A complete derivation of the phase accumulator truncation effects on the output spectrum is given in [10]. Portions of this material follow.

All DDS devices that use some variation of table lookup for the sine function can be mathematically analyzed using the model shown in Figure 7.21 where

W = binary word length of phase accumulator output used to address the table lookup
L = phase accumulator width in bits
D = binary word length of sine values stored in lookup table
F = frequency control word ($\Delta\theta$).

Three degradations from ideal DDS operation are also shown in this figure as described:

$\varepsilon_p(n)$ = truncation noise in addressing sine table with $W < L$ bits
$\varepsilon_T(n)$ = additive noise source from storing sine values with only finite precision
$\varepsilon_{DA}(n)$ = additive output noise arising from the nonideal D/A conversion process.

The subsequent analysis begins with consideration of the phase accumulator operation alone with truncation. Once this has been characterized, it is reasonably straightforward to determine the overall frequency spectrum at the DDS output.

The numerical period of the phase accumulator output sequence is defined as the minimum value of N for which the phase accumulator sample values repeat, i.e., $\theta(n) = \theta(n + N)$. In general, this is given by

$$P = \frac{2^L}{(F_r, 2^L)} \tag{7.28}$$

where $(F_r, 2^L)$ represents the greatest common divisor between F_r and 2^L and F_r is as defined earlier, the binary representation for $\Delta\theta$. One simple means for maximizing P is to simply make F_r an odd integer in which case $P = 2^L$ in all cases. As an additional consequence, the discrete Fourier frequency spectrum immediately prior to D/A conversion will also consist of P discrete lines.

The key element in the overall output spectrum derivation is computation of the spectral content of the phase error sequence represented by $\varepsilon_p(n)$. As stated ear-

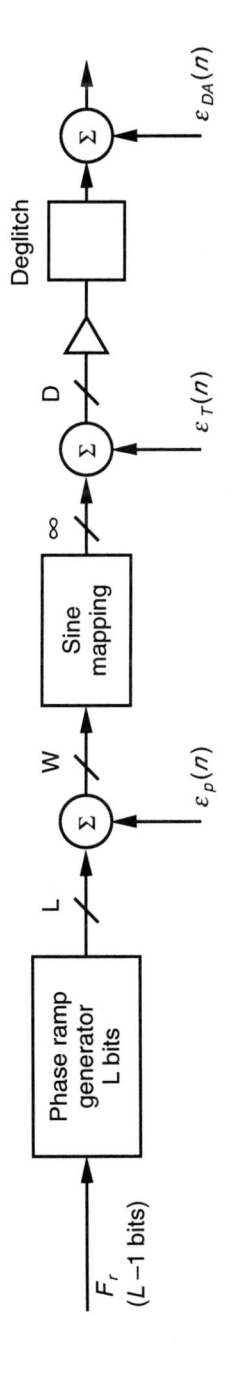

Figure 7.21 Direct digital synthesizer model considered in the text. (© 1987 IEEE; reprinted with permission). After: [10], Figure 6.

lier, in the ideal case of infinite precision in the lookup table and no phase truncation, the output sample sequence of the DDS is given by

$$s(n) = \sin\left(2\pi \frac{F_r}{2^L} n\right) \qquad (7.29)$$

If the phase accumulator value is now truncated to W bits prior to performing the table lookup operation, the output sequence must be modified as

$$s_t(n) = \sin\left(2\pi \frac{2^B}{2^L} \left\lfloor \frac{F_r}{2^B} n \right\rfloor\right) \qquad (7.30)$$

where the special brackets denote truncation to integer values. This may be rewritten as

$$s_t(n) = \sin\left\{\frac{2\pi}{2^L} [F_r n - \varepsilon_p(n)]\right\} \qquad (7.31)$$

where $\varepsilon_p(n)$ represents the error sequence introduced earlier. To proceed, it is fundamentally important to realize that this phase error sequence can be modeled as the continuous-time sampled waveform, which will be denoted by $\varepsilon_T(\ )$. The amplitude of this sawtooth waveform is 2^B and its frequency is $F_r 2^{-B}$ and is shown in Figure 7.22.

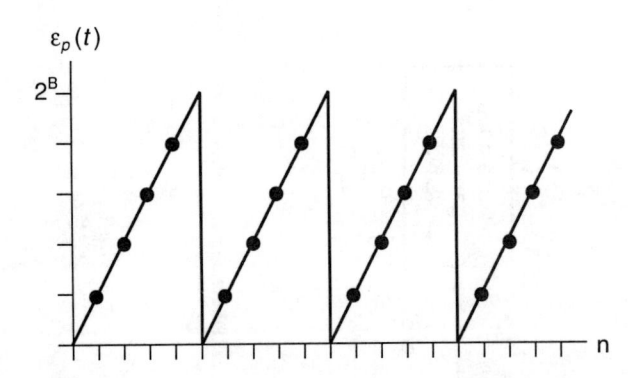

Figure 7.22 Phase accumulator error sequence. Source: [10], Figure 6. Reprinted with permission.

This sawtooth waveform is identical to the waveform that would be generated by a phase accumulator of word length B with an input frequency word of

$$\langle F_r \rangle_{2^B} \tag{7.32}$$

where the angle brackets denote taking the integer residue modulo 2^B.

Although not mentioned in [10], it is interesting to note what differences arise when phase accumulator rounding rather than truncation is assumed. (Normally, this is never done because the rounding operation would require additional hardware compared to simple truncation.) Purely graphical arguments can be used to show that this would remove the dc term in $\varepsilon_p(t)$ but leaves the waveform otherwise almost unchanged. The absence of the dc term would have simplified the remaining analysis, however, compared to the truncation case.

Continuing with the phase truncation case only, to characterize the spectral properties of the error sequence $\varepsilon_p(n)$, it is necessary to first compute the Fourier series representation for the error waveform shown in Figure 7.22. It is important to note that the error waveform is required to have a value of zero at each point of discontinuity, which is in violation of the required Dirichlet conditions for Fourier series representation. Dirichlet conditions require that a function take on its average value at any point of discontinuity, which is given by $2^B/2$ here.

Therefore, to represent the phase error sequence as a Fourier series, the error waveform must be reexpressed as the superposition of two Dirichlet valid waveforms, which also properly represent $\varepsilon_p(t)$. Such a set of waveforms is shown in Figure 7.23.

After fairly lengthy computations, it can be shown that

$$\varepsilon_p(t) = \sum_{k=1}^{\infty} \frac{2^B}{\pi k} \sin\left(2\pi k \frac{F_r}{2^B} t\right)$$
$$- \frac{2^B}{2\Lambda} \sum_{k=1}^{\infty} \frac{4\Lambda}{2\pi k} \sin\left(\frac{2\pi k}{4\Lambda}\right) \cos\left(2\pi k \frac{F_r}{2^B} t\right) \tag{7.33}$$

where

$$\Lambda = \frac{2^B}{2(F_r, 2^B)} \tag{7.34}$$

The first term in this result is due to the Dirichlet sawtooth waveform, whereas the second term is due to the pulse train correction waveform. Had phase rounding rather than truncation been employed, it can be shown that only the first term in this result

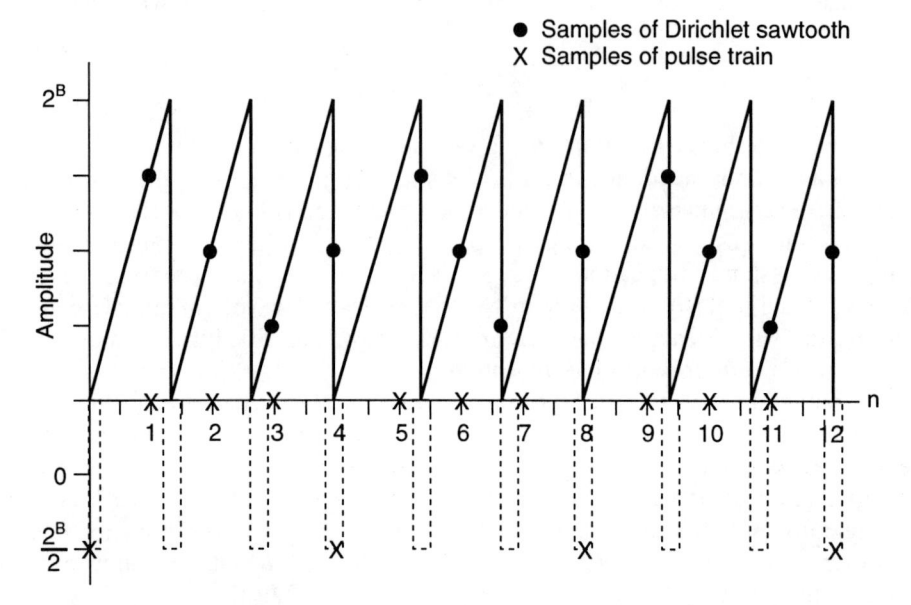

Figure 7.23 Representation of the error sequence as the superposition of two Dirichlet valid waveforms. (© 1987 IEEE; reprinted with permission). Source: [10], Figure 9.

would survive. Obviously, the sample sequence $\varepsilon_p(n)$ is computed by evaluating this result for each $t = n$.

Following a substantial amount of additional computation, the discrete spectrum for the phase error function $\varepsilon_p(n)$ can be summarized as follows [10]:

1. Find the number of discrete spurs from

$$\Lambda = \frac{2^{B-1}}{(F_r, 2^B)} \qquad (7.35)$$

where

B = number of accumulator bits truncated
F_r = integer representation for frequency control word

2. Divide the frequency range of 0 to $F_{\text{Clk}}/2(F_r, 2^B)$ into Λ equally spaced spur locations.

3. Find the spur number K for ζ_K from the sequential spur number F_n where
$$K = \langle F_n \Gamma^{\Lambda-1} \rangle_{2\Lambda} \qquad (7.36)$$
and
$$\Gamma = \frac{F_r}{(F_r, 2^B)} \qquad (7.37)$$

4. Compute the amplitude of the spur as
$$\zeta_K = \frac{2^B}{2\Lambda} \operatorname{cosec}\left(\frac{K\pi}{2\Lambda}\right) \qquad (7.38)$$

Given this result, and also additional computations to reflect this spurious content to the actual DDS output, the complete algorithm for determining the discrete output spectrum of the DDS in the presence of phase truncation can be summarized as follows [10]:

1. Divide the frequency range from 0 to $F_{\text{Clk}}/2$ into a sequence of $2^{L-1}/(F_r, 2^L)$ evenly spaced potential spur locations, each sequentially numbered. The sequential frequency number of a spur location is related to the actual analog spur frequency F_{sp} by
$$F_n = F_{sp} \frac{2^L}{F_{\text{Clk}}(F_r, 2^L)} \qquad (7.39)$$

2. Find the spur number K from the frequency number F_n as follows:
If both $F_n - \Lambda$ and $-F_n - \Lambda$ are not divisible by 2, then the magnitude of the spur at $F_n = 0$.
Else, if 2 does divide $F_n - \Lambda$, then
$$K = \left\langle \frac{F_n - \Gamma}{2^{L-B}} \Gamma^{\Lambda-1} \right\rangle_{2\Lambda} \qquad (7.40)$$
Else, if 2 divides $-F_n - \Lambda$, then
$$K = \left\langle \frac{-F_n - \Gamma}{2^{L-B}} \Gamma^{\Lambda-1} \right\rangle_{2\Lambda} \qquad (7.41)$$

3. The magnitude of the spurious line at F_n is given by
$$\zeta_{K\pm} = \frac{\pi 2^{B-L}}{2\Lambda} \operatorname{cosec}\left(\frac{K\pi}{2\Lambda}\right) \qquad (7.42)$$

The most important finding in this analysis is that the number of spurs and their magnitude only depend on F_r through the greatest common divisor factor $(F_r,$

2^B). As shown in [12], values of F_r which have the same value of $(F_r, 2^B)$ result in output spectra with the same number of spurs and with their respective amplitudes unchanged. Only the position of each spur in the output spectrum is altered. This result stems from several number theoretic arguments which are presented in [10,12].

Since ζ_K is a monotonically decreasing function of K in (7.42), the magnitude of the largest spur in the spectrum is given by

$$\zeta_{max} = 2^{B-L} \frac{\pi \frac{(F_r, 2^B)}{2^B}}{\sin\left[\pi \frac{(F_r, 2^B)}{2^B}\right]} \qquad (7.43)$$

For F_r values where $(F_r, 2^B) = 2^B$, no phase truncation spurs exist. Otherwise, the worst case truncation spur results when $(F_r, 2^B) = 2^{B-1}$ in which case

$$\zeta_{max} = 2^{B-L} \frac{\pi}{2} \qquad (7.44)$$

This is equivalent to a spur level of $6.02(B - L) - 3.92$ dBc.

An additional finding in [10] revealed that a fairly modest improvement in worst case phase truncation related performance could be delivered if a small modification were made to the traditional DDS architecture. This modification is shown in Figure 7.24. With this modification in place, all phase accumulator output sequences are forced to belong to the class for which $(F_r, 2^{L+1}) = 1$ for all F_r [11]. As a result, the worst case spurious response can be determined from only one simulation case, any dependency on the phase accumulator initial contents is eliminated, and the worst case spurious level is improved by 3.92 dB and is simply given by $6.02(B - L)$ dBc.

The illustration of some of these findings is straightforward with one simple example. Consider the situation where the phase accumulator width is 12 bits and 6 bits are truncated in the table lookup process. Two F_r values of 513 and 607 are considered. The DFT of the DDS output sequence for $F_r = 513$ is shown in Figure 7.25 where the worst base spur levels due to phase truncation appear to be approximately -36 dBc. Since $(F_r, 2^B) = 1$ for this case, the expected worst case spur value is $6.02(B - L) = -36.12$ dBc, which agrees very closely. The DFT spectrum for the second frequency case of $F_r = 607$ is shown in Figure 7.26. As predicted, since $(F_r, 2^B) = 1$ for this case as well, the worst case spur level is unchanged and only the position of the spurs has been permutated.

A compilation of worst case spurious performance resulting from all finite word length effects in the DDS is shown in Figure 7.27 [10]. This figure is invaluable for anyone contemplating the design of their own DDS core function.

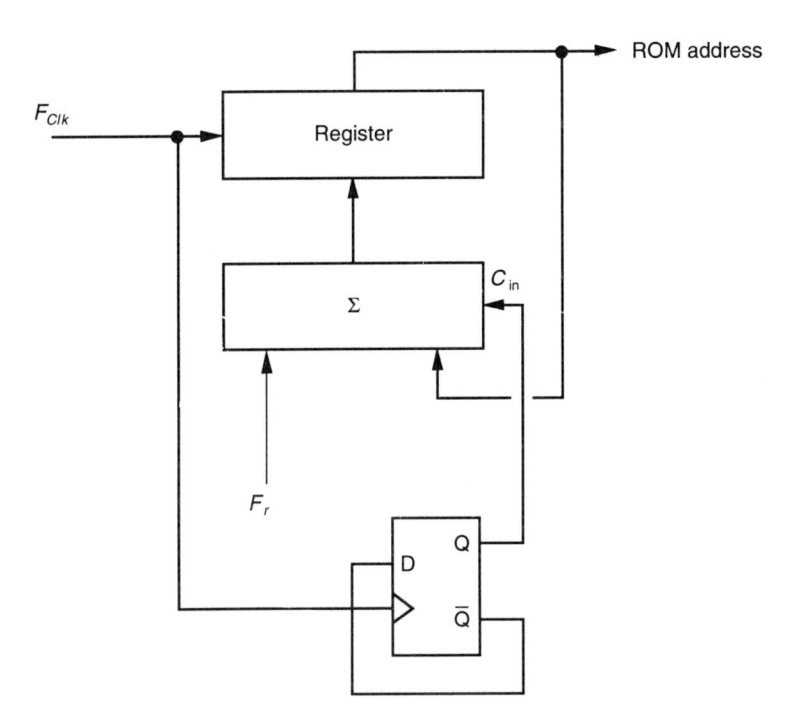

Figure 7.24 Modified direct digital synthesizer for improved spurious performance. (© 1987 IEEE; reprinted with permission). After: [10], Figure 13.

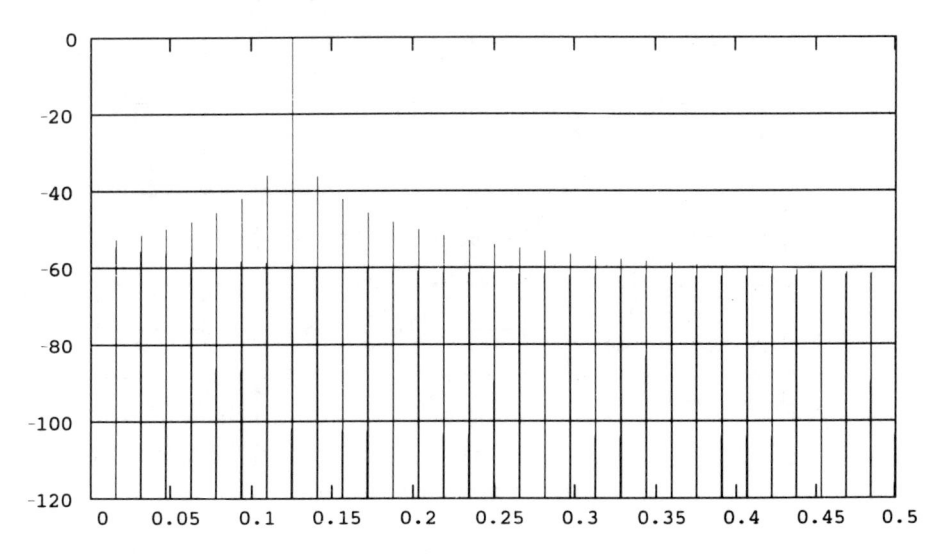

Figure 7.25 Discrete Fourier transform of the DDS output sample sequence for $L = 12$, $W = 6$, and $F_r = 513$.

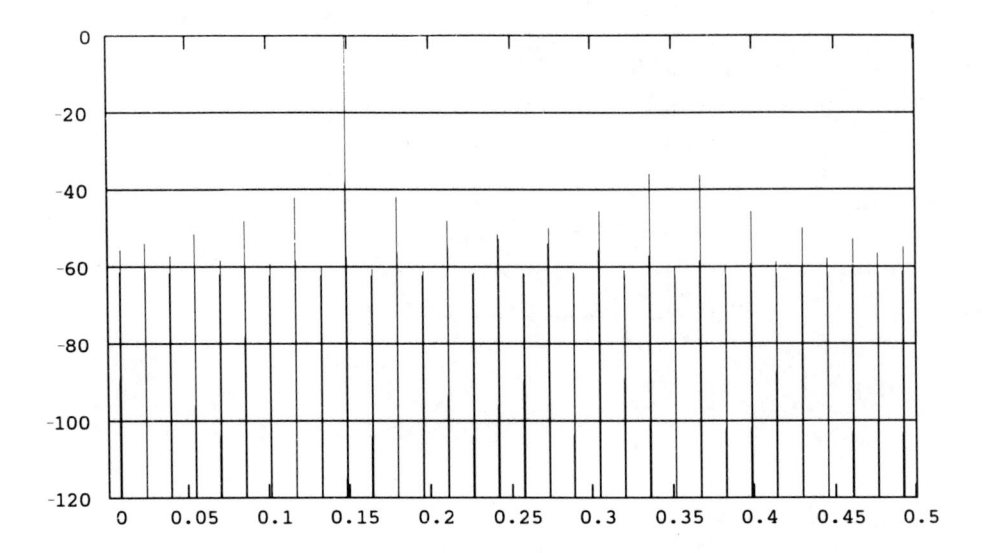

Figure 7.26 Discrete Fourier transform of the DDS output sample sequence for $L = 12$, $W = 6$, and $R_r = 607$.

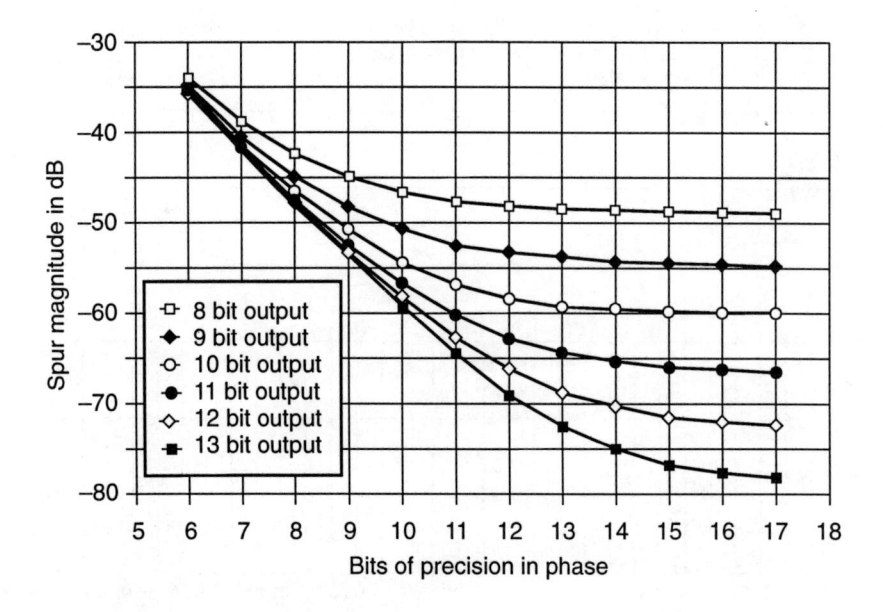

Figure 7.27 Worst case spurious response of a DDS for all finite word length effects. (© 1987 IEEE; reprinted with permission). Source: [10], Figure 14.

7.3.4 Lookup Table Finite Word Length Effects on Spectral Purity

Finite quantization in the sine lookup table sample values also leads to DDS output spectrum impairments. Equivalent errors are also introduced into the system by non-ideal D/A converter attributes. In this section, an ideal D/A converter is assumed and only sample finite quantization errors are considered.

It is straightforward to show that the sequence of table lookup quantization errors is periodic, repeating every P samples where

$$P = \frac{2^L}{(F_r, 2^L)} \tag{7.45}$$

Sample quantization effects on the DDS output may then be included by representing the DDS output by

$$s(t) = \sum_{n=-\infty}^{\infty} [\sin(2\pi f_o n T_c) + q(n \bmod P)] h(t - nT_c) \tag{7.46}$$

where $f_o = F_r F_{\text{Clk}}/2^L$ and the $q(k)$ values represent the repeating sequence of quantization errors for a given F_r and initial phase accumulator value. On simple substitution into (7.45),

$$s(t) = \sum_{n=-\infty}^{\infty} \left[\sin\left(\pi \frac{F_r}{2^{L-1}} n \right) + q(n \bmod P) \right] h(t - nT_c) \tag{7.47}$$

The noise-only portion of the output is clearly given by

$$V_n(t) = \sum_{n=-\infty}^{\infty} q(n \bmod P) h(t - nT_c) \tag{7.48}$$

The $q(k)$ values are in principle random error quantities which for $P \gg 1$ results in the errors being uniformly distributed from $-\text{LSB}/2$ to $\text{LSB}/2$. Depending on the specific nature of the $q(k)$ values and primarily P, widely variable output spectra can be observed due to the quantization error process.

In the case where $P = 2^L$ and the $q(k)$ can be taken to be statistically independent and uniformly distributed, the resulting noise spectrum contributes equal amounts of AM and PM noise at the DDS output, not unlike plain wideband noise. Computing the Fourier series coefficients of the error signal in this case,

$$V_n(f_k) = \frac{1}{PT_c} \int_{-PT_c/2}^{PT_c/2} \sum_n q(n \bmod P) h(t - nT_c) e^{-j2\pi f_k t} \, dt \tag{7.49}$$

for $f_k = k/PT_c$ with $0 \leq k < P$. From this,

$$|V_n(f_k)|^2 = \left(\frac{1}{PT_c}\right)^2 \int_{-PT_c/2}^{PT_c/2} dt \int_{-PT_c/2}^{PT_c/2} dt'$$

$$\sum_n \sum_m q(n \bmod P) q(m \bmod P) h(t - nT_c) h(t - mT_c) \, e^{-j2\pi f_k(t-t')} \quad (7.50)$$

The discrete power spectral density can then be found by taking the statistical expectation of (7.50), assuming statistical independence of the $q(k)$ for different values of k. Since the calculations are performed over one complete period, the mod P arguments can be dropped, finally leading to

$$S_n(f_k) = P\left(\frac{\sigma}{PT_c}\right)^2 \iint h(t - nT_c) h(t - mT_c) \, e^{-j2\pi f_k(t-t')} \, dt \, dt' \quad (7.51)$$

which can be further simplified to

$$S_n(f_k) = \frac{\sigma^2}{P} \left|\frac{\sin(\pi f_k T_c)}{\pi f_k T_c}\right|^2 \quad (7.52)$$

Therefore, the resulting output noise spectrum is simply a sinc-squared weighted line spectrum consisting of P equally spaced spectral lines each having an amplitude of $\sigma^2 = (LSB)^2/12$.

If the peak sine wave value is now assumed to have an amplitude of 2^{D-1} and the spectral "width" of each line is approximately given by F_{Clk}/P Hz, the carrier-to-noise ratio (CNR) at the DDS output is given by

$$\frac{C}{N_o} = \frac{\text{Carrier power}}{\text{Noise power in 1 Hz}} = \frac{(2^{D-1})^2}{2} \frac{F_{Clk}}{\sigma^2}$$

$$= 12 \, 2^{2D-3} \, F_{Clk} \quad (7.53)$$

For the case where $D = 10$ bits and $F_{Clk} = 40$ MHz, $CNR_o = 138$ dBc/Hz. In principle, following the DDS (after the anti-aliasing filter) with an ideal hard limiter would further improve the output CNR_o by roughly 3 dB since the AM noise would be eliminated.

In the general spurious case, $P < 2^L$ and roughly the same total amount of spurious energy is distributed between fewer discrete spectral lines. Worst case, the table errors can lead to a sinusoidal component of peak value $LSB/2$, which is precisely in phase-quadrature with the desired output signal. In this case all of the spur power is in the form of PM and the worst case spurious level is given by

$$S_T = 20 \log_{10}\left(\frac{\varphi_{pk}}{2}\right)$$

$$= -20 \log_{10}(2^D) \quad \text{dBc} \qquad (7.54)$$

Therefore, a D value of 10 bits leads to a worst case PM spur value of -60 dBc. Normally, the PM spur case is the most critical because AM spurs can in principle be removed by employing an output hard limiter.

The observed output spectra due to sample quantization effects can change substantially with only a small change in F_r. This point is underscored in Figures 7.28 through 7.32. In all cases, an accumulator length of 11 bits has been assumed with the sample values rounded and quantized to 8-bit values. In Figure 7.28, the DDS output sample sequence prior to anti-alias filtering is shown. The DFT of the output sample sequence follows in Figure 7.29. This choice of F_r (513) is relatively prime to 2^L, which results in a dense comb of fairly low-level spurious outputs. The value of F_r was changed to 524 in Figures 7.30 and 7.31, resulting in a substantially different interference pattern in Figure 7.30 and substantially fewer spurious lines in Figure 7.31. As shown, the output spectrum now contains several spurs in the vicinity of -70 dBc.

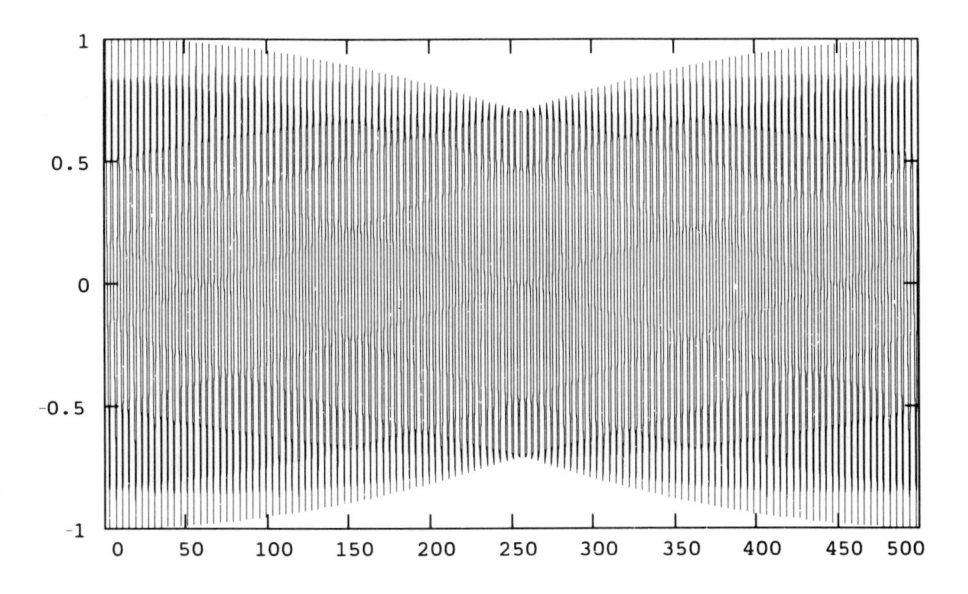

Figure 7.28 DDS output sample sequence for $L = 11$, $F_r = 513$, and 8-bit sample quantization.

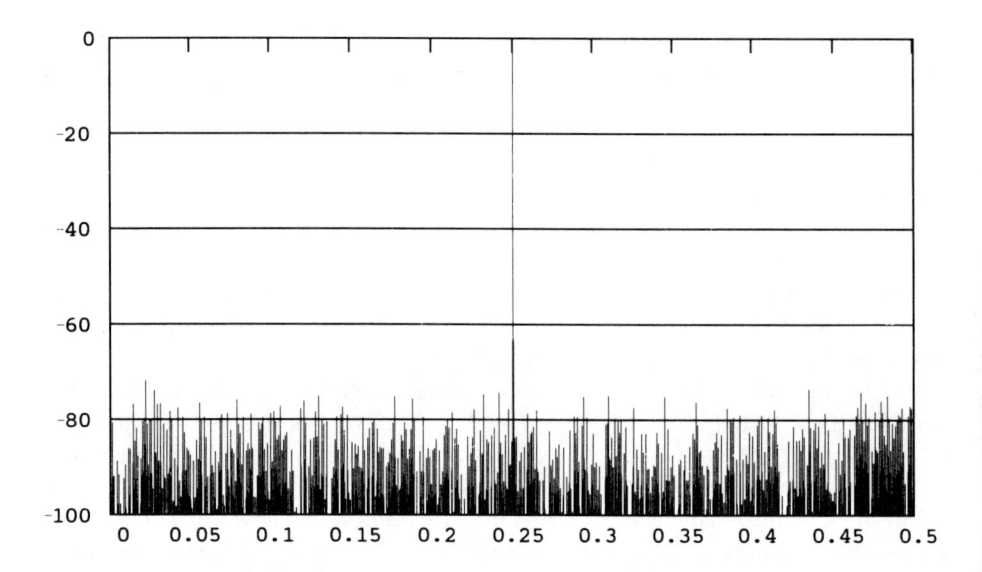

Figure 7.29 Discrete Fourier transform of the DDS output sample sequence corresponding to the time-domain waveform shown in Figure 7.28.

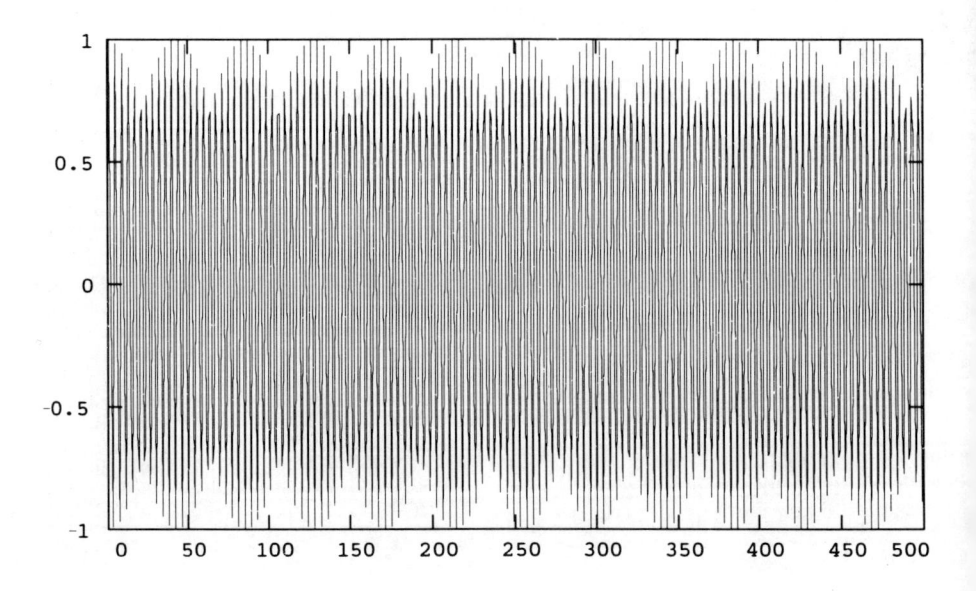

Figure 7.30 DDS output sample sequence for $L = 11$, $F_r = 524$, and 8-bit sample quantization.

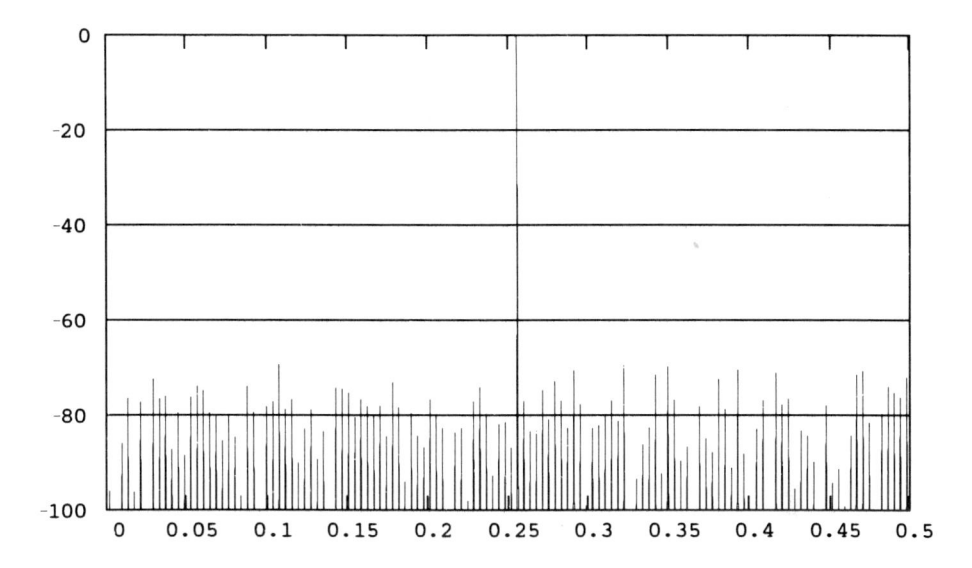

Figure 7.31 Discrete Fourier transform of the DDS output sample sequence corresponding to the time-domain waveform shown in Figure 7.30.

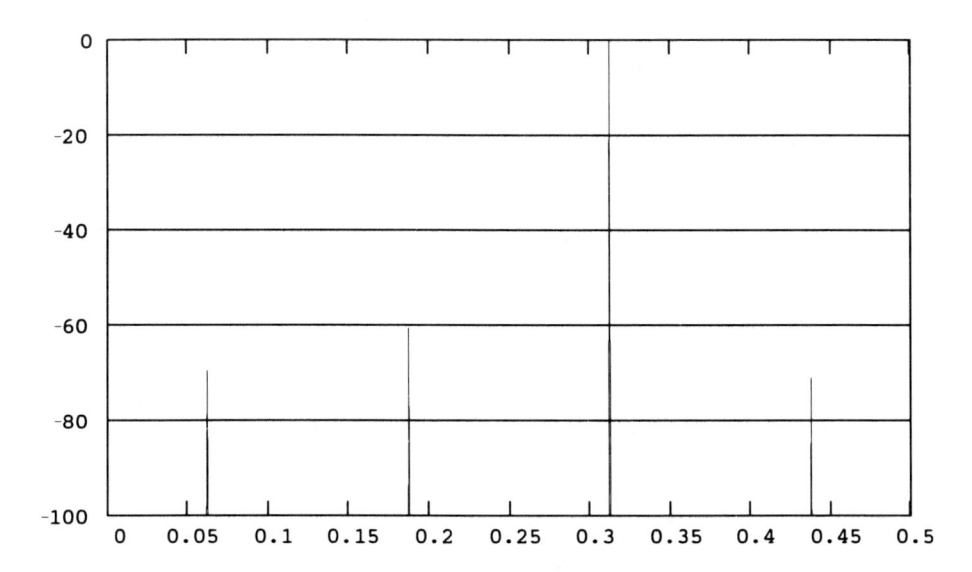

Figure 7.32 Discrete Fourier transform of the DDS output sample sequence for $L = 11$, $F_r = 640$, and 8-bit sample quantization.

In Figure 7.32, all parameters were left unchanged but $F_r = 640$ was selected to illustrate what can happen when $(F_r, 2^L) > 1$ occurs. In this case, the skewed spur levels indicate that elements of both AM and PM are present and the strongest spur component has increased to -60 dBc. This is substantially worse than that observed in the first two cases, which underscores the need to consider the worst case situation in actual DDS design.

7.3.5 Techniques for Mapping θ to $\mathrm{Sin}(\theta)$

The mapping technique [11] used for transforming the phase accumulator phase value θ to $\sin(\theta)$ is particularly crucial if table storage size is to be kept reasonable. Each additional bit used in the lookup table process potentially represents a doubling of the required table storage space.

The most simple means for compressed $\sin(\theta)$ storage is to exploit the symmetry of the function about $\pi/2$ and π. With proper manipulation of the phase and amplitude, lookup table samples need only be stored for phase values spanning the 0 to $\pi/2$ range. Such a scheme is shown schematically in Figure 7.33.

Normally, 2's-complement arithmetic is used in digital computing elements. However, this numerical representation presents some disadvantages wherever the negative of a number must be computed because formal negation involves first complementing each bit in the binary value followed by an addition of one to the quantity. In the table lookup case, if an LSB/2 offset is included in the number to be complemented, a simple 1's complementor can be used in place of the more complex 2's complementor without incurring additional error.

Sin(θ) Table Compression

One of the earliest methods used for sine table compression over the 0 to $\pi/2$ range is that of Sunderland et al. [12]. The subject architecture for this compression method is shown in Figure 7.34. In short, this technique allows the otherwise long table lookup ROM with 2^{A+B+C} storage locations to be replaced by two smaller ROMs having storage sizes of 2^{A+B} and 2^{A+C} locations. In the case where $A = B = C = 4$ and $D = 12$ bits, the noncompressed table would require 45056 bits (11×2^{12}). In contrast, even if full 12-bit-wide storage is used in both smaller ROMs, the total required bit storage is substantially less at 5632 bits ($11 \times 2^8 \times 2$) resulting in an 8:1 storage savings. The actual required word width for the second ROM in the present example is only 4 bits, which when factored in results in a total storage requirement of only 3840 bits ($11 \times 2^8 + 4 \times 2^8$), which is equivalent to an 11.73:1 storage reduction.

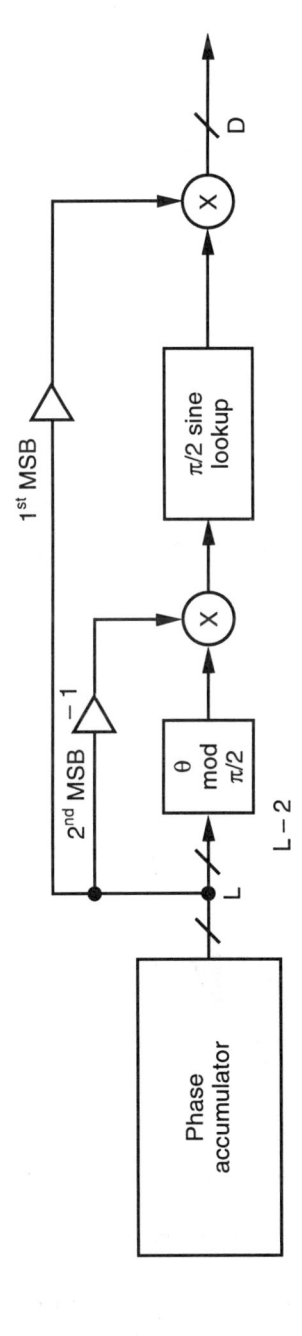

Figure 7.33 Exploitation of sin(θ) symmetries about $\pi/2$ and π for table storage compression. (© 1988 IEEE; reprinted with permission). After: [11], Figure 6.

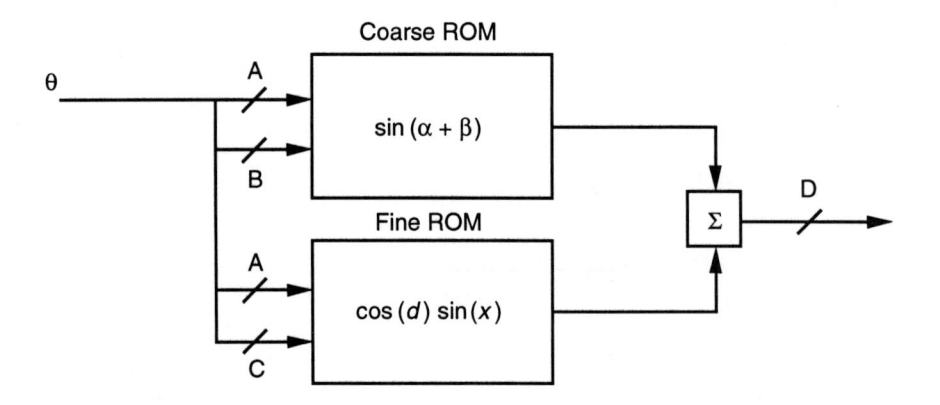

Figure 7.34 Sunderland architecture for compressed $\sin(\theta)$ table storage. (© 1988 IEEE; reprinted with permission). After: [11], Figure 7.

The Sunderland technique is based on simple trigonometric identities. The phase accumulator value θ is first reflected into the first quadrant by computing $\varphi = \theta$ mod $\pi/2$ and then further decomposed into a sum of three other angles as

$$\varphi = \alpha + \beta + \chi \tag{7.55}$$

where

$$\alpha < \frac{\pi}{2}$$

$$\beta < \frac{\pi}{2}2^{-A} \tag{7.56}$$

$$\chi < \frac{\pi}{2}2^{-(A+B)}$$

Using the trigonometric identity

$$\sin(\alpha + \beta + \chi) = \sin(\alpha + \beta)\cos(\chi)$$
$$+ \cos(\alpha)\cos(\beta)\sin(\chi) - \sin(\alpha)\sin(\beta)\sin(\chi) \tag{7.57}$$

and exploiting the small-angle approximations where possible,

$$\sin(\alpha + \beta + \chi) \approx \sin(\alpha + \beta) + \cos(\alpha)\sin(\chi) \tag{7.58}$$

The contents of the upper ROM in Figure 7.34 then contains the quantity $\sin(\alpha + \beta)$ and the lower ROM contains the quantity $\cos(\alpha)\sin(\chi)$. Since $\sin(\chi) \ll 1$, the width of the lower ROM can normally be made considerably smaller than the upper ROM.

An alternative means for calculating the ROM values based on numerical optimization is given in [11] also. Rather than be restricted to strictly a trigonometric interpretation, this approach permits each table value to be optimized independently. In the example where $A = B = C = 4$ and $D = 12$, this approach produced worst case spurious performance of approximately -84.2 dBc in comparison to -72.2 dBc using the Sunderland technique. The additional 12-dB performance improvement afforded by this numerical optimization technique is noteworthy.

Other optimizations and enhancements in the $\sin(\theta)$ compression area are certainly possible. One particular technique exploits the first-order Taylor series expansion for $\sin(\theta)$ by storing the alternate function

$$f(x) = \sin\left(\frac{\pi}{2}x\right) - \frac{\pi}{2}x \tag{7.59}$$

rather than $\sin(\theta)$ explicitly. Owing to the smaller dynamic range required to store $f(x)$, storage requirements can be further reduced. The required table lookup modification for use with this method is shown in Figure 7.35.

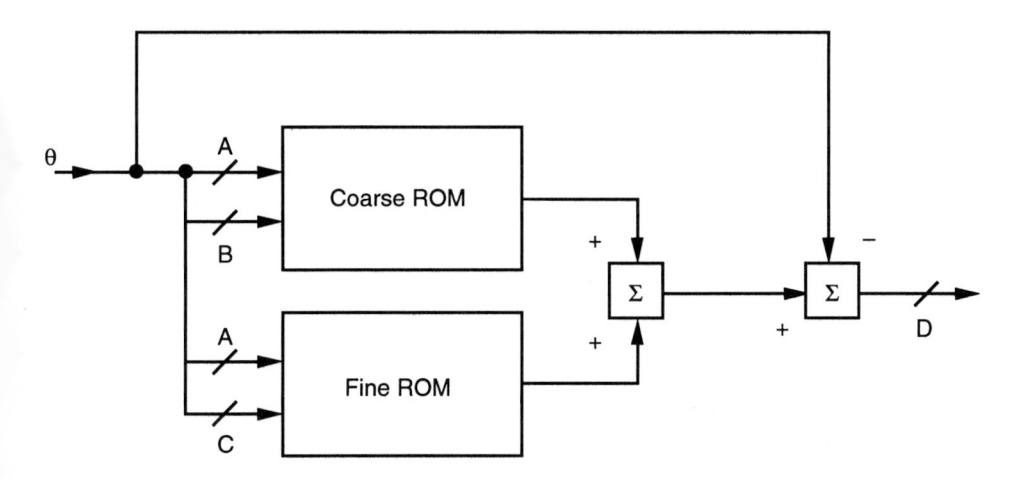

Figure 7.35 Modified architecture for $\sin(\theta)$ table compression using $f(\theta) = \sin(\theta) - \theta$. (© 1988 IEEE; reprinted with permission). After: [11], Figure 9.

A very interesting additional alternative for computation of sin(θ) is discussed at length in [13] where combinatorial binary logic is used to approximate the sine function. The basic algorithm for trigonometric function approximation follows these guidelines:

1. Express the function or a related function as a multiplication by
   (a) expressing the binary operands as polynomials,
   (b) taking the derivative of the corresponding inverse trigonometric function with the operands expressed as polynomials. The resulting equation is a multiplication of polynomials.
2. Expand multiplication into a partial products array.
3. Prohibit carry propagation between array columns for the most significant redundant digits of the unknown operand.
4. Express the resulting equation in a partial product array.
5. Reduce the array using Boolean and algebraic equivalences.

In the case of sin(θ), let Y equal the binary representation for sin(θ) and θ be the binary representation for the angle argument. Then for $0 \leq Y \leq 1$,

$$Y = \sum_{i=0}^{N} y_i 2^{-i} = \sum_{i=0}^{N} y_i x^i \qquad (7.60)$$

and for the angle θ where $0 \leq \theta \leq \pi/2$,

$$\theta = \sum_{i=0}^{N} \theta_i 2^{-i+1} = 2 \sum_{i=0}^{N} \theta_i x^i \qquad (7.61)$$

The polynomial representation is used because it can be differentiated. Following step 1 of the algorithm, the necessary formula are obtained as

$$\begin{aligned} \theta(x) &= \sin^{-1}[Y(x)] \\ \frac{d\theta(x)}{dx} &= \frac{1}{\sqrt{1 - Y^2(x)}} \frac{dY(x)}{dx} \\ Y(x) &= \sqrt{1 - Y^2(x)}\, \theta'(x) \\ [Y'(x)]^2 &= [1 - Y^2(x)][\theta'(x)]^2 \end{aligned} \qquad (7.62)$$

Following through the considerable amount of algebra as described in [13], the 12-bit estimate derived for sin(θ) is as shown in Figure 7.36 where each column entry is a logic minterm and the overbars indicate complement. This figure completely describes the combinatorial logic required to estimate sin(θ).

A comparison of this approach with more traditional methods is given in Figure 7.37. As shown, the 12-bit formulation is approximately equivalent to a third-order

Y1	Y2	Y3	Y4	Y5	Y6	Y7	Y8	Y9	Y10	Y11	Y12
1/2	1/4	1/8	1/16	1/32	1/64	1/132	1/256	1/512	1/1024	1/2048	1/4096
2	2	2	4	10	6	10	11	12	10	8	10 = 87
θ_1	$\theta_1\bar{\theta}_2$	$-\theta_1\theta_3$	$\theta_1\bar{\theta}_2$	$-\theta_2\theta_3$	$-\theta_1\theta_2\theta_3\bar{\theta}_4$	$-\theta_1\theta_2\theta_4$	$-\theta_1\theta_2\theta_3\theta_4$	$-\theta_1\theta_2\theta_4$	$-\theta_2\theta_3$	$-\theta_1\theta_3$	$-\theta_1\theta_3$
θ_2	θ_3	θ_4	$-\theta_1\theta_2\theta_3$	$-\theta_1\theta_2\theta_4$	$-\theta_1\theta_2\theta_5$	$-\bar{\theta}_1\theta_3\theta_4$	$\theta_1\theta_2\bar{\theta}_3\theta_5$	$-\theta_3\bar{\theta}_4$	$-\theta_3\theta_3\theta_4$	$\theta_1\theta_4$	$\theta_2\theta_3$
			$-\theta_1\theta_4$	$-\theta_1\theta_3\theta_4$	$\theta_1\theta_3\bar{\theta}_5$	$-\theta_2\theta_3\theta_5$	$\theta_2\theta_4\theta_5$	$-\theta_1\theta_2\theta_5$	$\theta_1\theta_4\theta_5$	$\theta_1\bar{\theta}_5$	$-\bar{\theta}_2\theta_4$
			$\bar{\theta}_2\theta_5$	$\theta_1\bar{\theta}_5$	$\bar{\theta}_1\theta_7$	$-\theta_1\theta_4\theta_5$	$\bar{\theta}_1\theta_2\theta_6$	$-\theta_1\theta_3\theta_5$	$\theta_1\theta_6$	$\theta_1\theta_2\bar{\theta}_5$	$\theta_1\theta_2\theta_4$
				$-\theta_1\theta_3$	$-\theta_2\bar{\theta}_5$	$-\bar{\theta}_1\theta_2\theta_6$	$-\theta_2\theta_3\theta_6$	$-\theta_3\theta_4\theta_5$	$-\bar{\theta}_1\theta_3\theta_6$	$\theta_1\theta_3\theta_5$	$\theta_1\theta_3\theta_5$
				$-\theta_2\theta_3$	$\theta_1\bar{\theta}_2\theta_6$	$-\theta_1\theta_3\theta_6$	$-\theta_1\theta_4\theta_6$	$-\theta_2\theta_4\theta_6$	$-\theta_1\bar{\theta}_2\theta_7$	$\theta_1\theta_2\theta_6$	$\theta_1\theta_4\theta_5$
				$-\theta_2\theta_4$		$\theta_1\bar{\theta}_2\theta_7$	$-\theta_1\theta_2\theta_7$	$-\theta_1\theta_6\theta_6$	$-\theta_2\theta_8$	θ_{12}	$-\bar{\theta}_1\theta_2\bar{\theta}_6$
				$\theta_1\bar{\theta}_6$		$\theta_1\theta_8$	$\theta_1\bar{\theta}_2\theta_8$	$\theta_2\theta_3\theta_7$	$-\theta_1\theta_{10}$	$\theta_1\theta_2\theta_6$	$\theta_1\theta_3\theta_6$
				$-\theta_2\bar{\theta}_5$		θ_2	$\theta_2\bar{\theta}_7$	$-\bar{\theta}_1\bar{\theta}_4\theta_7$	θ_{11}		$-\bar{\theta}_1\theta_2\theta_7$
				$\bar{\theta}_1\theta_2$		$\theta_1\theta_2\theta_3$	$-\theta_1\theta_4$	$-\theta_1\theta_3\theta_8$	$-\bar{\theta}_1\theta_2\theta_5$		$\theta_1\theta_8$
								$\theta_1\bar{\theta}_2\theta_9$			
								θ_{10}			

Figure 7.36 Reduced sine formulations for a 12-bit estimate. (© 1992 IEEE; reprinted with permission). After: [13].

(a)

Method	Bits or Order	Bits Correct Ave	Bits Correct Min.	Array Size Max Col.	Array Size Total Ele.	Latency Mpy	Latency Add
Prop.	8 b	7.14	3.78	5	26	< 1	0
Prop.	12 b	10.35	6.89	12	87	< 1	0
Prop.	16 b	13.52	9.49	33	250	< 1	0
Taylor	1 IRD	5.02	0.81	–	–	0	0
Taylor	3 ORD	10.88	3.73	–	–	3	1
Taylor	5 ORD	17.73	7.79	–	–	4	2
Cheby.	1 ORD	3.88	2.83	–	–	1	0
Cheby.	3 ORD	8.81	7.78	–	–	3	1
Cheby.	5 ORD	14.85	13.84	–	–	4	2

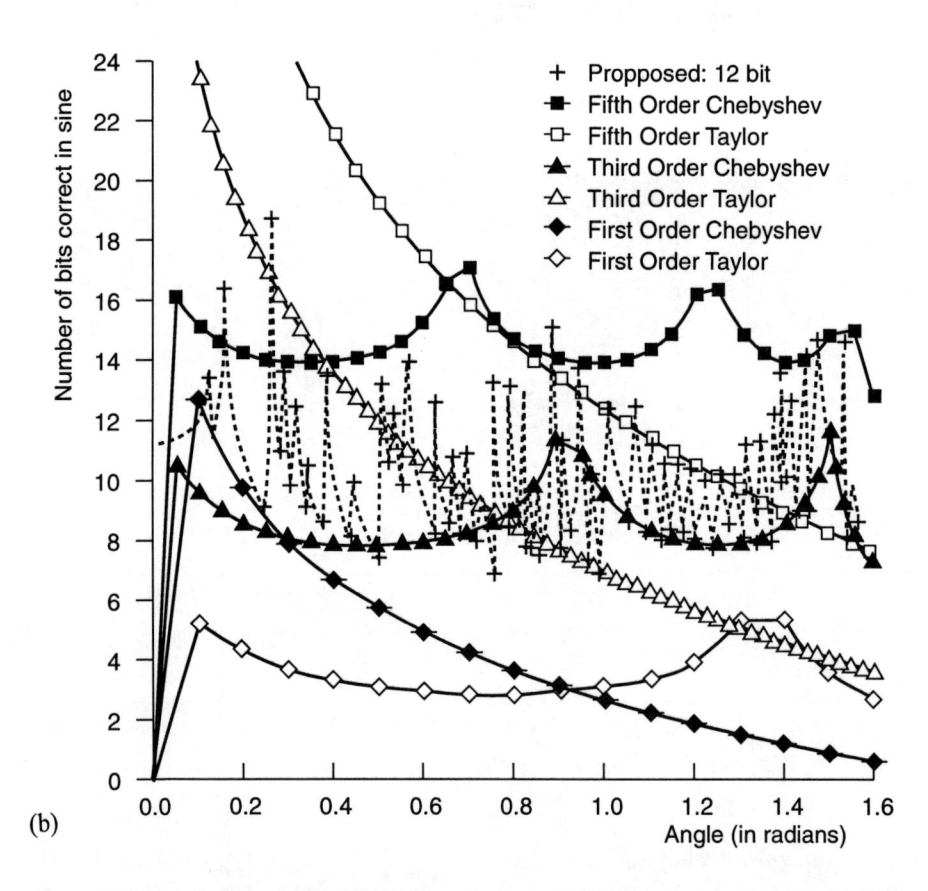

(b)

Figure 7.37 (a) Statistics of proposed, Taylor series, and Chebyshev polynomial methods for sine function. (b) Number of correct bits in proposed, Taylor, and Chebyshev methods for sine function. (© 1992 IEEE; reprinted with permission). Source: [13].

Chebyshev formulation. Although not directly comparable to the Sunderland approach described earlier, the underlying concepts appear very attractive for VLSI implementation.

7.3.6 Digital-to-Analog Converter Imperfections

The D/A converter is ultimately responsible for interfacing the digital world to the continuous RF/analog world. Although a number of performance measures have been standardized to quantify D/A converter imperfections, it is almost impossible to reflect these quantities to the DDS output and determine spectral purity. So, although these measures provide some evaluations guidelines for DACs, normally DDS spectral purity with a specific DAC must be evaluated on a case-by-case basis. A thorough introduction to this important aspect of DDS design can be found in [14–17].

7.3.7 Spurious Suppression in Direct Digital Synthesizers Using Numerical Techniques

Additional digital techniques may be incorporated into the generic DDS in order to largely eliminate the presence of discrete spurious signals at the DDS output. Normally, this results in a slight increase in wideband spectral noise but the impact can generally be made negligible.

From a standpoint of terminology, the earliest practitioners referred to these techniques as *dithering*[18] since randomization techniques are used to destroy the coherence of the undesired spurious components. Delta-sigma modulation techniques [19] based on digital signal processing techniques have spawned an alternative means for destroying these same discrete spectral components and are frequently referred to as *noise-shaping methods*. Unlike the randomization approaches that result in a white broadband output noise floor spectrum, these techniques normally produce close-in noise spectra, which are generally high-pass in nature in that phase noise very near the carrier is considerably better than at frequency offsets far removed from the carrier. Both of these methods are considered in this section.

A marriage between these two techniques is in principle possible as well. Rather than employ a randomization sequence in the dithering technique that is spectrally white, the random sequence could be processed by a digital high-pass filter prior to application in the dithering process. Although such a technique has to date not been observed in the literature, it is anticipated that performance similar to the noise-shaping approach could be achieved.

As of this writing, the author is only aware of one commercially available highly integrated DDS device that includes a means for discrete spurious suppression using numerical techniques [20]. It is anticipated that similar devices will appear in the marketplace in the near future.

Phase Dithering

One of the earlier literature references to phase dithering as applied to DDSs is [18], and although this paper deals with a square wave output DDS rather than a sinusoidal output, familiarity with this work is valuable.

The dithering technique used in [18] destroys the coherence of the undesirable discrete spurious components by periodically adding a random quantity to the DDS accumulator. The steps involved are shown in Figure 7.38.

As discussed in [18], a uniformly distributed random quantity x is added to the phase accumulator value, and the old value of x is also subtracted during each fundamental frequency output cycle. In effect, this addition process followed by the subtraction operation forms a digital high-pass filter and, as shown in [18], although any one period may be slightly in error, the average accumulator overflow event occurs at precisely the correct instant in time.

More significantly, the expected accumulator overflow time is not dependent on the phase value immediately prior to the overflow event, thereby creating adjacent accumulator resets that are statistically independent. Hence, all coherent spurious lines that originated from undesired correlations between overflow events are eliminated. Because the accumulator reset times can only occur synchronously with the edges of the reference clock F_{Clk}, at output frequencies that are not precise submultiples of F_{Clk}, the accumulator overflow events may occur as much as $\pm T_c$ in error with respect to the ideal overflow event. With the overflow event errors now randomized by the phase dithering process, the discrete spurious spectral components are eliminated from the DDS output at the expense of only a slightly higher output wideband noise floor.

As shown in [18], the spectrum of a square wave that exhibits a small amount of edge time jitter may be computed from the probability density function (pdf) of the time jitter. Further simplifications result if the time jitter is statistically independent from edge to edge. Let $p(\tau)$ represent the pdf of the actual overflow event with respect to the ideal event for a perfectly synthesized square wave. In the present context, τ can only take on one of two values, one corresponding to an overflow event occurring immediately prior to the ideal event and one corresponding to an event immediately following the desired ideal event. Given

$$F(\omega) = \int_{-\infty}^{\infty} p(\tau) e^{-j\omega\tau} d\tau \tag{7.63}$$

it can be shown [18] that the complete output square wave spectrum (less any antialias filtering) is given by

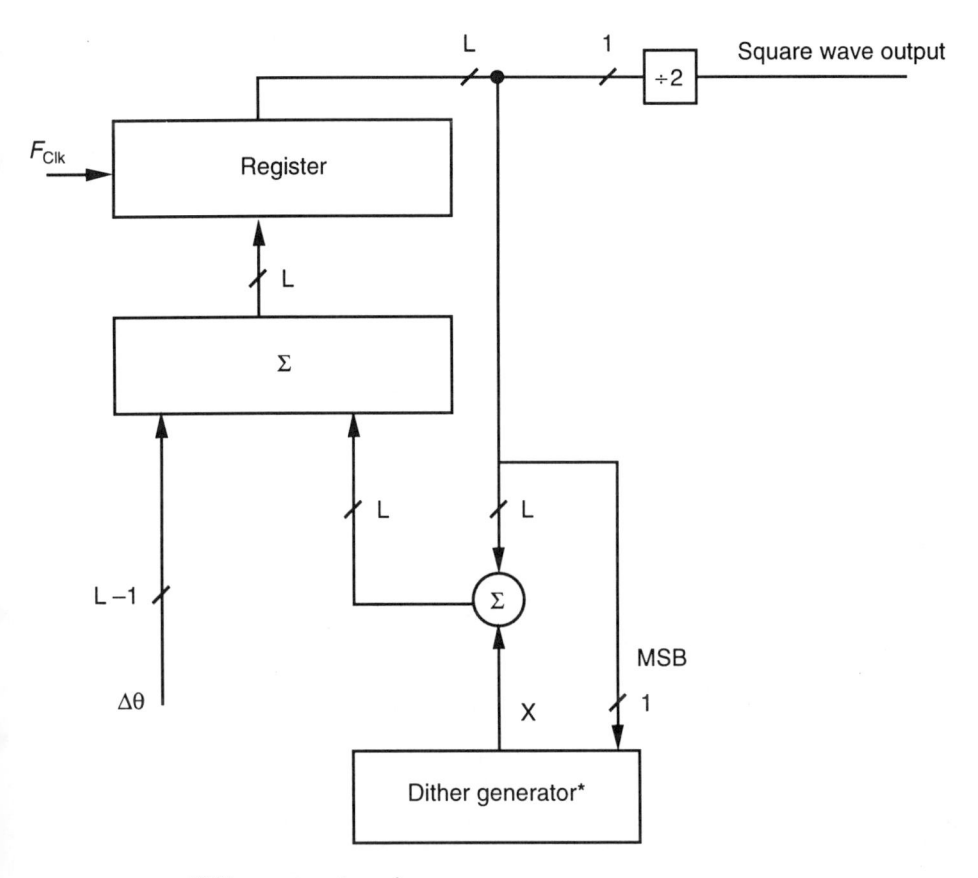

Figure 7.38 Dithered DDS square wave source. (© 1981 IEEE; reprinted with permission). After: [18].

$$S(\omega) = \frac{A^2}{\omega^2 T_o} \left[1 - |F(\omega)|^2 + \frac{2\pi}{T_o} |F(\omega)|^2 \sum_{m=-\infty}^{\infty} \delta\left(\omega - \frac{m\pi}{T_o}\right) \right] \quad (7.64)$$

where $(2T_o)^{-1} = f_o$, which is the frequency of the fundamental output square wave and A is the square wave peak-to-peak value.

In the case where the aforementioned phase accumulator dither is applied, it can be found that

$$|F(\omega)|^2 = 1 - 4\sin^2\left(\frac{\omega T_c}{2}\right)\left[\frac{\tau_e}{T_c} - \left(\frac{\tau_e}{T_c}\right)^2\right] \tag{7.65}$$

where $T_c = F_{\text{Clk}}^{-1}$ and τ_e is the error between the ideal and actual overflow event, which is uniformly distributed from 0 to T_c. The average value of $|F(\omega)|^2$ is then

$$\langle|F(\omega)|^2\rangle = \frac{1}{T_c}\int_0^{T_c}|F(\omega,\tau_e)|^2\,d\tau_e$$

$$= 1 - \frac{2}{3}\sin^2\left(\frac{\omega T_c}{2}\right) \tag{7.66}$$

Under the added assumption that $\omega_o = \pi T_o^{-1} \ll 2\pi T_c^{-1}$, this simplifies further to

$$\langle|F(\omega_o)|^2\rangle \approx 1 - \frac{1}{6}\left(\frac{\pi T_c}{T_o}\right)^2 \tag{7.67}$$

and on substitution into (7.64),

$$S(\omega_o) \approx \frac{(AT_c)^2}{6T_o} + 2\frac{A^2}{\pi}\left[\delta\left(\omega + \frac{\pi}{T_o}\right) + \delta\left(\omega - \frac{\pi}{T_o}\right)\right] \tag{7.68}$$

from which the noise density-to-carrier power ratio at the fundamental output frequency is given by

$$\frac{N_o}{C} = \frac{\pi^2}{3}\frac{f_o}{f_c^2} \tag{7.69}$$

Dithered Sinusoidal DDS

Although not addressed in [18], the same type of dithering may be applied to sinusoidal output DDSs with dramatically improved phase noise floor performance compared to the square wave DDS. Analysis of such a DDS can be found in [21].

As reiterated in [21], phase and amplitude representations are both quantized in any real-world DDS, and due to undesired periodicities in the subsequent error sequences, discrete spectral components result. Worst case spurious levels due to phase quantization typically improve 6 dB per phase bit, but since the sine lookup storage requirements normally increase exponentially with the phase word length, any reduction in this word length is highly desirable. In contrast, the dithering method described in [21] reports a worst case spurious reduction of 12 dB per phase bit.

Two quantities are actually of interest with this approach: (1) the level of the broadband output noise floor and (2) the level of the anticipated worst case discrete spurious signals. The subject phase dithered sinusoidal DDS architecture is shown in Figure 7.39.

The phase accumulator operations shown in Figure 7.39 are unaltered compared to the traditional DDS architecture shown in Figure 7.19. The phase dithering is imposed by adding a randomized B-bit binary quantity to the L-bit phase accumulator value and retaining only the W MSBs of the result to address the sine lookup table.

The phase dithering process may be analyzed by considering a continuous zero-mean, widesense stationary dithering sequence $z[n]$. The digital samples provided to the D/A converter may be represented ideally by

$$s[n] = \cos[\varphi(n)] \tag{7.70}$$

where

$$\varphi[n] = \frac{2\pi}{2^L}(F_r n) + \theta_o \tag{7.71}$$

with θ_o simply the initial phase. As discussed earlier, truncation of the full phase accumulator width from L bits to W bits results in truncation-related discrete spurs along with the desired output, which in total can be expressed mathematically by

$$s_t[n] = \cos\left(\frac{2\pi}{2^L} 2^B \left\lfloor \frac{F_r n}{2^B} \right\rfloor + \theta_o\right) \tag{7.72}$$

where the special brackets denote retainment of only the integer part and $B = L - W$. If left unmodified, periodicities in the truncation error sequences lead to the previously mentioned discrete spurious components.

If, on the other hand, a uniformly distributed random quantity $z[n]$ is added to the phase accumulator contents immediately prior to phase truncation, the DDS output waveform may be expressed as

$$s_{tr}[n] = \cos\left(\frac{2\pi}{2^L} 2^B \left\lfloor \frac{F_r n}{2^B} + \frac{z[n]}{2^B} \right\rfloor + \theta_o\right) \tag{7.73}$$

where it can be shown [22] that the phase truncation errors that now remain are now completely independent of the input frequency word F_r. The net effect is that the DDS output sample sequence may be written as

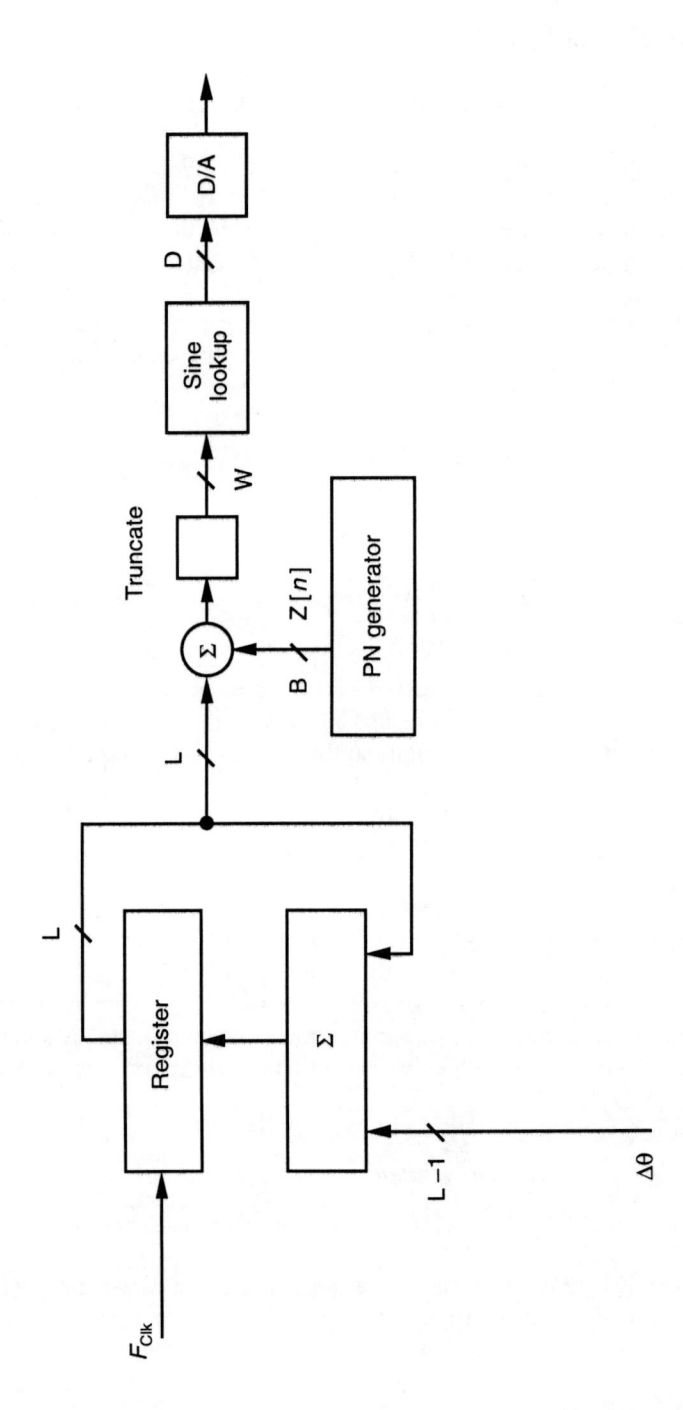

Figure 7.39 Block diagram for phase dithered sinusoidal output DDS. (© 1992; reprinted with permission). After: [21], Figure 3.

$$s_{tr}[n] = \cos\left(\frac{2\pi}{2^L} F_r n + \varepsilon[n] + \theta_o\right) \qquad (7.74)$$

where the phase error sequence $\varepsilon[n]$ can be represented as shown in Figure 7.40. Each value of $\varepsilon[n]$ is limited to the range of $[-\Delta/2, \Delta/2)$ where $\Delta = 2\pi/2^W$. The only restriction on W is that it must be large enough for the normal small-angle approximations to apply. Referring to Figure 7.40, assume that the dithering sequence $z[n]$ is uniformly distributed across the range $[-\Delta/2, \Delta/2)$. For any given high-precision phase accumulator value V_k, one of two coarse phase values may be selected as shown. The probabilities of selecting the value above and below each V_k may be summarized as follows:

Select value below V_k:

$\varepsilon[k] = -\Delta p[k]$ with probability = $\text{Prob}(z[k] \leq 0.5 - p[k])$.

Select value above V_k:

$\varepsilon[n] = \Delta(1 - p(k))$ with probability = $\text{Prob}(x[k] > 0.5 - p[k])$.

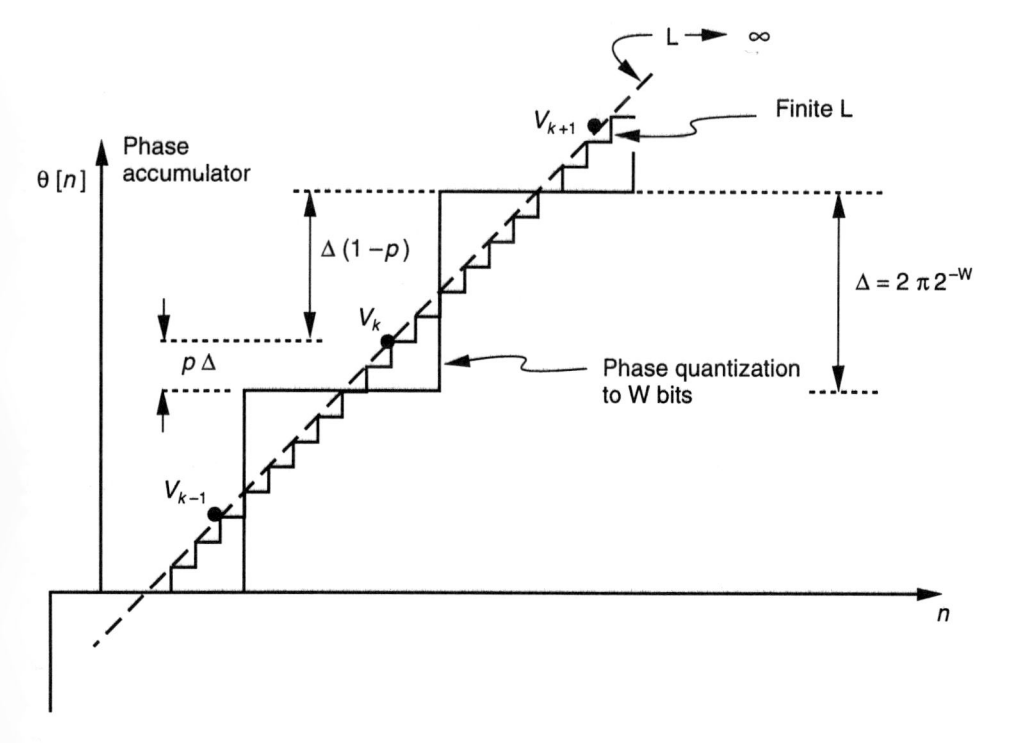

Figure 7.40 Phase quantization errors with respect to accumulator contents.

Based on this strategy, the mean value for every $\varepsilon[k]$ is zero and the variance for each $\varepsilon[k] = 4\pi^2 2^{-2W}/6$ rad^2.

Making use of the normal small-angle approximations, to first order,

$$s_{tr}[n] \approx \cos\left(\frac{2\pi}{2^L} F_r n + \theta_o\right) - \varepsilon[n] \sin\left(\frac{2\pi}{2^L} F_r n + \theta_o\right) \quad (7.75)$$

In this quadrature form, it should be clear that the phase error resulting from the dithering process at any time index n is simply $\varepsilon[n]$ as shown graphically in Figure 7.41.

Based on this reasoning, the output signal-to-noise ratio (SNR) is given by

$$\text{SNR} = \frac{1}{\text{var}\{\varepsilon[k]\}} \quad (7.76)$$

$$\text{SNR}_{\text{dB}} = -8.18 + 6.02W \quad \text{dB}$$

The one-sided noise power spectral density is then given by

$$N_o \approx -\left[10 \log_{10}\left(\frac{\text{SNR}}{2}\right) + 10 \log_{10}\left(\frac{F_{\text{Clk}}}{2}\right)\right]$$

$$= 5.18 - 6.02W - 10 \log_{10}\left(\frac{F_{\text{Clk}}}{2}\right) \quad \frac{\text{dBc}}{\text{Hz}} \quad (7.77)$$

which leads to a final output carrier-to-noise density ratio of

$$\frac{C}{N_o} \approx -8.17 + 6.02W + 10 \log_{10}\left(\frac{F_{\text{Clk}}}{2}\right) \quad \frac{\text{dBc}}{\text{Hz}} \quad (7.78)$$

Second-order analysis as reported in [21] reveals that the worst case discrete spurious

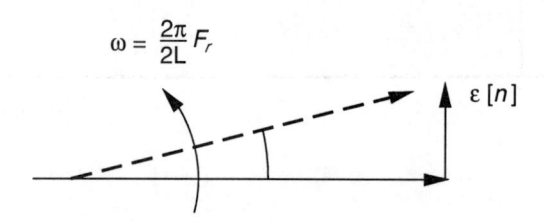

Figure 7.41 Quadrature representation for phase truncation error process.

components due to phase truncation can be reduced by 12 dB for each additional bit used in the sine table lookup function. This is a significant result, which can be used to reduce the lookup table size and or achieve additional improved spurious performance.

Improved DDS Spectral Performance Using Noise-Shaping Techniques

With the advent of very high-speed low-cost digital VLSI technology, it is only natural that digital signal processing techniques play an increasingly important role in what has traditionally been the RF/analog arena. This is particularly true in the case of DDS-applied noise shaping, which is really a direct growth out of sigma-delta data conversion techniques [19].

A representative DDS that makes use of noise shaping is reported on in [23] and makes use of the architecture shown in Figure 7.42. The additional signal processing that follows the phase accumulator in effect implements a digital high-pass filter, which helps to suppress any low-frequency truncation noise that might be present in the phase truncation error sequence. This additional processing (1) destroys the would-be correlation of phase truncation errors that is responsible for the undesired discrete spectral spurs and (2) pushes phase noise contaminants away from the desired carrier frequency in a high-pass manner. A typical output spectrum for such a DDS is shown in Figure 9.21. As pointed out in [23], higher order noise shaping is easily applied using the concepts provided in [24].

7.4 HYBRID PHASE-LOCKED LOOP/DDS FREQUENCY SYNTHESIZERS

It is only natural that DDS synthesis elements be incorporated into more sophisticated architectures such as that shown in Figure 7.1 or be used simply to provide a variable reference frequency for a following PLL. Since any multiplication of the reference frequency results in a degraded phase noise spectrum per the classical 20 log(N) rule, the latter technique might not deliver the desired performance unless N is quite small.

In most applications, particularly consumer ones, the high cost represented by the triple architecture in Figure 7.1(b) cannot be absorbed. Therefore, little else will be said concerning this topic. In contrast, the DDS-provided PLL reference technique has found fairly widespread use in many satellite modem applications. Typically, -30 dBc discrete spurious levels are adequate for that use and the 20 log(N) degradation can be tolerated. A thorough discussion of this synthesis technique can be found in [25] where the architecture considered there is shown in Figure 7.43.

Worst case spurious performance levels at the output of this synthesizer may be as large as -23 dBc.

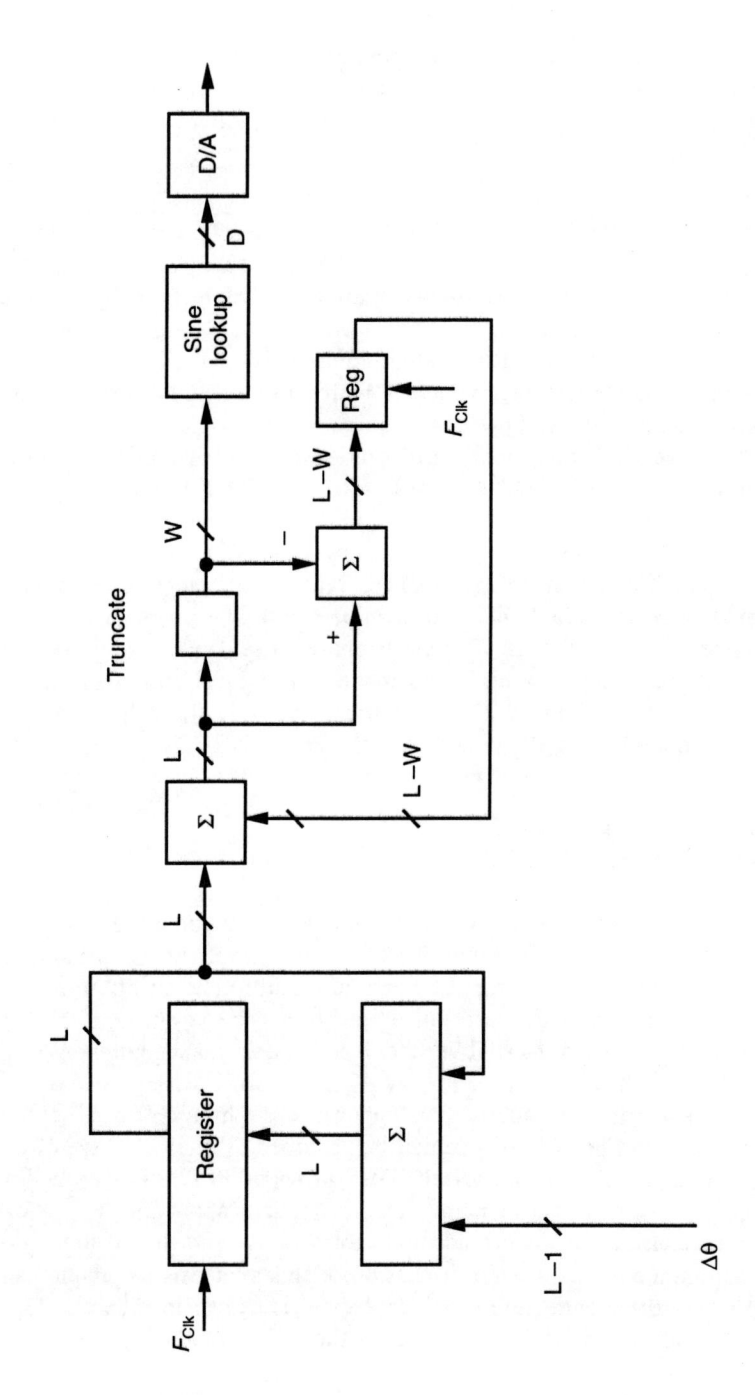

Figure 7.42 Sinusoidal DDS with noise shaping for improved spurious performance. (© 1991 IEEE; reprinted with permission). After: [23].

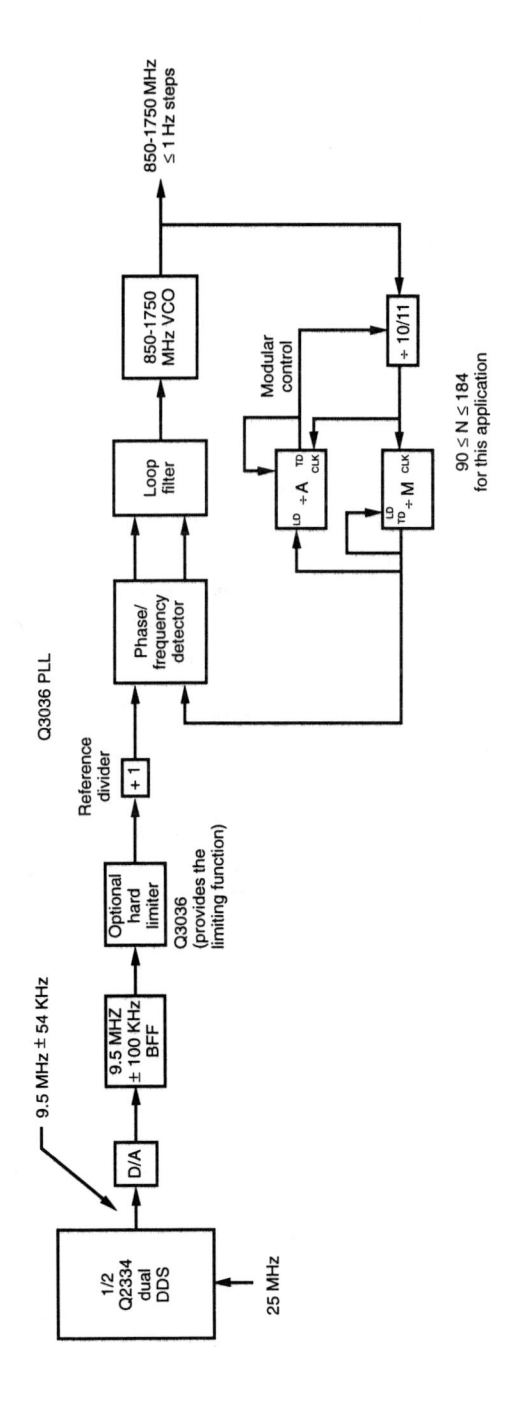

Figure 7.43 Block diagram of DDS-driven PLL synthesizer using QUALCOMM Q2334 DDS and Q3036 PLL. Source: [25]. Reprinted with permission.

7.5 PROS AND CONS OF DIRECT DIGITAL FREQUENCY SYNTHESIS

The advantages of direct digital frequency synthesis can be summarized as follows:

- An almost arbitrarily small channel step size is achievable.
- Extremely fast phase-continuous frequency switching is possible.
- DDS can be directly phase modulated with excellent linearity.
- Makes use of largely digital technology.
- Reasonably good spurious performance is achieved.
- Very good performance is achieved versus volume.

The disadvantages include:

- Performance is ultimately limited by the D/A converter.
- The cost is high, driven primarily by the D/A converter.
- Unless an extremely high F_{Clk} is used, the output must generally be either multiplied or heterodyned to higher frequencies before use.
- All circuitry runs at the maximum clock rate of F_{Clk}, thereby making power consumption high.
- Aliasing processes and clock leakage result in signal frequency components that are not integrally related to the fundamental output frequency. A substantial low-pass or bandpass filter is required for anti-alias filtering often followed by a hard limiter.
- The output bandwidth is often too restrictive unless a very high F_{Clk} used.
- It is not unusual for a separate PLL to be required to create the DDS high-frequency clock F_{Clk}.

In general, unless ultrafast switching speed and/or direct phase modulation is required, new VLSI techniques are beginning to appear that provide most of the DDS advantages while avoiding most of its disadvantages. This new category is discussed in the upcoming final section of this chapter and is called *advanced fractional-N synthesis*.

7.6 ADVANCED FRACTIONAL-*N* CONCEPTS

An entire chapter is provided in this text that addresses fractional-*N* synthesis from the traditional standpoint (Chapter 9). Historically, the negatives pertaining to fractional-*N* synthesis have included the following:

- High complexity in analog phase error correction circuitry
- Achievable spurious levels limited to roughly -70 dBc

Fractional-*N* displays a number of positive features, including these:

- Single phase-locked loop architecture (1 VCO)
- Output frequency and bandwidth fairly arbitrary

- No additional anti-aliasing filter required or creation of additional frequency components
- Low power consumption since most of the circuitry operates at a low reference frequency (e.g., 100 kHz)

One approach for eliminating the primary drawbacks of traditional fractional-N that is amenable to high integration is considered in this section.

7.6.1 Fractional-N Technique Based on Noise-Shaping Methods

As alluded to in Chapter 9, noise shaping can provide a means for reducing the analog complexity of fractional-N without incurring additional performance impairments. A very attractive approach for performing fractional-N synthesis using noise shaping (sigma-delta modulation) has been reported in [26–28] and is discussed here. Readers who are unfamiliar with general fractional-N concepts should read Chapter 9 before continuing.

The primary motivation behind fractional-N synthesis is to realize a feedback divider ratio having a nonzero fractional portion (e.g., $N.g$) without incurring the creation of subreference spurious phase sidebands. The concept presented here is based on the concepts of oversampled sigma-delta converters similar to the one shown in Figure 7.44. If the analog-to-digital converter (ADC) and digital-to-analog converter (DAC) here can be assumed to be perfect, the only error introduced by these components will be the quantization noise. This noise is generally distributed uniformly but is not necessarily spectrally white. In the closed-loop configuration shown, the output voltage closely tracks the input voltage for input frequency components that are within the closed-loop 3-dB bandwidth, and quantization noise is largely suppressed over this same frequency range. Normally, the sampling rate F_s is much greater than the closed-loop bandwidth used.

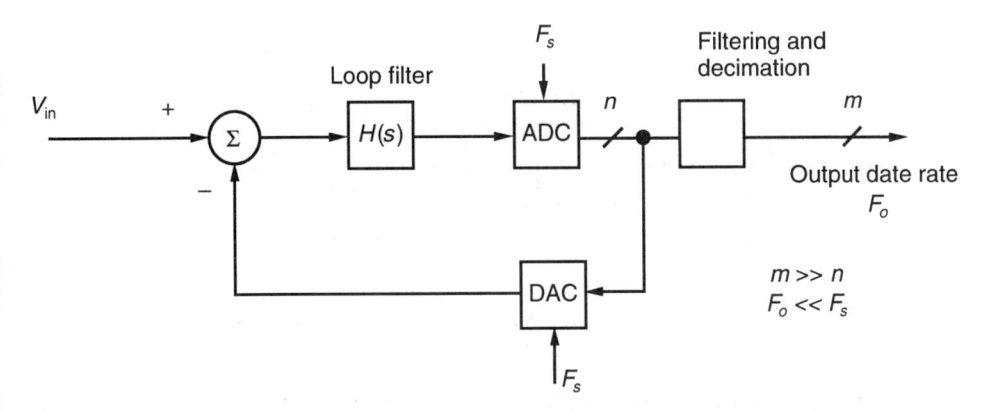

Figure 7.44 Simple representation for a generic oversampling sigma-delta converter. After: [26]. Reprinted with permission.

If it is recognized that the fractional-N divide ratio $N.g$ is analogous to the input voltage in Figure 7.44 and the feedback modulus control signal (N, $N + 1$ control for feedback divider prescaler) is analogous to the output of the 1-bit (quantizer) ADC, delta-sigma techniques can be applied directly to achieve the subreference spurious rejection needed in fractional-N synthesis.

In the fractional-N context, since the input signal is already in digital form ($N \cdot g$), the ADC and DAC are not required and only the fractional portion of the complete feedback divider ratio must be processed as shown in Figure 7.45. Here, the $(1 - z^{-1})^{-1}$ block represents a binary accumulator that sums all of the residual errors $e_q(k)$ resulting from the 1-bit quantization process and $y(k)$ represents the modulus control signal controlling the feedback divider prescaler. The quantization errors are normally taken to be uniformly distributed.

Based on Figure 7.45, the system transfer function is given by

$$y(k) = g(k) + e_{q1}(k)(1 - z^{-1}) \tag{7.79}$$

which represents a first-order digital high-pass processing of the quantization error sequence as desired. Any low-frequency components present in the error sequence are then reduced.

Nothing prevents similar processing from being performed on the residual quantization error sequence $e_{q1}(k)$ by a second stage of processing, or a third as shown in Figure 7.46. The net result is that the effective feedback divider ratio is given in the case of third-order processing by

$$N(z) = N.g(z) + (1 - z^{-1})^3 e_{q3}(z) \tag{7.80}$$

The second term represents frequency noise caused by the underlying quantization process.

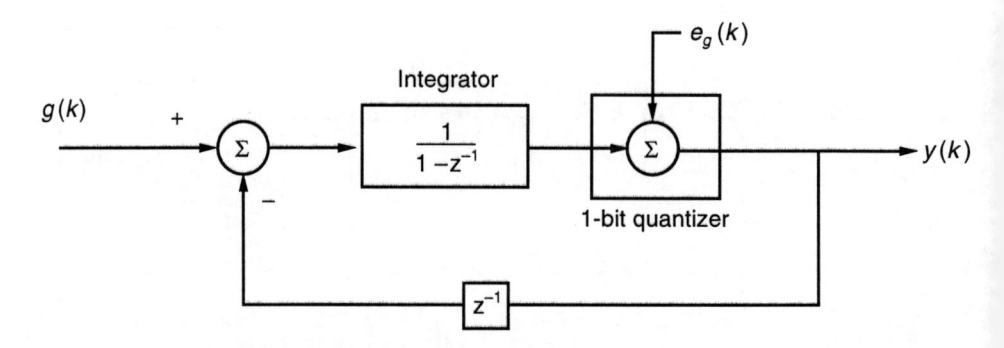

Figure 7.45 Sigma-delta modulator as applicable to fractional-N frequency synthesis. After: [26]. Reprinted with permission.

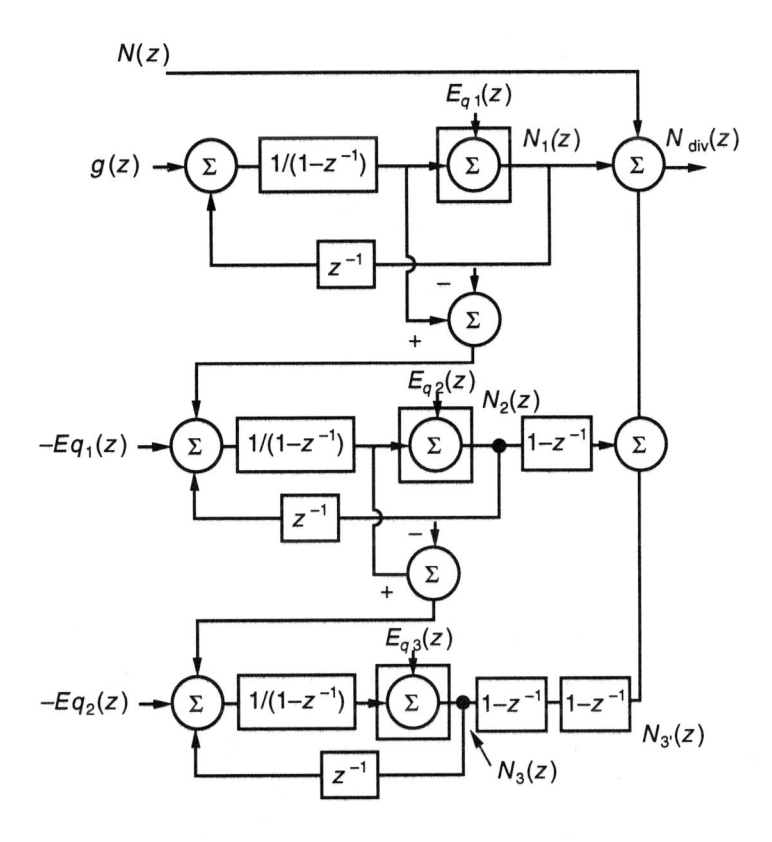

Figure 7.46 Multistage application of sigma-delta modulators to fractional-N frequency synthesis. Source: [26]. Reprinted with permission.

The quantization error sequence $e_{q3}(k)$ has a variance of $1/12$, which results in a two-sided power spectral density of $(12F_{ref})^{-1}$. The randomness of the feedback divider ratio represented by (7.80) results in frequency noise being created at the PLL output. Since we are primarily interested in phase noise, a conversion between frequency and phase noise measures must be done as discussed in Chapter 3. The required integration that was represented there in Laplace transforms as s^{-1} can be represented in z transforms here as $T_s/(1-z^{-1})$. The resulting single-sideband phase noise spectrum is then

$$\mathcal{L}(f) = \frac{(2\pi)^2}{12\,F_{ref}}\left[2\sin\left(\frac{\pi f}{F_{ref}}\right)\right]^{2(m-1)}\quad\frac{\text{rad}^2}{\text{Hz}} \qquad (7.81)$$

where m represents the number of sigma-delta processing stages employed.

Actual fabrication of such a sigma-delta based fractional-N PLL was also reported in [26]. The integrated circuit was fabricated in a 1.5-μm 5V CMOS, packaged in a 44-pin plastic chip carrier, and operated over an output frequency span of 500 MHz to 1 GHz. The prescaler consisted of a 32/33 dual-modulus prescaler preceded by a fixed divide-by-2 prescaler. A frequency reference of 200 kHz was used. Actual measured performance is shown in Figure 7.47.

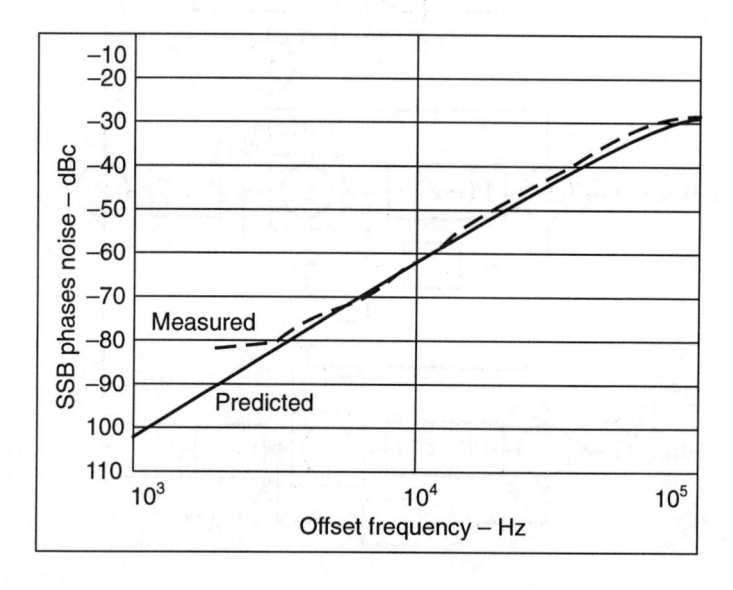

Figure 7.47 Output phase noise spectrum for design case sited in text. Source: [26]. Reprinted with permission.

The true elegance of this approach is that no phase error correction circuitry per se is required whatsoever since all correction is performed effectively in the frequency domain, and aside from perhaps the VCO and loop filter, the entire device is based on digital techniques. This type of frequency synthesis will do doubt have significant presence in the marketplace as it becomes more available.

It is interesting to note that a digital method that bears a similarity to first-order sigma-delta correction was reported in Chapter 7 of [4]. Although not discussed in the same context, the method was based on keeping the running phase error average zero by direct modulation of the dual-modulus prescaler.

REFERENCES

[1] Crawford, J.A., "Synthesizer Designs Minimize Phase Noise in Cellular Systems," *Microwaves & RF*, January 1993, pp. 69–78.
[2] Peterson, D.F., "Varactor Properties for Wide-Band Linear-Tuning Microwave VCOs," *IEEE Microwave Theory and Techniques*, Vol. MTT-28, February 1980, pp. 110–119.
[3] Haggai, T.F., "Phase Lock Receiver With a Constant Slope Network," U.S. Patent 3,551,829, December 29, 1970.
[4] Egan, W.F., *Frequency Synthesis by Phase Lock*, John Wiley & Sons, New York, 1981.
[5] Manassewitsch, V., *Frequency Synthesizers Theory and Design*, 2nd ed., John Wiley & Sons, New York, 1980.
[6] de Buda, R., "Frequency Synthesis Using Walsh Functions," *IEEE Trans. Electromagnetic Compatability*, Vol. EMC-21, August 1979, pp. 269–274.
[7] Furuno, K., "Design of Digital Sinusoidal Oscillators With Absolute Periodicity," *IEEE Trans. on Aerospace and Electronic Systems*, Vol. AES-11, November 1975, pp. 1286–1298.
[8] Fliege, N.J., J. Wintermantel, "Complex Digital Oscillators and FSK Modulators," *IEEE Trans. Signal Proc.*, February 1992, pp. 333–342.
[9] Reinhardt, V.S., Paper presented at 17th Annual Precise Time and Time Interval Applications and Planning Meeting (NASA/DOD), Washington, DC, December 3–5, 1985.
[10] Nicholas, H.T., H. Samueli, "An Analysis of the Output Spectrum of Direct Digital Frequency Synthesizers in the Presence of Phase-Accumulator Truncation," *41st Annual Frequency Control Symposium*, 1987, pp. 495–502.
[11] Nicholas, H.T., et al., "The Optimization of Direct Digital Frequency Synthesizer Performance in the Presence of Finite Word Length Effects," *42nd Annual Frequency Control Symposium*, 1988, pp. 357–363.
[12] Sunderland, D.A., et al., "CMOS/SOS Frequency Synthesizer LSI Circuit for Spread Spectrum Communications," *IEEE J. Solid State Circ.*, Vol. SC-19, August 1984, pp. 497–505.
[13] Schwarz, E.M., M.J. Flynn, "Approximating the Sine Function With Combinational Logic," *26th Asilomar Conference on Signals, Systems and Computers*, Pacific Grove, California, October 26–28 1992, IEEE Computer Society Press.
[14] Buchanan, D., "Choosing DACs for Direct Digital Synthesis," Analog Devices Application Note AN-237.
[15] Buchanan, D., "Choose DACs for DDS System Applications," *Microwaves & RF*, August 1992.
[16] Colotti, J.J., "Dynamic Evaluation of High Speed, High Resolution D/A Converters," *RF Design*, November 1990.
[17] Goodenough, F., "18-Bit Audio DACs Cut PCB Space Dramatically," *Electron. Design*, August 1990, pp. 51–54.
[18] Wheatley, C.E., III, "Spurious Suppression in Direct Digital Synthesizers," *Proc. 35th Annual Frequency Control Symposium*, 1981, pp. 428–435.
[19] Candy, J.C., G.C. Temes, *Oversampling Delta-Sigma Data Converters*, IEEE Press, New York, 1992.
[20] "Q2334 Dual Direct Digital Synthesizer," Qualcomm Data Sheet.
[21] Zimmerman, G.A., M.J. Flanagan, "Spur-Reduced Numerically-Controlled Oscillator for Digital Receivers," *Asilomar Conference on Signals, Systems and Computers*, Pacific Grove, California, 1992, pp. 517–520.
[22] Schuchman, L., "Dither Signals and Their Effects on Quantization Noise," *IEEE Trans. Comm.*, Vol. COM-12, December 1964, pp. 162–165.
[23] O'Leary, P., F. Maloberti, "A Direct Digital Synthesizer with Improved Spectral Performance," *IEEE Trans. Comm.*, Vol. COM-39, July 1991, pp. 1046–1048.

[24] Candy, J.A., A.N. Huynh, "Double Interpolation for Digital to Analog Conversion," *IEEE Trans. Comm.*, Vol. COM-34, Janurary 1984, pp. 77–81.
[25] "Hybrid PLL/DDS Frequency Synthesizers," Qualcomm Application Note, June 1990.
[26] Miller, B., "Technique Enhances the Performance of PLL Synthesizers," *Microwaves & RF*, January 1993.
[27] Miller, B., U.S. Patent, No. 5,038,117, August 6, 1991.
[28] Miller, B., R. Conley, "A Multiple Modulator Fractional Divider," *IEEE Trans. on Instrumentation and Measurement*, Vol. 40, No. 3, June 1991.

SELECTED BIBLIOGRAPHY

Jayant, N.S., L.R. Rabiner, "The Application of Dither to the Quantization of Speech Signals," *Bell Syst. Tech. J.*, Vol. 51, 1972, pp. 1293–1304.

Olsen, J.J., P.M. Fishman, "Truncation Effects in Direct Digital Frequency Synthesizers," IEEE 1987, pp. 186–190.

Sripad, A.B., D.L. Snyder, "A Necessary and Sufficient Condition for Quantization Errors to be Uniform and White," *IEEE Trans. Acoust. Speech Signal Proc.*, Vol. ASSP-25, October 1977, pp. 442–448.

Chapter 8
MACSET: A Computer Program for the Design and Analysis of Phase-Locked Loop Frequency Synthesizers

The acronym MACSET originated with the tool's primary purpose, which is the calculation of the macroscopic time-domain transient response of phase-locked loops that are used for frequency synthesis. The term *macroscopic* is emphasized here in that the program explicitly avoids modeling device details that are typically negligible in predicting a system's transient response (e.g., the rise time of a digital gate). On the other hand, device details that directly affect the transient performance such as op-amp slew-rate limitations and finite voltage supply rails are included because these are important high-level design considerations that are difficult to evaluate by means of purely analytical methods. It was also deemed important to be able to analyze fairly general cases, including such functions as elliptic low-pass filters and notch filters, thereby abandoning tightly constrained cookbook analysis approaches.

As the design of MACSET unfolded, it was clear that the program would have a complete loop description at its disposal, thereby making other analyses possible. Additional features were therefore added that allow the phase-locked loops to be analyzed in the frequency domain as well, addressing such important items as loop gain and phase margin, closed-loop gain peaking, and output phase noise performance.

In this chapter, the basic concepts used in MACSET will be discussed from a fairly general perspective in order to prevent the material from being too specific to MACSET. In Chapter 7, MACSET was used to create the design nomographs shown in Figures 7.16, 7.17, and 7.18. These figures are helpful for analyzing the traditional systems-level trade-offs that must be made between switching speed and spurious performance.

Before proceeding, it is necessary to point out a few of the limitations implicit in the MACSET discussion that follows. First, MACSET is limited to analyzing single-loop designs rather than multi-loop designs. Second, the present choice of phase detectors in MACSET is limited to (1) the simplified phase-frequency detector and (2) the zero-order sample-and-hold phase detector. Third, the time-domain analysis is restricted to only the transient response calculations involving a reference source phase step, or frequency step. Finally, a linear nodal admittance description is used throughout the program execution for speed and simplicity as compared to nonlinear hybrid techniques. The linearity assumption avoids the somewhat difficult convergence problems involved with nonlinear devices like diodes. This prevents some nonlinear design approaches from being thoroughly analyzed (e.g., shunting resistors with diodes) but this limitation is generally minimal except in very specific cases.

The first portion of the chapter is devoted to a discussion of time-domain simulation methods, in particular, the discrete-time models used in MACSET. This is followed by a discussion of the frequency-domain analysis tools used in MACSET including calculation of the output phase noise spectrum using Tellegen's theorem concepts [1].

8.1 TIME-DOMAIN MODELING

MACSET time-domain simulation techniques are based on well-grounded circuit analysis concepts that were developed in the 1960s and early 1970s. The interested reader is referred to [1–22] for further discussion. A helpful overview discussion and comparison of available techniques may be found in [2,5].

Two primary circuit equation formulations are available for use [5], the first is the familiar nodal analysis method and the second is the mixed or state-variable method. Quoting from [5]:

> The relative simplicity of the nodal network sharply contrasts the many manipulations needed in the state-variable approach. The primary reason for using the state-variable approach is that it yields a set of first-order linear or nonlinear differential equations in a minimal set of unknowns. The state-variable approach entails the fairly involved process of identifying trees and cut-sets within the network from which the state-variable equations may be derived. In all cases, the formulation procedures entail extensive matrix manipulations. The bookkeeping associated with these manipulations is extensive.

As a result of these strong statements, only nodal analysis methods are considered in this chapter.

Once the simultaneous equations that describe the network have been obtained, they may be solved in a variety of ways. Since the adopted approach in MACSET results in sets of linear equations, possible methods that can be used include Gaussian reduction, LU factorization, and sparse matrix methods [23]. If nonlinear network elements had been admitted into the analysis, this solution step would have been substantially more involved.

Network Equation Formulation

One of the most elegant methods available for extracting the simultaneous nodal equations that are needed to analyze a given network is based on discrete circuit element models. This approach is also called *companion modeling* by some. This technique provides an approach in which each circuit element, whether a resistor, capacitor, etc., can be represented by a subnetwork composed of only sources and conductances. With these subnetworks, all of the network node voltages may be easily computed. Perhaps equally important is the fact that this method automatically incorporates the needed numerical integration formulas directly into the nodal subnetwork descriptions. As seen shortly, the selection of a specific integration formula is very important.

8.2 NUMERICAL INTEGRATION FORMULAS

Although a complete discussion of numerical integration methods is well beyond the scope of this text, the subject is of central importance to the network modeling effort so some of the more relevant details are examined here. To facilitate this discussion, we start by examining the unit-step response of the simple second-order RLC circuit shown in Figure 8.1.

The voltage transfer function for the RLC circuit shown in Figure 8.1 is given in Laplace transform form by

$$H(s) = \frac{V_o(s)}{V_i(s)} = \frac{s/RC}{s^2 + s/RC + 1/LC} \tag{8.1}$$

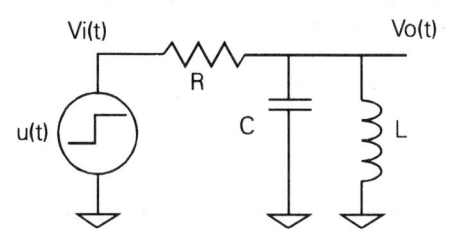

Figure 8.1. Simple RLC circuit used to examine numerical techniques.

which may be rewritten as

$$H(s) = \frac{2\zeta\omega_n s}{s^2 + 2\zeta\omega_n s + \omega_n^2} \tag{8.2}$$

where $\omega_n^2 = 1/LC$ and $2\zeta\omega_n = 1/RC$. The Laplace transform of the unit-step response is then given by

$$V(s) = \frac{2\zeta\omega_n}{s^2 + 2\zeta\omega_n s + \omega_n^2} \tag{8.3}$$

The time-domain response is easily found from (8.3) by taking the inverse Laplace transform [24], which gives

$$v(t) = \frac{2\zeta}{\sqrt{1-\zeta^2}} e^{-\zeta\omega_n t} \sin(\omega_n \sqrt{1-\zeta^2}\, t) \tag{8.4}$$

where $\zeta < 1$ and $t \geq 0$ restrictions apply. This exact result will be used to assess the precision of the numerical techniques, which are examined next.

From (8.3), the system response may be written in differential equation form as

$$\frac{dv(t)}{dt} + 2\zeta\omega_n v(t) + \omega_n^2 \int_0^t v(u)\,du = 2\zeta\omega_n \tag{8.5}$$

Let the system state variables s_1 and s_2 be given by

$$\begin{aligned} s_1 &= v(t) \\ s_2 &= \int_0^t v(u)\,du \end{aligned} \tag{8.6}$$

where the explicit time dependence has been dropped. Equation (8.5) may then be rewritten in state-variable form as

$$\begin{aligned} \dot{s}_1 &= -2\zeta\omega_n s_1 - \omega_n^2 s_2 + 2\zeta\omega_n \\ \dot{s}_2 &= s_1 \end{aligned} \tag{8.7}$$

If each state-variable derivative is now approximated as

$$\dot{s}(t) \approx \frac{s(t+h) - s(t)}{h} \tag{8.8}$$

where h is the incremental time step used in the numerical integration routine, the system of equations becomes

$$\begin{bmatrix} s_1(t+h) \\ s_2(t+h) \end{bmatrix} = \begin{bmatrix} s_1(t) \\ s_2(t) \end{bmatrix} + h \begin{bmatrix} 2\zeta\omega_n \\ 0 \end{bmatrix} + h \begin{bmatrix} -2\zeta\omega_n & -\omega_n^2 \\ 1 & 0 \end{bmatrix} \begin{bmatrix} s_1(t) \\ s_2(t) \end{bmatrix} \qquad (8.9)$$

Here, each $s(t+h)$ is defined explicitly in terms of the immediate time sample $s(t)$. This is a very important distinction that will become apparent in a moment. The numerical integration procedure used here by virtue of the approximation in (8.8) is known as the forward Euler method [5]. In practice, it is seldom used because of its low numerical accuracy [2,5] and poor stability [2].

Before considering the accuracy and stability attributes of the forward Euler method just introduced, a similar iterative solution will be developed making use of a slightly different approximation for the state-variable derivatives. Rather than (8.8), the derivatives at time $t+h$ could just as easily have been approximated as

$$\dot{s}(t+h) = \frac{s(t+h) - s(t)}{h} \qquad (8.10)$$

Employing this definition in (8.7) and continuing,

$$\frac{s_1(t+h) - s_1(t)}{h} = -2\zeta\omega_n s_1(t+h) - \omega_n^2 s_2(t+h) + 2\zeta\omega_n$$

$$\frac{s_2(t+h) - s_2(t)}{h} = s_1(t+h) \qquad (8.11)$$

These equations may be expressed in normal matrix form as

$$\begin{bmatrix} (1 + 2\zeta\omega_n h) & h\omega_n^2 \\ -h & 1 \end{bmatrix} \begin{bmatrix} s_1(t+h) \\ s_2(t+h) \end{bmatrix} = \begin{bmatrix} hs_1(t) + 2\zeta\omega_n h \\ s_2(t) \end{bmatrix} \qquad (8.12)$$

Solving (8.12) by means of the method of determinants, the final solution is given by

$$\begin{bmatrix} s_2(t+h) \\ s_2(t+h) \end{bmatrix} = \frac{\begin{bmatrix} 1 & -h^2\omega_n^2 \\ h & (1+2\zeta\omega_n h) \end{bmatrix} \begin{bmatrix} s_1(t) + 2\zeta\omega_n h \\ s_2(t) \end{bmatrix}}{1 + 2\zeta\omega_n h + h^2\omega_n^2} \qquad (8.13)$$

Solution in this manner using (8.10) is known as *backward Euler integration*. Because the unknowns are involved directly in the solution of (8.11), this integration method belongs to the class of methods known as *implicit integration* [5].

Although implicit integration methods are generally more complicated than explicit methods, the performance gains in terms of numerical precision and algorithm stability can be dramatic. MACSET is designed to utilize implicit integration methods for these reasons.

The forward and backward Euler integration methods are compared for a unit-step input excitation in Figures 8.2 and 8.3 using (8.4), (8.9), and (8.13) for the RLC circuit shown in Figure 8.1. Figure 8.2 shows the actual time-domain response calculated with these two numerical methods as compared to the closed-form exact response. The mean-squared error performance of the two integration methods is compared in Figure 8.3 using the weighted-error sum given by

$$\mathbf{E}_N = \sum_{n=1}^{N} [r(nh) - a(nh)]^2 (nh)^2 \qquad (8.14)$$

where $r(t)$ is the computed step response using each respective numerical method, and $a(t)$ is the exact step response given by (8.4). As shown in Figure 8.3, the forward Euler method becomes increasingly inaccurate (unstable) compared to the backward Euler method as the time step h is increased.

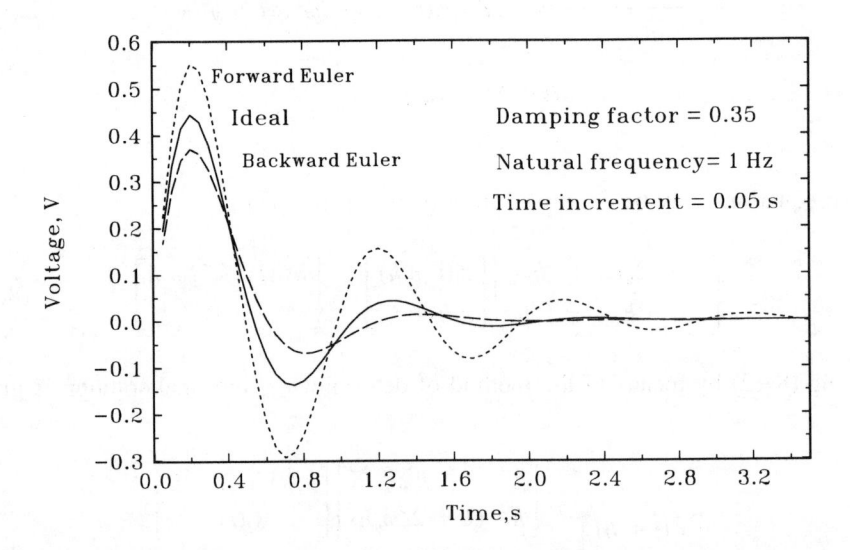

Figure 8.2. Comparison of forward and backward Euler methods for the step transient response of an RLC circuit.

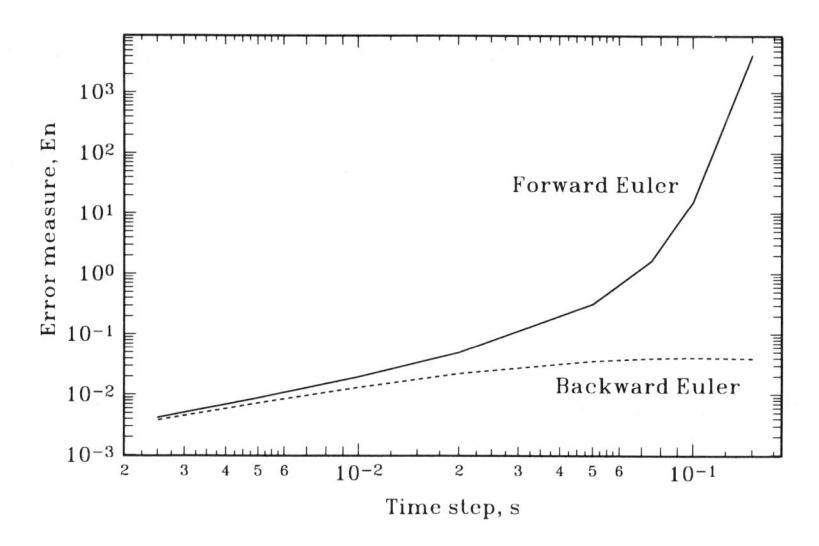

Figure 8.3. Computed error of the forward and backward Euler methods for the step transient response of an RLC circuit as a function of integration time step.

The numerical integration methods discussed thus far have had the general form [2]

$$x_{n+1} = \sum_{i=0}^{p} a_i x_{n-i} + h \sum_{i=-1}^{p} b_i \dot{x}_{n-i} \tag{8.15}$$

All of the a_i coefficients weight previous values of the state variable, whereas the b_i coefficients all weight previous values of the state-variable derivative except for b_{-1}, which weights the derivative value at index $(n+1)$. Methods for which the b_{-1} coefficient is zero are termed explicit methods (e.g., forward Euler), whereas integration formulas that have a nonzero b_{-1} coefficient are called implicit methods (e.g., backward Euler) [5]. As this previous example demonstrates, the performance differences between implicit and explicit methods can be substantial.

Generally, the $2p + 3$ coefficients in (8.15) are chosen such that any differential equation whose solution is a polynomial of degree $\leq K$ will be solved exactly. Recognizing that a Kth-order polynomial has $K + 1$ coefficients, this criterion requires that p be selected such that

$$2p + 3 \geq K + 1 \tag{8.16}$$

The exactness constraint for any polynomials of degree $\leq K$ results in restrictions on the a_i and b_i coefficients, which may then be found by considering the simply initial value problem given by

$$\dot{x} = f(x, t) \tag{8.17}$$

For the $K = 0$ case,

$$x = \alpha_0$$
$$f(x, t) = 0 \tag{8.18}$$

where α_0 is a constant. Substitution of (8.18) into (8.15), the exactness constraint for the zeroth-order case, can be obtained as

$$\sum_{i=0}^{p} a_i = 1 \tag{8.19}$$

Similarly, for the first-order case ($K = 1$),

$$x = \alpha_0 + \alpha_1 t$$
$$f(x, t) = \alpha_1 \tag{8.20}$$

which leads to the exactness constraint given by

$$\sum_{i=0}^{p} (-i) a_i + \sum_{i=-1}^{p} b_i = 1 \tag{8.21}$$

This procedure may be repeated for each order $j \leq K$ to obtain the full set of exactness constraints for the Kth-order multistep algorithm given as

$$\sum_{i=0}^{p} a_i = 1$$

$$\sum_{i=1}^{p} (-i)^j + j \sum_{i=-1}^{p} (-i)^{j-1} b_i = 1 \quad \text{for} \quad j = 1, 2, \ldots, K \tag{8.22}$$

This is a very important concept for anyone attempting to design their own numerical integration formula. Fortunately for us, this subject has been thoroughly studied by many researchers and it is only necessary to choose an existing algorithm that meets our specific needs.

The forward and backward Euler methods, which were examined here briefly, are actually members of two different algorithm families that have names worth mentioning. The forward Euler method is also known as the *first-order Adams-Bashforth method*. This family of multistep algorithms is characterized by the parameter choice given by

$$p = K - 1$$
$$a_1 = a_2 = a_3 = \ldots = a_{K-1} = 0 \quad (8.23)$$
$$b_{-1} = 0$$

Since $b_{-1} = 0$ in this family, the Adams-Bashforth methods are explicit in nature. The backward Euler method is also known as the *first-order member* of the Adams-Moulton algorithm family, which is characterized by the choice of parameters given by

$$p = K - 2$$
$$a_1 = a_2 = \ldots = a_{K-2} = 0 \quad (8.24)$$

which is a family of implicit methods since $b_{-1} \neq 0$.

More insight into these numerical methods may be gained if a specific numerical integration formula is derived as an example. In the case of the second-order Adams-Moulton formula, the unknown coefficients may be found by substituting (8.24) into the exactness constraints given by (8.22). Since $p = 0$, from (8.22), $a_0 = 1$. For the $j = 1$ and $j = 2$ elements of the exactness constraints, it must be true that

$$b_{-1} + b_0 = 1$$
$$2 b_{-1} = 1 \quad (8.25)$$

Upon solving these equations simultaneously, the solution is $a_0 = 1$, b_{-1} 1/2, and $b_0 = 1/2$ from which the complete numerical integration formula is given by

$$x_{n+1} = x_n + \frac{h}{2}(\dot{x}_{n+1} + \dot{x}_n) \quad (8.26)$$

This integration formula result is simple trapezoidal integration. An enumeration of the higher degree members of the Adams-Moulton family may be found in [2,3,10].

The trapezoidal integration method has been used in a number of circuit analysis programs in the past such as TRAC [5], CANCER, and SPICE1 [3]. Because it is an implicit method, it displays very good numerical stability.

Another family of implicit methods that is characterized by

$$p = K - 1$$
$$b_0 = b_1 = b_2 = \ldots b_{K-1} = 0 \tag{8.27}$$

is known as the *Gear family* of algorithms. The second-order Gear algorithm was used in CIRPAC, an early circuit analysis program used within Bell Telephone Laboratories [5]. The Gear algorithms were specifically designed to facilitate the solution of stiff differential equations [2], which are characterized by solutions containing both very fast and very slow components. Besides being implicit in nature, use of the second-order Gear algorithm leads to a slightly simpler circuit network equation formulation than does the trapezoidal method. This numerical algorithm is used within MACSET.

The second-order Gear algorithm may be derived from (8.27) and (8.22). As somewhat of a departure, this algorithm will be derived here for a variable time-step increment since this flexibility will be needed in MACSET. From (8.27) and (8.15),

$$x_{n+1} = a_0 x_n + a_1 x_{n-1} + T_1 b_{-1} f(x_{n+1}, t_{n+1}) \tag{8.28}$$

where it has been assumed that T_1 is the time increment between x_n and x_{n+1} and T_0 is the time increment between x_{n-1} and x_n. Assuming further that the second-order polynomial solution may be represented by

$$x(t) = \alpha_0 + \alpha_1 t + \alpha_2 t^2 \tag{8.29}$$

the derivative of the polynomial is given by

$$\dot{x}(t) = \alpha_1 + 2\alpha_2 t \tag{8.30}$$

Substituting (8.29) and (8.30) into (8.28) and collecting like terms, we obtain equality provided that

$$\begin{bmatrix} 1 & 1 & 1 \\ 0 & -\dfrac{T_0}{T_1} & 1 \\ 0 & \left(\dfrac{T_0}{T_1}\right)^2 & 2 \end{bmatrix} \begin{bmatrix} a_0 \\ a_1 \\ b_{-1} \end{bmatrix} = \begin{bmatrix} 1 \\ 1 \\ 1 \end{bmatrix} \tag{8.31}$$

Solving these simultaneous equations, the numerical integration formula coefficients are finally given by

$$b_{-1} = \frac{\gamma + 1}{\gamma + 2}$$

$$a_1 = \frac{-\gamma^{-1}}{\gamma + 2} \qquad (8.32)$$

$$a_0 = \frac{\gamma + 2 + \gamma^{-1}}{\gamma + 2}$$

where $\gamma = T_0/T_1$. For the situation where $\gamma = 1$ corresponding to a constant time increment between samples, these coefficients properly simplify to $b_{-1} = 2/3$, $a_0 = 4/3$, and $a_1 = -1/3$. This coefficient set is the set normally given for the constant time-step second-order Gear algorithm.

Before leaving the second-order Gear algorithm, it is insightful to examine the numerical accuracy of this algorithm compared to the backward Euler method. Using the constant time-step algorithm coefficients and (8.28) along with (8.7), the iterative numerical solution using the second-order Gear algorithm for the network step response is given by

$$\begin{bmatrix} s_1(t+h) \\ s_2(t+h) \end{bmatrix} = - \begin{bmatrix} -\dfrac{3}{2h} & -\omega_n^2 \\ -1 & 2\zeta\omega_n + \dfrac{3}{2h} \end{bmatrix}$$

$$\times \frac{\left\{ \begin{array}{l} 2\zeta\omega_n + \dfrac{3}{2h}\left(\dfrac{4}{3}s_1(t) - \dfrac{1}{3}s_1(t-h)\right) \\ -\dfrac{3}{2h}\left[\dfrac{4}{3}s_2(t) - \dfrac{1}{3}s_2(t-h)\right] \end{array} \right\}}{\dfrac{3}{2h}\left(2\zeta\omega_n + \dfrac{3}{2h}\right) + \omega_n^2}$$

(8.33)

The calculated step response is shown in Figure 8.4. The second-order Gear method clearly displays improved numerical accuracy for the selected time increment h compared to the first-order backward Euler method. The numerical error performance is

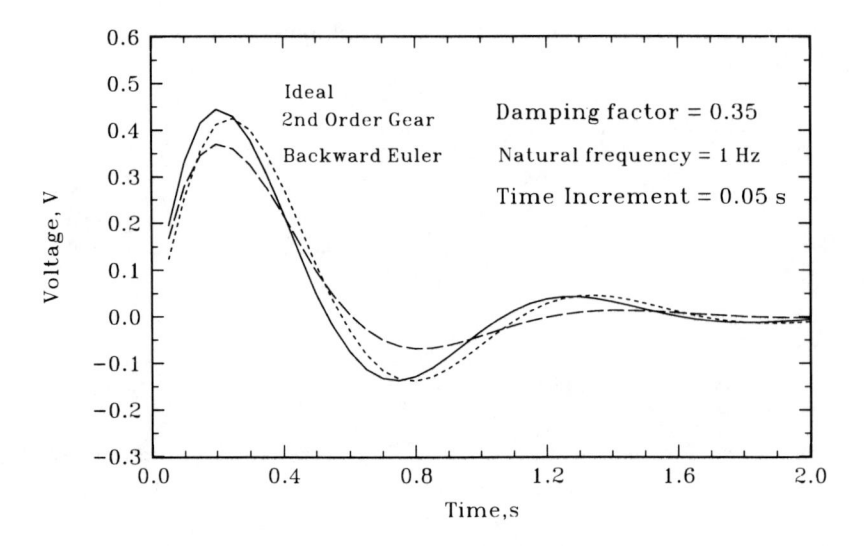

Figure 8.4. Comparison of the computed RLC circuit step response using the backward Euler and Gear methods with respect to ideal.

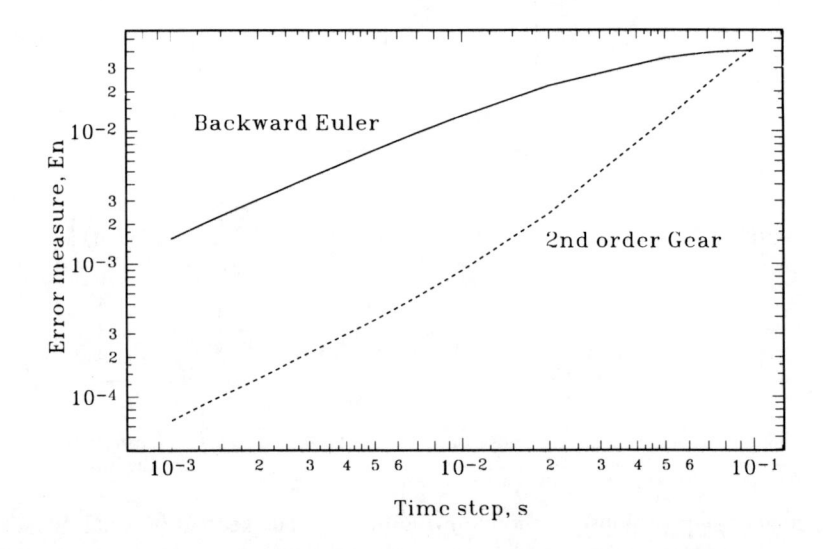

Figure 8.5. Comparison of the numerical error performance of the backward Euler and second-order Gear methods versus integration time step for the RLC circuit transient response.

further compared with the backward Euler method as a function of time step h in Figure 8.5 where the error measure given by (8.14) was again employed.

8.3 NUMERICAL INTEGRATION ALGORITHM STABILITY

Thus far, we have avoided any rigorous discussions concerning the numerical stability of the algorithms being employed because such discussions are in general beyond the scope of this text. Nonetheless, numerical stability is one of the primary motivations for favoring implicit methods over explicit integration methods. It is therefore necessary to make a brief detour into the subject of algorithm stability.

Stability of numerical integration algorithms is dealt with at length in [2]. The study of stability is carried out by substituting the differential equation of interest into the candidate numerical algorithm from which a difference equation results. The roots of the difference equation are then calculated, in some cases by means of z transforms. If the order of the difference equation exceeds the order of the original difference equation, which is normally assumed to be of first order, all but one of the equation solutions are considered to be parasitic. To have a stable well-behaved algorithm, all such parasitic solutions are required to decay quickly [5].

The issue of stability may be easily investigated for the numerical algorithms discussed thus far by considering the first-order initial value problem represented by

$$\dot{x} = -\lambda x \tag{8.34}$$

where $x(0) = 1$. For the forward Euler method, the difference equation that results from (8.8) is

$$\begin{aligned} x_{n+1} &= x_n + h\dot{x}_n \\ &= x_n + (-h\lambda)x_n \\ &= (1 - h\lambda)x_n \end{aligned} \tag{8.35}$$

which clearly diverges for $|1 - h\lambda| > 1$. The region of numerical stability for the forward Euler method is given by

$$|1 - h\lambda| < 1$$
$$(1 - x)^2 + y^2 < 1 \tag{8.36}$$

where $x = \text{Re}(h\lambda)$ and $y = \text{Im}(h\lambda)$. The region of algorithm stability is shown as the patterned area in Figure 8.6.

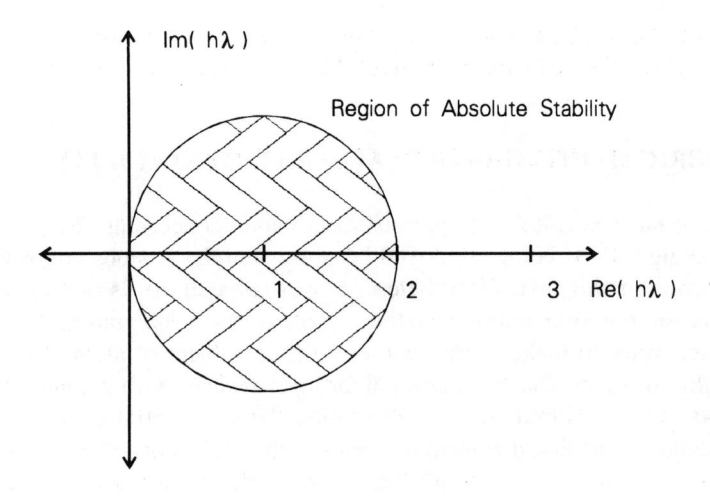

Figure 8.6. Stability region for the forward Euler integration algorithm.

In contrast, for the backward Euler method, the difference equation that results when solving (8.10) is given by

$$x_{n+1} = x_n + h\dot{x}_{n+1}$$

$$= x_n + (-h\lambda)x_{n+1}$$

$$= \frac{x_n}{1 + h\lambda} \qquad (8.37)$$

The solution diverges only for the cases where

$$(1 + x)^2 + y^2 < 1 \qquad (8.38)$$

which is the interior of a unit circle centered at the point (-1,0) as shown in Figure 8.7. The region of absolute stability for the backward Euler method is clearly much larger than for the forward Euler method.

Finally, for the second-order Gear algorithm, solution of the same first-order initial value problem leads to the difference equation

$$x_{n+1} = \frac{4}{3}x_n - \frac{1}{3}x_{n-1} + \frac{2}{3}\dot{x}_{n+1}$$

$$= \frac{4}{3}x_n - \frac{1}{3}x_{n-1} - \frac{2}{3}h\lambda x_{n+1} \qquad (8.39)$$

$$x_{n+1}\left(1 + \frac{2}{3}h\lambda\right) = \frac{4}{3}x_n - \frac{1}{3}x_{n-1}$$

Employing z transforms with (8.39),

$$x(z)\left(\frac{4}{3}z^{-1} - \frac{1}{3} - z^{-2} - \frac{2}{3}h\lambda z^{-2}\right) = 0 \qquad (8.40)$$

which leads finally to the characteristic equation given by

$$2z - \frac{1}{2}z^2 - \frac{3}{2} = h\lambda \qquad (8.41)$$

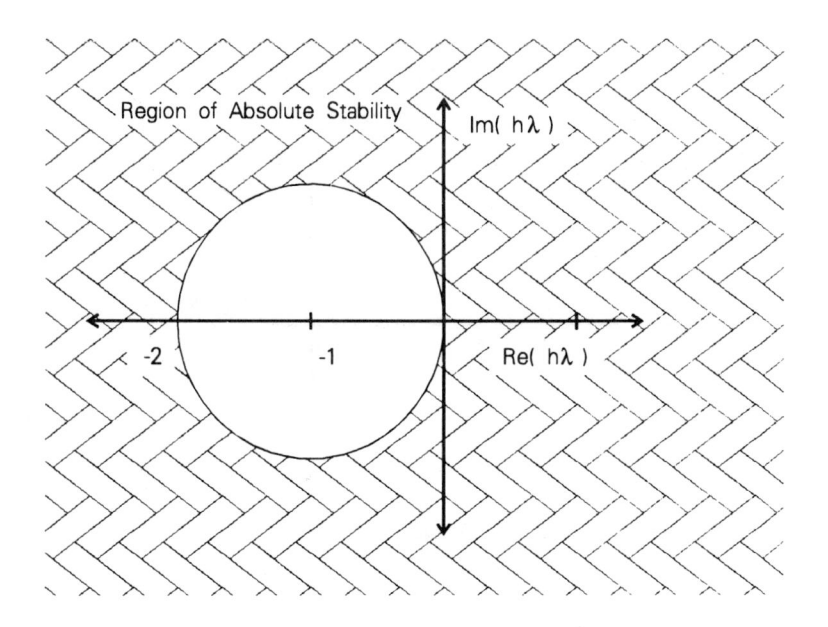

Figure 8.7. Region of stability for the backward Euler numerical integration formula.

If the algorithm is to be stable, all characteristic equation roots must lie in the interior of the unit circle in the complex z plane. Therefore, the range of $h\lambda$ for which this algorithm is stable can be identified by considering points along the unit circle given by

$$z = e^{j\theta} \quad \text{for} \quad 0 \le \theta < 2\pi \tag{8.42}$$

Substituting (8.42) into (8.41), the stable range of σ is given by

$$\sigma = h\lambda = 2e^{j\theta} - \frac{e^{j2\theta}}{2} - \frac{3}{2} \tag{8.43}$$

The region of absolute stability is shown for this case in Figure 8.8.

The interested reader is referred to [1] and [10] for more information on the subject of numerical integration methods. The variable time-step second-order Gear algorithm is utilized throughout the remainder of this chapter.

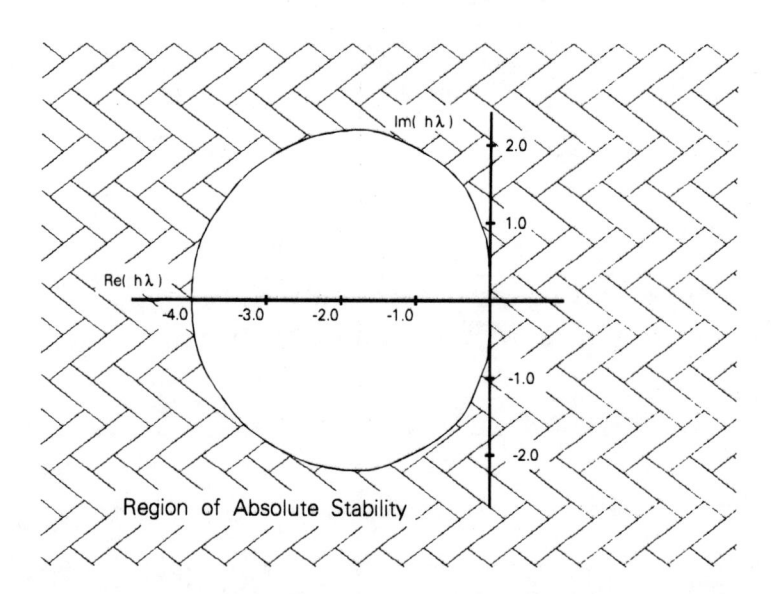

Figure 8.8. Region of stability for the second-order Gear numerical integration algorithm.

8.4 DISCRETE CIRCUIT MODELS

Armed with the variable time-step second-order Gear formula given by (8.32), it is now possible to derive the equivalent circuit models that will be used for the most common differential circuit elements, the capacitor and inductor. The underlying motivation for taking this approach is developed in parallel.

The voltage-current relationship for an ideal time-invariant capacitor is given by

$$C\frac{dV}{dt} = i \qquad (8.44)$$

where C is the capacitance value, and the voltage and current senses are as shown in Figure 8.9. The second-order Gear algorithm is given by

$$v_{n+1} = a_0 v_n + a_1 v_{n-1} + T_1 b_{-1} \dot{v}_{n+1} \qquad (8.45)$$

with coefficients given by (8.32). Recasting (8.44) where the dot notation implies differentiation with respect to time,

$$C\dot{v}_{n+1} = i_{n+1} \qquad (8.46)$$

and on substitution into (8.45),

$$i_{n+1} = \left(\frac{C}{T_1 b_{-1}}\right) v_{n+1} - \frac{C}{T_1 b_{-1}} (a_0 v_n + a_1 v_{n-1}) \qquad (8.47)$$

Recognizing that v_n and v_{n-1} have been previously computed and are therefore constant, (8.47) may be recast as the equivalent discrete circuit model shown in Figure 8.10. From this figure, it should be clear that this approach reduces the time-domain circuit analysis problem to an iterative dc analysis problem involving simple real admittances and constant sources at each time increment. This discrete model may

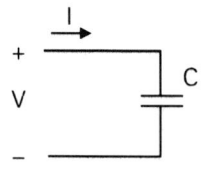

Figure 8.9. Ideal linear capacitor with voltage and current sense shown.

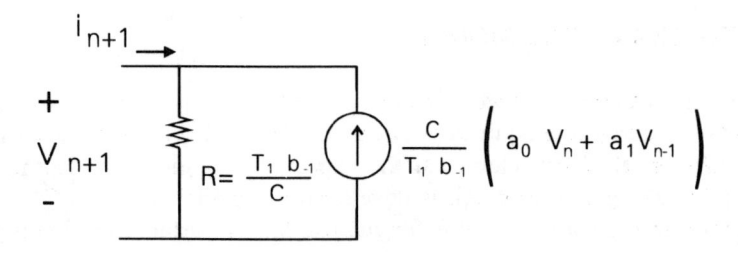

Figure 8.10. Discrete circuit model for an ideal capacitor using the second-order Gear numerical integration algorithm.

be substituted for every ideal capacitor that appears in the original circuit. Not only is this approach mechanically straightforward, every original node within the original circuit remains accessible.

The discrete model for an ideal inductor may be found in a similar manner. The voltage-current relationship for the ideal inductor is given by

$$L \frac{di}{dt} = v \tag{8.48}$$

where L is the inductance value and the voltage and current senses are as shown in Figure 8.11. Employing the second-order Gear algorithm for this case leads to the relationship

$$i_{n+1} = a_0 i_n + a_1 i_{n-1} + T_1 b_{-1} \left(\frac{di}{dt}\right)_{n+1} \tag{8.49}$$

where the dot-derivative notation has been dropped merely to avoid confusion. From (8.48), we have

$$L \frac{d}{dt} i_{n+1} = v_{n+1} \tag{8.50}$$

which leads finally to

$$i_{n+1} = (a_0 i_n + a_1 i_{n-1}) + \frac{T_1 b_{-1}}{L} v_{n+1} \tag{8.51}$$

This result may be similarly cast in a discrete model form as shown in Figure 8.12.

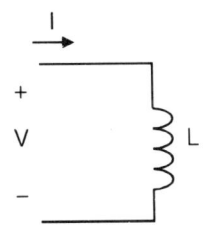

Figure 8.11. Ideal linear inductor with voltage and current sense shown.

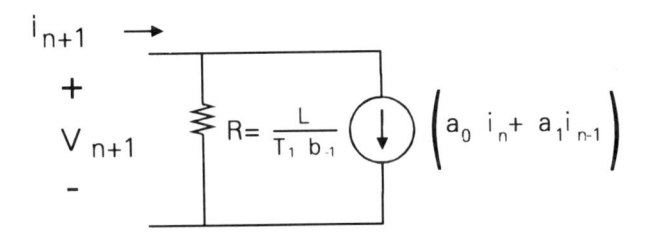

Figure 8.12. Discrete circuit model for an ideal inductor using the second-order Gear numerical integration formula.

The resistor value is $R = \dfrac{L}{T_1 b_{-1}}$ and the current source is $(a_0 i_n + a_1 i_{n-1})$.

To illustrate this discrete modeling technique and also gain more insight into using the second-order Gear algorithm, the circuit example shown in Figure 8.13 will be evaluated using this discrete modeling approach. The first step in the circuit analysis is to replace each circuit element in Figure 8.13 with its respective discrete circuit model. The input source is also replaced with its Norton equivalent since an admittance description will be used to analyze the network. These steps combined with the assumption of a constant time increment of h lead to the discrete system shown in Figure 8.14.

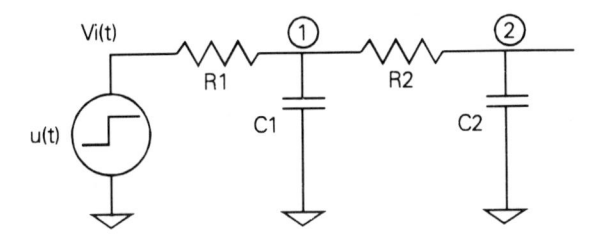

Figure 8.13. Simple RLC circuit for illustration of the second-order Gear algorithm discrete modeling approach.

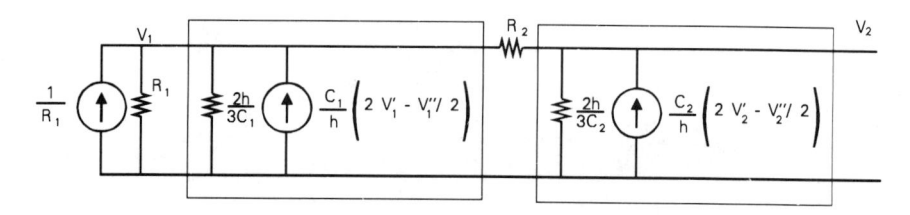

Figure 8.14. Discrete model representation for RLC circuit using the second-order Gear algorithm. Single primes denote sample time index n, whereas double primes denote a time index of $n-1$.

Applying Kirchoff's laws to Figure 8.14, the network is described by the matrix equation

$$\begin{bmatrix} \dfrac{1}{R_1} + \dfrac{3C_1}{2h} + \dfrac{1}{R_2} & -\dfrac{1}{R_2} \\ -\dfrac{1}{R_2} & \dfrac{3C_2}{2h} + \dfrac{1}{R_2} \end{bmatrix} \begin{bmatrix} v_1 \\ v_2 \end{bmatrix} = \begin{bmatrix} \dfrac{1}{R_1} + \dfrac{C_1}{h}\left(2v_1' - \dfrac{v_1''}{2}\right) \\ \dfrac{C_2}{h}\left(2v_2' - \dfrac{v_2''}{2}\right) \end{bmatrix} \qquad (8.52)$$

where the primes indicate different time indices. By means of the method of determinants, the solution to this matrix equation is given by

$$\begin{bmatrix} v_1 \\ v_2 \end{bmatrix} = \dfrac{\begin{bmatrix} \dfrac{3C_2}{2h} + \dfrac{1}{R_2} & \dfrac{1}{R_2} \\ \dfrac{1}{R_2} & \dfrac{1}{R_1} + \dfrac{3C_1}{2h} + \dfrac{1}{R_2} \end{bmatrix} \begin{bmatrix} \dfrac{1}{R_1} + \dfrac{C_1}{h}\left(2v_1' - \dfrac{v_1''}{2}\right) \\ \dfrac{C_2}{h}\left(2v_2' - \dfrac{v_2''}{2}\right) \end{bmatrix}}{\left(\dfrac{1}{R_1} + \dfrac{3C_1}{2h} + \dfrac{1}{R_2}\right)\left(\dfrac{1}{R_2} + \dfrac{3C_2}{2h}\right) - \dfrac{1}{R_2^2}} \qquad (8.53)$$

The node voltages at each time step may be calculated by repeated evaluation of (8.53) where each primed quantity is properly updated at each time step (e.g., $v' = v_{n-1}$, $v'' = v_{n-2}$). With this result and arbitrarily setting all of the component values to unity, the transient response at node 1 of the RC network as a function of time step h is shown in Figure 8.15. As evidenced by these curves, the Gear algorithm displays no evidence of overshoot even for fairly large time-step choices.

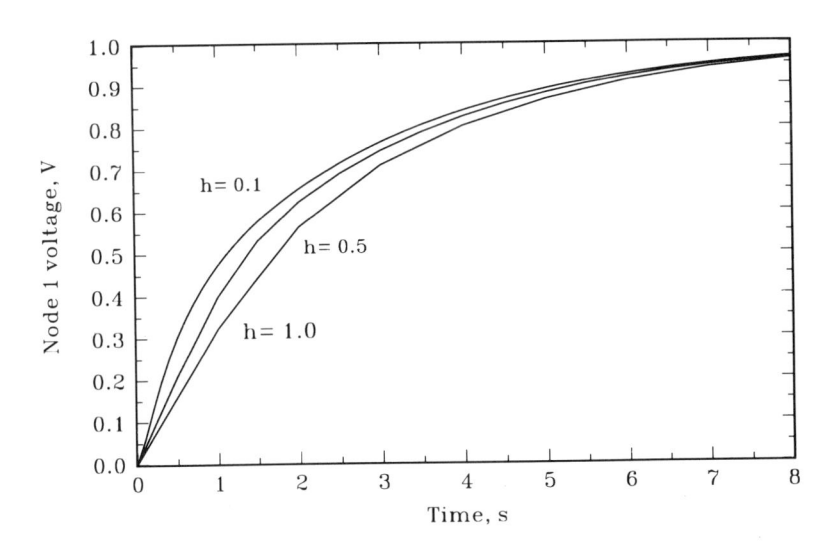

Figure 8.15. Discrete model step response computation for different numerical time integration quantities.

8.5 CIRCUIT MACROS

MACSET contains a number of circuit macros that permit a user to easily model more complicated circuit elements such as operational amplifiers without entering low-level equivalent circuits. The macros are specifically intended to capture device parameter effects that would directly affect the observed phase-locked loop characteristics while at the same time avoid unnecessary detail.

One of the simplest macros included is for the ideal voltage follower shown in Figure 8.16. It is ideal in that no reactive elements are included that would impair the device bandwidth. The model must necessarily have finite input and output impedances because of the admittance matrix formulation approach used within MACSET.

The ideal operational amplifier macro is closely related to the ideal voltage follower macro as evidenced in Figure 8.17. The dc voltage gain is given by the numerical parameter A_0.

A more realistic operational amplifier model, which includes finite slew-rate limitations and the first two amplifier frequency response poles, is shown in Figure 8.18. In this model, the finite voltage slew rate and dominant pole are characterized by the combination of the first current source and the first RC section. The limiting

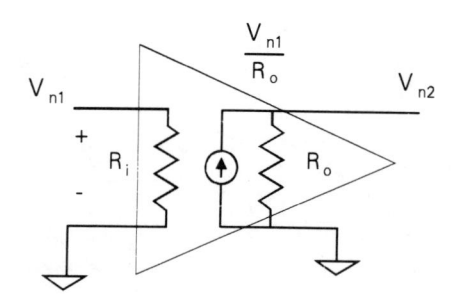

Figure 8.16. Ideal voltage follower macro.

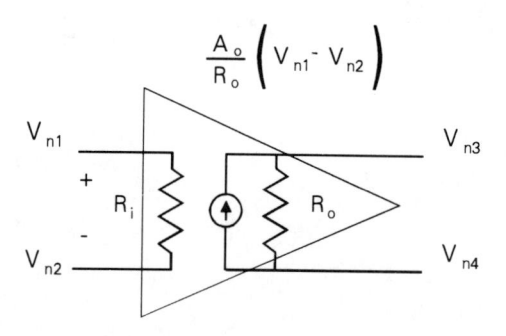

Figure 8.17. Ideal operational amplifier macro.

behavior of the I_m current source is shown in Figure 8.19 where $V = V_{n1} - V_{n2}$. From Figure 8.18, the dc differential voltage gain is given simply by

$$A_0 = g_m R_1 \tag{8.54}$$

Given that the maximum device slew rate is S volts/second, the slew rate and maximum source current are related as

$$S = \frac{I_{max}}{C_1} \tag{8.55}$$

where I_{max} and g_m are related as shown in Figure 8.18.

Continuing this definition phase further, the dominant pole for the operational amplifier is given by

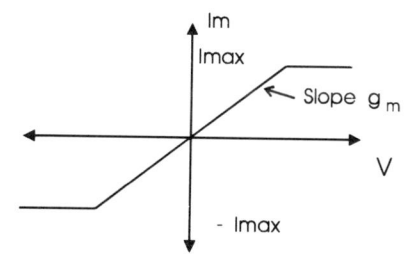

Figure 8.18. Nonlinear current source I_m for modeling finite slew-rate in an operational amplifier.

$$\omega_d = \frac{1}{R_1 C_1} \tag{8.56}$$

Using these defined quantities and arbitrarily setting $R_1 = 1\Omega$, the constitutive component elements of this macro model are given as

$$g_m = A_0$$

$$C_1 = \frac{1}{\omega_d} \tag{8.57}$$

$$I_{max} = \frac{S}{\omega_d}$$

Arbitrarily setting $R_2 = 1\Omega$ as well, the second pole represented by frequency ω_s defines the final capacitance value in Figure 8.18 as $C_2 = \omega_s^{-1}$.

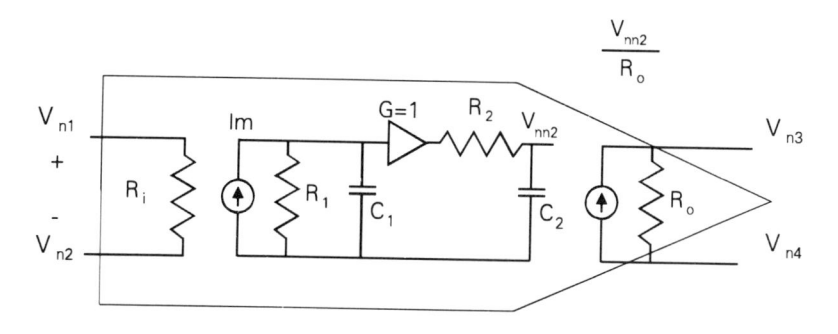

Figure 8.19. Nonideal operational amplifier macro.

Internal loop transport delays are modeled based on the discussion presented in Chapter 4, which used a cascade of first-order active all-pass sections. The form of a single all-pass section was shown in Figure 4.16. The only user inputs required for this macro are the number of all-pass sections to be cascaded (1 to 5) and the time-delay parameter to be modeled.

The phase detector macros that are included in MACSET are (1) the three-state simplified phase-frequency detector and (2) the zero-order sample-and-hold phase detector. The simplified phase-frequency detector is shown again in Figure 8.20 and is modeled precisely as shown in Figure 4.6 where the peak phase detector output voltage V_p is user definable. Owing again to the network admittance representation used within MACSET, the flip-flop output impedances must be nonzero (and are assumed to be 10Ω in MACSET.)

A schematic description of the zero-order sample-and-hold phase detector is shown in Figure 8.21. The finite sampling efficiency for the sampling switch of this detector is modeled as

$$V_{SH}(n) = V_{SH}(n-1) + \eta[V(n) - V_{SH}(n-1)] \qquad (8.58)$$

where $V_{SH}(n)$ is the phase detector output voltage following the nth sampling interval, η is the sampling efficiency ($0 < \eta \leq 1$), and $V(n)$ is the phase detector ramp voltage being sampled at the nth sampling instant, which is given mathematically by $K_d \theta_e(n)$. In this form, K_d is the phase detector gain in volts per radian and θ_e is the phase detector error in radians. This model is quite simplistic in that no time delays have been included, and the phase detector ramp resetting operation is assumed to be instantaneous.

Other circuit macros may of course be added to a program such as MACSET with little difficulty. The macros presented here represent the more important elements available within the program and also illustrate the adopted program methodology.

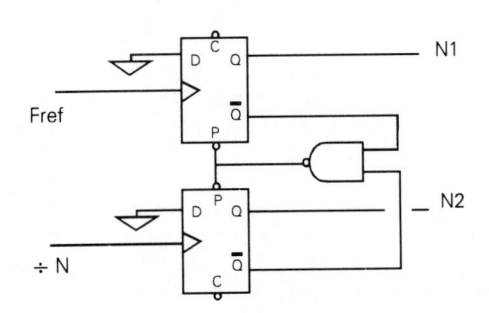

Figure 8.20. Simplified phase-frequency detector macro.

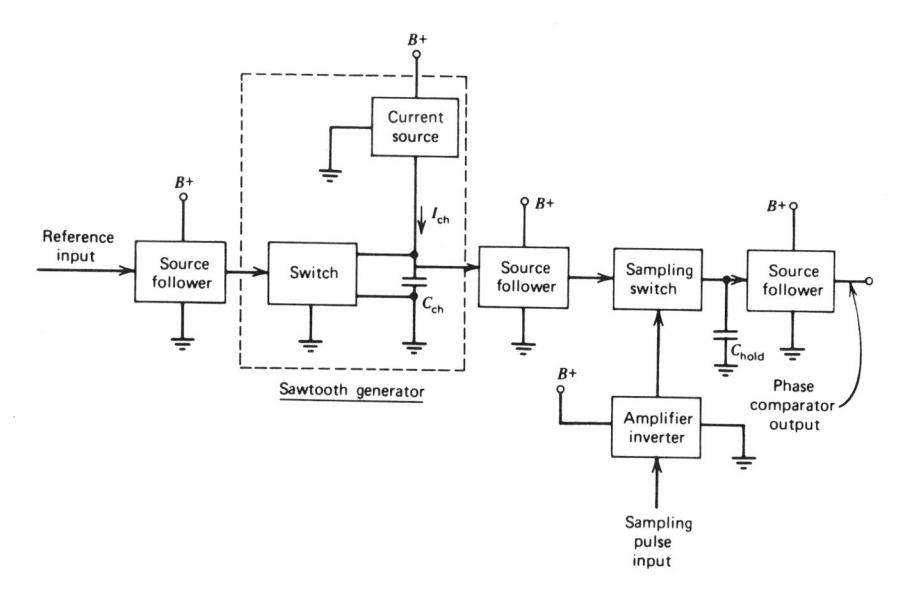

Figure 8.21. Sample-and-hold phase detector macro. Source: [25]. Reprinted with permission.

8.6 FREQUENCY-DOMAIN ANALYSIS

Analysis in the frequency domain is complicated by the discrete nature of the digital phase detectors that are generally used. Were it not for the presence of these digital devices, traditional Laplace transform methods would suffice as discussed in Chapter 4.

Based on the concepts that were developed in Chapter 5, the principle open-loop gain functions that must be calculated for a complete loop characterization are $G_{OL}(\omega)$ and $G_{OL}^*(\omega)$, the continuous and sampled open-loop gain functions, respectively. From (5.19),

$$G_{OL}^*(\omega) = \frac{1}{T} \sum_{n=-k}^{k} G_{OL}(\omega + n\omega_s) \qquad (8.59)$$

where as before k is a user-defined parameter that dictates the number of aliased gain terms to be included in the approximation for $G_{OL}^*(\omega)$. Once the two open-loop gain transfer functions have been calculated, the primary closed-loop transfer functions may be computed as

$$T_1(\omega) = \frac{N\, G_{OL}(\omega)}{1 + G_{OL}^*(\omega)} \qquad (8.60)$$

$$T_2(\omega) = \frac{1}{1 + G_{OL}^*(\omega)} \qquad (8.61)$$

The transfer functions represented by (8.59) through (8.61) are central to any calculations that will be performed in the frequency domain.

Assuming that $G_{OL}(\omega)$ has been calculated using methods that will be described shortly, a complete frequency-domain description of the candidate phase-locked loop may be found as summarized in Table 8.1.

Table 8.1
Frequency-Domain Characterization Summary

Analysis to be Performed	Reference for Method
Calculation of $G_{OL}^*(\omega)$	Section 5.3, (5.19)
System Stability	Section 5.3
Gain Peaking	Section 8.6
Output Phase Noise Spectrum	Sections 5.6, 8.7.

Calculation of $G_{OL}(\omega)$

Calculation of $G_{OL}(\omega)$ is central to all of the frequency-domain analyses of interest and it is therefore very important that the underlying concepts be understood well. The key to the computational approach is represented well by (5.2) where a clear separation between the system sampling aspects [e.g., $\varphi^*(s)$] and the continuous open-loop gain function [e.g., $H(s)G(s)$] has been made. The only factor missing in the open-loop gain function is the N^{-1} factor due to the $\div N$, which simply appends a multiplicative factor of N^{-1} to the gain function based on the discussions in Chapter 4. Rewriting (5.2) in the context of the present discussions,

$$\varphi_o(s) = \varphi_e^*(s) N G_{OL}(s) \qquad (8.62)$$

In short, all discrete sampling phenomena must be lumped into $\varphi_e^*(s)$, whereas all time-continuous transfer functions must be lumped into $NG_{OL}(s)$. Referring back to Figure 5.1, $G_{OL}(\omega)$ results from cascading all of the loop element transfer functions except the transfer function of the ideal impulse sampler, which is shown within the phase detector portion of the diagram. To be absolutely clear on the proper solution approach, the formulation of $G_{OL}(\omega)$ for each of the principal phase detector types in Figure 8.22 is summarized below.

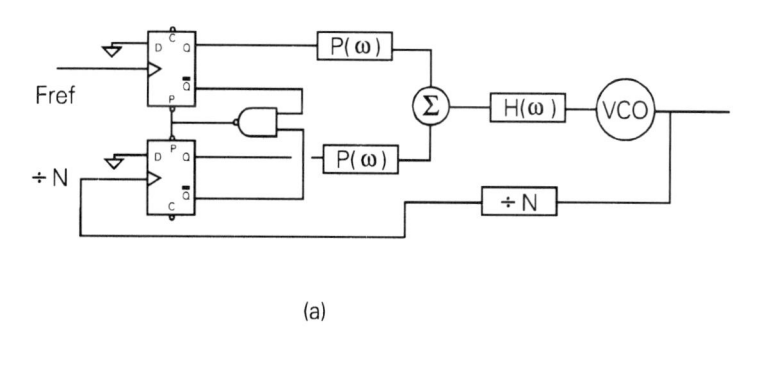

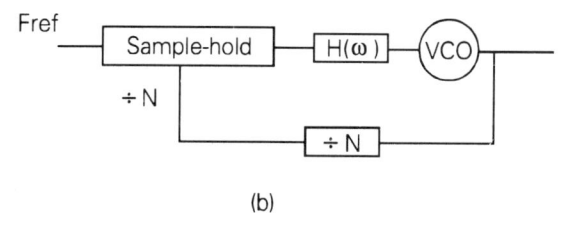

Figure 8.22. Summary of open-loop gain calculations involving different phase detector types: (a) Simplified phase-frequency detector case and (b) sample-and-hold detector case.

For the simplified phase-frequency detector case, the open-loop gain is given by

$$G_{\text{OL}}(\omega) = K_d \frac{P(\omega)H(\omega)K_v}{j\omega N} \tag{8.63}$$

where K_d is the phase detector gain in volts per radian, and K_v is the voltage-controlled oscillator (VCO) tuning sensitivity in radians per second per volt. For the ideal zero-order sample-and-hold phase detector case, the open-loop gain is given by

$$G_{\text{OL}}(\omega) = (e^{-j\omega T} - 1)\frac{H(\omega)K_d K_v}{\omega^2 N} \tag{8.64}$$

Finally, for the inefficient zero-order sample-and-hold case, the open-loop gain function is given by

$$G_{\text{OL}}(\omega) = K_d K_v \eta \frac{e^{-j\omega T} - 1}{1 - (1 - \eta)e^{-j\omega T}} \frac{H(\omega)}{\omega^2 N} \tag{8.65}$$

As shown here, most of the remaining analysis involves calculation of the embedded voltage transfer function $H(\omega)$, which requires nothing more than ordinary continuous-time network analysis concepts.

Concentrating now on calculation of $H(\omega)$, since a nodal admittance network description was used throughout the time-domain analysis, a similar representation will be adopted here for performing the frequency-domain analysis. Assume that the linear network represented by $H(\omega)$ may be described by

$$[Y]V = I \tag{8.66}$$

where Y is a complex admittance matrix, V is the vector of circuit node voltages, and I is the vector of external current inputs at each network node. To calculate $H(\omega)$, the input and output node voltages must be computed at each frequency using (8.66) and their complex ratio computed. Many computational methods are available for addressing this problem. Only one possible method is considered here in detail.

In the general case, all of the node voltages in (8.63) must be computed, but this is not necessary if only $H(\omega)$ is to be computed because this form implies a single input and a single output. In this special case, it is advantageous to use floating admittance matrix concepts [26] and reduce the $N \times N$ admittance matrix to an equivalent 2×2 matrix retaining only the input and output nodes as the network ports. This method is based on a generalization of Kirchhoff's current law in which the datum node voltage (which is normally taken as zero) is included in the equation formulation. The simple example that follows illustrates this computational technique.

The simplistic network example considered is shown in Figure 8.23. The admittance matrix for this network is given by

$$Y = \begin{bmatrix} G_1 + G_3 + G_5 & -G_1 & -G_3 & -G_5 \\ -G_1 & G_1 + G_2 + G_6 & -G_2 & -G_6 \\ -G_3 & -G_2 & G_2 + G_3 + G_4 & -G_4 \\ -G_5 & -G_6 & -G_4 & G_5 + G + 4 + G_3 \end{bmatrix} \tag{8.67}$$

where each conductance is given by $G_i = R_i^{-1}$. Unlike the usual admittance matrix representation, the datum node has been retained in (8.64) making Y a "floating" admittance matrix rather than one that is tied to a ground potential of 0V. A result of Kirchhoff's current law, each row and each column of Y must individually sum to zero. This fact may be used to reduce the dimensions of Y by eliminating embedded network nodes whose explicit solutions are not required. To this end, (8.64) may be rewritten as

$$\begin{bmatrix} g_{00} & g_{01} & g_{02} & g_{03} \\ g_{10} & g_{11} & g_{12} & g_{13} \\ g_{20} & g_{21} & g_{22} & g_{23} \\ g_{30} & g_{31} & g_{32} & g_{33} \end{bmatrix} \begin{bmatrix} V_0 \\ V_1 \\ V_2 \\ V_3 \end{bmatrix} = \begin{bmatrix} I_0 \\ I_1 \\ I_2 \\ I_3 \end{bmatrix} \qquad (8.68)$$

Assume now that node 2 is considered to be an internal inaccessible node and it is therefore to be eliminated from any explicit appearance in (8.68). From the matrix equation (8.68) for node 2,

$$g_{20}V_0 + g_{21}V_1 + g_{22}V_2 + g_{23}V_3 = I_2 = 0 \qquad (8.69)$$

Solving now for V_2,

$$V_2 = -\frac{1}{g_{22}}[g_{20}V_0 + g_{21}V_1 + g_{23}V_3] \qquad (8.70)$$

Substituting this result into (8.68), the reduced matrix description for the network is given by

$$\begin{bmatrix} \left(g_{00} - \dfrac{g_{02}g_{20}}{g_{22}}\right) & \left(g_{01} - \dfrac{g_{02}g_{21}}{g_{22}}\right) & \left(g_{03} - \dfrac{g_{02}g_{23}}{g_{22}}\right) \\ \left(g_{10} - \dfrac{g_{12}g_{20}}{g_{22}}\right) & \left(g_{11} - \dfrac{g_{12}g_{21}}{g_{22}}\right) & \left(g_{13} - \dfrac{g_{12}g_{23}}{g_{22}}\right) \\ \left(g_{30} - \dfrac{g_{32}g_{20}}{g_{22}}\right) & \left(g_{31} - \dfrac{g_{32}g_{21}}{g_{22}}\right) & \left(g_{32} - \dfrac{g_{32}g_{23}}{g_{22}}\right) \end{bmatrix} \begin{bmatrix} V_0 \\ V_1 \\ V_2 \end{bmatrix} = \begin{bmatrix} I_0 \\ I_1 \\ I_2 \end{bmatrix} \qquad (8.71)$$

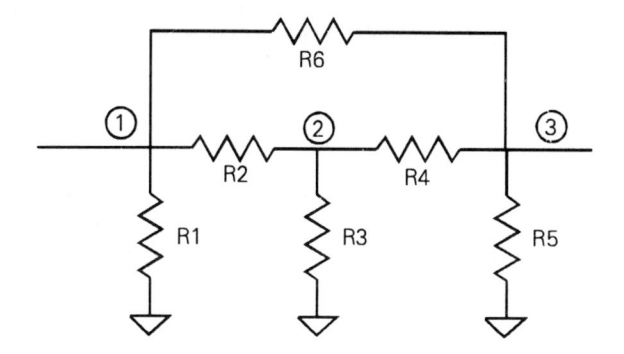

Figure 8.23. Simple resistive network for illustrating the floating admittance matrix concept.

which is itself a floating admittance matrix. For larger networks, this node elimination method may be used repeatedly as necessary to reduce the network description to an equivalent two-port one. Ignoring the datum node in (8.71) by setting $V_0 = 0$, the final 2×2 admittance matrix for this example is given as

$$\begin{bmatrix} \left(g_{11} - \dfrac{g_{12}g_{21}}{g_{22}} \right) & \left(g_{13} - \dfrac{g_{13}g_{23}}{g_{22}} \right) \\ \left(g_{31} - \dfrac{g_{32}g_{21}}{g_{22}} \right) & \left(g_{33} - \dfrac{g_{32}g_{23}}{g_{22}} \right) \end{bmatrix} \quad (8.72)$$

In the more general case where the solution of all node voltages is required, a number of computational techniques are available including the Cholesky and Gaussian elimination methods [2,27]. Sparse matrix methods utilizing a Cholesky solution approach are used within MACSET for improved execution speed. Although beyond the scope of this text, these methods are very important for dealing with large network problems.

8.7 PHASE NOISE COMPUTATIONS

Although the methods explored in Chapters 4 and 5 provide an avenue for calculating the phase noise spectrum at the output of a phase-locked loop, the discussions assumed that all of the underlying noise sources had been previously equivalenced with phase noise sources at either the frequency reference or VCO ports. In this section, we discuss the noise arising from individual discrete components such as resistors and operational amplifiers in the loop filter electronics and how these may be efficiently referred back to one of these two points.

The noise arising in an individual resistor R may be represented by the equivalent open-circuit model shown in Figure 8.24. The random noise voltage source e_n is represented in the frequency domain as a uniformly flat noise spectrum having a spectral density of

$$e_n = \sqrt{4kTR}, \quad \text{Vrms}/\sqrt{\text{Hz}} \quad (8.73)$$

Noise arising from an operational amplifier may be similarly modeled as shown in Figure 8.25 [28] where the e_n and i_n noise source spectra must be defined by the user. All of the noise sources are assumed to be statistically independent.

Given these simple noise model equivalents, the contribution of each noise source at one of the two aforementioned reference points must still be calculated. While it is possible to calculate the transfer function from each noise source to the reference point for every frequency of interest, and to combine the results, network

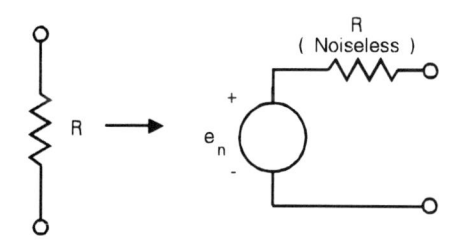

Figure 8.24. Equivalent noise circuit for a physical resistor.

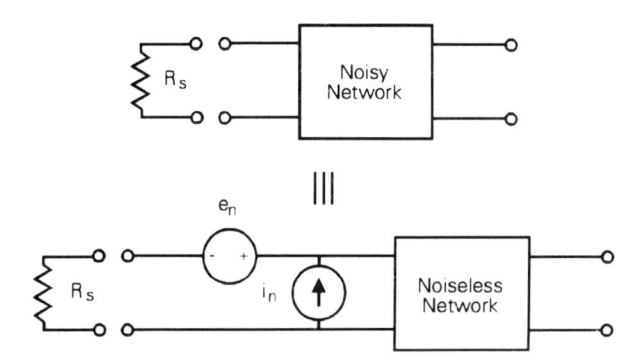

Figure 8.25. The noise sources present in a noisy linear two-port may be equivalanced using suitable input voltage and current noise generators.

interreciprocity concepts may be used to perform the computations in a much faster manner [2,29]. The proof of these concepts follows directly from Tellegen's theorem.

Assume that all of the network noise sources are represented by statistically independent noise current sources I_k, $k = 1, 2, \ldots, N$. Further assume that all of the noise contributors are to be referred forward to the tuning port of the noiseless VCO in the phase-locked loop being analyzed. If the transimpedances from each noise current source to the VCO tuning port Z_k are known, the overall noise contribution at the VCO tuning port is given by

$$V_T = \sqrt{\sum_{k=1}^{N} |I_k Z_k|^2}, \quad \text{Vrms}/\sqrt{\text{Hz}} \tag{8.74}$$

Since the I_k and Z_k are both functions of frequency, this computation must be performed at each frequency of interest.

The Z_k transimpedances may be calculated by using the interreciprocity relationships just referenced. The first step in the solution process is to form the admittance matrix for the adjoint network. This matrix turns out to be simply the transpose of the original circuit admittance matrix [29]. A 1-A current source is then connected to the port corresponding to the VCO tuning port and all of the complex network node voltages V_j are computed. Having followed these steps, the transimpedance for the kth current noise source connected between circuit nodes i and j is simply given by $Z_k = V_i - V_j$. Since one such adjoint network analysis permits all of the transimpedance values to be computed at once, this approach is much faster than the brute force method where each noise source transfer function is individually computed for every frequency of interest. Further computational details are available in [29].

8.8 SUMMARY

The basic elements of a computer program suitable for analyzing the transient response of fairly general phase-locked loops have been discussed. A number of results developed elsewhere in the text have been brought together to make this design and analysis method viable.

SPICE computer methods can in many respects be used to duplicate elements of the MACSET design tool. Depending on the circuit elements available for modeling, however, low-level details that are not pertinent to the macroscopic behavior of the system can lead to long simulation times. In addition, computation of $G_{OL}^*(\omega)$ will normally not be straightforward, nor is the calculation of the important functions given by (8.60) and (8.61).

REFERENCES

[1] Cuthbert, T.R., *Circuit Design Using Personal Computers*, John 1 Wiley & Sons, New York, 1983.

[2] Chua, L.O., P.M. Lin, *Computer-Aided Analysis of Electronics Circuits*, Prentice-Hall, Englewood Cliffs, NJ, 1975.

[3] Nagel, L.W., "SPICE2: A Computer Program to Simulate Semiconductor Circuits," Memorandum ERL-M520, Electronics Research Laboratory, University of California, Berkeley, May 9, 1975.

[4] Adby, P.R., *Applied Circuit Theory: Matrix and Computer Methods*, Ellis Horwood, Chichester, West Sussex, UK, 1980.

[5] McCalla, W.J., D.O. Pederson, "Elements of Computer-Aided Circuit Analysis," *IEEE Trans. CT*, Vol. CT-18, No. 1, January 1971, pp. 14–26.

[6] Fidler, J.K., C. Nightingale, *Computer-Aided Circuit Design*, John Wiley & Sons, New York,

[7] Weeks, W.T., et al., "Algorithms for ASTAP—A Network Analysis Program," *IEEE Trans. CT*, Vol. CT-20, No. 6, November 1973, pp. 628–634.

[8] Pederson, D.O., "Historical Review of Circuit Simulation," *IEEE Trans. CAS*, Vol. CAS-31, No. 1, January 1984, pp. xxx–111.

[9] Hachtel, G.D., A.L. Sangiovanni-Vincentelli, "A Survey of Third-Generation Simulation Techniques," *IEEE Proc.*, Vol. 69, No. 10, October 1981, pp. 1264–1280.

[10] Smith, J.M., *Mathematical Modeling and Digital Simulation for Engineers and Scientists*, 2nd ed., John Wiley & Sons, New York, 1987.
[11] Branin, F.H., Jr., "The Relations Between Kron's Method and the Classical Methods of Network Analysis," *IRE Wescon Conference Record*, Part 2, August 1959, pp. 3–28.
[12] Kron, G., "A Set of Principles to Interconnect the Solutions of Physical Systems," *J. Appl. Phys.*, Vol. 24, No. 8, August 1953, pp. 965–980.
[13] Kuh, E.S., R.A. Rohrer, "The State Variable Approach to Network Analysis," *IEEE Proc.*, July 1965, pp. 672–686.
[14] Branin, F.H., Jr., "Computer Methods of Network Analysis," *IEEE Proc.*, Vol. 55, No. 11, November 1967, pp. 1787–1801.
[15] Pinel, J.F., M.L. Blostein, "Computer Techniques for the Frequency Analysis of Linear Electrical Networks," *IEEE Proc.*, Vol. 55, No. 11, November 1967, pp. 1810–1819.
[16] Idleman, T.E., et al., "SLIC—A Simulator for Linear Integrated Circuits," *IEEE J. Solid-State Circuits*, Vol. SC-6, No.4, August 1971, pp. 188–203.
[17] Manaaaaktala, V.K., "A Versatile Small Circuit-Analysis Program," *IEEE Trans. CT*, September 1973, pp. 583–586.
[18] Root, C.D., "Circus Means Versatility as a CAD Program," *Electronics*, February 2, 1970, pp. 86–96.
[19] Jenkins, F.S., S.P. Fan, "Time—A Nonlinear DC and Time-Domain Circuit Simulation Program," *IEEE J. Solid-State Circuits*, Vol. SC-6, No. 4, August 1971, pp. 182–188.
[20] Chinichiaan, M., C.T. Fulton, "Software for Closed-Form Solutions of Linear Time-Invariant Differential Systems," *IEEE Circuits Dev. Magazine*, May 1986, pp. 25–32.
[21] Schreiber, H.H., "Digital Simulation of Analysis Systems for a Restricted Class of Inputs," *IEEE Circuits Syst. Magazine*, December 1982, pp. 4–9.
[22] Saunders, J.H., "Place 2.0- An Interactive Program for PLL Analysis and Design," *AT&T Tech. J.*, Vol. 64, No. 5, May–June 1985, pp. 1101–1133.
[23] Press, W.H., et al., *Numerical Recipes in C*, 2nd ed., Cambridge University Press, New York, 1990.
[24] Dorf, R.C., *Modern Control Systems*, 2nd ed., Addison-Wesley Publishing, Reading, MA, 1974, p. 384.
[25] Manassewitsch, V., *Frequency Synthesizers Theory and Design*, 2nd ed., John Wiley & Sons, New York, 1980.
[26] Daruvala, D.J., "Unify Two-Port Calculations with a Single Analysis Technique—the Indefinite Matrix," *Electronic Design*, Vol. 1, January 1974, pp. 112–116.
[27] Gerald, C.F., P.O. Wheatley, *Applied Numerical Analysis*, 3rd ed., Addison-Wesley Publishing, Reading, MA, 1970.
[28] Gray, P.R., R.G. Meyer, *Analysis and Design of Analog Integrated Circuits*, 2nd ed., John Wiley & Sons, New York, 1984.
[29] Rohrer, R., et al., "Computationally Efficient Electronic-Circuit Noise Calculations," *IEEE J. Solid-State Circuits*, Vol. SC-6, No. 4, August 1971, pp. 204–213.

SELECTED BIBLIOGRAPHY

"1620 Electronic Circuit Analysis Program [ECAP][1620-EE-02X] User's Manual," IBM Application Program File H20–0170–1, 1965.

Chapter 9
Fractional-N Frequency Synthesis

Fractional-N frequency synthesis was conceived to circumvent the long-standing limitation that constrained indirect synthesis methods to integral multiples of the reference frequency employed. In principle, fractional-N synthesis loops also permit frequency switching speeds that are considerably faster than their traditional indirect loop counterparts. The trade-offs involved in obtaining this added flexibility include potentially poorer spurious and phase noise performance, and increased circuit complexity. The fractional-N concept is closely related to other techniques that have appeared in the literature including Digiphase [1].

As the name implies, fractional-N techniques can be used to modify classical indirect phase-locked loops in such a way as to effectively permit the feedback divider ratio N to take on noninteger values. For the sake of the discussions here, assume that the nonintegral divider ratio is represented by the quantity $N.F$ where N and F are the integer and fractional portions, respectively. A hypothetical phase-locked loop (PLL) that facilitates this discussion is shown in Figure 9.1 where an ideal voltage-controlled oscillator (VCO) and linear phase detector have been assumed. For simplicity, we will assume that the PLL is a time-continuous system. If the VCO is operating at a frequency of precisely $N F_{\text{ref}}$, the phase detector output will be a dc voltage with a value that is proportional to the phase difference between $\theta_o(t)/N$ and $\theta_r(t)$, which can be taken as zero with no loss of generality. If instead the VCO output frequency is now changed to $N.F\, F_{\text{ref}}$ while maintaining the feedback divider ratio as N, the phase detector phase error takes on a sawtooth shape as shown in Figure 9.2 where the slope is given by

$$\frac{\Delta \theta}{\Delta t} = \frac{N.F\, F_{\text{ref}}\, T}{NT} 2\pi - \frac{2\pi}{T} F_{\text{ref}} T \qquad (9.1)$$

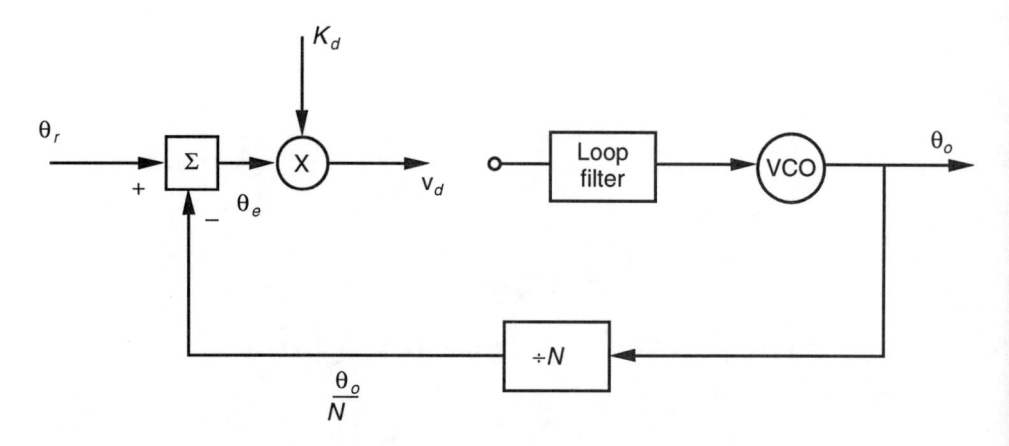

Figure 9.1. An ideal phase-locked loop may be used to visualize the fractional-N synthesis concept.

This may be simplified to give

$$\frac{\Delta \theta}{\Delta t} = \left[\frac{(2\pi F_{\text{ref}}\, 0.F)}{N} \right] \text{ modulo } 2\pi \qquad (9.2)$$

The so-called "beat note," shown in Figure 9.2 as a sawtooth, arises because the VCO output frequency is not an integral multiple of the reference frequency F_{ref}.

In simple terms, fractional-N synthesis is achieved by means of two (or more) integral feedback divider ratios with different duty cycles such that a weighted av-

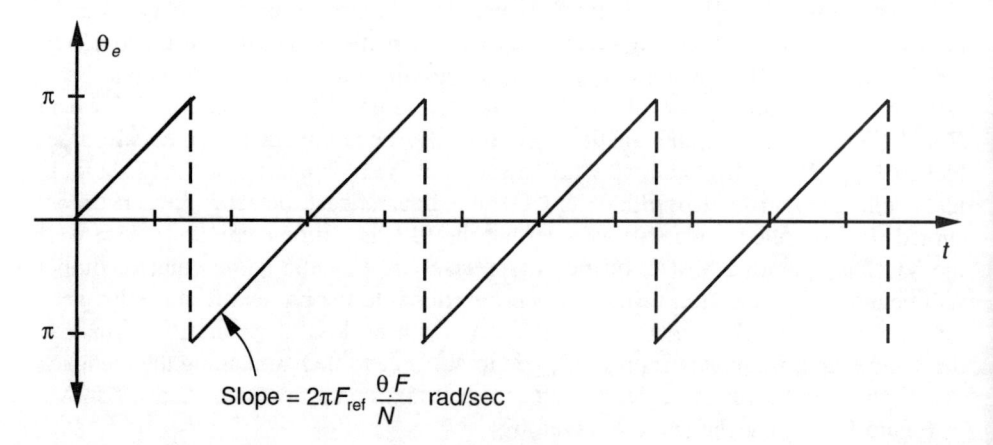

Slope = $2\pi F_{\text{ref}} \dfrac{0.F}{N}$ rad/sec

Figure 9.2. The phase detector output contamination or "beat note" takes the form of a sawtooth wave in fractional-N synthesis.

erage of the divider ratios is created. Since this average is formed over more than one reference time period, this gives rise to the undesired beat note phenomenon mentioned above.

In order to illustrate this perspective further, assume that the feedback divider ratio is N for n reference periods and $(N + 1)$ for m reference periods. The phase-locked loop although contaminated with the beat note presence attempts to operate in a normal fashion regardless. Over the $(n + m)$ reference periods, the average output frequency is therefore just the weighted average of the two synthesized output frequencies $N F_{ref}$ and $(N + 1) F_{ref}$ and is given by $N F_{ref} + F_{ref} m/(n + m)$. This is clearly a powerful synthesis technique provided that the spurious effects of the fractional-N beat note can be effectively eliminated.

The basic phase-locked loop shown in Figure 9.1 may be modified in order to eliminate the phase detector beat note output by including a phase error function that precisely cancels this contaminant as shown in Figure 9.3. Since the beat note phase error is completely deterministic based on the fractional portion being synthesized, in principle this cancellation approach is very straightforward. In actuality, these are only idealistic representations and actual circuit implementations are considerably more complex.

Another view of fractional-N synthesis that may be helpful is to visualize the $\div N$ function as a dual-modulo counter that divides the VCO output frequency by N until the relative phase error at the VCO equals or exceeds 2π radians with respect

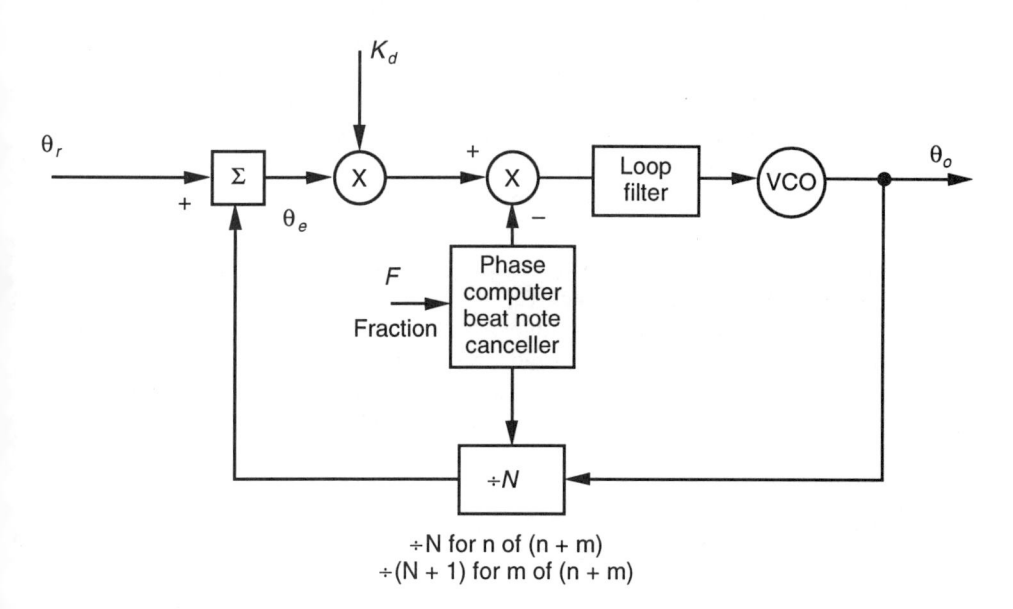

Figure 9.3. Conceptual cancellation of the beat note in fractional-N synthesis.

to $N\ F_{\text{ref}}$. At this instant, the $\div N$ ratio is changed to $N + 1$ for the next reference period in order to remove 2π radians of the additional phase that had accumulated. From this perspective, the maximum phase error ever seen at the phase detector is $2\pi/N$ radians.

Although a clear deviation from normal practice, nothing prevents us from allowing the extra VCO phase to accumulate to $2\pi P$ radians before P additional divider counts are removed by altering the feedback divider ratio. In this case, the maximum phase error seen at the phase detector is increased to $2\pi P/N$ radians. This concept may be useful for reducing the effects of nonideal beat note cancellation in some systems since it results in the fractional-N spurious frequencies being reduced by a factor of P. For example, under normal conditions with $P = 1$, assume that the first fractional-N spur occurs at a particularly sensitive offset frequency of 5 kHz due to nonideal cancellation. By making $P = 16$, the first spurious sideband could be moved to 5 kHz/16 or 312.5 Hz, which may well eliminate the problem. Although certainly adding complexity, P values could be randomly selected from a predetermined set of values to further destroy the coherence of any nonideal cancellation errors present. As briefly discussed later in this chapter, dithering techniques in use with modern direct digital synthesizer (DDS) technology may be effectively used to eliminate coherence in the residual error without adding complication to the analog portions of the fractional-N correction circuitry. In actual practice, the nonideal cancellation factors should be weighed carefully with respect to the system requirements before any complicated measures for dealing with them are implemented. More recently, completely digital methods are beginning to simplify the beat note cancellation problem, as discussed in Chapter 7. Only the more traditional fractional-N techniques are discussed within this chapter.

Thus far, we have only been talking in terms of fractional-N synthesis where the $N.F$ notation is explicit. The general fractional-N concept is shown in Figure 9.4. In this implementation, only the fractional portion (N_2) of the divide ratio is accumulated and, once an accumulator carry occurs, an additional VCO count (equivalent to 2π radians of phase) is removed thus effectively removing at least a portion of the excess phase that has accumulated [2].

The fractional-N concept is intuitively comfortable for most frequency synthesizer designers because they are generally familiar with dual-modulus counters in the context of swallow-counters [2]. By means of, for instance, a dual-modulus counter with moduli 10 and 11, it is possible to divide the VCO output by 103 by dividing by 11 three times and dividing by 10 seven times within the confines of one reference time period. Since these divisions all occur within one reference period, the effective divide-by-N ratio is 103 and no beat note is observed. If, on the other hand, we had divided by 11 three times followed by a division by 10 seven times and this had been performed within the confines of two reference time periods rather than one, the average divide-by-N ratio would have been $(3 \times 11 + 7 \times 10)/2 = 51.5$ and, in effect, fractional division has been performed. Generalizing, the divider action in

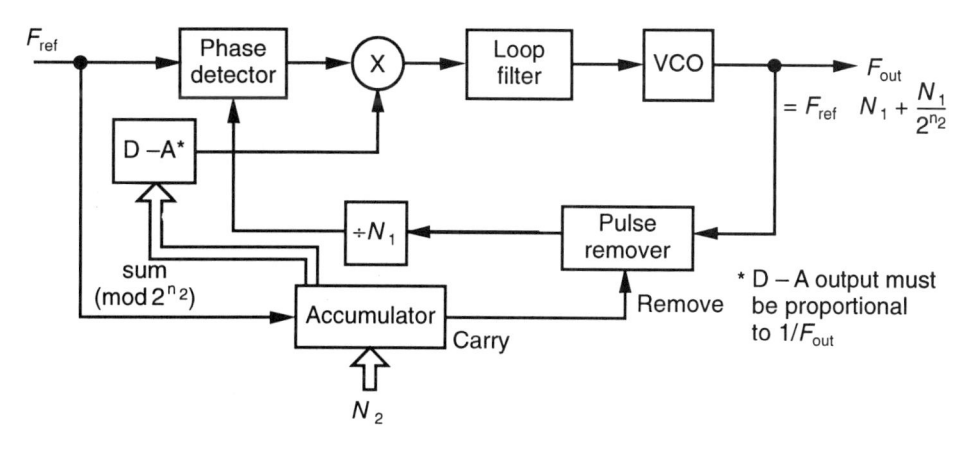

Figure 9.4. Fractional-N concept. (Copyright © 1981 John Wiley & Sons, Inc. Reprinted by permission [2].)

fractional-N may be thought of as a traditional swallow-counter where the period of operation extends over more than one reference time period.

A second approach, which is closely related to fractional-N synthesis and is really responsible for introducing this synthesis method in general, is called Digiphase® [1]. This concept was first discussed at length by Gillette [3] in a fairly classic paper. A generalized block diagram of this concept is provided in Figure 9.5. Compared to fractional-N synthesis, this approach accumulates both the integer and fractional portions of the frequency control word and therefore requires no explicit divide-by-N block. For RF frequency synthesis, the fractional-N approach has clear advantages in this respect since only the dual modulus prescaler must operate at the VCO clock frequency.

In a perhaps purer sense, fractional-N frequency synthesis may also be viewed as a PLL that incorporates a DDS to perform phase interpolation [4]. Used in this manner, no trigonometric sine lookup table is required within the DDS nor is an anti-aliasing filter required. This interpretation is shown in Figure 9.6. It is this same interpretation that will be drawn on for calculation of fractional-N spurious performance later in this chapter.

Phase interpolation can be implemented using a precision variable time delay or variable phase shifter as conceptually suggested in Figure 9.7 [4]. A single-sideband mixer can also perform the function of a variable phase shifter as shown in Figure 9.8. The spurious performance of this approach is primarily driven by the performance limitations of the single-sideband (SSB) mixer. If the carrier leakage through the SSB mixer and unwanted sideband fall well outside the closed-loop bandwidth (including aliasing effects introduced by the feedback divider action), excellent spurious performance is attainable. Since the spurious frequency components that arise are again deterministic, avenues for further correction at the phase detector

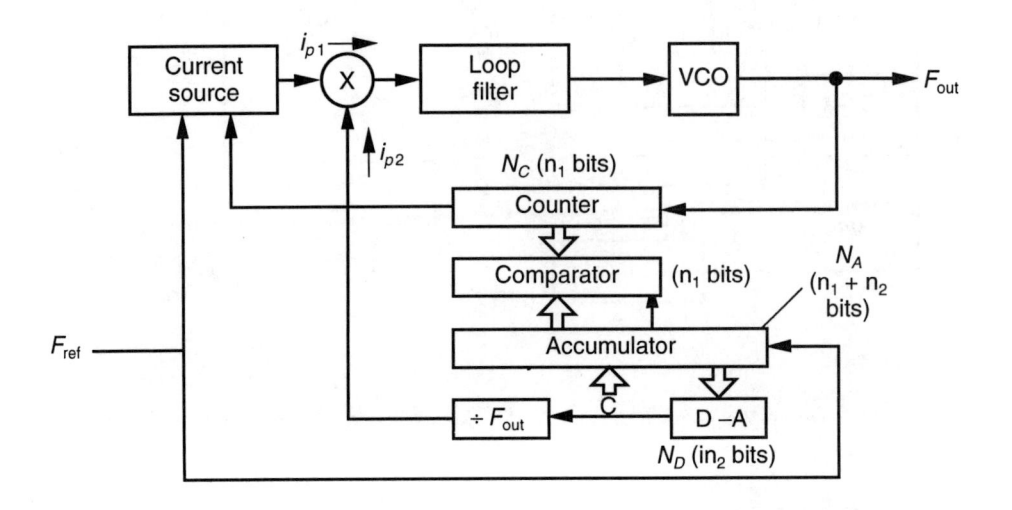

Figure 9.5. Digiphase concept. (Copyright © 1981 John Wiley & Sons, Inc. Reprinted by permission [2].)

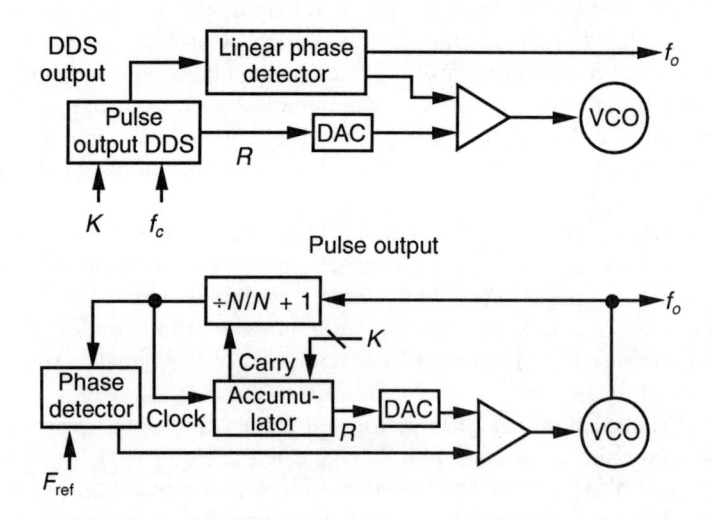

Figure 9.6. Phase-locked loop type of phase interpolation using DDS. (© 1986 IEEE; reprinted with permission). Source: [4].

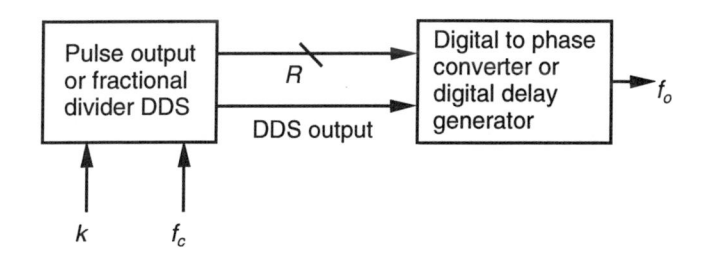

Figure 9.7. Direct output type of phase interpolation DDS. (© 1986 IEEE; reprinted with permission). Source: [4].

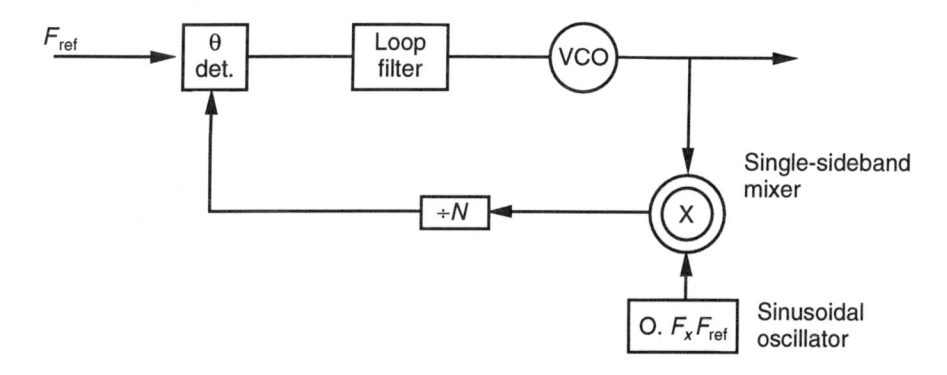

Figure 9.8. Phase interpolation using an SSB mixing approach.

using feedforward techniques can be considered. Additional references for these techniques and others are provided in [4].

The fractional-N concept may also be used without performing any phase error compensation provided that the beat note frequencies are well outside the closed-loop bandwidth where they may be easily filtered. Fractional parts such as 1/2 and 1/3 would normally qualify in this context, whereas relatively prime factors would still result in close-in spurious components that generally cannot be filtered out.

9.1 SOME PROS AND CONS BEHIND FRACTIONAL-N SYNTHESIS

Researchers over the years have attempted to devise methods for frequency synthesis that are more flexible than the standard PLL where the output frequencies are restricted to be integer multiples of the reference frequency [3–7]. This is in large part the motivation behind modern DDS in which the output frequencies may be represented by

$$f_o = \frac{k}{2^M} F_{\text{ref}} \qquad (9.3)$$

where M is the accumulator width in bits and k is representative of the frequency control word and $0 < k < 2^{M-1}$.

In most frequency synthesis cases involved in RF communications, the DDS output frequency span must be translated to a higher frequency range using other techniques in order to get sufficient spurious performance and bandwidth. In this respect, fractional-N combines the best features of the DDS and indirect PLLs into one package. In addition, the fractional-N approach does not require an additional output anti-aliasing filter, unlike the DDS since the loop filter and VCO double in this capacity. Due to this somewhat higher degree of integration, lower total power consumption is also more obtainable with fractional-N synthesis.

The fractional-N approach is not without its disadvantages and limitations, however. In general, fractional-N loops are considerably more tedious to design and implement, particularly unless gate arrays of some kind are utilized. The worst case raw spurious performance is generally limited to roughly -70 dBc unless special techniques (e.g., multiple reference frequencies utilized) are additionally employed. Compared to the DDS, only phase modulation well within the closed-loop bandwidth is possible and the output is strictly constant envelope (i.e., no amplitude modulation).

Fractional-N synthesis is a staple ingredient in Hewlett-Packard frequency synthesis products [5,8–11]. The HP8662 synthesizer incorporates both corrected and uncorrected fractional-N loops within its architecture [5]. Other manufacturers are also using the advantages afforded by fractional-N synthesis [3,12].

Fractional-N techniques combined with modern device technologies can be used to create the ultimate in monolithic pseudo-DDS building blocks. One semiconductor manufacturer is planning to integrate several novel phase interpolation concepts with their 20-GHz process to realize a single-chip monolithic synthesis block that spans the range of 5 to 10 MHz with a step size <1 Hz, spurious < -80 dBc, and close-in phase noise < -150 dBc/Hz. The fact that PLL-based techniques do not require an additional output anti-aliasing filter is a very important consideration in achieving high functionality within a single monolithic device. A simplified block diagram of this device is shown in Figure 9.9. Since the large output division factor of 128 reduces the VCO phase noise and spurious components 42 dB, the rather poor phase noise performance of the on-chip VCO is quite adequate and the open-loop phase interpolation technique must provide only 40 dB of spurious reduction in order to realize -80-dBc spurs. Proprietary techniques within the phase interpolation block permit this architecture to use a high reference frequency (5 MHz), which further disperses the residual phase noise arising in the interpolation process over an appreciably large bandwidth compared to most fractional-N implementations.

The fractional-N frequency synthesis concept has also been used for military applications such as the Hughes PRC-104 HF manpack radio with good success. In

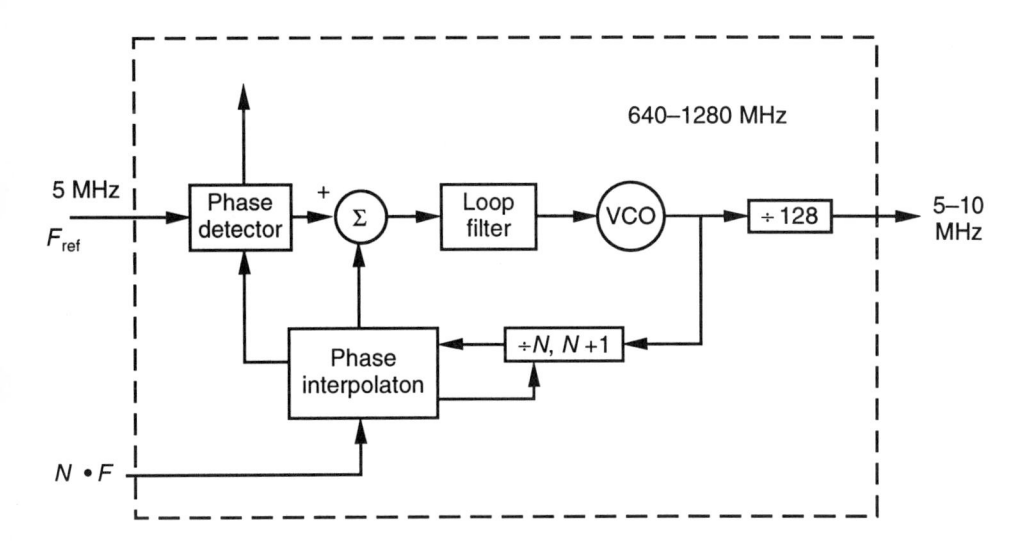

Figure 9.9. Top-level block diagram for the monolithic synthesis component described in the text.

this case, a complete fractional-N loop covering 77 to 105 MHz in 100-Hz steps was achieved that consumed only 1.8 W of power, achieving less than -65-dBc spurious noise, with a switching time of ≤ 0.4 ms.

Through the remainder of this chapter, the fractional-N discussions will largely parallel the architecture used in the HP3325 function generator [8,11]. We first consider some of the high-level system issues involved with fractional-N synthesis such as the degree of beat note cancellation required at the phase detector. The remainder of the chapter addresses mainly some of the many detailed implementation issues involved with fractional-N.

9.2 FRACTIONAL-N SYNTHESIS: GENERAL CONCEPTS

A number of detailed discussions of the fractional-N concept have appeared in the literature previously [2,5,8,13]. The discussions we pursue in this section are based on the block diagram shown in Figure 9.10 [8].

As mentioned earlier, the primary feature that distinguishes the fractional-N concept from the Digiphase® concept is that only the fractional portion of the divide ratio is accumulated in fractional-N synthesis.

For the time being, the analog phase interpolation (API) circuitry shown in Figure 9.10 will be disregarded, which is equivalent to setting the fractional portion of $N.F$ to zero. The phase detector is assumed to be an ultralinear first-order sample-and-hold type in which a linear voltage ramp is created by integrating a precision

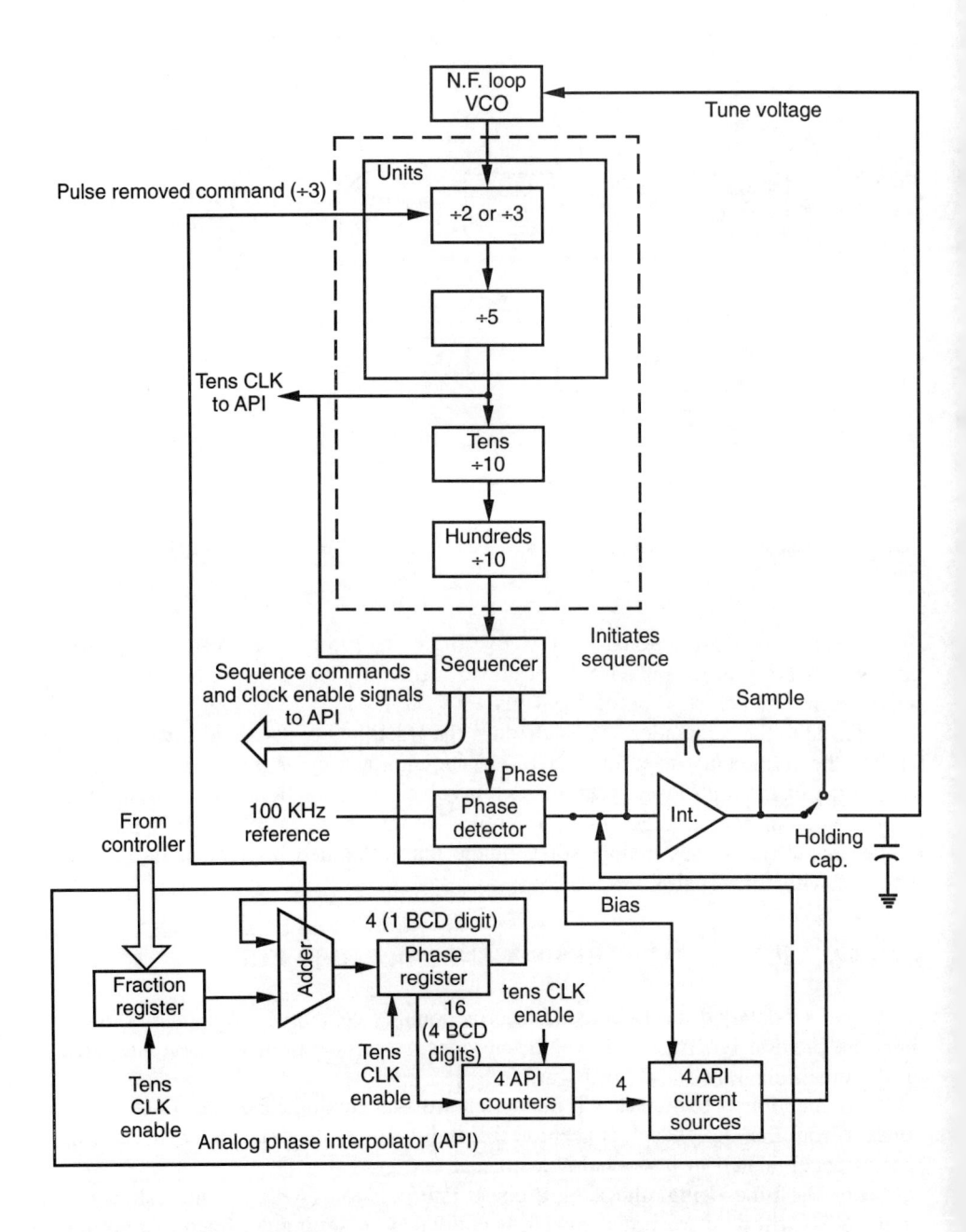

Figure 9.10. Basic block diagram for a fractional-N PLL. (Source: [17]. Figure 3–19. Reprinted with permission.)

reference current I_r in a capacitor C. Assuming that the reference frequency is $F_{ref} = T^{-1}$, the phase detector gain K_d is given by

$$K_d = \frac{I_r T}{C 2\pi}, \quad \text{V/rad} \tag{9.4}$$

Following the same line of discussion as earlier, now assume that the fractional portion of the frequency control word F is nonzero and that the VCO is operating at precisely $N.F\ F_{ref}$. Since the VCO is operating as a slightly higher frequency than $N.F_{ref}$, the feedback divider will accumulate its $2\pi N$ radians of phase causing an output in slightly less than one reference period. To examine this further, let $N.F$ be represented as

$$N.F = N_1 + \frac{N_2}{2^M} \tag{9.5}$$

where M is the length of the phase register in Figure 9.9. The feedback divider output precedes the reference clock output by an amount Δt each reference period where

$$\Delta t = \frac{1}{F_{ref}} - \frac{N_1}{F_{VCO}} \tag{9.6}$$

where F_{VCO} is the VCO frequency, which is given as $N.F\ F_{ref}$. As developed in [2], this time increment may be expressed equivalently as

$$\Delta t = \frac{1}{F_{VCO}} \left(\frac{F_{VCO}}{F_{ref}} - N_1 \right)$$

$$= \frac{1}{F_{VCO}} \left[\left(N_1 + \frac{N_2}{2^M} \right) - N_1 \right]$$

$$= \frac{N_2}{2^M} \frac{1}{F_{VCO}} \tag{9.7}$$

which is well suited for hardware implementation as developed shortly. The steady decrease in the phase detector output voltage that results if the no beat note cancellation is provided is shown graphically in Figure 9.11.

In order to make the general fractional-N concept viable, the steadily decreasing phase detector output voltage must be compensated for in order to remove the sawtooth component. Each reference period, the sample-and-hold output voltage is reduced an additional voltage amount ΔV, which may be equivalenced to a charge ΔQ

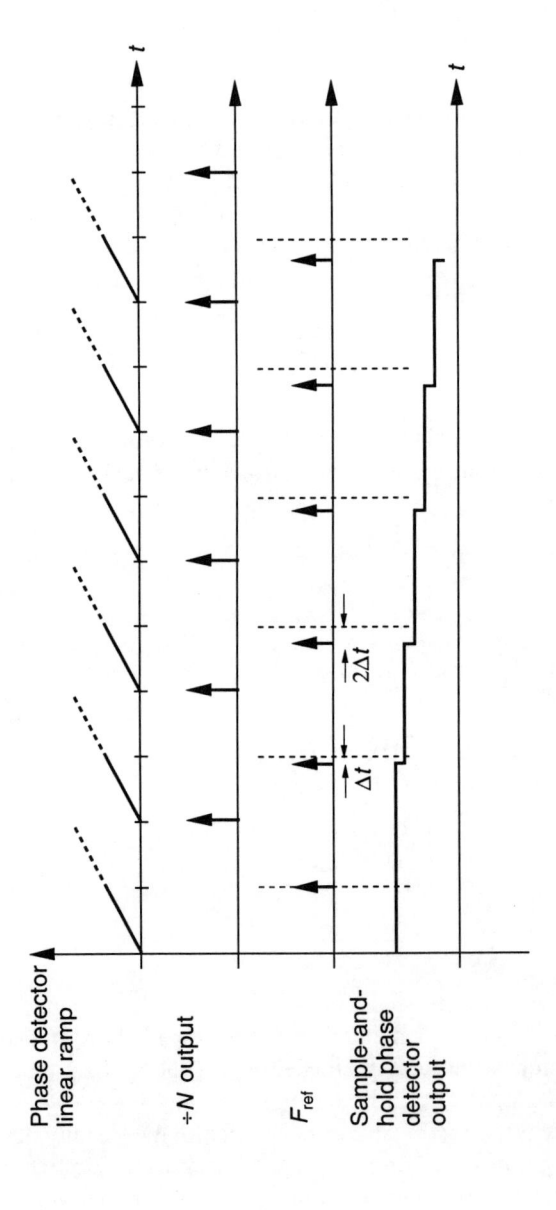

Figure 9.11. The fractional-N beat note is manifested as a sawtooth error at the phase detector output.

missing in the current integration process that is used to form the linear ramp. This charge increment, which must be gated into the current integration process for each ΔV, is given by

$$\Delta Q = I_r \Delta t \qquad (9.8)$$

This same charge correction may also be performed by directing a different precision current source I_p of proper polarity into the current integration process for a time period T_c as long as

$$\Delta Q = I_p T_c = \Delta t I_r \qquad (9.9)$$

Since Δt is inversely proportional to the output frequency F_{VCO}, which of course changes with $N.F$, T_c must also change proportionally with Δt. This is most easily accomplished by making

$$T_c = \frac{M}{F_{\text{VCO}}} \qquad (9.10)$$

where M is an integer constant.

From (9.9), it should be clear that any combination of I_p and T_c values may be selected as long as the overall relationship for ΔQ is satisfied. For implementation simplicity, normally T_c is fixed for a given $N.F$ ratio and either binary-coded-decimal (BCD) or decade-weighted current sources are summed to form I_p. This is precisely the role of the four analog current sources shown in Figure 9.10.

If we assume that a decade-based phase register is used in Figure 9.10, the required total charge correction at reference index m may be implemented as

$$\begin{aligned}\Delta Q(m) &= \left(I_c p_3 + \frac{I_c}{10} p_2 + \frac{I_c}{100} p_1 + \frac{I_c}{1000} p_0\right) \frac{M}{F_{\text{VCO}}} \\ &= I_c \left(p_3 + \frac{p_2}{10} + \frac{p_1}{100} + \frac{p_0}{1000}\right) \frac{M}{F_{\text{VCO}}} \end{aligned} \qquad (9.11)$$

where I_c is the precision source and (p_3, p_2, p_1, p_0) represents the phase register value in decade form at time index m. Examining (9.11) further, in effect at index m, a current source of magnitude I_c is gated into the charge integrator for $p_3 M$ cycles of the VCO frequency. Add to this a second precision current source of magnitude $I_c/10$, which is simultaneously gated into the same charge integrator for $p_2 M$ cycles of the VCO, and so on. Schematically, this may represented as shown in Figure 9.12. The comparator implementation for gating the current sources on and off in this figure was first conceived by Fountain [14]. If the range of $N.F$ values used is

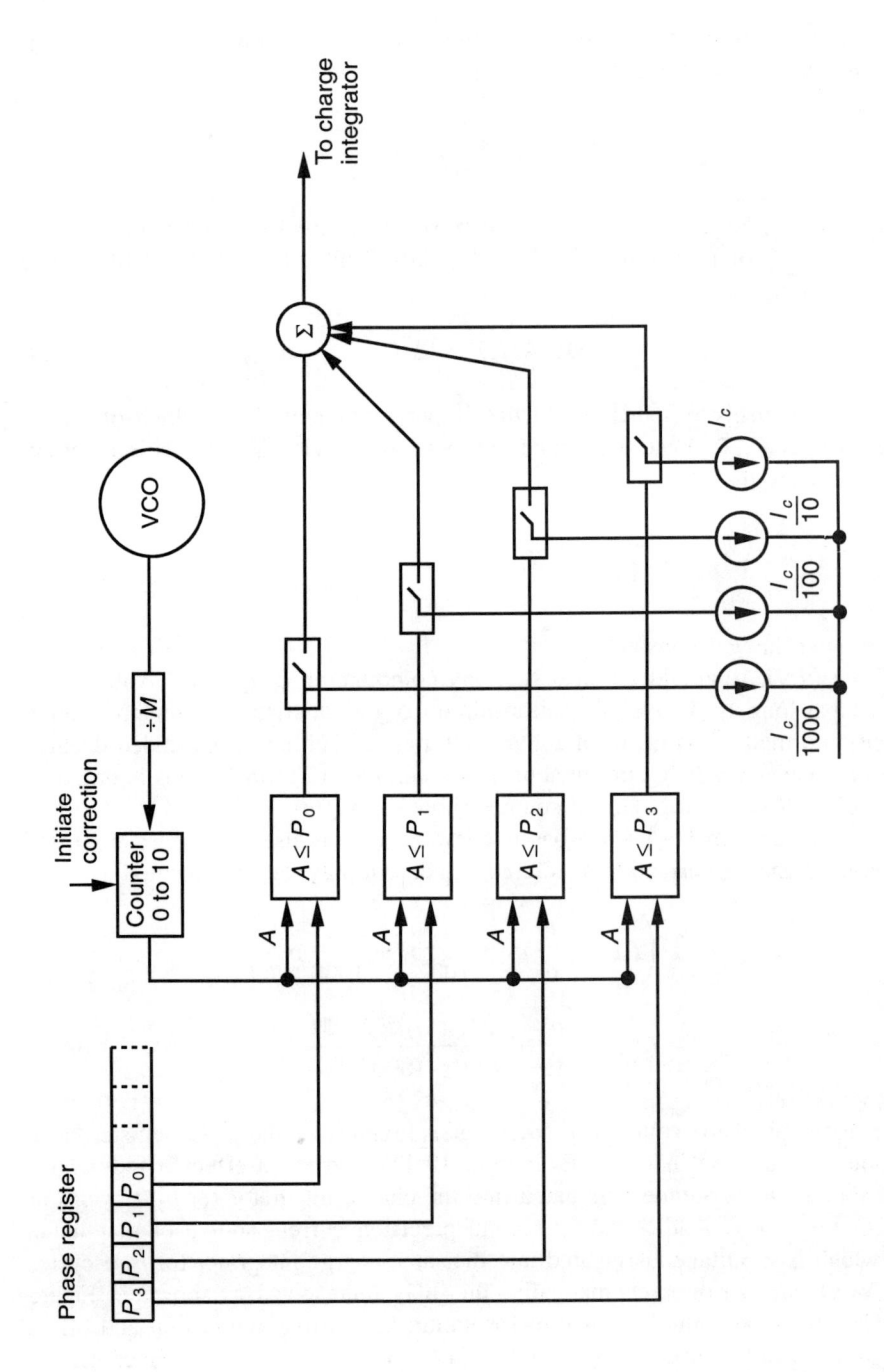

Figure 9.12. Fractional-N beat note cancellation.

substantially restricted, the required compensation can be performed quite accurately with a simple digital-to-analog (D/A) converter as advocated in [6].

9.3 FRACTIONAL-N SYSTEMS ASPECTS

The systematic issues involved with fractional-N frequency synthesis are rarely dealt with in the open literature. Some effort will be expended here to deal with a number of systems issues as well as propose several alternative viewpoints that may be helpful in specific design situations.

9.3.1 Spectral Performance: Discrete Spurious

Discrete spurious contaminants in the fractional-N synthesizer output result primarily from nonideal beat note cancellation. Although a general methodology has been discussed that addresses how this cancellation may be performed, no guidelines have been presented that outline to what degree this beat note cancellation must be performed in order to attain a specific degree of spectral purity. This issue is addressed in this section.

Unlike the DDS spectrum examined in Chapter 7, the fractional-N output spectrum is completely void of any AM spurious components since the VCO output is assumed to be constant envelope. The discrete spurious signals that do occur can only be a result of either (1) phase accumulator truncation in applying the beat note correction or (2) nonideal cancellation effects resulting from nonlinearities or imprecision in applying the correction.

Based on the following helpful viewpoint [15], it can be shown that the required fractional-N beat note cancellation is fairly independent of the particular fractional portion being synthesized. Assume first that a fractional part of 0.1 is to be synthesized, which results in a fundamental beat note frequency of 0.1 F_{ref}. Ignoring any filtering action by the phase-locked loop, the beat note that must be cancelled at the phase detector is as shown in Figure 9.13 where the phase error is referenced to the VCO output phase. The situation is quite similar for a fractional part of 0.9 as shown in Figure 9.14. In this case, the beat note frequency would be expected to occur at 0.9 F_{ref}, but due to aliasing about harmonics of F_{ref}, the beat note fundamental appears at $(1 - 0.9)F_{ref}$ instead. These waveforms are essentially identical to the signals observed in Chapter 7.

It is insightful to take the beat note sample values from Figure 9.13 and perform a discrete Fourier transform (DFT) in order to examine the spectral makeup of the beat note. Given the convenient sawtooth waveform in Figure 9.13, the magnitude of the Fourier components are given by $C(k) = 2/k$ rad as shown in Figure 9.15. The fundamental beat note frequency has the largest Fourier coefficient magnitude, which in this case is 2 rad.

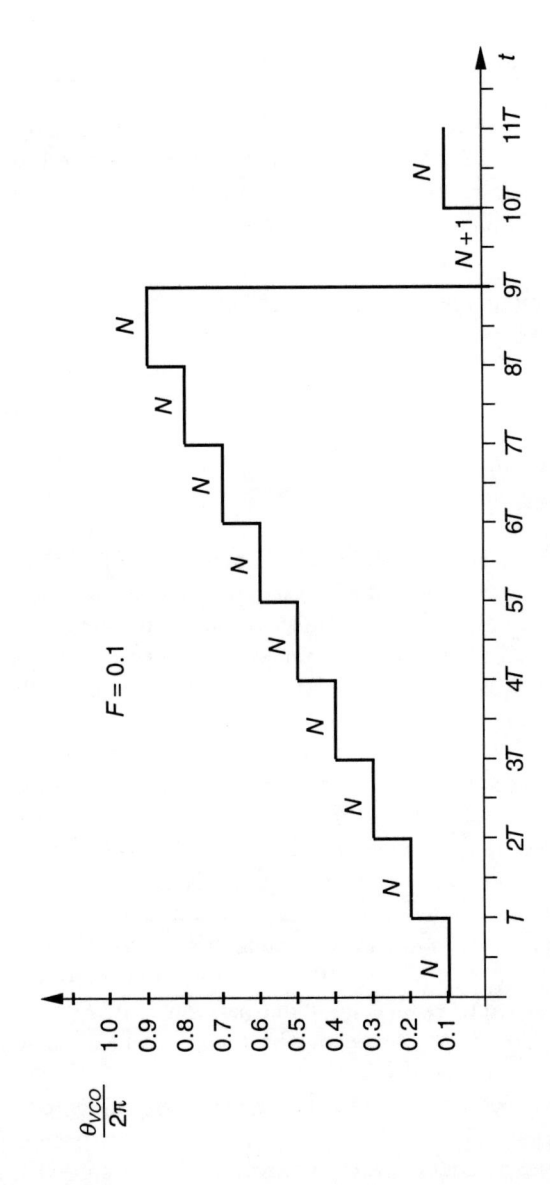

Figure 9.13. Fractional-N beat note for a fractional part of $0.1\ F_{\text{ref}}$.

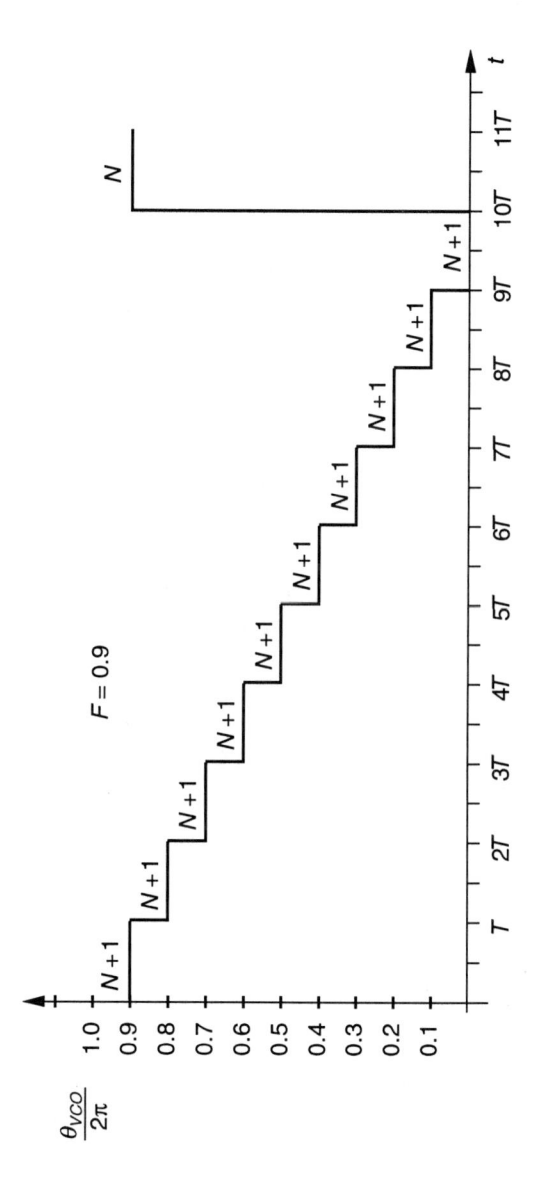

Figure 9.14. Fractional-N beat note for a fractional part of 0.9 F_{ref}.

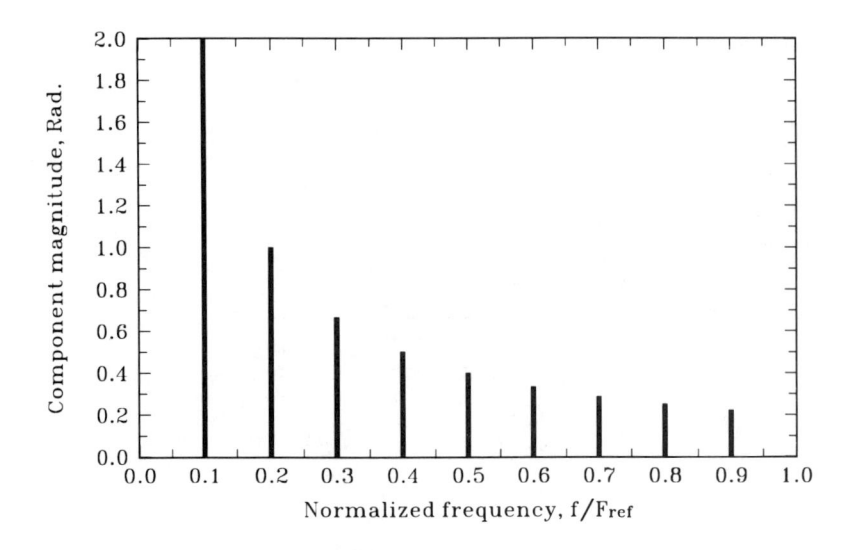

Figure 9.15. Beat note Fourier components for a fractional part of 0.1.

The DFT calculation may be performed for an arbitrary fractional portion such as $F = 0.3$. The time-domain error for this case is shown in Figure 9.16 whereas the DFT coefficient magnitudes are plotted in Figure 9.17. Once again, the largest DFT coefficient occurs for the fundamental beat note frequency having a magnitude of 1.743 rad in this case. It can be shown that under worst case conditions, the maximum $C(k)$ value for any fractional part F occurs at the fundamental beat note frequency with an approximate magnitude of 2 rad. Since the resulting discrete spur level given a sinusoidal phase modulation is given approximately by

$$L = 20 \log_{10}\left(\frac{\Delta\theta}{2}\right) \tag{9.12}$$

where $\Delta\theta$ is the peak phase deviation in radians, and the maximum phase deviation that results with fractional-N is 2 rad, we may conclude that there is a decibel-for-decibel relationship between the fractional-N beat note cancellation performed and the discrete output spurious levels observed. More specifically, if the residual phase detector beat note voltage after cancellation is reduced x dB with respect to its uncorrected form, the discrete spurious at the VCO output will be no higher than approximately $-x$ dBc. Based on this reasoning, only three decades of correction (60 dB) are normally implemented because further levels of correction would generally be masked by other nonideal implementation features.

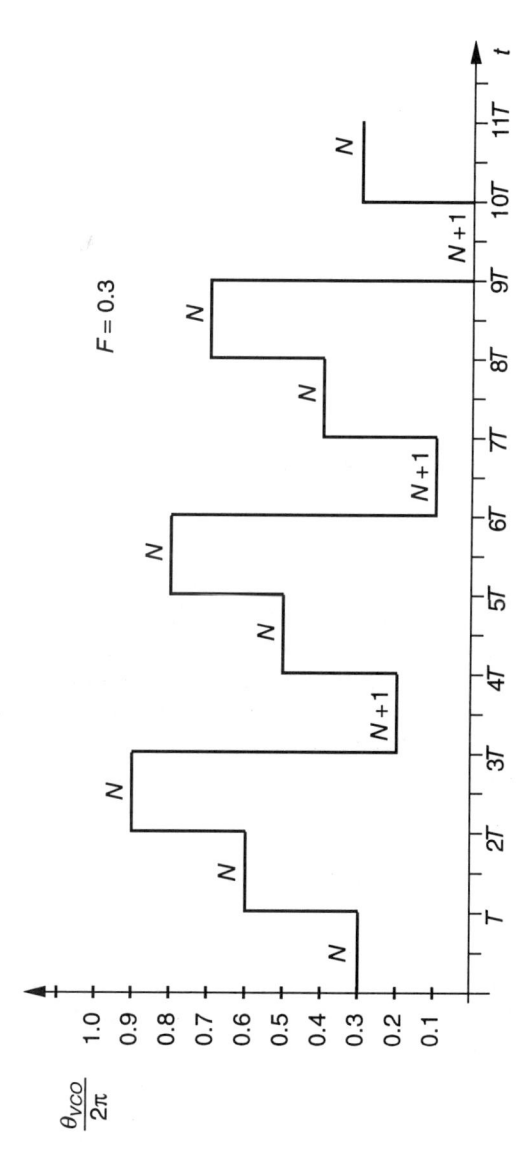

Figure 9.16. Fractional-N beat note for a fractional part of 0.3 F_{ref}.

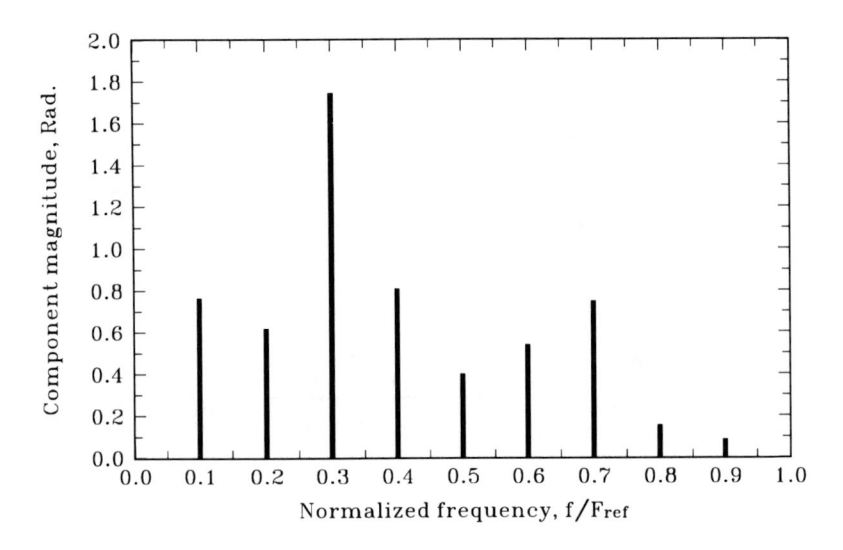

Figure 9.17. Beat note Fourier components for a fractional part of 0.3.

As mentioned earlier, since the phase-locked VCO output is by definition constant envelope, only PM discrete spurs can occur at the output. These spurious sidebands arise from (1) inaccurate beat note cancellation primarily in the analog electronics and (2) precision truncation in going from the phase accumulator to the analog correction. This latter factor is identical to the phase error truncation effects that occur in DDSs, and we can therefore apply the results of Chapter 7 to determine the location of these spurs. Generally, prediction of discrete spurs resulting from nonideal analog circuit performance must be done on a case-by-case basis.

9.3.2 Loop Parameter Limitations

Implementation of the fractional-N concept carries with it several limitations pertaining to the selection of loop parameters if the type of correction shown in Figure 9.11 is employed. Several of these limitations are best illustrated by considering the sample-and-hold phase detector signals shown in Figure 9.18.

Considering Figure 9.17 without fractional-N correction for the moment, clearly the loop must be designed such that after the divide-by-N pulse occurs, adequate time is left to (1) take an efficient sample of the linear ramp and (2) completely reset the linear ramp voltage prior to the beginning of the next reference period. Any residual charge left on the ramp generating capacitor from the previous reference period could otherwise seriously affect the fractional-N correction efforts. In a type

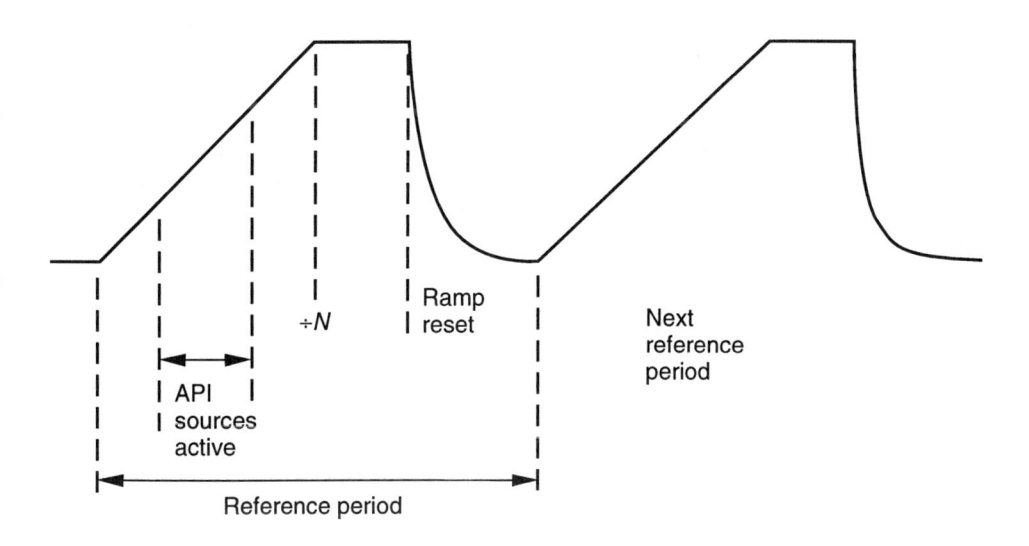

Figure 9.18. Fractional-N phase detector timing relationships.

2 system where the steady-state phase detector phase error is zero, it is often convenient to center the linear ramp about the steady-state operating point (i.e., ground), thereby providing a nearly symmetric phase error dynamic range. In type 1 systems, however, since the steady-state phase error is not zero, the operating point effectively slides up and down the linear ramp as a function of output frequency, thereby compressing the time available within each reference period for performing the two aforementioned steps. (If VCO presetting is used, this problem can be substantially reduced.)

The timing constraints depicted in Figure 9.17 are further complicated if fractional-N cancellation is also to be accommodated, particularly in type 1 systems. Since the API current gating period is inversely proportional to the VCO output frequency as shown in (9.10), unless different values of M are used in (9.11), the ratio of the minimum API correction window time width to the maximum is given by the ratio of the minimum VCO output frequency to maximum. This is an undesirable complication considering that normally it is best to reserve as much of the reference period as possible to perform the API correction.

9.3.3 An Alternative for Dealing With Truncation-Related Discrete Spurs

A direct correlation exists between the discrete spurs resulting from phase accumulator truncation in fractional-N and the similar phenomenon observed in DDSs as we mentioned earlier. Unlike in DDSs, however, the output spurious level from the fractional-N loop can also be substantially enhanced by the loop filter transfer function used within the PLL for spurious frequencies falling well outside the loop bandwidth. This degree of freedom is simply not available in the DDS.

If the worst case discrete spurious signals are due to primarily phase accumulator truncation error in applying the analog correction via the gated API current sources, noise-shaping techniques may be employed to improve the phase noise performance close to the carrier and to largely eliminate discrete spectral contaminants [1]. This technique is very effective for small fractional values that result in a synthesized cancellation waveform that is substantially oversampled. The oversampling effectively causes interpolated API corrections to be applied, thereby reducing the maximum peak phase error experienced by the system. Conceptually, this form of noise shaping is very similar to a fractional-N cancellation approach discussed in [2].

The noise-shaping concept is most easily discussed in the context of a DDS, the traditional architecture appearing as shown in Figure 9.19. First-order noise shaping may be added to this architecture as shown in Figure 9.20. A typical output phase noise spectrum for the modified architecture is shown in Figure 9.21. Since the spectrum is high pass in nature, the low-pass nature of the loop filter in a fractional-N loop can be very helpful in further enhancing the final output spectrum. Furthermore, if the oversampling rate is large for all components within the loop bandwidth, this approach could also be used to trade somewhat poorer phase noise performance for a reduced number of API sources in Figure 9.12 or eliminate them entirely.

Since fractional-N does not involve the use of a sine table compared to the DDS, application of noise-shaping theory is simplified compared to the DDS case. Second-order noise-shaping techniques are briefly discussed in [16] also.

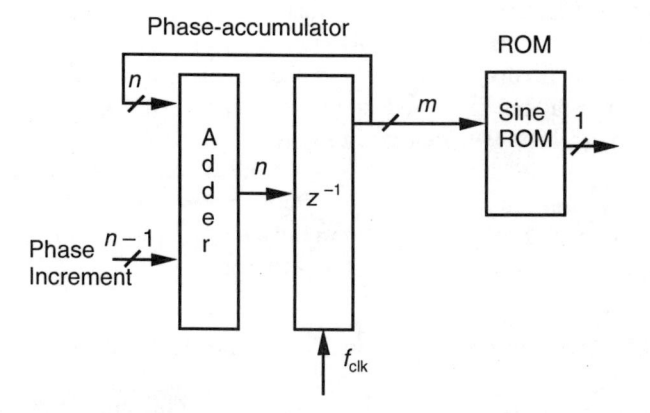

Figure 9.19. Conventional direct digital frequency synthesizer. (© 1991 IEEE; reprinted with permission). Source: [16], Figure 1.

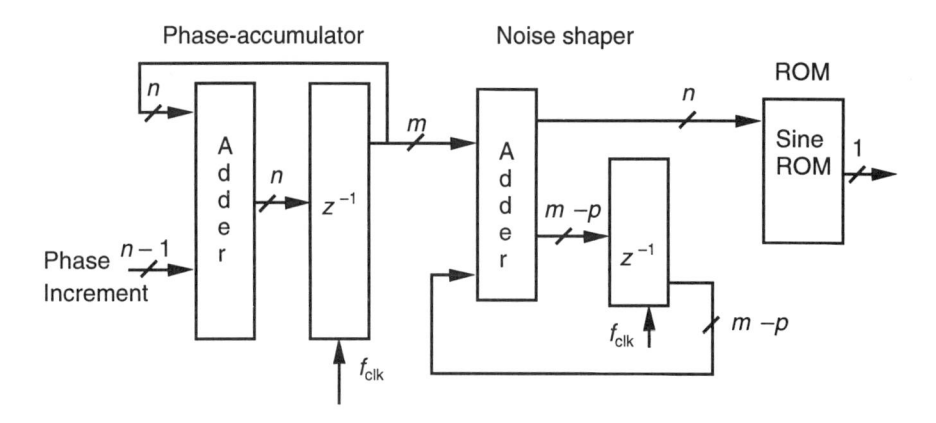

Figure 9.20. New direct digital frequency synthesizer architecture with noise shaping. (© 1991 IEEE; reprinted with permission). Source: [16], Figure 4.

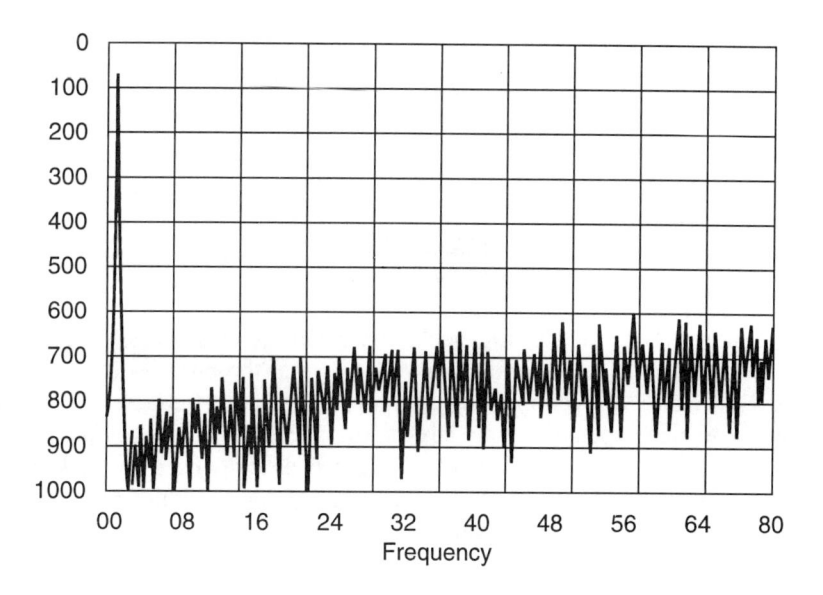

Figure 9.21. Representative output phase noise spectrum for the new synthesizer architecture. (© 1991 IEEE; reprinted with permission). Source: [16], Figure 5.

9.3.4 System Noise Floor Degradations Arising in Fractional-*N* Synthesis

The fractional-*N* output spectrum also contains a continuous noise spectrum contribution that arises due to nonideal beat note cancellation. Ideally, any residual cancellation errors lead to a completely random error sequence at the sample-and-hold phase detector output [13], in which case the effective noise contribution may be represented by

$$n(t) = \sum_{n=-\infty}^{\infty} a_n \sqcap (t - nT) \qquad (9.13)$$

where the a_n represent the random phase cancellation errors occurring during each reference period in radians and

$$\sqcap(t) = \begin{cases} 1 & \text{for } |t| < \dfrac{T}{2} \\ 0 & \text{otherwise} \end{cases}$$

In the case where the a_n are statically independent and mean-zero, the power spectral density corresponding to (9.12) is given by

$$\begin{aligned} S_n(f) &= \lim_{M \to \infty} \frac{1}{2MT} \mathbf{E}\{|\mathbf{F}[n_M(t)]|^2\} \\ &= \frac{\sigma^2}{F_{\text{ref}}} \left[\frac{\sin(\pi f/F_{\text{ref}})}{(\pi f/F_{\text{ref}})} \right]^2 \text{ rad}^2/\text{Hz} \end{aligned} \qquad (9.14)$$

where σ^2 is the variance of the a_k cancellation errors in squared radians.

The primary point to observe from (9.14) is that the phase noise floor due to imperfect fractional-*N* correction is lowered as the reference frequency is increased. It is therefore advantageous to use the largest loop reference frequency possible. The phase detector noise floor is effectively increased as evidenced by (9.13) due to nonideal fractional-*N* correction leading to potentially poorer output phase noise performance compared to a nonfractional loop.

9.3.5 API Gating Period Inaccuracies

Each reference period, the API current sources are gated on for a period of time given by (9.10), which is based on the VCO output frequency. During this time period, which must be reasonably small compared to a reference period, the VCO is essentially free-running. Therefore, VCO self-noise well outside the loop bandwidth can also contribute errors to the cancellation process. The time period in question is shown in Figure 9.22.

Due to the VCO self-noise, the time gating period may be slightly in error. The variance of the time gating error may be evaluated in terms of the VCO phase noise by examining the excess VCO phase accumulated over the maximum API gating interval, which is taken to be τ seconds by employing the relationship

$$\Delta t = \frac{\Delta \theta}{2\pi F_{\text{VCO}}} \tag{9.15}$$

where F_{VCO} is the nominal VCO frequency. The ratio of the time error Δt to the nominal gating period τ is given by

$$\frac{\Delta t}{\tau} = \frac{\Delta \theta}{2\pi M} \tag{9.16}$$

The variance of the accumulated phase $\Delta \theta$ may be computed from

$$\sigma^2 = \mathbf{E}\{[\theta(t - \tau) - \theta(t)]^2\} \tag{9.17}$$

where $\mathbf{E}$ denotes statistical expectation and the VCO phase noise process is assumed to be widesense stationary. Equation (9.16) may be further simplified to

$$\sigma^2 = 4 \int_{-\infty}^{\infty} S_\theta(f) \sin^2(\pi f \tau) \, df \tag{9.18}$$

Take the VCO phase noise spectrum to be low pass in nature and given as

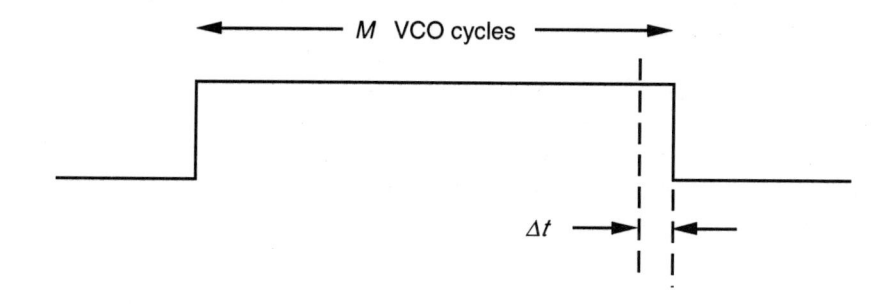

Figure 9.22. API current source gating period inaccuracies.

$$S_\theta(f) = \frac{\alpha}{1 + (f/f_c)^2} \quad \text{rad}^2/\text{Hz} \tag{9.19}$$

Substituting (9.18) into (9.17) finally gives

$$\sigma^2 = 4\alpha \int_{-\infty}^{\infty} \frac{\sin^2(\pi f \tau)}{1 + (f/f_c)^2} df$$

$$\text{var}\left(\frac{\Delta t}{\tau}\right) = \frac{\sigma^2}{(2\pi F_{\text{VCO}} \tau)^2} \tag{9.20}$$

Generally, this noise-induced gating accuracy is a secondary effect that is rarely considered. It should nonetheless be evaluated when either low reference frequencies or very noisy VCOs are used.

Another potentially more serious inaccuracy that arises in gating the API current sources involves the disparity between current switch turn-on and turn-off times. Since a device's turn-on time is in general unequal to its turn-off time, beat note cancellation errors result. The situation is illustrated further in Figure 9.23 where API gating periods of 0, $3T$, and $6T$ are illustrated.

Since the API current sources are integrated in order to form the beat note cancellation signal, the area under each curve is the quantity of principal interest. It should be clear from Figure 9.22 that precise cancellation results only if area $A_1 = A_2$. If $A_1 \neq A_2$, the cancellation signal is in error by a factor proportional to $|A_1 - A_2|$. This inaccuracy naturally becomes more pronounced as high reference frequencies are used with the same device technology because the turn-on and turn-off times become an appreciable fraction of the total possible API time gating period.

The turn-on/turn-off asymmetry problem with the API current sources can be virtually eliminated by a very simple technique. The problem with forming the cancellation signal as shown in Figure 9.23 stems from the desire to synthesize a zero cancellation value (top trace Figure 9.23). If instead at each reference period we synthesize a correction quantity $B + C(t)$ where B is a fixed dc offset and $C(t)$ is the previously used cancellation signal, all of the API current sources may be gated on for one T interval in Figure 9.23 every reference period. Since the (potentially unequal) areas A_1 and A_2 are now present for all of the API sources every reference period, even though the cancellation signal has a small dc bias, the incremental beat note cancellation signal is precise. In the context of (9.11), the charge offset producing the B bias is simply given by

$$I_c\left(1 + \frac{1}{10} + \frac{1}{100} + \frac{1}{1000}\right)\frac{M}{F_{\text{VCO}}}$$

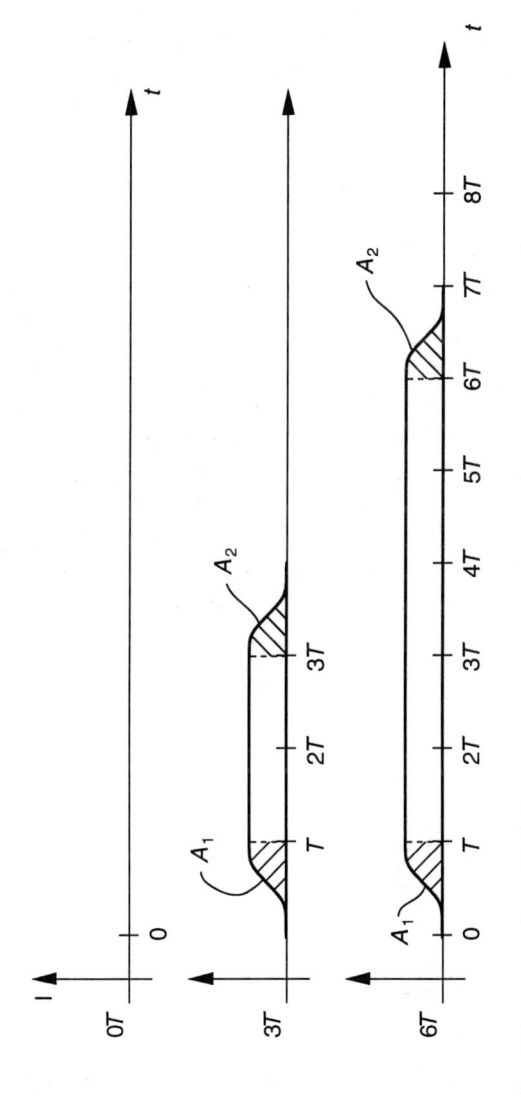

Figure 9.23. API source gating with finite current gate switching speed.

The desired bias may be implemented by incrementing each p_i before applying them to the API control circuitry shown in Figure 9.12 or by simply modifying the magnitude comparators slightly.

Modification of the API current source gating in this manner should make it possible to utilize the same current switching technology at considerably higher reference frequencies, thereby obtaining better phase noise performance. In general, with this technique, CMOS technology should be adequate for the API gating circuitry for reference frequencies well into the low-megahertz regime.

9.3.6 Phase Detector Ramp Nonlinearities

The need for thorough analog design for fractional-N frequency synthesis cannot be understated because many factors can cause degraded performance if not anticipated. Nonlinearities in the phase detector ramp signal can, for instance, contribute to substantial beat note cancellation errors as is now briefly explored.

In high-performance fractional-N synthesizers such as the HP3325, the linear phase detector voltage ramp is formed by integrating a precision current source in an operational-amplifier configuration such as that shown in Figure 9.24. In an effort to simplify the circuitry as well as build on more traditional sample-and-hold phase detector designs, at least one company has designed the linear ramp forming circuitry as shown in Figure 9.25. In this case, the phase detector ramp voltage is formed by integrating a precision current source directly in a holding capacitor. As shown momentarily, the finite output impedance of the current source and input resistance of

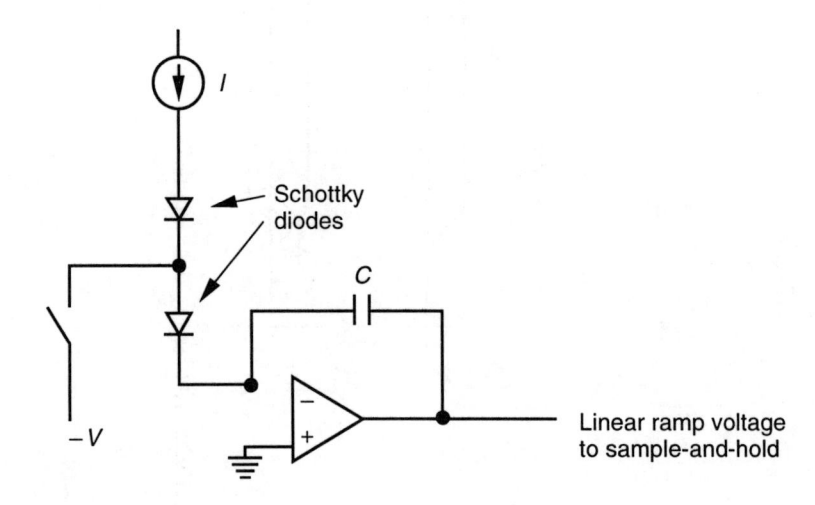

Figure 9.24. Linear ramp voltage for sample-and-hold phase detector input.

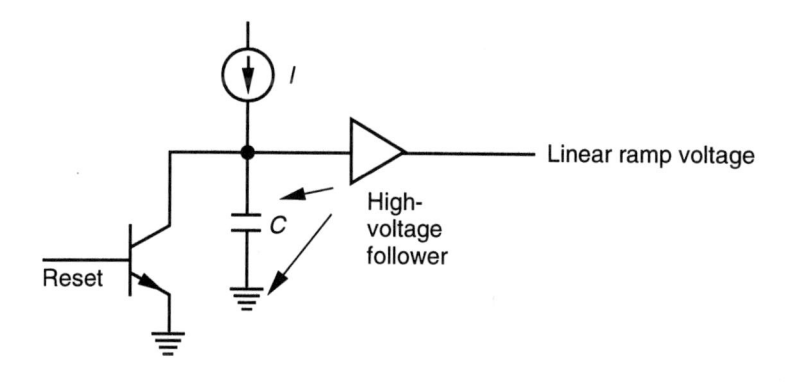

Figure 9.25. Linear ramp voltage generation using a simplified ramp generator.

the voltage follower in Figure 9.25 easily results in sufficient phase detector ramp curvature so as to cause a severe impact in beat note cancellation performance.

In order to study the impact of these finite impedances on the ramp linearity, consider the simple phase detector ramp circuit model shown in Figure 9.26. Due to the finite output impedance of the source, the linear phase detector ramp is instead exponential as depicted in Figure 9.27.

Due to the exponential decay of charge with time on the holding capacitor, additional charge ΔQ intended to cancel the fractional-N beat note is not faithfully held until the time sampling instant.

Since the beat note induced error voltage and fractional-N applied correction are both affected by this ramp curvature problem, the residual error with fractional-N correction applied is a bit more complex than might be expected. Assume for purposes of illustration that the fractional portion to be synthesized is 0.001. Under ideal operation, the beat note induced phase error is a simple linear phase ramp. Due to the finite source output impedance, however, the observed beat note voltage appearing at the phase detector output is instead given by

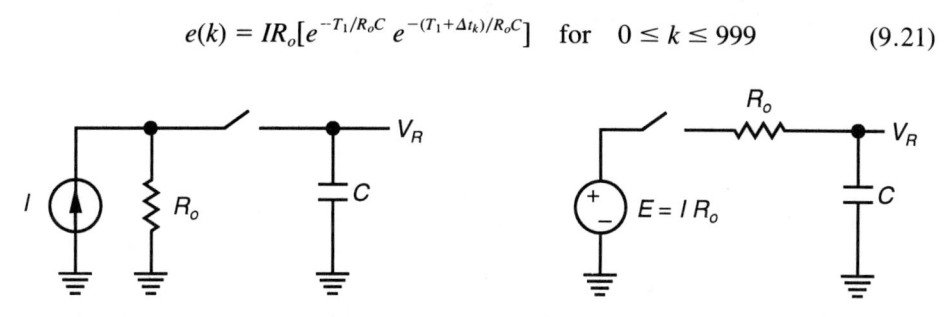

$$e(k) = IR_o[e^{-T_1/R_oC} e^{-(T_1+\Delta t_k)/R_oC}] \quad \text{for} \quad 0 \le k \le 999 \tag{9.21}$$

Figure 9.26. Linear ramp generator model including finite output impedance.

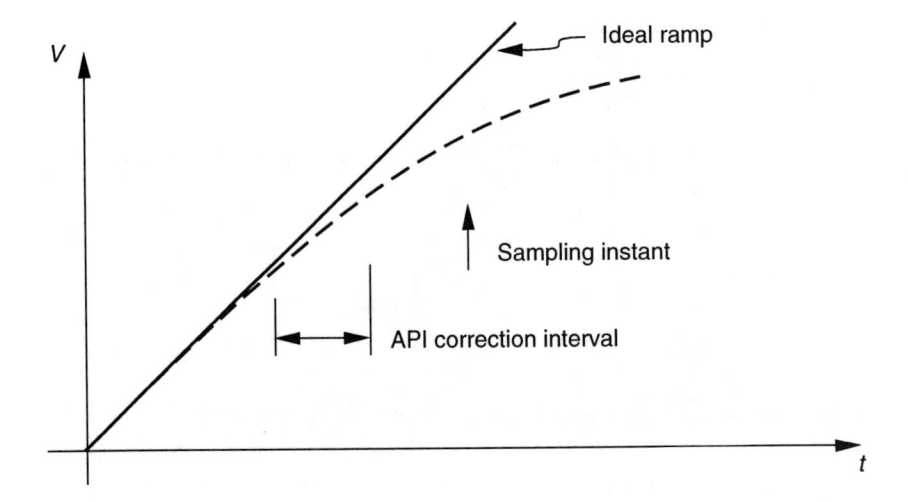

Figure 9.27. Curvature of the otherwise ideal linear ramp voltage resulting from finite source output impedance.

where

$$T_1 = \frac{1}{2 F_{\text{ref}}} \quad \text{(nominal)}$$

$$\Delta t_k = \frac{k}{F_{\text{ref}}} \frac{0 \cdot F}{N + 0 \cdot F}$$

This situation is more clearly illustrated in Figure 9.28.

Not surprisingly, as the quantity $R_o C \to \infty$, the fractional-N induced voltage error at the phase detector output (without correction) tends toward

$$e(k) = \frac{I \Delta t_k}{C} \tag{9.22}$$

which is the linear result that would be expected without the ramp curvature impairment.

The API correction currents are applied during the time period from $t = T_o$ to $t = T_f$ as shown in Figure 9.28. Considering a single API current source I_1 gated on for the time period starting at $t = T_f$ for a length of time given by m counts of $N.F$

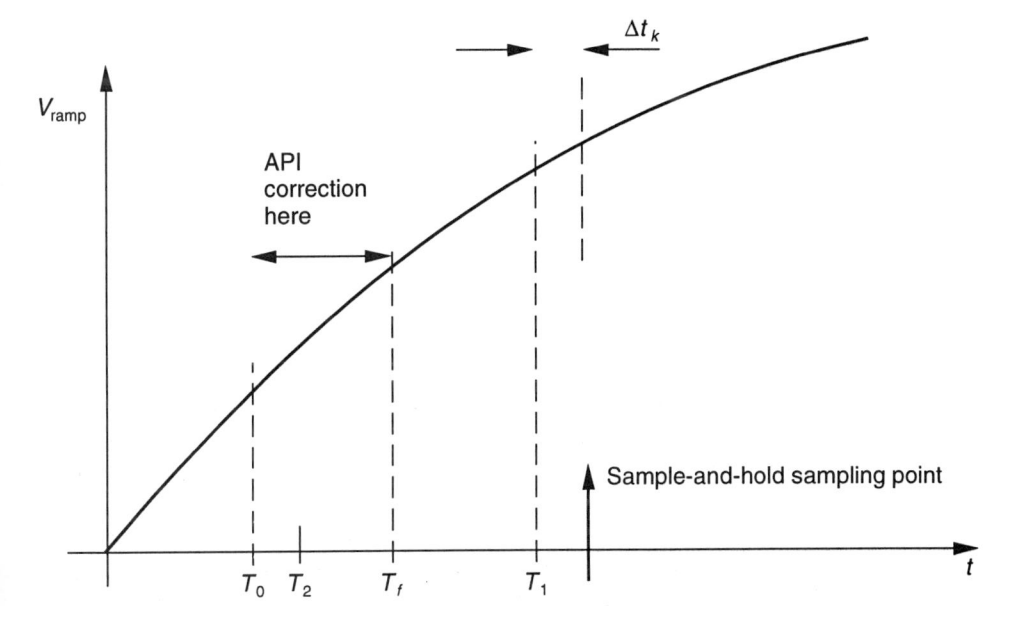

Figure 9.28. Illustration of the API gating period with respect to the finite ramp curvature.

$\times F_{\text{ref}}/L$ (i.e., VCO output divided by L), the applied fractional-N correction imparted at the sampling instant $T_1 = \Delta t_k$ is given by

$$C(m,k) = I_1 R_o [1 - e^{-(T_2-T_o)/R_oC}] e^{-(T_1+\Delta t_k-T_2)/R_oC} \qquad (9.23)$$

where

$$T_2 = T_o + m \frac{N.FF_{\text{ref}}}{L} \qquad (9.24)$$

In general, multiple API current sources are involved, each being gated on for various time durations. Their effects may nonetheless be characterized in this same manner.

With this simple introduction serving as background, the quantity of interest is the residual error remaining after applying the API corrections (e.g., $e(k) - c(m,k)$) with the phase detector ramp curvature present. The case chosen to demonstrate the ramp curvature impairment is given as follows:

F_{ref}	100 kHz
T_1	5 μsec
N	801
$0 \cdot F$	0.001
I	2.0 mA
API I_1	10.0 μA
I_2	1.0 μA
I_3	0.1 μA
C	2000 pF
L	20
T_o	1 μsec

The finite output impedance of the ramp current source results in a small magnitude error in the fractional-N correction, which is easily compensated for and is therefore not of further concern here. Much more serious is the appearance of a cusp-shaped error waveform as shown in Figure 9.29, which was calculated for the case parameters just given with $R_o = 10$ kΩ. For this case, the peak cancellation error is approximately 150 μV, which, depending on the VCO tuning sensitivity, can produce substantial spurious sidebands since it more than likely falls well within the

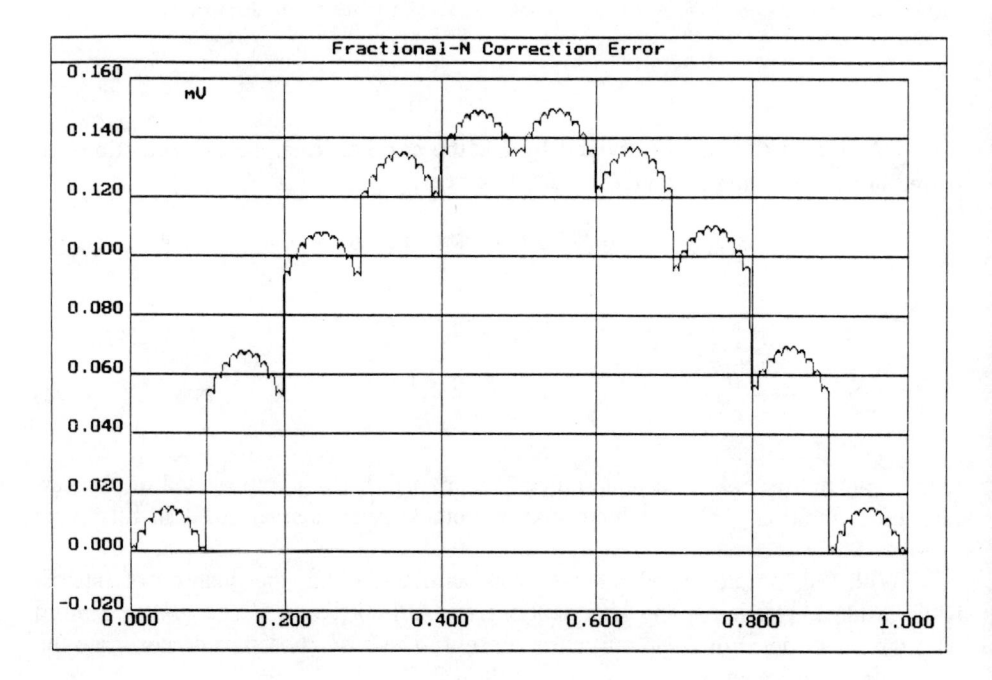

Figure 9.29. Residual error for the output impedance case with $R_o = 10$ kΩ.

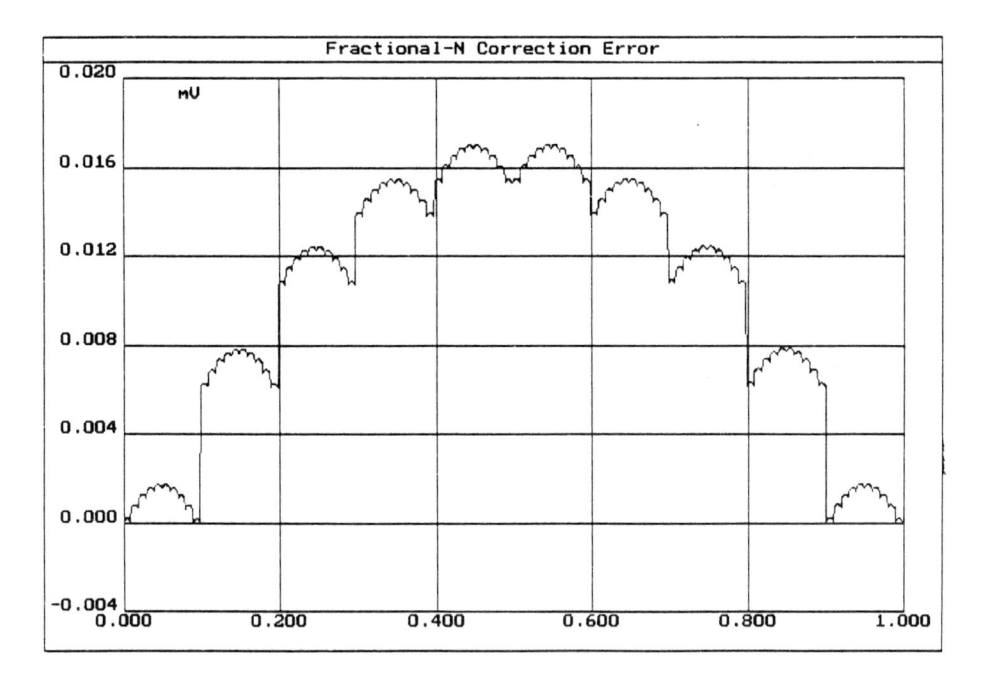

Figure 9.30. Residual error for the output impedance case with $R_o = 100$ k$\Omega \cdot$ H.

closed-loop bandwidth of the system. An x-fold increase in the current source output impedance results in approximately the same degree of improvement in the residual error as shown in Figure 9.30 where R_o has been increased to 100 kΩ. This brief analysis clearly accentuates the need for thorough design analysis when designing a fractional-N synthesizer.

REFERENCES

[1] Dana Series 7000 Digiphase, Publication 980428 (Manual). Dana Laboratories, 2401 Campus Drive, Irvine, CA 92664, 1973.
[2] Egan, W.F., *Frequency Synthesis by Phase Lock,* John Wiley & Sons, New York, 1981.
[3] Gillete, G.C., "Digiphase Principle," Frequency Technology, 1969.
[4] Reinhardt, V., et al., "A Short Survey of Frequency Synthesizer Techniques," *40th Annual Frequency Control Symposium,* 1986, pp. 355–365.
[5] Hassun, R., "A High-Purity, Fast-Switching Synthesized Signal Generator," *Hewlett-Packard J.,* February 1981.
[6] Bjerede, B., "A New Phase Accumulator Approach to Frequency Synthesis," *Proc. IEEE NAECON,* 1976.
[7] Messerschmitt, D.G., "A New PLL Frequency Synthesis Structure," *IEEE Trans. Comm.,* Vol. COM-26 No. 8, August 1978, pp. 1195–1200.

[8] Rohde, U.L., "Low-Noise Frequency Synthesizers Using Fractional-N Phase-Locked Loops," *RF Design,* January/February 1981, pp. 20–34.
[9] Faulkner, T.R., et al., "Signal Generator Frequency Synthesizer Design," *Hewlett-Packard J.,* December 1985, pp. 24–31.
[10] Aken, M.B., W.M. Spaulding, "Development of a Two-Channel Frequency Synthezier," *Hewlett-Packard J.,* August 1985, pp. 11–18.
[11] Danielson, D.D., S.E. Froseth, "A Synthesized Signal Source with Function Generator Capabilities," *Hewlett-Packard J.,* Vol. 30, No. 1, January 1979.
[12] Browne, J., "Miniature RF Synthesizer Generates Giant Performance," *Microwaves & RF,* December 1984, pp. 135–6.
[13] Hassun, R., "The Common Denominators in Fractional N," *Microwaves & RF,* June 1984, pp. 107–110.
[14] Fountain, E., Hughes Ground Systems Group, Fullerton California (personal communication).
[15] Frey, G., Hughes Ground Systems Group, Fullerton, California (personal communication).
[16] O'Leary, P., F. Maloberti, "A Direct Digital Synthesizer with Improved Spectral Performance," *IEEE Trans. on Comm.,* Vol. 39, No. 7, July 1991, pp. 1046–1048.
[17] Rohde, U.L., *Digital PLL Frequency Synthesizers Theory and Design,* Prentice-Hall, Englewood Cliffs, NJ, 1983.

About the Author

James A. Crawford is founder and president of Comfocus Corporation, a research and development firm in San Diego, California, that specializes in wire and wireless communications. Founded in 1990, Comfocus has played an active role in developing cellular telephone technology as well as high-speed modem technology for digital cable systems. In previous years, Mr. Crawford has held staff engineering positions at TRW, M/ACOM Linkabit, and Hughes Aircraft Company. While at Hughes Ground Systems Group in the early 1980s, he became group leader for their synthesized frequency sources activities. Mr. Crawford holds three patents in the area of frequency synthesis with additional patents and patents pending in digital signal processing pertaining to TDMA SATCOM modem work and completely digital receiver technology. Mr. Crawford studied toward the PhD in communication systems for approximately two years at USC beginning in 1985, and earned an MSEE degree from this same institution in quantum electronics in 1979. Mr. Crawford received his BSEE degree from the University of Nebraska–Lincoln in 1976. He grew up in Sidney, Iowa, a rural farming community in southwest Iowa with a population of approximately one thousand.

Index

8-PSK
  gray-coded, 128
    BER for, 134
  symbol error probability, 130
  two-channel direct detection receiver for, 132
16-QAM, 114
  BER performance, 120
  channel cutoff rate for, 137
  constellation symmetries, 124
  gray-coded, 123
    symbol error rate, 126
  phase noise effects, 120–27
  signal constellation, 123
  waveform sensitivity, 125
64-QAM, 114

Adams-Bashforth method. *See* Forward Euler integration
Adams-Moulton algorithm family, 361
  first-order member. *See* Backward Euler integration
  higher degree members of, 361
Additive noise, 64
Additive white Gaussian noise (AWGN) channel, 117
  16-QAM signals over, 123
  BPSK demodulation with, 120
  QPSK demodulation with, 120
Admittance matrix, 380
  floating, 380–82
  illustrated, 382
Advanced fractional-N synthesis, 346–50
  See also Fractional-N frequency synthesis
Analog frequency discriminator, 17
Analog hold function, 189
Analog phase interpolation (API), 395

  correction currents, 416
  current source gating modification, 414
  gating period inaccuracies, 410–14
    with finite gate switching speed, 413
    illustrated, 411
  illustrated gating period, 417
Analog-to-digital converter (ADC), 347
Autoregressive modeling, 69

Backward Euler integration, 357–58
  comparison with forward Euler, 358
  computed error for, 359
  computed step response for, 364
  numerical error performance using, 364
  stability, 366
    illustrated, 367
  See also Numerical integration formulas
Bandpass filters, BPF2, 249
Bayes rule, 118
Beat note, 388
  cancellation, 400
  conceptual cancellation of, 389
  Fourier components for fractional part of 0.1, 404
  Fourier components for fractional part of 0.3, 406
  for fractional part of 0.1F, 402
  for fractional part of 0.3F, 405
  for fractional part of 0.9F, 403
  illustrated, 388
  in phase detector output, 398
  See also Fractional-N frequency synthesis
Bilinear transformation techniques, 66
Bilinear transform method, 196
  of Laplace transform inversion, 197
Binary phase-shift-keyed (BPSK) systems, 114
  BER for, 121
  phase noise effects of, 120

Bit error rates (BERs), 41
   for 8-PSK with gray coding, 134
   16-QAM and, 120
   BFSK with coherent processing vs., 119
   BPSK and, 120
      illustrated, 121
   coherent demodulation, 117
   MPSK and, 127
   QPSK and, 120
      illustrated, 122
Blackbody radiation, 41
Butterworth filtering, 53

Carrier-to-noise ratio (CNR), 115
   at DDS output, 324
Cauchy residue theorem, 164
Cauchy-Riemann differential equations, 198
CD4046 phase/frequency comparator, 17
   illustrated, 18
Channel cutoff rate, 136
   for 19-QAM, 137
   illustrated, 137
Charge pumps, 23
Chebyshev filter, 34
   idealized, 34
Chebyshev polynomial method, 334
Cholesky elimination method, 382
Chua's second-order determining equation, 102
Circuit equivalence method, 182
   defined, 188
Circuit macros, 373–77
   ideal operational amplifier, 374
   ideal voltage follower, 374
   nonideal operational amplifier, 375
   sample-and-hold phase detector, 377
   simplified phase-frequency detector, 376
   See also MACSET computer program
Classical residue method, 197
Closed-loop gain functions, 174
Closed-loop transfer function, 151, 171
Colpitts topology, 92
Companion modeling, 355
Complementary metal-oxide semiconductor
     (CMOS), 83
Contamination
   of adjacent frequency channels, 41
   signal in phase-locked loops, 31–33
Continuous linear systems, 147–88
   frequency-domain analysis, 174–82
   time delays in, 171–74
   transient response evaluation for, 182–88
      classical residue method, 197
      Corrington method, 194–97

      Ross method, 192–94
Continuous phase modulation (CPM), 135
   minimum (Euclidean) distance, 136
Control loop
   damping factor, 150
   frequency, 150
   stability, 179–82, 188
   type 1, 250
   See also Phase-locked loops (PLLs)
Corrington method, 194–97
   bilinear method and, 196
   example, 196
   of Laplace transform inversion, 196

D/A converter. *See* Digital-to-analog converter
Damping factor, 175
Data windowing, 58
DBMs. *See* Double-balanced mixer phase detectors
DDSs. *See* Direct digital synthesizers
Delta-sigma modulation techniques, 335
Design equation, 3–8
   frequency switching speed, 4–5
   interface related noise and contamination, 8
   power supply contamination budgets, 7–8
   RF/analog and digital techniques, 5–6
   system design, 6–8
Design tools, 104
Differential phase-shift-keyed (DPSK), 63
Digiphase, 391, 395
   illustrated, 392
Digital frequency dividers, 71
   divide-by-20, 73
   divide-by-N, 72
   ECL test set, 76
      in characterization efforts, 79
      measurement setups, 80
   feedback, 73
   phase noise in, 71–81
      comparison, 76
      illustrated, 75
      measurement, 73
      model for assessing, 72
   PLL design and, 78
   residual noise of, 77
      illustrated, 81
      normalized to 10 GHz, 78
   TTl test set, 76
      in characterization efforts, 79
      measurement setups, 80
Digital signal processing (DSP), 6
Digital-to-analog converter, 335, 347
   imperfections, 335
Dirac delta function, 157

Direct digital frequency synthesis, 346
    with noise shaping, 409
Direct digital synthesizers (DDSs), 1, 6
    D/A output, 311
    dithered
        sinusoidal, 338–43
        square wave source, 337
    finite word length effects, 320
    frequency synthesizers, 343–45
    fundamentals, 308–43
    general concepts, 309–11
    history of, 308–9
    ideal output spectrum, 311–13
    illustrated architecture, 310
    illustrated model, 315
    improved spurious performance
            modification, 321
    lookup table, 323–28
    mapping techniques, 328–35
    operation degradations, 314
    output
        CNR, 324
        CNR, density, 342
        DFT, 321, 322, 326, 327
        frequency, 311
        phase dithered sinusoidal, 340
        quantization effects, 323
        sample sequence, 325, 326
        SNR, 342
        square wave, 336
    phase accumulator, 316
    phase truncation related spurious effects, 313–22
    PLL synthesizer, 345
    sampling process, 312
    sine function table lookup, 314
    sinusoidal, with noise shaping, 344
    spurious response, worst case, 322
    spurious suppression, 335–43
    using noise shaping techniques, 343
Dirichlet sawtooth waveform, 317
    error sequence, 318
Discrete circuit models, 369–72
    for ideal capacitor, 370
    for ideal inductor, 371
    for RLC circuit, 371, 372
    step response computation, 373
Discrete Fourier transform (DFT), 48–50, 401
    of DDS output, 321, 322, 326, 327
    equation development, 49
Discrete sampled systems
    computation example, 51–58
    creation methodology, 53
    PSD for, 50–51
    calculations in, 48–60
Discrete spurs, 318–19
    amplitude computation, 319
Dithered sinusoidal DDS, 338–43
Dithering, 335
    phase, 336–38
Double-balanced mixer phase detectors, 81–84
Dual-loop frequency synthesis, 289–308
    frequency channel spacing, 289
    gain compensation strategies, 292–302
    offset mixer considerations, 289–91
    phase noise performance, 289
    proper sideband selection, 302–4
    signal isolation, 291–92
    switching speed, 304–8

Elliptic filter, 305
    components, 307
    type 2 PLL with, 306
Equation roots, calculating, 197–99
Equivalent circuit
    for noisy resistor, 270
    for physical resistor, 383
Ergodicity, 50

Fast switching, 247
    frequency synthesizer design, 247–85
Fast switching speed
    figure of merit, 247
    optimum frequency, 248
Feedback dividers, 4, 27–31, 168–70
    changing ratio, 251
    integral, ratios, 388–89
    operation, 168
        illustrated, 169
    operator notation, 151
    role of, 27–31
    sampled-data PLLs and, 204
    sampling action, 170
    as zero-crossing selector gate, 31
Feldtkeller energy equation, 86
Field-effect transistors (FETs), 89
Filter noise, lossless two-port network and, 85
First-order active all-pass network, 173
First-order Adams-Bashforth method, 361
    See also Forward Euler integration
Flicker noise, 44
Forward error correction (FEC), 114, 135
Forward Euler integration, 357, 358
    comparison with backward Euler, 358
    computed error for, 359
    stability, 365

Forward Euler integration *(cont.)*
    illustrated, 366
    See also Numerical integration formulas
Fourier frequency, 60
Fourier transforms
    continuous-time, 48–50
    discrete-time, 48–50
Fractional-N frequency synthesis, 2, 387–419
    advanced concepts, 346–50
    API gating period inaccuracies, 410–14
        with finite switching speed, 413
        illustrated, 411
    concept visualization, 388
    discrete spurious contaminants, 401–6
    general concepts, 395–401
    illustrated, 391
    loop parameter limitations, 406–7
    multistage application to, 349
    noise floor degradations in, 410
    noise-shaping methods, 347–50
    phase detector output voltage and, 397
    phase detector ramp nonlinearities, 414–19
    phase detector timing relationships, 407
    PLL block diagram, 396
    pros and cons, 393–95
    spectral performance, 401–6
    system aspects, 401–19
    truncation-related discrete spurs and, 407–9
    uses, 387, 394–95
    See also Beat note
Frequency channel spacing, 289
Frequency dividers. *See* Digital frequency dividers
Frequency-domain analysis
    continuous systems, 174–82
    MACSET, 377–82
    sampled-data control systems, 208–9
Frequency-domain representation, 205–6
Frequency-locked loops, 31
    See also Phase-locked loops (PLLs)
Frequency switching speed, 4–5
    achievable, 304–8
    elliptic filter attenuation vs., 305
    frequency error 5 MHz, phase error 0 deg., 269
    initial phase error impact on, 264–69
    initial phase error of 50 deg., 265
    initial phase error of 100 deg., 266
    initial phase error of 360 deg., 267
    initial phase error of 720 deg., 268
    limits on, 250
    open-loop transfer function attenuation vs., 307
    optimum frequency, 248
    systematic gain compensation and, 302

Frequency synthesis
    architecture, complex, 8
    building blocks for, 3–39
    direct digital
        advantages, 346
        disadvantages, 346
        with noise-shaping, 409
    dual-loop, 289–308
        frequency channel spacing, 289
        gain compensation strategies, 292–302
        offset mixer considerations, 289–91
        phase noise performance, 289
        sideband selection, 302–4
        signal isolation, 291–92
        switching speed, 304–8
    fractional-N, 2, 387–419
        advanced concepts, 346–50
        API gating period inaccuracies, 410–14
        based on noise-shaping methods, 347–50
        concept visualization, 388
        discrete spurious contaminants, 401–6
        general concepts, 395–401
        illustrated, 391
        loop parameter limitations, 406–7
        multistage application to, 349
        noise floor degradations in, 410
        phase detector ramp nonlinearities, 414–19
        phase detector timing relationships, 407
        phase detector voltage output and, 397
        PLL block diagram, 396
        pros and cons, 393–95
        spectral performance, 401–19
        system aspects, 401–19
        truncation-related discrete spurs and, 407–9
        uses, 387, 394–95
        See also Beat note
    using sampled-data control systems, 203–44
Frequency synthesizers
    DDS, 343–45
    direct digital, with noise shaping, 409
    fast-switching, 247–85
    high-performance, 1
    output phase noise spectrum for, 409
    SNR in, 147
    system performance, 114–37
Frequency warping, 58

Gain compensation, 251, 252
    dual-loop strategies, 292–302
    need for, 292
    phase detector, 295–96
        using current steering method, 298
    systematic, 296–302

VCO, 292–95
Gain functions
   closed-loop, 174
   open-loop, 175, 189
      approximation error and, 179
      exact, 191
      ideal type 2, 225
      MACSET calculations, 379
      Monte Carlo analysis, 260
      percentage bandwidth, 190
      pseudo-continuous vs., 177, 178, 191
      sampled, 209
      type 1 with inefficient gain detector, 216–17
      type 1 with internal delay, 221
      type 2 with inefficient phase detector, 228
      type 2 with internal delay, 231
   pseudo-continuous, 176
      open-loop vs., 177, 178, 191
      stability and, 179–82
Gain imbalance, 26
Gaussian elimination method, 382
Gaussian noise, discrete model, 54
Gaussian noise source
   low-pass filtering of, 51
   modeled in time domain, 53
Gaussian random number generator, 54
Gear family of algorithms, 362
Gear method, 362
   computed step response, 364
   second-order, 362–63
      for network step response, 363
      numerical error performance and, 364
   stability, 366–68
      illustrated, 368
Gray coding
   8-PSK, 128
   16-QAM, 126
   256-QAM, 125
   constellations, 133
   for information bit assignment, 131
   three-tuples (LB, MB, RB), 132
Grounding, 33–39
   cost-effective, 38
   effectiveness, 35
   guideline #1, 35–37
   guideline #2, 37–38
   guideline #3, 38–39
   Ohm's law and, 33–39
   shielding and, 37

Harmonic sampling phase detectors, 13–15
   sampling gate interval, 13
   sampling inefficiency and, 15

Heaviside operator notation, 151
Hybrid phase-locked loops, 287–350
   architectures, 287
      illustrated, 288
   DDS frequency synthesizers, 343–45
   DDS fundamentals, 308–43
   dual-loop, 289–308
   See also Phase-locked loops (PLLs)

Ideal linear capacitor, 369
   discrete circuit model for, 370
IF frequencies, 290
Implicit integration, 358
Inefficient sample-and-hold phase detector, 165–68
   attenuation and, 167
   phase argument and, 168
   with unit-step input, 166
   See also Sample-and-hold phase detectors
Injection locking, 108–9
Integration methods
   backward Euler, 357, 358
   forward Euler, 357, 358
   implicit, 358
   numerical, 355–65
Interface related noise, 8
Interpolated zero-crossing, 212

Johnson noise, 41, 44

Kirchoff's laws, 372
   current law, 380
Kurokawa theory, 99

Laplace transform
   arbitrary rational, 186
   bilinear transform method of, inversion, 197
   classical residue method of, inversion, 197
   Corrington method, inversion, 196
   general quadratic rational, 186
   Ross method, inversion, 194
   of transient error response, 192
Lead-lag loop filters
   feedback elements, 298
   illustrated, 88, 152
   noise sources for, 87–89
   operator notation, 151
   system gain compensation modification, 301
Leeson's phase noise model, 94–97
   comparison, 97
   extensions and implications, 97–101
   illustrated, 94
   single-sideband phase noise spectrum, 94
Linear phase detector, operator notation, 151
Linear ramp generator, 415

Linear ramp voltage
   curvature, 416
   generation, 415
   for sample-and-hold phase detector input, 414
Load pulling, 109–11
   baseband-modulation-induced, 109
   defined, 109
   oscillator assessment, 110
Local oscillator (LO)
   phase noise impairment, 135
   port, 87
Loop bandwidth, 189–99
Loop filters, 23–27
   concepts, 24
   synthesize, structure, 24
   transfer function, 150
Loop phase error, 149
Lossless two-port network, 85
   linear, 86
   passive, 85
Low-noise oscillators, 91–114
   design example, 111–14
      methodologies, 112–14
      parameters, 112
   impairments, 107
   injection locking, 108–9
   Leeson's phase noise model, 94–97
      extensions and implications, 97–101
      nonlinear oscillator theory and, 101–2
   load pulling, 109–11
   negative resistance, 91–93
   nonlinear analysis, 103–4
   nonlinear effects in, 106–7
   pushing, 111
   van der Pol model, 102–3
   varactor diode nonlinear effects, 104–5
Low-pass filter
   additional, 394
   of continuous Gaussian noise source, 51
   discrete model for, 54
   impulse response of, 52
   low-order filters, 394
   type 1 PLL with single-pole, 189
   unit delay implementation for, 185
LPC-10 speech compression techniques, 70

MACSET computer program, 26, 353–84
   circuit macros, 373–77
   companion modeling, 355
   discrete circuit models, 369–72
   frequency-domain analysis, 377–82
   frequency-domain characterization summary, 378
   history of, 353
   limitations, 353–54
   network equation formulation, 355
   numerical integration formulas, 355–65
   open-loop gain calculations, 379
   phase noise computations, 382–84
   time-domain modeling, 354–55
   uses, 353
Mapping techniques, 328–35
   sine(0) table compression, 328–35
M-ary phase-shift-keyed modulation (MPSK), 127–34
   BER performance, 127
   degradation of, 139
   demodulator, 127–28
Microstrip
   ground length, 35
   width, 36
Minimum-shift-keyed (MSK) waveform, 125
Mixer phase detectors, 9–13
   characteristic, 10
   double-balanced, 81–84
      mixer termination for, 11
   mixer sensitivities, 12
   with output lowpass filter, 9
   output voltage, 10
   performance guidelines, 11
   uses, 9
Monte Carlo loop analysis, 260–63
   open-loop gain function, 260
   transient phase error response, 260–61
      for type 1 system, 262
      for type 2 system, 263
Motorola
   IN821, generated noise, 90
   IN935, generated noise, 90
   IN941, generated noise, 89
   MC12040 phase/frequency comparator, 17
      illustrated, 19
MPSK. *See* M-ary phase-shift-keyed modulation
M-PSK trellis-coded modulation, 114
M-QAM systems, 138
   degradation of, 138

Negative resistance generator, 92
Negative resistance oscillator, 91–93
   illustrated, 91
   impedance conditions, 100
Newton-Raphson method, 198
NLO-100 test equipment, 63, 67
   desired output, 65
   output for different loop parameters, 67
   output noise spectrum, 68

top-level block diagram, 67, 69
Nodal analysis method, 354
Noise. *See* Phase noise; Specific noise
Noise floor
   degradations in fractional-N synthesis, 410
   with integrated phase-locked loop chips, 84
   at phase detector output, 272
   for phase noise, 82
Noiseless biasing, 112
Noise power, 85
   incident and reflected terms, 86
Noise-shaping methods, 335
   direct digital frequency synthesizer and, 409
   fractional-N, 347–50
Nonlinear oscillators
   analysis methods, 103–4
   design guidelines, 101–2
   pulling, 110
   varactor diode, 104–6
Numerical integration formulas, 355–65
   RLC circuit for examination, 355
   stability of, 365–68
   See also Backward Euler integration; Forward Euler integration
Nyquist stability criterion, 179
   complex path of integration, 180
   example, 180
      illustrated plot, 181
   stability assessment using, 210
   type 2 PLL with internal delay and, 232

Offset mixing
   for dual loop architecture, 289–91
      illustrated, 291
Ohm's law, 33–39
   grounding and, 35
Open-loop gain function, 175, 189
   approximation error and, 179
   ideal type 2, 225
   MACSET calculations, 379
   Monte Carlo analysis, 260
   percentage bandwidth, 190
   pseudo-continuous vs., 177, 178, 191
   sampled, 209
   type 1 with inefficient phase detector, 216–17
   type 1 with internal delay, 221
   type 2 with inefficient phase detector, 228–31
   type 2 with internal delay, 231
   type 2 with simplified phase-frequency detector, 273
Open-loop transfer function
   attenuation vs. loop overshoot, 309
   attenuation vs. phase margin, 308
   attenuation vs. switching speed, 307

magnitude after systematic compensation, 300
magnitude vs. frequency, 299
phase after systematic compensation, 300
phase vs. frequency, 299
Operational amplifier noise modeling, 87
Oscillators
   design example, 111–14
      methodologies, 112–14
      parameters, 112
   feedback of, 95
   impairments, 107
   injection locking and, 108–9
   Leeson's model, 94–97
   linear, analysis, 94
   load pulling and, 109–11
   local (LO), 87
      with phase noise impairment, 135
   low-noise design and characterization, 91–114
   negative resistance, 91
   nonlinear
      analysis methods, 103–4
      design guidelines, 101–2
      effects in, 106–7
      pulling, 110
      varactor diode, 104–6
   phase noise, 98
      optimum performance, 101
   post-tuning drift, 282
   pushing, 111
   sideband noise, 99
   van der Pol, 102–3
   See also Voltage-controlled oscillators (VCOs)

Padgve approximates, 172–73
   first-order, 183
Partial fraction expansion, 199
PDF
   of computed phase estimate, 128
      illustrated, 130
   Tikhonov phase error, 131
Phase accumulator, 316
   error sequence, 316
   mapping technique, 328–35
   rounding, 317
   truncation, 317
   unaltered, operations, 339, 340
Phase detectors, 8–23
   analog output voltage, 238
   differential arm balance, 26
   finite pulse widths and, 19
   fractional-N frequency synthesis
      output voltage and, 397
      ramp nonlinearities, 414–19

Phase detectors *(cont.)*
    timing relationships, 407
    gain compensation, 295–96
    gain control, 298
    harmonic sampling, 13–15
    inefficient sample-and-hold, 165–68
    input zero-crossing, 213
    linear, operator notation, 151
    mixer, 9–13
        double-balanced, 81–84
    output, 149
        voltage, 23
    phase-frequency, 17–23, 153
    phase noise performance, 81–84
    pseudo-continuous, 152–70
    sample-and-hold, 250–51, 282, 284–85, 414
        inefficient, 165–68
        noise, 270–73
    three-state, 19, 21
    transfer function, 149
    types of, 9
    zero-order sample-and-hold, 15–16, 163–65
Phase dithering, 336–38
    process analysis, 339
    sinusoidal output DDS, 340
Phase error, 252
    function, 318
        continuous, 157
    impact on switching speed, 264–69
    sample sequence, 157
    transient, 183
Phase fluctuations, 60
    one-sided PSD of, 60
Phase-frequency detectors, 17–23, 153
    advantages, 285
    characteristics of, 17
    flip-flop, outputs, 281
    gain behavior, 159
    loop bandwidth considerations using, 189–99
    quad-D, 22
    refinements, 280–82
    sample-and-hold detectors vs., 282
        choosing between, 284–85
    simplified, 19, 21, 158–62
        gain imbalances, 26
        logic states, 158–59
        macro, 376
        timing, 159
    topologies, 281
    tri-state, 153–58
        analysis, 154
        equivalent circuit, 153

    illustrated, 153
    signal timing relationships, 154
    type 2 PLL with, 273–82
        gain margin contours, 274, 275–76
        gain/phase margin contours overlay, 279
        illustrated, 274
        open-loop gain function, 273
        phase margin contours, 274, 277–78
        transient error response optimization, 280
        transient response, 279
    zero phase error dead zone, 17
        avoiding, 17–19
Phase interpolation, 391
    direct output type of, 393
    illustrated, 392
    using SSB mixing, 393
Phase-lock concept, 1
    loop, 1
    noise floor and, 84
Phase-locked loops (PLLs), 3
    with acquisition circuitry, 25
    analysis for continuous linear systems, 147–88
    basics of, 148–52
    chip families, 71
    closed-loop bandwidth, 301
    components, 8–31, 148
        illustrated, 149
    DDS-driven synthesizer, 345
    equivalent model, 170
    feedback dividers, 27–31
    frequency dividers and, 78
    hybrid, 287–350
        architectures, 287–88
        DDS frequency synthesizers, 343–45
        DDS fundamentals, 308–43
        dual-loop, 289–308
    including modeling for dominant noise
            sources, 65
    to investigate phase detector arm balance, 26
    loop filters, 23–27
    loop order, 148
    loop type, 148
    noise corrupted model, 67
    phase detectors, 8–23
    phase noise and, 44
    pseudo-continuous, 147
    purpose for using, 27
    sampled-data, 203
        frequency-domain representation for, 205–7
        ideal type 1, 214–16
        ideal type 2, 225–28
        illustrated, 204

stable, 211–12
transient response of, 211
type 1 with inefficient phase detector, 216–20
type 1 with internal delay, 320–24
type 2 with inefficient phase detector, 228–31
type 2 with internal delay, 231–37
signal contamination in, 31–33
SNR, 133
   for frequency synthesis, 148
type 1, 171
   design example, 249
   high-speed, 248
   post-tuning drift, 264
type 2, 148
   with elliptic filter, 306
   with phase-frequency detector, 273–82
   post-tuning drift, 264, 282–84
   second-order, 152
   third-order, 175, 177
voltage controlled oscillators, 27
Phase noise, 41–139
   arbitrary spectra, creating, 69–71
   in components, 89–90
   correlation, measurement system, 83
      illustrated, 83
   in devices, 71–114
   in digital frequency dividers, 71–81
      illustrated, 75
      measurement, 73
      model for assessing, 72
      performance comparison, 76
   DPSK impact, 63
   dual-loop frequency synthesis and, 289
   filters and, 84–87
   first principles, 45–46, 58–60
   Gaussian, statistics, 65
   illustrated impairments, 43
   impairments, 115–17
      for advanced waveforms, 135–37
      LO, 135
   in lead-lag loop filters, 87–89
   low-noise electronic design and, 84–87
   MACSET computations, 382–84
   noise floor, 82
   oscillator, 98
      optimum performance, 101
   performance
      of divide-by-20 components, 73
      frequency divider comparison, 76
      phase detector, 81–84
      system, 114–37
   in phase detectors, 81–84

PLL components and, 44
as primary design issue, 285
PSD concept, 46–48
PSDs, creating in lab, 63–68
reference, 238–42
   output phase noise PSD, 240
   system design and, 241–42
requirements, 42
sample-and-hold phase detector, 270–73
sampled-data control systems, 237–44
   reference, 238–42
   VCO-related, 242–44
spectrum output, 350
standardized quantities, 60–63
template description, 62
theory, 44–48
undesirable levels of, 41–42
VCO-related, 242–44
   equivalent phase error process, 244
   nonband-limited, 244
   output phase noise PSD, 243
worst case bounding for, 137–39
Phase noise effects
   with coherent demodulation of 16-QAM, 120–27
   with coherent demodulation of binary
      FSK, 117–19
   with coherent demodulation of BPSK, 120
   with coherent demodulation of QPSK, 120
   See also Phase noise
Phase quantization errors, 341
Phase truncation, 317
   in DDS, 313–22
Pole-zero pair, 24
Postdetection filtering, 19
   RC time constant for, 21
Post-tuning drift, 264
   oscillator, 282
   with type 2 systems, 282–84
   VCO, 264, 282
      evaluation, 284
      parameters, 283
Power law noise sources, 44
Power spectral density (PSD), 44
   calculations, 48–60
   for case k=2048, 57
   computation using DFTs, 55
   concept, 46–48
   creating accurate phase noise, 63–68
   definition, 46–47
   for discrete sampled systems, 50–51
   modified, calculation technique, 55
   for noise-corrupted sample sequence, 272

Power spectral density *(cont.)*
  one-sided, 59
    of phase fluctuations, 60
  at PLL output, 238
  for reduced variance for case k,128, 56
  for smoothing case K=2, 56
  two-sided, 59
    defined, 61
Power spectrum estimation theory, 69
Power supply, 7–8
  contamination levels, 32
PSD. *See* Power spectral density
Pseudo-continuous, 147
Pseudo-continuous gain function, 176
  open-loop vs., 177, 178, 191
  stability and, 179–82
Pseudo-continuous phase detectors, 152–70
  model accuracy, 161
  model usage, 161
  types of, 152
Pseudo-continuous transfer function, 158, 160, 163
  for inefficient sampling operation, 166–67
  for phase-frequency detector, 158, 160
  for sample-and-hold phase detector, 165

Quad-diode bridge, 256
  illustrated, 257
Quad-D phase-frequency detectors, 22
  illustrated, 22
Quadrature amplitude modulation (QAM), 114
Quadrature-based sideband selection, 302
  illustrated, 303
  modified, 304
Quadrature phase-shift-keyed (QPSK), 4, 114
  BER for, 122
  phase noise effects, 120
Quasi-static diode bridge, 256
  illustrated, 258

Ramp current
  generator, 270
  representative current source for, 270
RC low-pass filter, 21–22
Receiver desensitization, 42
Reference frequency, 249
Reference phase noise, 238–42
  output phase noise PSD, 240
  system design and, 241–42
Residual errors, 418–19
Residual FM, 61
  computation, 61–62, 63
Residual noise
  frequency divider, 77

  illustrated, 81
  normalized to 10 GHz, 78
Reverse isolation, 291
RLC circuit
  discrete circuit model, 371, 372
  illustrated, 355
  numerical error performance comparison, 364
  state-variable derivative, 356
  step response comparison, 364
  time-domain response, 356
  unit-step response, 355
  voltage transfer function, 355
Rms phase jitter, 61
ROM values, 331
Ross method, 192–94
  example calculation, 193
  of Laplace transform inversion, 194
  time-domain transient response
    calculation and, 193
Routh-Hurwitz criterion, 179

Sample-and-hold phase detectors
  advantages, 284–85
  design features, 250
  inefficient, 165–68
  internal timing relationships, 251
  linear ramp voltage for, 414
  macro, 377
  noise, 270–73
  phase-frequency detectors vs., 282
    choosing between, 284–85
  See also Zero-order sample-and-hold
    phase detector
Sampled-data control systems, 203–44
  basics, 203–8
  closed-form, 214–37
    case 1, 214–16
    case 2, 216–20
    case 3, 220–24
    case 4, 225–28
    case 5, 228–31
    case 6, 231–37
  complex path of integration for, 210
  frequency-domain analysis, 208–9
  phase noise, 237–44
    reference, 238–42
    VCO-related, 242–44
  stability, 209–11
  time-domain analysis, 211–13
  transfer functions, 207
Sampled-data phase-locked loops, 203
  frequency-domain representation for, 205–7
  ideal type 1, 214–16

gain and phase margins, 215
open loop gain, 214
use, 216
ideal type 2, 225–28
   gain margin, 225
   gain margin contours, 225
   open-loop gain function, 225
   phase margin, 226
   phase margin contours, 227
   transient phase error response, 226
illustrated, 204
stable, 211–12
transient response of, 211
type 1 with inefficient phase detector, 216–20
   loop bandwidth and, 219
   loop optimum K factor, 220
   phase error, 219
   phase margin, 217–18
   transient response, 218
type 1 with internal delay, 220–24
   gain margin, 221
   gain margin contours, 222
   loop bandwidth and, 224
   open-loop gain function, 221
   phase margin contours, 223
   system phase margin, 222–23
   transient error responses, 224
   transient response, 223
type 2 with inefficient phase detector, 228–31
   gain margin, 228–29
   loop parameters, 231
   open-loop gain, 228
   phase-locking speed, 231
   phase margin, 229
   phase margin contours, 229
   transient response, 229
type 2 with internal delay, 231–37
   gain margin contours, 233–34
   Nyquist criterion and, 232
   open-loop gain, 231
   phase-locking speed, 236
   phase margin contours, 235
   system gain margin, 232
   transient error response, 233
See also Phase-locked loops (PLLs)
Sampling bridge
   balancing of, 254
      illustrated, 255
   design considerations, 254–60
   diode, 255
      illustrated, 256
      low reference spurs and, 259

quad, 256, 257
quasi-static, 256, 258
Schottky diode
   barrier, 81
   bridge, 254, 256
   quad, 257
   quasi-static, 258
   reverse bias capacitance, 256
Schottky TTL gate, 280
Shielding, 37
Shot noise, 44
Sideband selection, 302–4
   quadrature-based, 302
   illustrated, 303
   modified, 304
Sideband signal contamination, 32
Sigma-delta converter, 347
   illustrated, 348
   multistage application, 349
Signal contamination
   in phase-locked loops, 31–33
   sideband, 32
Signal isolation, 291–92
Signal-to-noise ratio (SNR), 64
   DDS output, 342
   in frequency synthesizers, 147
      illustrated, 148
   limiting, 64
   of synthesizer phase-locked loop, 133
Simplified phase-frequency
   detectors, 19, 21, 158–62
   gain behavior, 159
   gain imbalances, 26
   logic states, 158–59
   timing, 159
Sine formulations, reduced, 333
Sine function, statistics for, 334
Sine lookup table
   compression, 328
      illustrated, 329
      Sunderland architecture for, 330
   finite quantization in, 323
   function, 314
Single-sideband (SSB) mixer, 391
   phase interpolation using, 393
Sinusoidal PM, first-order spectrum, 45
SPICE computer methods, 384
Stability, 179–82
   backward Euler method, 366
      illustrated, 367
   control loop, 188
   for forward Euler method, 365

Stability *(cont.)*
    illustrated, 366
    Gear algorithm, 366–68
        illustrated, 368
    of linear time-invariant systems, 179
    numerical integration algorithm, 365–68
    Nyquist, 211
    sampled-data control systems, 209–11
State-variable method, 354
Sunderland technique, 330–31
    illustrated, 330
Switching speed, 4–5
    achievable, 304–8
    elliptic filter attenuation vs., 305
    frequency error 5Mhz, phase error 0 deg., 269
    initial phase error impact on, 264–69
    initial phase error of 50 deg., 265
    initial phase error of 100 deg., 266
    initial phase error of 360 deg., 267
    initial phase error of 720 deg., 268
    limit on, 250
    open-loop transfer function attenuation vs., 307
    optimum frequency, 248
    systematic gain compensation and, 302
Symbol error probability, 131
Systematic gain compensation, 296–302
    switching speeds and, 302

Taylor series expansion, 331
    statistics for, 334
Tellegen's theorem, 354
Thermal noise, 44
Thevenin equivalent circuit, 42
Three-state phase detector, 19
    illustrated, 21
Time delays
    approximation, 173
        all-pass network delay, 174
    in continuous systems, 171–74
    type 1 PLL, 220–24
Time-domain
    MACSET modeling, 354–55
    partial fractions inverse-transformed into, 199
    response to exponentially decaying frequency, 283
    sampled-data control system analysis, 211–13
    system impulse response, 172
        using Ross method, 193
    transient response calculations, 182
Timing initialization, 253
Timing-sample-hold (TS&H) hybrid, 249, 250
    timing diagram, 251
Transfer function, 65

closed-loop, 151, 171
generalized rational quadratic, 186
loop filter, 150
open-loop
    attenuation vs. loop overshoot, 309
    attenuation vs. phase margin, 308
    attenuation vs. switching speed, 307
    magnitude after systematic compensation, 300
    magnitude vs. frequency, 299
    phase after systematic compensation, 300
    phase vs. frequency, 299
phase detector, 149
pseudo-continuous, 158, 160, 163
    for inefficient sampling operation, 166–67
    for phase-frequency detector, 158, 160
    for sample-and-hold phase detector, 165
sampled-data, 207
in Tustin method, 183
voltage, 355
zero-order sample-and-hold phase detector, 16
Transient response evaluation, 182–88
    classical residue method, 197
    Corrington method, 194–97
    ideal type 2 PLL, 227–28
    Monte Carlo analysis, 260–61
        for type 1 system, 262
        for type 2 system, 263
    Ross method, 192–94
    of sampled-data PLLs, 211
    type 1 PLL with inefficient phase detector, 218
    type 1 PLL with internal delay, 223
    type 2 PLL with inefficient phase detector, 229
    type 2 PLL with internal delay, 233
    type 2 PLL with phase-frequency detector, 279–81
Transimpedances, 383–84
Transportation delay, 220–24
    transient response performance and, 220
Trigometric function approximation, 332
Tri-state phase-frequency detectors, 153–58
    analysis, 154
    equivalent circuit, 153
    illustrated, 153
    signal timing relationships, 154
Tustin method, 182
    transfer function in, 183
    transformation with, 184
    unit delay elements and, 187
    using, 183

Two-channel coherent receiver
  for 8-PSK, 132
  illustrated, 116
Unit delay implementation, 185
  for generalized second-order system, 187

van der Pol oscillator model, 102–3
Varactor diode, 104–6
  back-to-back configuration, 104
  current, 106, 107
  designing, 105
  harmonic distortion, 105
  hyper-abrupt, 294
  nonlinear effects, 104–6
VCO-related phase noise, 242–44
  equivalent phase error process, 244
  nonband-limited, 244
  output phase noise PSD, 243
Very large scale integrated (VLSI), 5–6
  digital, 6
Voltage-controlled oscillators (VCOs), 4, 27
  design parameters, 112
  frequency slew rate, 25
  gain compensation, 292–95
  as ideal integrator of phase, 149
  negative resistance source, 292
    illustrated, 293
  operator notation, 151
  phase-locked, output, 406
  pole-zero pair and, 24
  post-tuning drift, 264, 282
    evaluation, 284
    parameters, 283
  pretuning errors, 250
  self-noise, 411
  spectral purity in, 98
  state-of-the-art, 113
  tuning curve, 292
    illustrated, 293
    normalized, 296–97
    slope, 294
Voltage transfer function, 355
Volterra series techniques, 101, 104

Waveforms
  detection loss for, 127
    illustrated, 127
  MPSK, 127–34
  phase noise impairments for, 135–37
Wiener-Khinchine relation, 50, 59
Wiener-Khinchine theorem, 70

Zero-crossings, 64, 212–13
  interpolated, 212
  phase detector input, 213
  selector gate, 31
Zero-order sample-and-hold phase
  detector, 15–16, 163–65
  advantages, 15
  for continuous input waveform, 163
  disadvantages, 15
  high-speed performance, 16
  implementation, 15
  inefficient, 165–68
    type 1 PLL, 216
  transfer function, 16
  See also Sample-and-hold phase detector
Zero phase error dead zone, 17
  avoiding, 17–19

The Artech House Microwave Library

Acoustic Charge Transport: Device Technology and Applications,
   R. Miller, C. Nothnick, and D. Bailey

*Advanced Automated Smith Chart Software and User's Manual,
   Version 2.0,* Leonard M. Schwab

Algorithms for Computer-Aided Design of Linear Microwave Circuits, Stanislaw
   Rosloniec

Analysis, Design, and Applications of Fin Lines, Bharathi Bhat and Shiban K. Koul

Analysis Methods for Electromagnetic Wave Problems, Eikichi Yamashita, editor

Automated Smith Chart Software and User's Manual, Leonard M. Schwab

*C/NL2 for Windows: Linear and Nonlinear Microwave Circuit Analysis and
   Optimization, Software and User's Manual,* Stephen A. Maas and Arthur Nichols

Capacitance, Inductance, and Crosstalk Analysis, Charles S. Walker

Design of Impedance-Matching Networks for RF and Microwave Amplifiers, Pieter L. D.
   Abrie

Dielectric Materials and Applications, Arthur von Hippel, editor

Dielectrics and Waves, Arthur von Hippel

Digital Microwave Receivers, James B. Tsui

Electric Filters, Martin Hasler and Jacques Neirynck

Electrical and Thermal Characterization of MESFETs, HEMTs, and HBTs, Robert Anholt

E-Plane Integrated Circuits, P. Bhartia and P. Pramanick, editors

Feedback Maximization, Boris J. Lurie

Filters with Helical and Folded Helical Resonators, Peter Vizmuller

Frequency Synthesizer Design Toolkit Software and User's Manual, Version 1.0, James
   A. Crawford

Fundamentals of Distributed Amplification, Thomas T. Y. Wong

GaAs MESFET Circuit Design, Robert A. Soares, editor

GASMAP: Gallium Arsenide Model Analysis Program, J. Michael
   Golio *et al.*

Handbook of Microwave Integrated Circuits, Reinmut K. Hoffmann

Handbook for the Mechanical Tolerancing of Waveguide Components,
W. B. W. Alison

HELENA: HEMT Electrical Properties and Noise Analysis Software and User's Manual,
Henri Happy and Alain Cappy

HEMTs and HBTs: Devices, Fabrication, and Circuits, Fazal Ali, Aditya Gupta, and
Inder Bahl, editors

High-Power Microwave Sources, Victor Granatstein and Igor Alexeff, edtiors

High-Power GaAs FET Amplifiers, John Walker, editor

High-Power Microwaves, James Benford and John Swegle

Introduction to Microwaves, Fred E. Gardiol

Introduction to Computer Methods for Microwave Circuit Analysis and Design, Janusz
A. Dobrowolski

Introduction to the Uniform Geometrical Theory of Diffraction, D. A. McNamara, C. W.
I. Pistorius and J. A. G. Malherbe

LOSLIN: Lossy Line Calculation Software and User's Manual, Fred E. Gardiol

Lossy Transmission Lines, Fred E. Gardiol

Low-Angle Microwave Propagation: Physics and Modeling, Adolf Giger

Low Phase Noise Microwave Oscillator Design, Robert G. Rogers

MATCHNET: Microwave Matching Networks Synthesis, Stephen V. Sussman-Fort

*Matrix Parameters for Multiconductor Transmission Lines: Software and User's
Manual*, A. R. Djordjevic et al.

Microelectronic Reliability, Volume I: Reliability, Test, and Diagnostics, Edward B.
Hakim, editor

Microelectronic Reliability, Volume II: Integrity Assessment and Assurance, Emiliano
Pollino, editor

Microwave and Millimeter-Wave Diode Frequency Multipliers, Marek T. Faber,
Jerzy Chamiec, Miroslaw E. Adamski

Microwave and RF Circuits: Analysis, Synthesis, and Design, Max Medley

Microwave and RF Component and Subsystem Manufacturing Technology, Heriot-Watt
University

Microwave Circulator Design, Douglas K. Linkhart

Microwave Engineers' Handbook, 2 Volumes, Theodore Saad, editor

Microwave Materials and Fabrication Techniques, Second Edition, Thomas S.
Laverghetta

Microwave MESFETs and HEMTs, J. Michael Golio *et al.*

Microwave and Millimeter Wave Heterostructure Transistors and Applicatons, F. Ali,
editor

*Microwave and Millimeter Wave Phase Shifters, Volume I: Dielectric and Ferrite Phase
Shifters*, S. Koul and B. Bhat

Microwave and Millimeter Wave Phase Shifters, Volume II: Semiconductor and Delay Line Phase Shifters, S. Koul and B. Bhat

Microwave Mixers, Second Edition, Stephen Maas

Microwave Transmission Design Data, Theodore Moreno

Microwave Transition Design, Jamal S. Izadian and Shahin M. Izadian

Microwave Transmission Line Couplers, J. A. G. Malherbe

Microwave Tubes, A. S. Gilmour, Jr.

Microwaves: Industrial, Scientific, and Medical Applications, J. Thuery

Microwaves Made Simple: Principles and Applicatons, Stephen W. Cheung, Frederick H. Levien *et al.*

MMIC Design: GaAs FETs and HEMTs, Peter H. Ladbrooke

Modern GaAs Processing Techniques, Ralph Williams

Modern Microwave Measurements and Techniques, Thomas S. Laverghetta

Monolithic Microwave Integrated Circuits: Technology and Design, Ravender Goyal *et al.*

Nonuniform Line Microstrip Directional Couplers, Sener Uysal

PC Filter: Electronic Filter Design Software and User's Guide, Michael G. Ellis, Sr.

PLL: Linear Phase-Locked Loop Control Systems Analysis Software and User's Manual, Eric L. Unruh

RF Design Guide: Systems, Circuits, and Equations, Peter Vizmuller

Scattering Parameters of Microwave Networks with Multiconductor Transmission Lines: Software & User's Manual, A. R. Djordjevic *et al.*

Solid-State Microwave Power Oscillator Design, Eric Holzman and Ralston Robertson

Terrestrial Digital Microwave Communications, Ferdo Ivanek *et al.*

Time-Domain Response of Multiconductor Transmission Lines: Software and User's Manual, A. R. Djordjevic *et al.*

Transmission Line Design Handbook, Brian C. Waddell

Yield and Reliability in Microwave Circuit and System Design, Michael Meehan and John Purviance

For further information on these and other Artech House titles, contact:

Artech House
685 Canton Street
Norwood, MA 02062
617-769-9750
Fax: 617-769-6334
Telex: 951-659
email: artech@world.std.com

Artech House
Portland House, Stag Place
London SW1E 5XA England
+44 (0) 171-973-8077
Fax: +44 (0) 171-630-0166
Telex: 951-659
bookco@artech.demon.co.uk